Methods in Enzymology

Volume XXXIX

HORMONE ACTION

Part D

Isolated Cells, Tissues, and Organ Systems

METHODS IN ENZYMOLOGY

EDITORS-IN-CHIEF

Sidney P. Colowick   Nathan O. Kaplan

Methods in Enzymology

Volume XXXIX

Hormone Action

Part D

Isolated Cells, Tissues, and Organ Systems

EDITED BY

Joel G. Hardman

DEPARTMENT OF PHYSIOLOGY
VANDERBILT UNIVERSITY SCHOOL OF MEDICINE
NASHVILLE, TENNESSEE

Bert W. O'Malley

DEPARTMENT OF CELL BIOLOGY
BAYLOR COLLEGE OF MEDICINE
TEXAS MEDICAL CENTER
HOUSTON, TEXAS

1975

ACADEMIC PRESS New York San Francisco London
A Subsidiary of Harcourt Brace Jovanovich, Publishers

ACADEMIC PRESS, INC.
111 Fifth Avenue, New York, New York 10003

United Kingdom Edition published by
ACADEMIC PRESS, INC. (LONDON) LTD.
24/28 Oval Road, London NW1

Library of Congress Cataloging in Publication Data

Hardman, Joel G., joint author
Hormone Action.

(Methods in enzymology ; v. 36-)
Includes bibliographical references and indexes.
CONTENTS: pt. A. Steroid hormones.—pt. C. Cyclic nucleotides.—pt. D. Isolated cells, tissues, and organ systems.—pt. E. Nuclear structure and function.
1. Enzymes. 2. Hormones. 3. Cyclic nucleotides. I. O'Malley, Bert W. II. Title.
III. Series: Methods in enzymology ; v. 36 [etc.]
[DNLM: 1. Cell nucleus. 2. Hormones. 3. Nucleotides, Cyclic. W1 ME9615K v. 40 / QH595 H812]
QP601.C733 vol. 36, etc. [QP601] 574.1'925'08s
[574.1'92] 74-10710
ISBN 0–12–181939–6 (v. 39) (pt. D)

PRINTED IN THE UNITED STATES OF AMERICA

Table of Contents

Section I. Kidney

Section II. Liver

Section III. Heart

Section IV. Skeletal and Smooth Muscle

Section V. Endocrine and Reproductive Tissue

Section VI. Neural Tissue

Section VII. Exocrine Tissue

Section VIII. Vascular Tissue

Section IX. Unicellular Organisms

Section X. Substrate and Ion Fluxes

Contributors to Volume XXXIX

Article numbers are in parentheses following the names of contributors. Affiliations listed are current.

S. Adolfsson (11), *Department of Physiology, University of Göteborg, Göteborg, Sweden*

Kurt Ahrén (11, 19), *Department of Physiology, University of Göteborg, Göteborg, Sweden*

E. Alsat (21), *Laboratoire de Chimie Hormonale, Maternité de Port-Royal, Paris, France*

A. Arvill (11), *Department of Physiology, University of Göteborg, Göteborg, Sweden*

C. Wayne Bardin (38), *Department of Medicine, Division of Endocrinology, Milton S. Hershey Medical Center, The Pennsylvania State University Hershey, Pennsylvania*

Michael J. Berridge (40), *Agricultural Research Unit of Invertebrate Chemistry and Physiology, Department of Zoology, University of Cambridge, Cambridge, England*

John D. Biggers (25), *Department of Physiology, and Laboratory of Human Reproduction and Reproductive Biology, Harvard Medical School, Boston, Massachusetts*

M. W. Bitensky (41), *Department of Pathology, Yale University School of Medicine, New Haven, Connecticut*

Andre B. Borlé (44), *Department of Physiology, University of Pittsburgh School of Medicine, Pittsburgh, Pennsylvania*

R. H. Bowman (1), *State University of New York, Upstate Medical Center, Veterans Administration Hospital Syracuse, New York*

Edward Bresnick (5), *Department of Cell and Molecular Biology, Medical College of Georgia, Augusta, Georgia*

Gary Brooker (7), *Department of Pharmacology, University of Virginia Medical School, Charlottesville, Virginia*

Kurt Bürki (5), *Pathologisches Institut, Universität Bern, Bern, Switzerland*

Leslie P. Bullock (38), *Department of Medicine, Division of Endocrinology, Milton S. Hershey Medical Center, The Pennsylvania State University, Hershey, Pennsylvania*

Ian M. Burr (32), *Department of Pediatrics, Vanderbilt University School of Medicine, Nashville, Tennessee*

Daniel P. Cardinali (33), *I.L.A.F.I.R., Universidad del Salvador, San Miguel, Argentina*

K. J. Catt (22), *Section on Hormonal Regulation, Reproduction Research Branch, National Institute of Child Health and Human Development, National Institutes of Health, Bethesda, Maryland*

L. Cedard (21), *Laboratoire de Chimie Hormonale, Maternité de Port-Royal, Paris, France*

Cornelia P. Channing (18), *Department of Physiology, University of Maryland School of Medicine, Baltimore, Maryland*

Jorge E. Cidre (20), *Department of Biochemistry, University of Miami School of Medicine, Miami, Florida*

M. L. Dufau (22), *Reproduction Research Branch, National Institute of Child Health and Human Development, National Institutes of Health, Bethesda, Maryland*

Kristen B. Eik-Nes (23), *Department of Biophysics, N.T.H., University of Trondheim, Trondheim, Norway*

John H. Exton (4), *Department of Physiology, Vanderbilt University School of Medicine, Nashville, Tennessee*

Rudolph E. Fanska (31), *Metabolic Research Unit, University of California, San Francisco, California*

Judah Folkman (30), *The Department of Surgery, The Children's Hospital Medical Center, and the Harvard Medical School, Boston, Massachusetts*

Michael A. Gimbrone, Jr. (30), *National Cancer Institute, National Institutes of Health, Bethesda, Maryland*

ALFRED L. GOLDBERG (10), *Department of Physiology, Harvard Medical School, Boston, Massachusetts*

GEROLD M. GRODSKY (31), *Metabolic Research Unit and Department of Biochemistry and Biophysics, University of California, San Francisco, California*

LARS HAMBERGER (35), *Department of Physiology, University of Göteborg, Göteborg, Sweden*

HANS HERLITZ (35), *Department of Physiology, University of Göteborg, Göteborg, Sweden*

B. J. HOFFER (36), *Laboratory of Neuropharmacology, IRP, National Institute of Mental Health, St. Elizabeths Hospital, Washington, D.C.*

RAGNAR HULTBORN (35), *Department of Neurobiology, University of Göteborg, Göteborg, Sweden*

SIRET DESIREE JAANUS (27), *Southern California College of Optometry, Fullerton, California*

PER OLOF JANSON (19), *Department of Obstetrics and Gynecology, University of Göteborg, Göteborg, Sweden*

L. S. JEFFERSON (9), *Department of Physiology, The Milton S. Hershey Medical Center, The Pennsylvania State University, Hershey, Pennsylvania*

J. J. KEIRNS (41), *Department of Pathology, Yale University School of Medicine, New Haven, Connecticut*

PETER O. KOHLER (13), *Departments of Medicine and Cell Biology, Baylor College of Medicine, Texas Medical Center, Houston, Texas*

MICAH I. KRICHEVSKY (42), *National Institute of Dental Research, National Institutes of Health, Bethesda, Maryland*

MARTIN J. KUSHMERICK (10), *Department of Physiology, Harvard Medical school, Boston, Massachusetts*

FLORENCE LEDWITZ-RIGBY (18), *Department of Physiology, University of Pittsburgh School of Medicine, Pittsburgh, Pennsylvania*

BARRY D. LINDLEY (29), *Division of Molecular and Cellular Endocrinology, Department of Physiology, Case Western Reserve University, Cleveland, Ohio*

LESLIE L. LOVE (42), *National Institute of Dental Research, National Institutes of Health, Bethesda, Maryland*

JOHN M. MARSH (20), *Department of Biochemistry, University of Miami School of Medicine, Miami, Florida*

SUSAN B. MARTEL (10), *Department of Physiology, Harvard Medical School, Boston, Massachusetts*

STEVEN E. MAYER (8), *Department of Medicine, University of California, San Diego, School of Medicine, La Jolla, California*

ANDREW M. MICHELAKIS (3), *Department of Pharmacology, Vanderbilt University School of Medicine, Nashville, Tennessee*

A. JOSEPHINE MILNER (16), *Department of Pathology, Cambridge University, Cambridge, England*

IRENE MOWSZOWICZ (38), *Faculte de Médecine, Pitie-Salpetriere, Service de Biochimie, Paris, France*

J. R. NEELY (6), *Department of Physiology, The Milton S. Hershey Medical Center, The Pennsylvania State University, Hershey, Pennsylvania*

WENDELL E. NICHOLSON (28). *Department of Medicine, Vanderbilt University School of Medicine, Nashville, Tennessee*

TAKAMI OKA (37), *National Institute of Arthritis, Metabolism, and Digestive Diseases, National Institutes of Health, Bethesda, Maryland*

ANDRE PEYTREMANN (28), *Endocrine Unit, Department of Medicine, Harvard Medical School, and Massachusetts General Hospital, Boston, Massachusetts*

JOHN C. PORTER (17), *Cecil H. and Ida Green Center for Reproductive Biology Sciences, Departments of Obstetrics and Gynecology and Physiology, University of Texas Health Science Center at Dallas, Southwestern Medical School, Dallas, Texas*

William T. Prince (40), *Agricultural Research Unit of Invertebrate Chemistry and Physiology, Department of Zoology, University of Cambridge, Cambridge, England*

Howard Rasmussen (2, 40), *Department of Biochemistry, University of Pennsylvania School of Medicine, Philadelphia, Pennsylvania*

David M. Regen (43), *Department of Physiology, Vanderbilt University Nashville, Tennessee*

Adalgisa Rojo (20), *Department of Biochemistry, University of Miami School of Medicine, Miami, Florida*

M. J. Rovetto (6), *Department of Physiology, Jefferson Medical College, Philadelphia, Pennsylvania*

Ronald P. Rubin (27), *Department of Pharmacology, Medical College of Virginia, Richmond, Virginia*

George Sayers (29), *Division of Molecular and Cellular Endocrinology, Department of Physiology, Case Western Reserve University, Cleveland, Ohio*

Michael Schramm (39), *Department of Biological Chemistry, The Hebrew University of Jerusalem, Jerusalem, Israel*

Dennis Schulster (26), *Biochemistry Group, School of Biological Sciences, University of Sussex, Brighton, England*

Steven Seelig (29), *Division of Molecular and Cellular Endocrinology, Department of Physiology, Case Western Reserve University, Cleveland, Ohio*

Zvi Selinger (39), *Department of Biological Chemistry, The Hebrew University of Jerusalem, Jerusalem, Israel*

Gunnar Selstam (19), *Department of Physiology, University of Göteborg, Göteborg, Sweden*

Shail K. Sharma (12), *Department of Biochemistry, All India Institute of Medical Sciences, New Delhi, India*

Robert A. Sharp (32), *Department of Pediatrics, Vanderbilt University School of Medicine, Nashville, Tennessee*

Harvey M. Shein (34), *Department of Psychiatry, Harvard Medical School, Boston, Massachusetts, and McLean Hospital, Belmont, Massachusetts*

G. R. Siggins (36), *Laboratory of Neuropharmacology, IRP, National Institute of Mental Health, St. Elizabeths Hospital, Washington, D.C.*

Dianne Moore Smith (25), *Department of Physiology, and Laboratory of Human Reproduction and Reproductive Biology, Harvard Medical School, Boston, Massachusetts*

Anna Steinberger (14, 24), *Program in Reproductive Biology and Endocrinology, University of Texas Medical School at Houston, Houston, Texas*

James T. Stull (8), *Department of Medicine, University of California, San Diego, School of Medicine, La Jolla, California*

Sylvia A. S. Tait (26), *Department of Physics as Applied to Medicine, The Middlesex Hospital Medical School, London, England*

G. P. Talwar (12), *Department of Biochemistry, All India Institute of Medical Sciences, New Delhi, India*

Yale J. Topper (37), *National Institute of Arthritis, Metabolism, and Digestive Diseases, National Institutes of Health, Bethesda, Maryland*

Roberto Vitale (15), *Department of Cell Biology, Baylor College of Medicine, Houston, Texas*

Barbara K. Vonderhaar (37), *National Institute of Arthritis, Metabolism, and Digestive Diseases, National Institutes of Health, Bethesda, Maryland*

R. C. Wagner (41), *Department of Biology, University of Delaware, Newark, Delaware*

Richard J. Wurtman (33), *Laboratory of Neuroendocrine Regulation, Department of Nutrition and Food Science, Massachusetts Institute of Technology, Cambridge, Massachusetts*

Preface

Our aim in compiling this volume was to provide a source of descriptions and critical evaluations of currently useful intact cell systems and general techniques applicable to the study of hormone action on biochemical and biophysical parameters of intact cells. Many of these systems and techniques will be useful for studying the actions of a variety of hormones, neurotransmitters, and pharmacological agents. Others will be suitable primarily for investigators whose interests are confined to actions of a single hormone or class of hormones.

The hormone-responsive systems described represent most mammalian organ systems and include tissue slices, isolated cells, isolated perfused and nonperfused organs and organ fragments, cell and organ cultures, and *in situ* organ preparations. The general techniques described, for example those involving measurement of substrate and calcium fluxes, generally involve processes that are sensitive to modulation by hormones or neurotransmitters in a variety of cell types.

Omissions have inevitably occurred—some because potential authors were overcommitted, some because of editorial oversight, some because of the timing of new developments relative to the publication deadline. Some apparent omissions, particularly those involving the isolation and culture of cells, can be found in Volume XXXII.

We thank Drs. S. P. Colowick and N. O. Kaplan who originated the idea for and encouraged the compilation of this volume. We thank the staff of Academic Press for their help and advice. We especially thank the contributing authors for their patience and full cooperation and for carrying out the research that made this volume possible.

Joel G. Hardman
Bert W. O'Malley

METHODS IN ENZYMOLOGY

EDITED BY

Sidney P. Colowick and Nathan O. Kaplan

VANDERBILT UNIVERSITY
SCHOOL OF MEDICINE
NASHVILLE, TENNESSEE

DEPARTMENT OF CHEMISTRY
UNIVERSITY OF CALIFORNIA
AT SAN DIEGO
LA JOLLA, CALIFORNIA

METHODS IN ENZYMOLOGY

EDITORS-IN-CHIEF

Sidney P. Colowick Nathan O. Kaplan

VOLUME XX. Nucleic Acids and Protein Synthesis (Part C)
Edited by KIVIE MOLDAVE AND LAWRENCE GROSSMAN

VOLUME XXI. Nucleic Acids (Part D)
Edited by LAWRENCE GROSSMAN AND KIVIE MOLDAVE

VOLUME XXII. Enzyme Purification and Related Techniques
Edited by WILLIAM B. JAKOBY

VOLUME XXIII. Photosynthesis (Part A)
Edited by ANTHONY SAN PIETRO

VOLUME XXIV. Photosynthesis and Nitrogen Fixation (Part B)
Edited by ANTHONY SAN PIETRO

VOLUME XXV. Enzyme Structure (Part B)
Edited by C. H. W. HIRS AND SERGE N. TIMASHEFF

VOLUME XXVI. Enzyme Structure (Part C)
Edited by C. H. W. HIRS AND SERGE N. TIMASHEFF

VOLUME XXVII. Enzyme Structure (Part D)
Edited by C. H. W. HIRS AND SERGE N. TIMASHEFF

VOLUME XXVIII. Complex Carbohydrates (Part B)
Edited by VICTOR GINSBURG

VOLUME XXIX. Nucleic Acids and Protein Synthesis (Part E)
Edited by LAWRENCE GROSSMAN AND KIVIE MOLDAVE

VOLUME XXX. Nucleic Acids and Protein Synthesis (Part F)
Edited by KIVIE MOLDAVE AND LAWRENCE GROSSMAN

VOLUME XXXI. Biomembranes (Part A)
Edited by SIDNEY FLEISCHER AND LESTER PACKER

VOLUME XXXII. Biomembranes (Part B)
Edited by SIDNEY FLEISCHER AND LESTER PACKER

VOLUME XXXIII. Cumulative Subject Index Volumes I–XXX
Edited by MARTHA G. DENNIS AND EDWARD A. DENNIS

Methods in Enzymology

Volume XXXIX

HORMONE ACTION

Part D

Isolated Cells, Tissues, and Organ Systems

Section I

Kidney

[1] The Perfused Rat Kidney

By R. H. BOWMAN

The perfused kidney has had limited use as an experimental organ system. This has arisen not from a lack of interest in, nor potential usefulness of, such a preparation, but rather from technical difficulties that have arisen in the perfusion attempts. Several groups have reported studies in which they developed biochemically viable[1] and functional[2-4] perfused kidneys. However, there has been a reluctance by physiologists to study kidneys that do not have the capacity to reabsorb 98–99% of filtered water and sodium, and this performance is difficult to attain *in vitro*.[5] In actuality, it is not at all necessary to have such renal efficiency in order to investigate profitably a variety of problems involving the kidney. It is of importance to have a system composed of simple apparatus, which at the same time yields a preparation that gives consistent results, approaching physiological normalcy. The technique described herein has been used successfully in the author's laboratory for several years, and has proved useful for both biochemical[6,7] and physiological investigation.[8,9]

Procedure

Apparatus. Perfusion of the rat kidney is performed with apparatus that is essentially identical to that employed for perfusing the rat heart by the Langendorff method.[10] A detailed description of that apparatus, including the specific glassware, is given elsewhere in this volume.[11] Figure 1 shows a diagram of the apparatus arranged for kidney perfu-

[1] J. M. Nishiitsutsuji-Uwo, B. D. Ross, and H. A. Krebs, *Biochem. J.* **103,** 852 (1967).

[2] J. Bahlman, G. Giebisch, B. Ochwadt, and W. Schoeppe, *Amer. J. Physiol.* **212,** 77 (1967).

[3] A. Nizet, Y. Cuypers, P. Deetjen, and K. Kramer, *Pfluegers Arch. Gesamte Physiol. Menschen Tiere* **296,** 179 (1967).

[4] W. H. Waugh and T. Kubo, *Amer. J. Physiol.* **217,** 277 (1969).

[5] H. Franke, H. Huland, and C. Weiss, *Z. Gesamte Exp. Med.* **156,** 268 (1971).

[6] R. H. Bowman, *J. Biol. Chem.* **245,** 1604 (1970).

[7] R. Walter and R. H. Bowman, *Endocrinology* **88,** 189 (1973).

[8] R. H. Bowman and T. Maack, *Amer. J. Physiol.* **222,** 1499 (1972).

[9] R. H. Bowman, J. Dolgin, and R. Coulson, *Amer. J. Physiol.* **224,** 416 (1973).

[10] H. E. Morgan, M. J. Henderson, D. M. Regen, and C. R. Park, *J. Biol. Chem.* **236,** 253 (1961).

[11] J. R. Neely and J. J. Rovetto, this volume, [6].

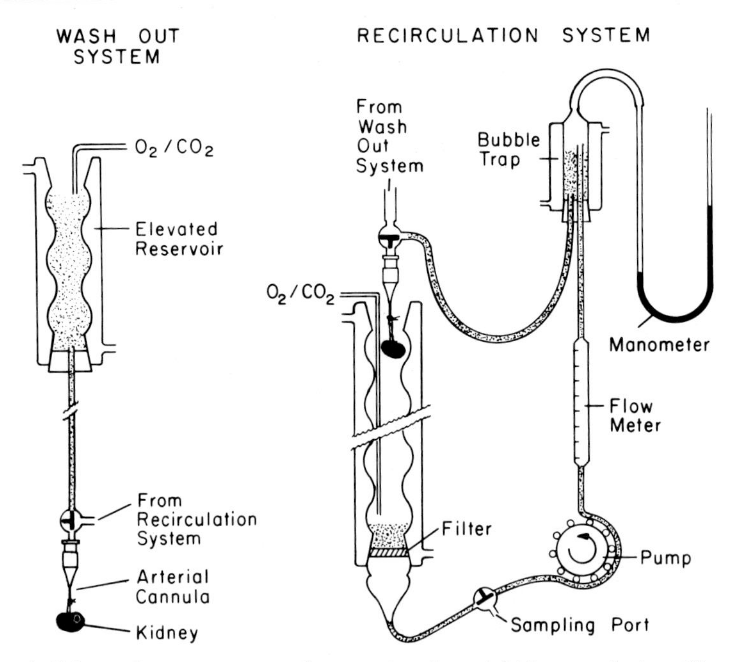

FIG. 1. Schematic arrangement of apparatus for rat kidney perfusion. The reservoir shown on the left is positioned approximately 85 cm above the chamber used in the recirculation system. See Article [6] in this volume for details of the glassware depicted in this figure, and see text for description of its use for the perfused rat kidney.

sion; the reader should refer to this volume [6][11] for further explanation. Modifications from the heart perfusion system are (1) a three-way stopcock positioned in the effluent line from the perfusion chamber to allow perfusate sampling; (2) a flowmeter (Arthur H. Thomas, 5083-D35) inserted in the line between the pump and the bubble trap; (3) a spring clamp (Arthur H. Thomas, 2835-L15) for holding the renal artery cannula (Fig. 2); and (4) use of an 18-gauge hypodermic needle for the arterial cannula. The latter is cut to a length of approximately 2.5 cm, slightly beveled, and filed to a smooth tip. The tubing leading from the bubble trap to the arterial cannula must be long enough to permit its extension when cannulating the artery; unlike cannulation of the heart for perfusion, the renal artery is cannulated while the organ is *in situ*. The ureter is catheterized with PE-10 tubing, and urine is collected in preweighed plastic vials with conical internal dimensions, such as those employed with the Technicon AutoAnalyzer. The vial is held in place by a spring clamp (Arthur H. Thomas, 2835-L20). A detailed drawing

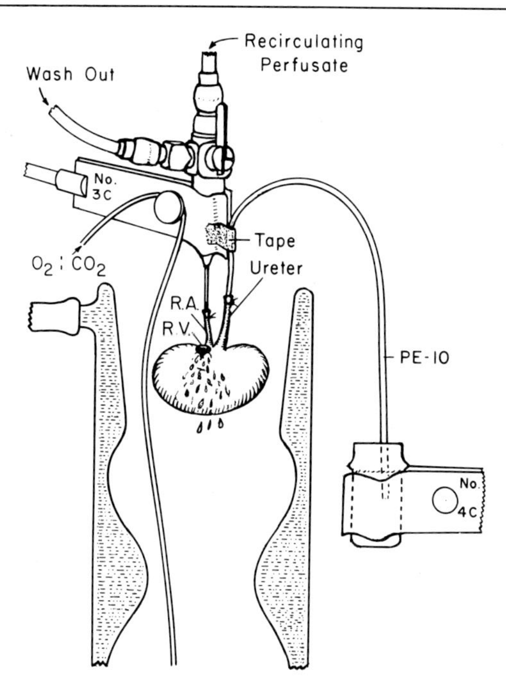

FIG. 2. View of the kidney as it is positioned in the perfusion chamber. R.A., renal artery; R.V., renal vein. See text for description.

showing the kidney suspended in the perfusion chamber and the arrangement of the adjacent apparatus is depicted in Fig. 2. As with the perfused heart, the temperature of the system is maintained by recirculating 37° water through the double-walled glassware.

The tubing used in the perfusion system is Tygon formulation S-50-HL (U.S. Stoneware, Inc.). Except where required for making connections to glassware and other parts of the perfusion apparatus, the size of the tubing is 0.093 × 0.156 inch (i.d. × o.d.). A finer diameter tubing, 0.05 × 0.08 inch (American Instrument Co., No. 5-8963), is used for gas delivery to the perfusion chamber. Note that this tubing extends to the lower portion of the chamber, but not below the level of the perfusate; with care, foaming of the perfusate can be avoided.

The perfusion pump can be any of the roller-bearing variety (other types of peristaltic pumps may be suitable but have not been tried). It is important to be able to control finely the flow through the pump. We use a Brunswick Instrument Co. model PA (less its motor) to which a variable speed motor has been attached, and drive the pump at approximately 180 cycles per minute. A series of such pumps can be mounted side by side and driven by a shaft connected to a single motor. It is convenient to use a variable speed motor, although perfusate flow is controlled by adjusting the pump yoke rather than by the speed of the pump.

Perfusate. Perfusate consists of Krebs-Henseleit bicarbonate buffer containing bovine serum albumin, fraction V, at a concentration of 7.5 g/100 ml. The perfusate is dialyzed at 2° against an approximate 10-fold excess of Krebs bicarbonate buffer (without albumin) for 18–24 hours prior to use. A longer or shorter (or no) dialysis may prove satisfactory for some studies. To curtail bacterial growth, antibiotics are added to the dialyzate; a satisfactory combination is 4.8 mg penicillin and 4 mg streptomycin per liter of dialyzate. Before the perfusate is removed from the dialysis sacs, the contents of the dialysis jar are equilibrated with $O_2:CO_2$ (95:5) for 30 minutes. The perfusate then is emptied from the dialysis tubing, and any substrates or other substances (e.g., glucose, inulin, creatinine, urea, drugs) are added; if necessary, the pH is adjusted with NaOH to approximately 7.45.

The dialysis tubing is boiled for a number of hours in frequent changes of distilled water before it is used. It is then stored in dilute HCl at 2°, and is well rinsed before filling with perfusate. All water used in constituting the perfusate is glass-distilled.

The desired amount of perfusate (25 ml minimum) is added to the perfusion chamber and recirculated through the system while the kidney is being prepared for perfusion. $O_2:CO_2$ (95:5) is delivered to the perfusion chamber during this time, permitting the perfusate to become fully oxygenated prior to actual perfusion. The overhead reservoir is filled with oxygenated perfusate, and a stream of $O_2:CO_2$ is delivered over the top of this fluid. Tubing leads from the bottom of the reservoir to the three-way stopcock, which also receives the circulating perfusate and to which is attached the arterial cannula (see Figs. 1 and 2).

Operative Procedure. Male rats weighing 350–400 g are anesthetized with sodium pentobarbital (approximately 5 mg/100 g body weight). A midline incision is made, and the major abdominal blood vessels are exposed. Two ligatures are passed around the left renal vein into which heparin (200 units/rat) is injected. This blood vessel is then tied off. A loose ligature is passed around the right renal artery, and distal and proximal ligatures around the mesenteric artery. These blood vessels are exposed previously by gentle dissection. It is important to contact the right renal artery as little as possible. If urine is to be collected, the ureter is catheterized next. To avoid undue resistance to flow from use of the PE-10 catheter, it should be kept as short as possible.

Cannulation. The foregoing procedure is carried out at some suitable worktable with the rat secured to a small operating tray. After preparation of the rat, the tray containing the animal is transferred to a convenient platform adjacent to the perfusion apparatus. Before the artery is cannulated, the three-way stopcock holding the cannula is turned to allow flow from the overhead chamber. After any air bubbles are expelled from this tubing, it is occluded by attaching a hemostat. By adjusting the perfusion pump, the pressure in the recirculating system is brought to 120–140 mm Hg with zero flow, air bubbles also being eliminated from the delivery tubing. To prepare for cannulation, the distal mesenteric ligature is tied, and a small serrefine is placed on the mesenteric artery close to the aorta. A cut is made in the mesenteric artery, just proximal to the tied distal ligature, and the cannula is inserted into the vessel up to the level of the serrefine; the serrefine is removed, and the cannula is manipulated across the aorta and into the renal artery.[1] As the cannula enters the renal artery, the hemostat holding back the perfusate from the overhead reservoir is released, and kidney perfusion is begun. The proximal mesenteric ligature and the ligature around the renal artery are tied, securing the cannula in place. The inferior vena cava is rapidly severed, and the kidney is dissected free from the animal and trimmed of adhering tissue. During the time that the arterial cannula is being tied into place and the kidney trimmed, the gravity perfusion system serves to flush blood from the renal vessels and to maintain oxygen delivery to the tissue. This procedure takes about 1 minute. Perfusate flow is then switched so as to be delivered from the recirculating system, the cannula is secured in the spring clamp (Fig. 2), and the perfusion pressure is adjusted.

The vasculature will dilate over the first 5–10 minutes of perfusion with the result that the pressure will fall unless flow is increased via the pump. After 10 minutes of perfusion, the renal resistance will usually stabilize, and a perfusate flow of 27–32 ml per minute per kidney will result when the perfusion pressure is held between 90 and 100 mm Hg.

Functional Behavior

Glomerular filtration is estimated by the clearance of exogenous creatinine (initial perfusate concentration: 1.5 mg/ml). The adequacy of creatinine for measuring filtration in the perfused kidney has been assessed by using perfusate containing both creatinine and inulin, with or without [^{3}H]inulin.[12] Creatinine and chemically determined inulin give similar clearance values. [^{3}H]Inulin often is cleared at less than the rate of chemically determined inulin or creatinine, perhaps owing to contamination of the [^{3}H]inulin with smaller molecular weight, absorbable, triti-

[12] M. E. Trimble, unpublished observations.

ated compounds. Probenecid, at a concentration (0.9 mM) which inhibits p-aminohippuric acid secretion, is found not to alter the clearance of creatinine (1.5 mg/ml, initial concentration), indicating that secreted creatinine is not a significant component of the total amount excreted.[12] An advantage of creatinine as an indicator of filtration rate is its solubility (compared to inulin) and its convenience of assay, using the Technicon AutoAnalyzer.

FUNCTIONAL BEHAVIOR OF THE ISOLATED PERFUSED RAT KIDNEY[a]

Parameter	Perfusate composition		
	5% Albumin,[b] no ADH (4)[c]	7.5% Albumin[b]	
		No ADH (10)	ADH[d] (8)
Perfusate flow (ml/min)	35 ± 4[e]	31 ± 1.6	30 ± 0.3
Glomerular filtration rate (μl/min)	831 ± 84	492 ± 67	449 ± 50
Urine flow (μl/min)	104.1 ± 6.4	38.9 ± 3.5	18.8 ± 1.8
Fractional excretion (% of filtered load)			
Na^+	10.6 ± 1.3	5.1 ± 0.87	2.24 ± 0.45
K^+	62.9 ± 13.7	55.3 ± 3.0	46.1 ± 5.9

[a] All kidneys perfused at a pressure of 100 ± 10 mm Hg.
[b] Concentration of albumin in the perfusate.
[c] Numbers in parentheses indicate the number of kidneys perfused under each condition.
[d] The antidiuretic hormone (ADH) employed was synthetic arginine vasopressin.
[e] Mean and standard error of the mean. The data are from final 10-minute clearance periods obtained after prior perfusion for 50 minutes. Kidney wet weights were approximately 1.7 g.

The table lists several functional parameters that have been measured with the perfusion system described. They were obtained with a perfusion pressure of 90–100 mm Hg, and at two different albumin concentrations. The data show that the filtration rate can be increased nearly 2-fold by using 5% albumin rather than 7.5%, but the reabsorptive efficiency then decreases. However, for many studies it is more desirable to have a higher filtration rate and a greater urine output, and in such cases a lower albumin concentration should be considered. If the perfusion pressure is set at 60 mm Hg or below, very little glomerular filtration will occur, but the perfusion flow rate will not be greatly reduced. In experiments in which it is desired to study biochemical function in the absence of urine formation, the kidney may be perfused at a lower pressure. At any perfusion pressure the perfusate flow rate is unphysiologically high. In practice this presents no difficulty, and, in fact, is responsible for allowing adequate oxygenation in the absence of red cells. The drawback to the high flow rate is that the medullary solute concentration becomes diluted, and this mediates against obtaining hypertonic urine. It does not, however, prevent adequate reabsorption of filtered water and electrolytes (see the table).

Biochemical Parameters

Kidneys perfused according to the foregoing procedure utilize oxygen at a rate of 1–4 μmoles per minute per gram wet weight, depending upon

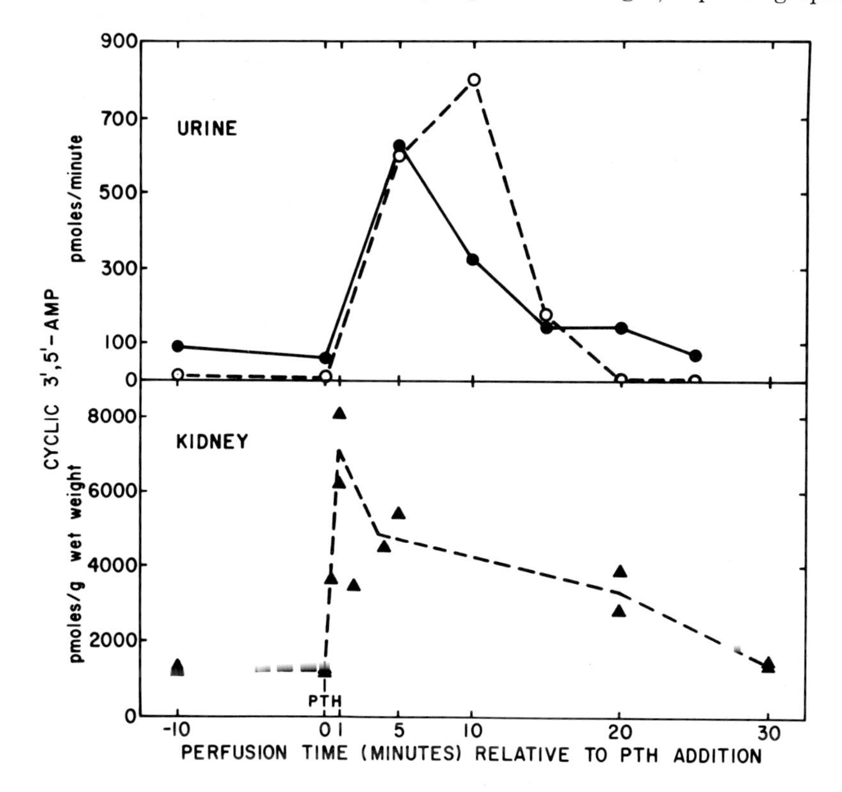

FIG. 3. Effect of parathyroid extract (PTH) on cyclic AMP levels in the kidney and on excreted cyclic AMP. PTH was injected into the perfusate via the tubing immediately adjacent to the arterial cannula (Fig. 2). Bottom panel: Each symbol represents a kidney frozen either before or after PTH addition (see time scale). The dashed line is arbitrarily drawn between the points. Top panel: Urinary cyclic AMP excretion rates in two kidneys before and after PTH addition to the perfusate.

the exogenous substrate.[12] The ATP level is well maintained and will be found to be 6–8 μmoles/g, dry weight, at any time the kidney is rapidly frozen during an hour or two of perfusion. Evidence of the biochemical viability of the perfused kidney is derived from studies of glucose synthesis. At saturating concentrations of pyruvate (20 m*M*), glucose is formed by the perfused kidney at a rate of approximately 1 μmole per minute per gram, wet weight.[6] Each micromole of glucose formed requires the expenditure of 6 μmoles of ATP, which is four times as much as is present in the whole kidney at any one moment. Since there are other demands on energy production, and since the tissue ATP level does not diminish with time, it must be concluded that the energy-generating system is well maintained during perfusion.

Hormone Responses

Responses of the perfused kidney to hormones have been investigated only superficially. The table shows that arginine vasopressin causes an antidiuresis in the absence of a decrease in glomerular filtration rate. The hormone, therefore, causes an increase in both fractional and absolute reabsorption of water. Contrary to the usual mammalian response to vasopressin, there is also an increased reabsorption of sodium in kidneys perfused with antidiuretic hormone (see the table). The significance of this observation remains to be investigated, and the effect of vasopressin on the perfused kidney may or may not reflect the same mechanism of action as occurs *in vivo*.

Parathyroid hormone added to the perfusate, as the extract (Lilly) or as purified acetone powder (Wilson Laboratories), increases phosphate excretion by the perfused kidney,[13] and large amounts of intrarenal cyclic AMP can be generated in such experiments.[14,15] Figure 3 (bottom panel) shows a time course of the change in tissue cyclic AMP concentration after perfusing kidneys with parathyroid extract. The upper panel of Fig. 3 shows that the rise in tissue cyclic AMP is reflected in increased excretion of the nucleotide.

General Comment

The technique of kidney perfusion is probably best learned by using rather large rats (450 g or heavier) and employing a small arterial cannula (No. 19 rather than No. 18 hypodermic needle). Considerable prac-

[13] R. H. Bowman, unpublished observations.
[14] R. W. Butcher, R. H. Bowman, and E. W. Sutherland, unpublished observations.
[15] R. Coulson, unpublished observations.

tice is required before one can rapidly pass the cannula across the aorta and into the renal artery. Regardless of one's facility at this procedure, two persons are required in order to promptly initiate perfusion. It is necessary to have the hemostat on the wash-out system released as soon as the cannula enters the renal artery, and to have the cannula held while the kidney is cut free from the animal.

The renal artery is extremely sensitive to tactile stimulation, and any unnecessary handling of the artery may lead to its constriction. If constriction does occur, as manifested by a low perfusate flow rate, elevation of the perfusion (pump) pressure will usually overcome the arterial resistance and result in gradual dilatation. For adequate oxygenation it is desirable to maintain a perfusion flow of at least 15 ml per minute.

[2] Isolated Mammalian Renal Tubules[1]

By Howard Rasmussen

The classic methods for the study of renal intermediary metabolism have employed either homogenates or slices of renal cortex.[1a,2-5] However, one of the unique features of renal tubular cells not shared by most other mammalian cells is their permeability to Krebs cycle intermediates.[6-8] Because of this property, renal cortical slices contain significant concentration of these metabolites in their interstitial fluids, and these acids also accumulate in the incubation media. Most studies employing such slices have measured intermediate concentrations in blotted slices. However, this method may give variable results, depending upon the amount of interstitial fluid present and the thickness of the slice. A number of these difficulties can be obviated by employing preparations of isolated

[1] Supported by a grant from the United States Public Health Service (AM 00650).

[1a] H. A. Krebs, D. A. Bennett, P. de Gasquet, T. Gascoyne, and T. Yoshida, *Biochem. J.* **86**, 22 (1963).

[2] J. Z. Rutman, L. E. Meltzer, J. R. Kitchell, R. J. Rutman, and P. George, *Amer. J. Physiol.* **213**, 967 (1967).

[3] A. D. Goodman, R. E. Fuisz, and G. F. Cahill, Jr., *J. Clin. Invest.* **45**, 612 (1966).

[4] D. E. Kamm, R. E. Fuisz, A. D. Goodman, and G. F. Cahill, Jr., *J. Clin. Invest.* **46**, 1172 (1967).

[5] R. Rognstad, *Biochem. J.* **116**, 493 (1970).

[6] D. P. Simpson, *Amer. J. Physiol.* **206**, 875 (1964).

[7] J. J. Cohen and E. Wittmann, *Amer. J. Physiol.* **204**, 795 (1963).

[8] L. Runeberg and W. D. Lotspeich, *Amer. J. Physiol.* **211**, 467 (1966).

renal tubules, prepared by the enzymatic digestion of the renal cortex from a suitably prepared experimental animal.[9-14]

Preparation of Rat Renal Tubules

Isolated renal cortical nephrons are prepared from young adult rats, using a mixture of collagenase and hyaluronidase.[12] The method is adapted from that developed by Howard and Pesch[15] for the preparation of intact liver cells.

Male Wistar rats weighing 150–200 g are anesthetized with urethane (0.16–0.2 g/100 g body weight) and the abdominal aorta exposed through a midline abdominal incision. The celiac artery, the abdominal aorta, and the inferior vena cava are all ligated just below the diaphragm. The kidneys are then perfused via the abdominal aorta and the renal arteries with 10 ml of an enzyme solution containing, per 100 ml, 40 mg of collagenase (Worthington CLS), 100 mg of hyaluronidase (Sigma type I), 25 mg of streptomycin sulfate, penicillin, and 180 mg of glucose dissolved in Ca^{2+}-free Hanks' solution. The kidneys are rapidly extirpated, and the medulla are excised. The cortices from 4–6 perfused kidneys are combined, minced finely with surgical scissors, and suspended in 10 ml of the same enzyme solution in a 250-ml Erlenmeyer flask. The flask was incubated at 37° under O_2–CO_2 (95:5, by volume) for 60 minutes with shaking in a metabolic shaker at a rate of 60–70 oscillations per minute. At the end of this time, the tubules are dispersed by gentle pipetting with a broad-tipped 10-ml serological pipette. The contents of the flask are then filtered through a single layer of nylon stocking and then through nylon mesh with a pore size of 60 μm. Filtration is aided by addition of cold, Ca^{2+}- and glucose-free Hanks' solution to a final volume of 50 ml. This filtrate is centrifuged at 50 *g* for 60 seconds at 4° in an International Centrifuge (PRII). The supernatant is removed by suction, and the resulting loose pellet is resuspended and washed twice with cold Ca^{2+}- and glucose-free Hanks' solution, and centrifuged at 30 *g* for 60 seconds. The final loosely packed pellet contains approximately 60–70 mg of protein per milliliter and is the preparation used for incubations *in vitro*. Light and electron microscopic examination reveals a preparation con-

[9] C. Arnaud and H. Rasmussen, *Endocrinology* **75,** 277 (1964).
[10] M. B. Burg and J. Orloff, *Amer. J. Physiol.* **203,** 327 (1962).
[11] N. Nagata and H. Rasmussen, *Biochim. Biophys. Acta* **215,** 1 (1970).
[12] N. Nagata and H. Rasmussen, *Proc. Nat. Acad. Sci. U.S.* **65,** 368 (1970).
[13] H. Rasmussen and N. Nagata, *Biochim. Biophys. Acta* **215,** 17 (1970).
[14] A. Struyvenberg, R. B. Morrison, and A. S. Relman, *Amer. J. Physiol.* **214,** 1155 (1968).
[15] R. B. Howard and L. A. Pesch, *J. Biol. Chem.* **243,** 3105 (1968).

taining elements of both proximal and distal tubule with basement membranes largely intact and a normal arrangement and configuration of mitochondria. The proximal tubular elements represent approximately 80% of the total preparation.

Comments. The use of urethane as anesthetic has proved to be the most reliable and least harmful to the metabolic properties of the isolated tissue.[16–18] Amytal anesthesia is to be avoided, because it depresses the respiratory activity of the isolated tubes. Ether anesthesia, for reasons that are not completely clear, invariably gives a tubule preparation that will not maintain its integrity for more than 25–30 minutes *in vitro*.

An alternative procedure, in which kidney cortex is removed directly from the animal without intraarterial infusion of enzyme solution, and minced and incubated with enzyme solution *in vitro,* gives a very low yield of isolated tubules.

Functional Intactness of the Tissue

This tissue depends primarily upon oxidative metabolism. Citrate, α-ketoglutarate, pyruvate, lactate, and fatty acids are the main substrates which these cells utilize *in vivo*. Hence, to test functional intactness of the tubule preparation, two different kinds of measurements can be made[12]: (1) measurement of the rate of glucose production from either malate, lactate, or α-ketoglutarate; or (2) measurement of adenine nucleotide concentrations. Representative rates of glucose formation from these substrates in typical preparations are shown in Table I. Particularly important is the fact that in the usual preparation of tubules, the rate of glucose formation is linear for 60–90 minutes, and the ATP concentration is maintained during this same time interval. When gluconeogenesis declines upon prolonged incubation, this is almost invariably associated with a fall in adenine nucleotide concentration in the tubules, and the appearance of adenine nucleotides, principally ADP, in the incubation medium.

Comments. The limitation in the use of isolated tubules is that experiments lasting longer than 1.5 hours are not possible.

Standard Incubation Conditions

For the study of renal gluconeogenesis and its ionic and hormonal control, it is essential that the buffer be bicarbonate rather than phos-

[16] K. Kurokawa and H. Rasmussen, *Biochim. Biophys. Acta* **313**, 17 (1973).
[17] K. Kurokawa and H. Rasmussen, *Biochim. Biophys. Acta* **313**, 42 (1973).
[18] K. Kurokawa and H. Rasmussen, *Biochim. Biophys. Acta* **313**, 59 (1973).

TABLE I
RATES OF GLUCONEOGENESIS FROM VARIOUS SUBSTRATES BY SLICES OF RAT RENAL CORTEX AND BY ISOLATED RENAL TUBULES[a,b]

Substrate	Nanomoles per milligram of protein per hour	
	Slices[c]	Tubules
Pyruvate	151 ± 19	296 ± 10
Lactate	173 ± 24	272 ± 11
Succinate	—	220 ± 8
α-Ketoglutarate	141 ± 30	179 ± 9
Oxaloacetate	99 ± 15	198 ± 7
Malate	—	181 ± 9
Dihydroxyacetone	360 ± 60	594 ± 26
Glycerol	—	151 ± 6

[a] This table was taken from N. Nagata and H. Rasmussen, *Biochim. Biophys. Acta* **215,** 1 (1970).
[b] Krebs-Ringer bicarbonate buffer (pH 7.4), 10 m*M* substrate, and 2.5 m*M* $CaCl_2$ were used.
[c] These values are reported as mean ± SE and are very similar to those reported by D. E. Kamm, R. E. Fuisz, A. D. Goodman, and G. F. Cahill, Jr., *J. Clin. Invest.* **46,** 1172 (1967).

phate and that the medium contain 2% bovine serum albumin and 0.5 m*M* palmitate.[11-13,16-19] The standard incubation conditions employ a Krebs-Ringer bicarbonate buffer containing standard amounts of all salts except calcium chloride. The standard K-R buffer employs 2.5 m*M* $CaCl_2$. However, the calcium ion concentration *in vivo* is actually closer to 1.0 m*M* (cf. Rasmussen[20]). Thus, 1.0 m*M* $CaCl_2$ is a more physiological concentration for *in vitro* studies. This is of particular importance with the isolated renal tissues, because changes in extracellular [Ca^{2+}] alter the rate of glucose formation in this tissue. Likewise, renal gluconeogenesis is controlled by the extracellular [H^+], hence it is important to adjust buffer pH very accurately in order to obtain reproducible results. The effects of Ca^{2+} and H^+ are interrelated, and are also modified by the phosphate concentration, so phosphate concentration must also be carefully controlled.

In order to study the independent effects of [H^+] and [Ca^{2+}] upon renal glucose production, it is best to use Ca^{2+}-EGTA buffers.[21] The [Ca^{2+}] with these buffers is dependent upon pH, so that any change in [H^+] requires a different Ca^{2+}-EGTA buffer. [H^+] concentration, [HCO_3^-], and pCO_2 can be controlled independently by proper use of

[19] K. Kurokawa, T. Ohno, and H. Rasmussen, *Biochim. Biophys. Acta* **313,** 32 (1973).
[20] H. Rasmussen, *Amer. J. Med.* **50,** 567 (1971).
[21] C. N. Reilley, *Fed. Proc., Fed. Amer. Soc. Exp. Biol.* **20,** Suppl. 10, 22 (1961).

differing [HCO_3^-] and O_2:CO_2 gas mixtures. Incubations are carried out for 5–20 minutes at 37° with constant shaking. At the end of the incubation either trichloroacetic acid or $HClO_4$ extracts are prepared. Duplicate flasks can be run. In one set, the trichloroacetic acid or $HClO_4$ is added to the total contents of the flask. In the other, the contents of the flask are first subjected to centrifugation for 1 minute at 600 *g* or filtered through fine (0.5-μm pore size) Millipore filters, and the acids are added to the supernatant, or pellet. This is done so that the total amount of various intermediates and their concentrations in the medium can be determined. Glucose and most other metabolites are assayed on $HClO_4$ extracts using the fluorometric enzymatic methods described by Maitra and Estabrook[22] and by Williamson and Corkey.[23] Protein is determined by the biuret method,[24] cyclic AMP by the method of Gilman.[25]

The rate of substrate decarboxylation can be estimated by measuring the rate of decarboxylation of specifically labeled substrates; $^{14}CO_2$ is collected and counted by the method of Fain *et al.*[26]

This tubule preparation responds to both epinephrine and parathyroid hormone (PTH), with an increase in rate of glucose formation from a variety of substrates.[13,27] No special precaution is necessary for use of epinephrine, but is with use of PTH. Best results are obtained if the hormone is made up in small aliquots in a 10 m*M* acetic acid containing 0.5% bovine serum albumin, and then frozen at —20° until used. If stored in this way or as a dry lyophilized powder under N_2, the hormone is stable for several years. However, a particular aliquot of dissolved frozen hormone should not be used more than twice, because repeated freezing and thawing lead to a loss of biological activity.

Addition of hormone to the incubation flask should be made after addition of the tissue preparation, because, if added before the tissue, the hormone tends to bind to glass and lose activity, even in the presence of bovine serum albumin. It is also sensitive to oxidative inactivation.

Measurement of Metabolite Concentrations and Comparison of Whole Organ, Slices, and Tubules

The most striking feature of the renal tubule preparations is their permeability to the Krebs cycle intermediates and to pyruvate, lactate, glutamate, aspartate, and glutamine. When tubules are incubated in high concentrations of one of these, e.g., 10 m*M* citrate, there is a rapid ac-

[22] P. K. Maitra, and R. W. Estabrook, *Anal. Biochem.* **7**, 472 (1964).

[23] J. R. Williamson and B. E. Corkey, this series, Vol. 13, p. 434.

[24] A. G. Gornall, C. S. Bardawill, and M. M. David, *J. Biol. Chem.* **177**, 751 (1949).

[25] A. G. Gilman, *Proc. Nat. Acad. Sci. U.S.* **67**, 305 (1970).

[26] J. N. Fain, R. O. Scow, and S. S. Chernick, *J. Biol. Chem.* **238**, 54 (1963).

[27] K. Kurokawa and S. G. Massry, *J. Clin. Invest.* **52**, 961 (1973).

cumulation of all the others in the medium.[12] In contrast, the glycolytic intermediates and adenine nucleotides do not accumulate in the medium (see below).

Comparison of the rates of gluconeogenesis by isolated tubules with those obtained by the slice technique is shown in Table I. The rates of gluconeogenesis in the tubules are 127–202% of that seen in slices incubated under similar conditions with similar substrates, using milligrams of protein as the basis of comparison.

The amounts of adenine nucleotides, Krebs cycle, and glycolytic intermediates in the whole organ (quick frozen *in vivo*) and in slices or tubules incubated *in vitro* are also compared. These data are reported in terms of nanomoles per milligram of protein. When reported in this way, there are very marked differences in the concentration of Krebs cycle and glycolytic intermediates between whole organ and tubules. Slices had intermediate values (Table II). However, the values of adenine nucleotides are comparable in the three preparations.

The reason for these marked differences depend upon three factors: (1) the different manner in which the experiments are carried out with the three different preparations; (2) the manner of reporting the

TABLE II
RELATIVE CONCENTRATIONS OF VARIOUS METABOLIC INTERMEDIATES IN QUICK-FROZEN WHOLE RAT KIDNEY, RENAL CORTICAL SLICES, AND ISOLATED RENAL TUBULES[a,b]

Intermediate	Nanomoles per milligram of protein		
	Whole organ	Cortical slices	Isolated tubules
ATP	7.55 ± 0.21	6.05 ± 0.30	7.95 ± 0.16
ADP	2.54 ± 0.07	4.91 ± 0.06	2.16 ± 0.05
AMP	0.43 ± 0.02	2.18 ± 0.09	0.34 ± 0.03
Phosphoenol pyruvate	0.19 ± 0.03	0.45 ± 0.06	1.10 ± 0.11
Pyruvate	0.27 ± 0.05	1.74 ± 0.13	2.40 ± 0.05
Glucose-6-P	0.79 ± 0.06	0.25 ± 0.04	Barely detectable
Fructose-6-P	0.17 ± 0.03	0.04 ± 0.02	Barely detectable
3-Phosphoglyceraldehyde	0.42 ± 0.04	0.04 ± 0.01	Barely detectable
Citrate	2.41 ± 0.07	10.2 ± 0.2	50.3 ± 1.5
Isocitrate	0.21 ± 0.03	0.62 ± 0.06	2.43 ± 0.06
α-Ketoglutarate	2.10 ± 0.11	3.21 ± 0.06	69.7 ± 2.1
Malate[c]	1.18 ± 0.07	1.56 ± 0.12	10.2 ± 0.8

[a] This table was taken from N. Nagata and H. Rasmussen, *Biochim. Biophys. Acta* **215**, 1 (1970).

[b] Both glucose 10 mM and malate 10 mM were used as substrates in Krebs-Ringer bicarbonate (pH 7.4), and 2.5 mM $CaCl_2$.

[c] Measured with only glucose as substrate.

data; and (3) the difference in the ability of phosphorylated and nonphosphorylated intermediates to diffuse into and out of renal cells.

The differences in experimental technique are as follows: In the case of whole organ, the tissue is rapidly frozen between precooled aluminum tongs upon evulsion of the organ from the animal. In this organ, the total intracellular fluid volume is probably on the order of 50% of the total tissue volume. In the case of slices, they are removed from the incubation vessel, blotted, then frozen and extracted. The total intracellular fluid volume is probably 30 40% of the tissue volume. In the case of the renal tubules, tissue *plus* medium is treated with acid, and the entire extract is analyzed. The total intracellular fluid volume is approximately 2.5% of the total volume analyzed, approximately 0.026 ml per 7 mg of tubular protein in a total volume of 1.1 ml.

The data have all been reported in terms of nanomoles of intermediate per milligram of protein for comparative purposes and because of the difficulty of obtaining an exact measure of intracellular fluid volume in these tubules. When reported in this way, they are not equivalent to concentrations in preparations with a variable ratio of intracellular to extracellular fluid unless all the intermediates are confined solely to the intracellular space. This is the case with the adenine nucleotides (Tables II and III) and the phosphorylated glycolytic intermediates, but not with pyruvate, lactate, and the Krebs cycle intermediates. In fact, the data in Table III indicate that these intermediates appear in the medium. This is to be expected upon the basis of the known behavior of the renal tubule *in vivo* because it has been shown to be permeable to these intermediates,[28-30] and it is known that the kidney is one of the major organs that utilizes citrate from blood.

The concentration of these intermediates in the medium and intracellular fluids can be calculated in terms of nanomoles per milliliter. Based upon the fact that tubules representing 7 mg of tubular protein occupy a volume of 0.031 ml when spun in a special hemocrit tube, and assuming the intracellular fluids represent 80% of this volume, the volume of intracellular fluid in each incubation is 0.025 ml. Assuming an equal concentration of citrate, for example, in intra- and extracellular fluid, the total citrate is 50 nmoles per milligram of protein $\times$ 7 mg of protein = 350 nmoles/1.1 ml or 320 nmoles/ml in the extra- and intracellular fluid of the tubules. Since the latter is 0.025 ml, the amount of citrate in it is 0.025 ml $\times$ 320 nmoles/ml = 8 nmoles, or 1.1 nmoles of intracellular

[28] S. Balagura and W. J. Stone, *Amer. J. Physiol.* **212**, 1319 (1967).

[29] A. P. Grollman, H. C. Harrison, and H. E. Harrison, *J. Clin. Invest.* **40**, 1290 (1961).

[30] P. Vishwakarma, *Amer. J. Physiol.* **202**, 572 (1962).

TABLE III

CONCENTRATION OF VARIOUS METABOLITES IN TOTAL FLASK CONTENTS (MEDIUM *plus* TUBULES) AND IN MEDIUM ALONE AFTER THE INCUBATION OF ISOLATED RENAL TUBULES[a,b]

Intermediate	Nanomoles per milligram of protein: Medium and tubules	Medium alone
ATP	7.85 ± 0.26	0.15 ± 0.11
ADP	2.09 ± 0.06	—
AMP	0.34 ± 0.03	—
Phosphoenol pyruvate	1.10 ± 0.11	0.22 ± 0.04
Pyruvate	2.40 ± 0.05	2.34 ± 0.06
Citrate	50.1 ± 1.7	48.8 ± 2.1
α-Ketoglutarate	66.9 ± 2.1	68.7 ± 3.2
Isocitrate	2.4 ± 0.06	2.45 ± 0.05
Malate[c]	10.2 ± 0.8	10.4 ± 0.8

[a] This table was taken from the paper by N. Nagata and H. Rasmussen, *Biochim. Biophys. Acta* **215**, 1 (1970).

[b] The incubation was done under conditions similar to those described in Table II. The values obtained in both medium *plus* tubules and medium alone were referred to the protein content obtained by measuring amount of protein in $HClO_4$ precipitate of an aliquot of isolated tubules.

[c] Value when glucose alone was substrate.

citrate per milligram of tubular protein. In contrast, the value for citrate in the whole organ is 2.41 nmoles per milligram of protein. It is difficult to estimate the intracellular citrate values in the whole organ, but these values appear to be in the same range as seen *in vitro*. Furthermore, the citrate value in the tubular medium, 320 nmoles/ml, is similar to the value in mammalian blood plasma of 200 nmoles/ml. Similar calculations for malate show a value of 61 nmoles/ml in tubular medium and 70 nmoles/ml in blood plasma, and for α-ketoglutarate 420 nmoles/ml in tubular medium and 200 nmoles/ml in plasma. Thus the concentrations of citrate and the other Krebs cycle intermediates bathing the tubules *in vitro* are in the same range as those that were obtained *in vivo*.

Insofar as it is possible to accurately determine the distribution of the Krebs cycle intermediates in the medium and intracellularly, their concentrations appear to be equal, and equilibration is rapid (within 1 minute), so that it is possible to measure their concentrations in the total incubation flask rather than separating tubule from medium, using $[^{14}C]$inulin to estimate extracellular H_2O contained in tubule pellet, and measuring only intermediate concentrations in the tissue itself.

This unique feature of the tubules means that it is quite easy to measure the concentrations of Krebs cycle intermediates, and to measure decided changes in their concentrations under different metabolic conditions. The only difficulty that arises in the interpretation of these data is that it is not possible to distinguish between control of a certain metabolic step by changing the activity of a particular mitochondrial enzyme, e.g., succinic dehydrogenase, or by altering the permeability of the tissue to succinate and fumarate. However, this limitation is not as serious as it may seem, because it is possible to carry out similar studies of metabolic control properties in isolated renal mitochondria.[16] A comparison of these results with those obtained with intact tubules allows one to assess the probable role of the cellular transport of these intermediates in the regulation of cell metabolism. In the one careful study in which such comparisons have been made, there was little evidence that regulation of transport of these intermediates across the cell membrane was an important aspect of metabolic control in this tissue.[16] On the other hand, it was clear that changes in $[Ca^{2+}]$, $[H^+]$, and $[HCO_3^-]$ did regulate gluconeogenesis by altering the activity of key mitochondrial and cytosolic enzymes.

A second aspect of the unique permeability properties of these tubules is the types of experiments which can be carried out to define metabolic control points in this metabolic sequence. For example, it is possible to study the rate of conversion of each of the Krebs cycle intermediates to glucose, and the influence of ionic and hormonal factors on each. This allows one to identify possible control steps in this metabolic sequence, and to corroborate data obtained by measuring rates of substrate decarboxylation and concentrations of metabolic intermediates.

A disadvantage of this tissue preparation is that the phosphorylated glycolytic intermediates (with the exception of phosphoenol pyruvate) do not diffuse out of the cell. Hence, their concentrations are extremely low in the usual total flask extract made with these tubule preparations. However, this may not be a serious limitation, because glycolytic rates in cortical nephrons are very low, and the major sites of metabolic control in the gluconeogenic sequence are in its early steps.

Disadvantages of Tubule Preparation

The two most serious disadvantages of this preparation are (1) the inability to measure accurately the intracellular H_2O content, and therefore to measure ion fluxes in this tissue,[12-14] and (2) the inability to correlate the changes in carbohydrate and energy metabolism with changes in physiological functions in this tissue.

Measurements of intracellular H_2O, using standard methods such as $[^{14}C]$insulin as a marker of extracellular water, have been reported. However, there is at least a 25% error in such measurements. It seems likely that this large error is due to a unique and variable extracellular space in this preparation, i.e., the extracellular fluid contained in the collapsed tubules. When viewed microscopically, the tubular segments vary greatly in length and in the degree of collapse of their lumens. Hence, it seems likely that there is a great difference in the amount of extracellular fluid contained in the luminal space of individual tubular segments, and the rate at which the constituents of this luminal fluid exchange with those in the bulk ECF. This does not appear to be a serious limitation for metabolic studies, but has made it difficult to follow ion uptakes and exchanges.

The isolated tubules suffer from the serious disadvantage that it is not possible to correlate metabolic with physiological changes. Thus, for example, when parathyroid hormone (PTH) is added to tubules *in vitro*, there are striking changes in rates of gluconeogenesis.[13] Likewise, when it is infused into thyroparathyroidectomized rats *in vivo*, it induces changes in renal carbohydrate metabolism,[31] but also has marked effects upon calcium, magnesium, sodium, potassium, and phosphate transport. However, there are no simple techniques by which to study these transport functions in bulk preparations of isolated renal tubules. On the other hand, perfused segments of renal cortical nephron have been employed successfully to study transport functions *in vitro*.[32] Studies of the metabolic events in such nephrons has not yet been reported, but should be possible because of the unique property of these cells, i.e., their permeability to pyruvate, lactate, and the Krebs cycle intermediates. Because of this property, these intermediates should accumulate in the incubation medium surrounding the perfused segment of nephron, and should, therefore, be measurable.

[31] N. Nagata and H. Rasmussen, *Biochemistry* **7**, 3728 (1968).
[32] M. B. Burg and J. Orloff, *Amer. J. Physiol.* **219**, 1714 (1970).

[3] The Preparation of Renal Cell Suspensions for the Study of Renin Production and Other Renal Functions[1]

By Andrew M. Michelakis

A method for the preparation of a renal cell suspension which was found to be suitable for the study of renin production *in vitro* is described. This cell suspension system has the advantages that it can be used to

[1] Supported by Grants HL-12683, HL-14359, and GM-15431 from the U.S. Public Health Service, and Grant 72-1045 from the American Heart Association.

study renal cell physiology in the absence of hemodynamic or neurogenic influences.

Although valuable information on kidney functions has been gained utilizing renal slices, such preparation might have the difficulty that relatively few of its cells are in direct contact with the incubation medium. While cell-free homogenates of renal tissue overcome this difficulty, they are obviously unsuitable for studies of metabolic functions which require cellular integrity. The renal cell suspension system described here combines the advantages of cellular integrity with those of exposure of cell surfaces and could, therefore, be used as a versatile tool in physiological and metabolic studies of the kidney.[2-4]

Procedure

Medium. To each 100 ml of sterile MEM Eagle, Spinner modified medium without glutamine (Baltimore Biological Laboratory, Baltimore, Maryland), 1.3 ml of 200 mM L-glutamine sterile solution (Robbin Laboratories, Chapel Hill, North Carolina) was added. The resulting solution was used as the medium in the procedure. The pH was 7.2. The collagenase used was a product of Nutritional Biochemicals Corporation, Cleveland, Ohio.

Methods. Young mongrel dogs weighing 12–20 kg were anesthetized with pentobarbital; through an abdominal incision, both kidneys were excised and immediately placed in ice. A needle was tied in place in each renal artery, and the kidney was perfused with normal saline followed by medium until the fluid coming out through the renal vein was clear. The renal vein was occluded with ligatures, and then a freshly prepared solution of 0.3% collagenase in medium was forcibly injected into the kidney via the renal arterial needle until fluid escaped through the capsule of the kidney. The renal cortex was then removed and minced into pieces less than 3 mm in size. These fragments were washed with a small volume of 0.3% collagenase solution in medium and then placed in 50 ml of the same solution for digestion. This mixture was slowly stirred for 1 hour with a magnetic stirrer under an atmosphere of 95% oxygen and 5% carbon dioxide. After digestion, the mixture obtained was diluted to 100 ml with medium and then filtered in the cold through a double layer of 20 × 12 mesh cotton gauze. The filtrate was then centrifuged in the cold for 3 minutes at 50 *g*, and the supernatant was discarded.

[2] A. M. Michelakis, J. Caudle, and G. W. Liddle, *Proc. Soc. Exp. Biol. Med.* **130,** 748 (1969).

[3] A. M. Michelakis, *Proc. Soc. Exp. Biol. Med.* **135,** 13 (1970).

[4] A. M. Michelakis, *Proc. Soc. Exp. Biol. Med.* **138,** 1106 (1971).

The sediment was then washed with medium and centrifuged as above, three times, in order to remove collagenase. Finally, the cells were suspended in medium for the preparation of the suspension. Microscopic examination of the suspension revealed intact cells which stained with methylene blue. Some tubular fragments were found, but intact glomeruli were not identified.

Several experiments were performed utilizing this system in order to study the effect of catecholamines and cyclic AMP on renin production. In each experiment, 5-ml aliquots of cell suspension were incubated with epinephrine, norepinephrine, or cyclic AMP for 1 hour in a 37° water bath with slow, constant shaking under an atmosphere of 95% oxygen and 5% carbon dioxide.[2] The samples were run in duplicate. At the end of the incubation period the samples were homogenized and assayed for renin by the method of Boucher and associates[5] with minor modifications described by Gordon *et al.*[6] using autologous nephrectomized plasma as substrate. Since the method we employed in studying the effect of various agents on renin production does not provide information concerning the rate of synthesis and the rate of destruction as independent processes, only their algebraic sum is known, i.e., "net production of renin." Suspension incubated with epinephrine (0.5 μg/ml of suspension, added at the beginning and every 5 minutes of the first 0.5 hour of incubation) had 30–55% more renin than the incubated controls. Suspension incubated with norepinephrine (1 μg/ml of cell suspension, added at the beginning and every 5 minutes of the first 0.5 hour of incubation) had 35–65% more renin than the incubated controls. Similarly, in suspension incubated with cyclic AMP (2–4 mg per milliliter of cell suspension, added at the beginning and at the middle of incubation period) there was a 30–50% rise of renin over the incubated controls. The results suggest that the renal cell suspension system might prove to be a versatile tool in renal physiology.

[5] R. Boucher, R. Veyrat, J. De Champlain, and J. Genest, *Can. Med. Ass. J.* **90,** 194 (1964).

[6] R. Gordon, L. Wolfe, D. P. Island, and G. W. Liddle, *J. Clin. Invest.* **45,** 1587 (1966).

Section II

Liver

[4] The Perfused Rat Liver

By JOHN H. EXTON

The perfused liver preparation offers many advantages in the study of intermediary metabolism. It preserves the control systems associated with intact cells and provides a milieu similar to that in the intact animal while retaining the advantages of an *in vitro* system, i.e., strict control of substrate and hormone concentrations and elimination of complicating interactions with other tissues. The technique of liver perfusion was first utilized extensively by Embden and his co-workers in their studies of lipid and carbohydrate metabolism at the beginning of this century. It was later used by Lundsgaard and his students in investigations of hormone effects on liver metabolism. In more recent years, the technique has come back into vogue due to the efforts of Miller[1] and Mortimore.[2] The method of liver perfusion described in this chapter is based on that of Mortimore. Although developed originally for rats[3] it is also being used for mice with minor modifications.[4]

Apparatus. The perfusion apparatus described in this section was originally designed by Dr. Howard E. Morgan, then of the Physiology Department, Vanderbilt University School of Medicine, and Mr. Bailey F. Moore, head of the Apparatus Shop at the Vanderbilt University School of Medicine. The prototype has been modified several times, but the basic system of perfusion remains unchanged. The latest model will be described.

As shown in Figs. 1–4, the apparatus consists of a Plexiglas box (120 cm long, 40 cm high, 40 cm deep) placed on a Formica-coated wooden base containing two light bulbs for heating (150 W, severe service), the blowers for circulation of air (Dayton 2C782, W. W. Grainger, Inc., Chicago, Illinois), the motor for rotating the oxygenation chambers (Bodine 2246 E-30, Bodine Electric Co., Chicago, Illinois), the temperature control system (Thermistemp model 63Ra, Yellow Springs Instrument Co., Yellow Springs, Ohio), and switches. The top of the Formica base can be removed for servicing the motors and heating system, and the front panel of the Plexiglas box can be readily removed to allow the

[1] L. L. Miller, C. G. Bly, M. L. Watson, and W. F. Bale, *J. Exp. Med.* **94**, 431 (1951).

[2] G. E. Mortimore, *Amer. J. Physiol.* **200**, 1315 (1961).

[3] J. H. Exton and C. R. Park, *J. Biol. Chem.* **242**, 2622 (1967).

[4] F. Assimacopoulos-Jeannet, J. H. Exton, and B. Jeanrenaud, *Amer. J. Physiol.* **225**, 25 (1973).

various components of the perfusion system proper to be dismantled (Fig. 1). The air in the Plexiglas box is circulated continuously during perfusion and is kept at 38.5° by a thermostat, the probe (model 621, Yellow Springs Instrument Co.) of which is placed near the livers.

Each of the two oxygenation chambers (Fig. 2) consists of a Plexiglas cylindrical drum (9 cm internal diameter), the external face of which is removable and allows penetration of a Teflon-sealed metal shaft carrying four stainless steel tubes for influent and effluent perfusion media and gases (Figs. 1 and 2). The drum rotates around the metal shaft at 60 rpm being driven by a gear system (model 133, Crown Gear Co., Worcester, Massachusetts) coupled to the Bodine motor described above.[5]

The animals whose livers are to be perfused *in situ* are placed on stainless steel trays (13 cm × 27 cm) which can be slid out of the Plexiglas box through removable windows to facilitate cannulation and other procedures involved in setting up the perfusion (Figs. 1 and 3). The trays have two blocks to which the cannulas and tubing can be anchored by adhesive tape during perfusion (Fig. 3).

Perfusion medium is pumped by means of a peristaltic pump (model 1201, Harvard Apparatus Co., Millis, Massachusetts) from the oxygenation chamber through a filter assembly (Fig. 4), containing nylon gauze (100 mesh), and a bubble trap (Fig. 5) to the liver. Tygon tubing (⅛ o.d. × 1/16 i.d., Formula S50 HL, Norton Plastics and Synthetics Division, Akron, Ohio) is used throughout the perfusion circuit except in the pumps, where Silastic tubing (ST 430, Extracorporeal Medical Specialties, Inc., King of Prussia, Pennsylvania) is used. The size of the pump tubing can be varied together with the pump rate to give the desired flow rate (generally 1–2 ml per minute per gram of liver). For purposes of sampling or infusing substrates or hormones, 2.5-cm lengths of latex tubing (3/16 o.d. × ⅛ i.d.) are inserted into the perfusion circuit proximal to the pump (sampling) and proximal to the portal cannula (infusion).

The medium is equilibrated with humidified O_2–CO_2 (95:5) which enters the oxygenation chamber at a rate of about 200 ml per minute. In experiments with ^{14}C-labeled substrates in which $^{14}CO_2$ is to be measured, the gas mixture may be withdrawn through a series of 3 tubes, each containing 25 ml of 2 *M* NH_4OH. Aliquots from each tube are mixed with scintillation fluid and counted.[3] An alternative method of measuring $^{14}CO_2$ formation is described later.

The cannulas have short bevels and may be of stainless steel tubing

[5] Other fabrication and specification details of the perfusion apparatus are available upon request from Mr. Bailey F. Moore, The Apparatus Shop, Vanderbilt University, School of Medicine, Nashville, Tennessee 37232.

(Eisele Co., Nashville, Tennessee) or Teflon tubing (available from Sargent-Welch Scientific Co., Skokie, Illinois). The portal cannulas may be gauge 18 for rats or gauge 20 for mice. The vena cava cannulas are notched for tying in place and may be gauge 12 for rats or gauge 17 for mice.

Perfusion Media. The media used in the author's laboratory generally contain red blood cells because the increased O_2 carrying capacity permits low flow rates, which are more desirable in experiments using nonrecirculating medium. In some experiments (e.g., studies of ion fluxes, lactate transport or glucose transport), it is preferable to omit red cells and use high flow rates (4–6 ml per minute per gram of liver). Bovine erythrocytes are routinely used in the author's studies, although rat, aged human, or sheep cells may be employed.[6] Beef blood (approximately 5 liters) obtained from the jugular vein of stunned animals at a commercial slaughterhouse is immediately mixed with 0.2 volume of anticoagulant solution consisting of 13.2 g of glucose, 13.2 g of sodium citrate, 4.4 g of citric acid, and 150 mg of chloramphenicol per liter. The anticoagulant solution is prepared on the day of collection and the anticoagulated blood is kept at 2°C before being used within 5 days. The blood is centrifuged, and the plasma is discarded. The cells are washed three times in 2 volumes of saline (9 g of NaCl per liter) and twice in 2 volumes of bicarbonate buffer[7] containing 1 g of bovine serum albumin per liter. The white cell layer is removed by aspiration at each wash step. The packed red cells are stored at 2° and used within 2 days.

Perfusion medium is made up by resuspending the red cells in oxygenated bicarbonate buffer[7] containing 30 g of bovine serum albumin (Cohn fraction V)[8] per liter. Sufficient cells are added to give a hematocrit of 18–22% (for experiments employing recirculating medium) or 36–40% (for experiments with nonrecirculating medium). The pH of the medium is then adjusted to 7.4. The required volume of medium (50–70 ml if recycled and 130 ml or more if not recycled) is pumped into the oxygenation chamber and is recirculated through the system for 15–20 minutes to bring the pH, O_2 tension and temperature to the required values (7.4, 250 m*M* Hg, and 37.5°, respectively). In the case of perfusion at slow flow rates, it may be necessary to pass the medium which is returning from the pumps through a coiled length of tubing immersed in water at 37.5°.

[6] Bovine cells are preferred, especially in studies of hepatic carbohydrate metabolism, because of their low rate of glycolysis.

[7] H. A. Krebs and K. Henseleit, *Hoppe-Seyler's Z. Physiol. Chem.* **210**, 33 (1932).

[8] Albumin purchased from any of the major supply houses is usually satisfactory. Human serum albumin may also be used if available.[4]

FIG. 1. *Upper panel:* General view of liver perfusion apparatus. The circuit tubing, filter assemblies and bubble traps have been omitted for clarity. The left-

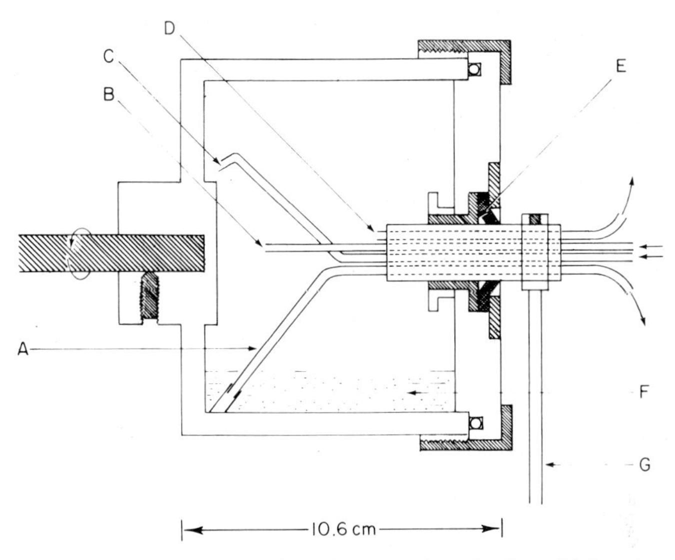

Fig. 2. Cross section of oxygenation chamber. A, tube for withdrawing medium; B, gas inlet tube; C, tube for returning medium. D, gas outlet tube; E, Teflon seal; F, medium; G, stainless steel arm to keep tube cluster fixed while chamber is rotating.

Technique of Perfusion. The procedure of Mortimore[2] is followed. The rat (90–150 g) is injected intraperitoneally with 6 mg of sodium pentobarbital per 100 g. The anesthetized animal is placed on the mobile tray and fastened in place using adhesive tape (Fig. 3). The abdomen is opened widely and the gut is reclined to the left to expose the portal vein and inferior vena cava. Using surgical silk (4/0) mounted on a curved needle, three loose ligatures are set: (a) around the portal vein just before its entry into the liver (inclusion of the bile duct must be avoided); (b) around the superior mesenteric and celiac arteries; (c) around the inferior vena cava above the right renal vein (Fig. 3).

Circulation of medium through the perfusion system is continued dur-

hand animal tray (A) has been slid out of Plexiglas box through the removable window to its position during the cannulation procedures; the right-hand tray (B) is in its normal position during perfusion. The Plexiglas oxygenation chambers (C) and the gear system (D) which drives them are shown. The four front inlet ports (E) to the blowers for circulation of air are visible, the four rear outlet ports are obscured. The temperature control system is housed behind panel F. *Lower panel:* View of exposed base of perfusion apparatus. Visible are the electric motor (A) which rotates the oxygenation chambers, the blowers (B) for circulation of air, the temperature control system (C), the switch and cooling grille (D) for the electric motor, and the switch (E) for the temperature control system.

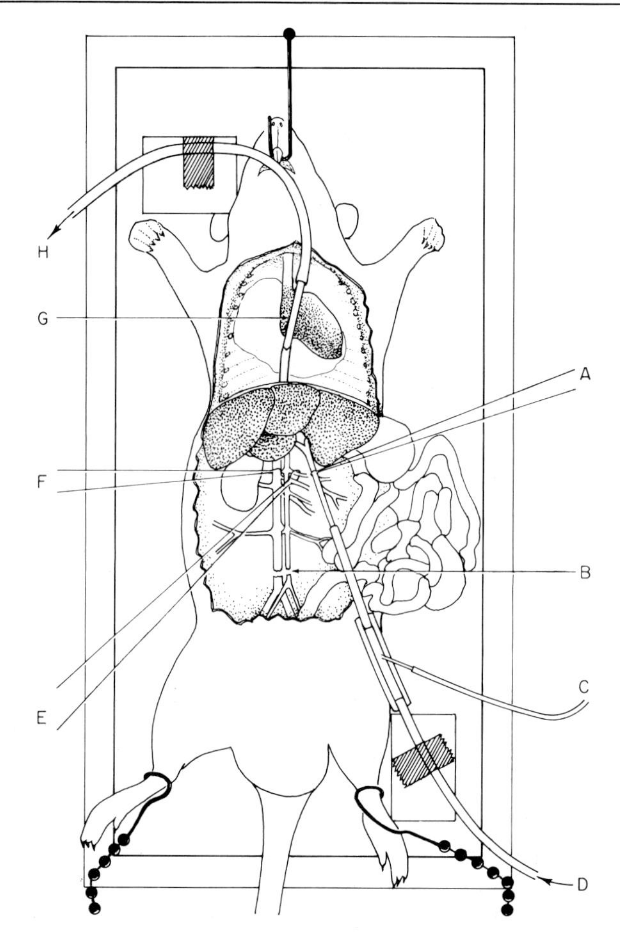

FIG. 3. Diagram depicting cannulation and perfusion procedures. A, ligature tying portal vein cannula and placed above entry of splenic vein; B, inferior vena cava and abdominal aorta sectioned after insertion of portal cannula; C, infusion tubing and cannula inserted into rubber tubing insert in inflow line; D, inflow line taped to aluminum block to hold portal cannula in place; E, ligature around celiac axis and superior mesenteric arteries, tied off after insertion of portal cannula; F, ligature around inferior vena cava above R renal vein, tied off after insertion of vena cava cannula; G, vena cava cannula inserted through right atrium into inferior vena cava, not tied; H, outflow line taped to aluminum block to hold vena cava cannula in place.

ing the above procedures by temporarily connecting portal and vena cava cannulas. It is then stopped by turning off the pump and clamping the tubing proximal to the portal cannula and distal to the vena cava cannula. The cannulas are then separated and the portal cannula is inserted

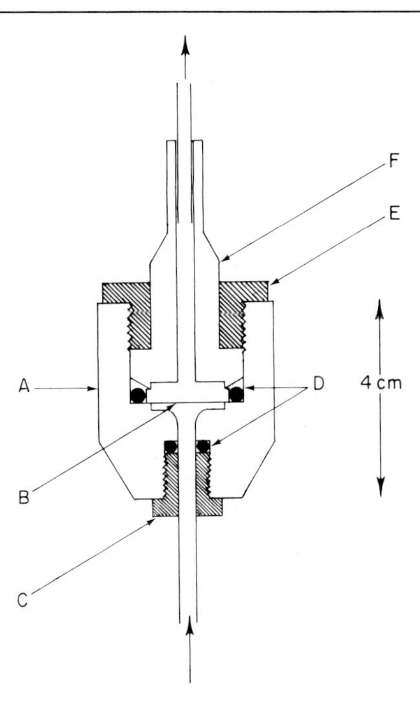

FIG. 4. Cross section of filter assembly. A, Plexiglas outer casing; B, nylon gauze (100 mesh); C, brass screw carrying stainless steel inflow tube; D, rubber O-ring seals; E, anodized aluminum screw for holding outflow assembly in place; F, Teflon outflow assembly carrying stainless steel outflow tube.

into the portal vein. Immediately following this, the proximal clamp is removed, and the pump is restarted. Speed is essential at this stage in order to minimize interruption of blood flow to the liver. The vena cava is then sectioned below the set ligature so that perfusion medium flowing through the liver may escape via this route. The portal cannula is fixed in place with adhesive tape and the ligatures around the portal vein and the arteries are tied.

The chest is then opened widely and the heart is exposed. The vena cava cannula is inserted through the right atrium into the inferior vena cava. The ligature around the lower part of the inferior vena cava is tightened, thus closing the circuit (Fig. 3). In experiments with recirculating medium, the perfusate leaving the liver flows by gravity back to the oxygenation chamber, which is 15 cm below the liver.[9] In experiments with nonrecirculating medium, commencement of the flow of effluent

[9] A negative hydrostatic pressure must be maintained in the portal cannula to prevent swelling and deterioration of the liver.

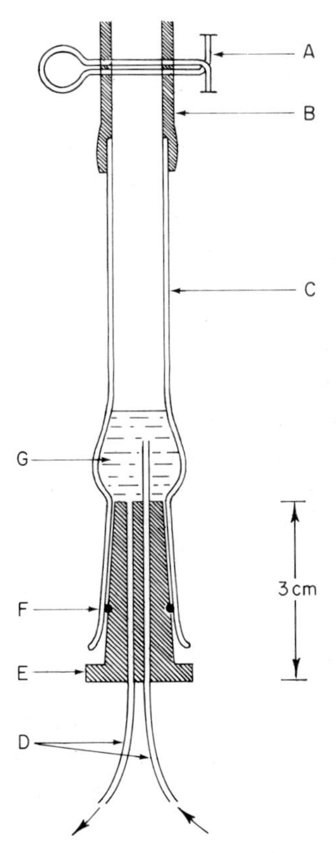

FIG. 5. Cross section of bubble trap assembly. A, clamp; B, rubber tubing, C, glass outer ground joint, standard taper 10/30; D, stainless steel inflow and outflow tubes; E, Teflon inner ground joint carrying inflow and outflow tubes; F, rubber O-ring seal; G, medium.

medium from the liver is facilitated by mild suction. The effluent medium is collected in sequential samples and is not returned to the oxygenation chamber. If $^{14}CO_2$ production is to be measured in this type of experiment, samples (1 ml) of effluent perfusate are drawn directly from the tubing and injected into Erlenmeyer flasks (25 ml) sealed with rubber stoppers (Kontes Glass Co., Vineland, New Jersey) and equipped with plastic center wells (Kontes) containing filter paper wicks impregnated with 0.3 ml of 1 *M* NaOH. To measure $^{14}CO_2$, 0.5 ml of 30% $HClO_4$ is injected into the flasks which are then incubated with shaking for 2 hours at 37°. The center wells are transferred to scintillation fluid and counted.[3]

To prevent the liver from drying during perfusion, it is routinely covered with a gauze pad soaked in warm saline. This is removed at intervals for inspection of the liver.

In experiments with recirculating medium, samples are withdrawn from the oxygenation chambers before including the liver in the perfusion circuit and usually at 20-minute intervals thereafter. Since loss of perfusate during the cannulation procedures is variable, the volume of recirculating medium is measured by pumping the medium out of the system into a measuring cylinder at the end of perfusion and adding to this volume the residual volume (usually 1 ml) and the volume of the samples withdrawn during perfusion. In experiments with nonrecirculating medium, it is necessary to know the flow rate of the medium accurately. This is measured at the beginning of each perfusion experiment.

Cleaning. Bacterial contamination is a complication of all perfusion studies and must be vigorously searched for and prevented if possible. Some bacterial infection of blood is almost inevitable during collection at the slaughterhouse. This is minimized by routine sterilization of the container, inclusion of chloramphenicol in the anticoagulant solution, and instruction of the personnel at the slaughterhouse. Spot checks of bacterial contamination and of the antibiotic sensitivity of contaminating organisms are also undertaken. If sterility is important, aged human blood may be used.

At the end of a single perfusion, the circuit is cleaned by pumping through it at least 500 ml of saline at the rate of 30 ml per minute. At the end of a day of perfusion, the perfusion circuit, including all tubing and all components, is completely dismantled and soaked in bactericidal detergent (Dreft, Proctor and Gamble, Cincinnati, Ohio). Tubing is also routinely discarded after 1 month of use.

General Considerations. Livers perfused by this technique for up to 2 hours show no deterioration, as evidenced by their gross and electron microscopic appearance.[3,4] There is negligible edema as shown by the low water content of livers perfused for 1 hour (0.765 ± 0.003 ml per gram, wet weight)[10] and by the absence of an increase in liver weight during perfusion for 2 hours.[3] Bile production is maintained and release of K^+ into the medium is minimal.[11] Release of enzymes (alkaline phosphatase, glutamate oxaloacetate transaminase, lactate dehydrogenase) into the perfusate is absent or negligible. Oxygen consumption is steady[3,12] and levels of ATP are maintained.[4,13] Steady rates of glycoge-

[10] L. E. Mallette, J. H. Exton, and C. R. Park, *J. Biol. Chem.* **244,** 5713 (1969).

[11] T. F. Williams, J. H. Exton, N. Friedmann, and C. R. Park, *Amer. J. Physiol.* **221,** 1645 (1971).

[12] J. H. Exton, J. G. Corbin, and S. C. Harper, *J. Biol. Chem.* **247,** 4996 (1972).

[13] J. H. Exton and C. R. Park, *J. Biol. Chem.* **244,** 1424 (1969).

nolysis, gluconeogenesis, ureogenesis, and steady levels of cyclic AMP are attained within 15 minutes of completion of the perfusion circuit and are maintained for at least 60 minutes.[3,4,10,12,14] Steady rates of utilization and oxidation of substrates such as lactate and alanine are also observed.[3,4,10,12] The rates of most metabolic processes appear to be similar to those occurring in the intact animal[3,4] with the exception of glycogen synthesis which is low[3,15] and proteolysis which is increased after 30 minutes of perfusion without added amino acids or insulin.[16] Livers perfused by this technique also respond rapidly to physiological levels of hormones[4,10,11,14,16-18] or to changes in the substrate concentration within the physiological range.[3,4,10,13,19]

The perfusion technique[2] described here has been successfully employed in a wide variety of studies of hepatic metabolism. The following is a list of selected references to such studies. Carbohydrate metabolism,[3,4,10-15,17-26] protein metabolism,[10,12,16,26] lipid metabolism,[3,12,27-30] glucagon, catecholamine, or cyclic AMP action,[4,11-14,16,17,19,20,23-25,29,31,32] insulin action,[2,11,12,18,19,21,25,31-33] steroid action,[34-36] membrane transport,[22,37] ion fluxes,[2,11,18,25,30,31,38] and hormone assay.[14,39,40]

[14] J. H. Exton, G. A. Robison, E. W. Sutherland, and C. R. Park, *J. Biol. Chem.* **246**, 6166 (1971).

[15] The low rate of glycogen synthesis observed in earlier studies[3] was probably due to the low level of glucose in the perfusion medium since this substrate markedly affects the rate of glycogen deposition (H. Buschiazzo, J. H. Exton, and C. R. Park, *Proc. Nat. Acad. Sci. U.S.* **65**, 383 (1970); W. Glinsmann, G. Pauk, and E. Hern, *Biochem. Biophys. Res. Commun.* **39**, 774 (1970).

[16] C. E. Mondon and G. E. Mortimore, *Amer. J. Physiol.* **212**, 173 (1968).

[17] J. H. Exton and C. R. Park, *J. Biol. Chem.* **243**, 4189 (1968).

[18] G. E. Mortimore, *Amer. J. Physiol.* **204**, 699 (1963).

[19] J. H. Exton, J. G. Corbin, and C. R. Park, *J. Biol. Chem.* **244**, 4095 (1969).

[20] J. H. Exton, S. B. Lewis, R. J. Ho, G. A. Robison, and C. R. Park, *Ann. N.Y. Acad. Sci.* **185**, 85 (1971).

[21] G. E. Mortimore, E. King, Jr., C. E. Mondon, and W. Glinsmann, *Amer. J. Physiol.* **212**, 179 (1967).

[22] T. F. Williams, J. H. Exton, C. R. Park, and D. M. Regen, *Amer. J. Physiol.* **215**, 1200 (1968).

[23] T. H. Claus, D. M. Regen, J. H. Exton, and C. R. Park, *J. Biol. Chem.* Submitted for publication.

[24] M. Ui, T. H. Claus, J. H. Exton, and C. R. Park, *J. Biol. Chem.* **248**, 5344 (1973).

[25] W. Glinsmann and G. E. Mortimore, *Amer. J. Physiol.* **215**, 553 (1968).

[26] K. H. Woodside and G. E. Mortimore, *J. Biol. Chem.* **247**, 6474 (1972).

[27] J. D. McGarry and D. W. Foster, *J. Biol. Chem.* **246**, 1149 (1971).

[28] J. D. McGarry and D. W. Foster, *J. Biol. Chem.* **246**, 6247 (1971).

[29] D. M. Regen and E. B. Terrell, *Biochim. Biophys. Acta* **170**, 95 (1968).

[30] R. L. Hamilton, D. M. Regen, M. E. Gray, and V. S. LeQuire, *Lab. Invest.* **16**, 305 (1967).

In most instances, perfusion with nonrecirculating medium is preferred. With this method, substrates and hormones can be maintained at constant physiological concentrations, and rapid changes in substrate utilization or product formation can be readily measured. Complications due to the transformation of substrates or products in the medium itself are also avoided with this method because substrates can be infused into the influent medium just prior to the portal cannulas and effluent medium can be immediately deproteinized. The disadvantages are that more medium is needed and that some metabolic changes are too small to produce measurable alterations in substrates or products during a single passage through the liver.

In this description, attention has been focused on the use of the perfused liver for the study of short-term aspects of hormonal control of metabolism. It has not as yet proved feasible to perfuse livers satisfactorily for more than 3 hours with the unsupplemented medium described in this chapter. The reasons for the deterioration of the liver during longer perfusions are not known, but it is suspected that the lack of certain plasma constituents and excessive bacterial growth may be factors. It has been found in the author's laboratory that livers can be perfused satisfactorily for up to 6 hours with medium consisting of equal parts of rat blood collected under sterile conditions and bicarbonate buffer. Inclusion of antibiotics in this medium might result in satisfactory perfusions of even longer duration. Such perfusions would be desirable in the study of the mechanisms of action of slowly acting hormones, of the induction of enzymes and of the regulation of the synthesis and release of proteins such as albumin and lipoproteins.

---

[31] L. S. Jefferson, J. H. Exton, R. W. Butcher, E. W. Sutherland, and C. R. Park, *J. Biol. Chem.* **243,** 1031 (1968).

[32] J. H. Exton, J. G. Hardman, T. F. Williams, E. W. Sutherland, and C. R. Park, *J. Biol. Chem.* **246,** 2658 (1971).

[33] J. H. Exton, S. C. Harper, A. L. Tucker and R. J. Ho, *Biochim. Biophys. Acta* **329,** 23 (1973).

[34] J. H. Exton, N. Friedmann, E. H. A. Wong, J. P. Brineaux, J. D. Corbin, and C. R. Park, *J. Biol. Chem.* **247,** 3579 (1972).

[35] J. H. Exton, S. C. Harper, T. B. Miller, Jr., and C. R. Park, *Biochim. Biophys. Acta.* Submitted for publication.

[36] J. H. Exton, S. C. Harper, A. L. Tucker, J. L. Flagg, and C. R. Park, *Biochim. Biophys. Acta* **329,** 41 (1973).

[37] T. H. Claus, D. M. Regen, and A. Roos, *Amer. J. Physiol.* Submitted for publication.

[38] N. Friedmann and C. R. Park, *Proc. Nat. Acad. Sci. U.S.* **61,** 504 (1968).

[39] R. H. Unger, A. Ohneda, I. Valverde, A. M. Eisentraut, and J. H. Exton, *J. Clin. Invest.* **47,** 48 (1968).

[40] J. Marco, G. R. Faloona, and R. H. Unger, *J. Clin. Invest.* **50,** 1650 (1971).

Addendum. A recent study[41] indicates that fluorocarbons are satisfactory substitutes for erythrocytes in perfusion experiments. Livers perfused with the fluorocarbon FC-47 (3M Company, St. Paul, Minnesota) emulsified with bicarbonate buffer[7] and Pluronic polyol F-68 (BASF Wyandotte Corp., Wyandotte, Michigan) were virtually identical in their metabolic behavior to livers perfused with bicarbonate buffer containing 15% erythrocytes.

[41] M. N. Goodman, R. Parrilla, and C. J. Toews, *Amer. J. Physiol.* **225**, 1384 (1973).

[5] The Fetal Rat Liver Explant as a Model System in the Study of Hormone Action[1]

By Edward Bresnick and Kurt Bürki[2]

Any study of the mechanism of action of a particular hormone in the intact animal may be complicated by different effects upon different tissues, environmental influences, varying rates of absorption from the site of entry, etc. Consequently, it is sometime necessary to isolate a particular tissue for a detailed study of its biochemistry as influenced by hormonal administration. In this regard, organ or tissue culture has been of value as a biological model system. Although many techniques are available for maintenance of tissue explants, they depend upon growth on either solid or fluid media. The latter have been modified such that the tissue is grown at an interphase between the gas and liquid phases; the method of Trowell[3] has been useful in the investigation on liver tissue.

Liver is a particularly difficult tissue to study by the explant technique partly because of its heterogeneity with regard to types of cells. Adult liver is composed of connective tissue elements, a variety of reticuloendothelial cells including Kupffer cells, as well as the hepatic parenchymal cells, i.e., hepatocytes. Under most circumstances, the parenchymal cells of adult liver tissue will not proliferate under culture conditions and, in fact, die. However, fetal liver adapts quite well to primary explanation, i.e., cultivation of pieces of tissues fresh from the organism. The fetal liver explant system has several other desirable properties which make it well suited for studying problems of development, cell

[1] The work presented in this article was supported by grants from the National Institutes of Health (GM 18623 and CA 12609), the National Science Foundation, Pharmaceutical Manufacturers Association, and Brown-Hazen Fund.

[2] Present address: Institute of Pathology, University of Bern, Bern, Switzerland.

[3] O. A. Trowell, *Exp. Cell Res.* **6**, 246 (1954).

differentiation, and specific effects of hormones on metabolism. The fetuses are present in a sterile container, which minimizes bacterial contamination. Fetal liver is easily obtained and manipulation of the tissue is easy. The tissue has a homogeneous parenchymatous structure and is composed of relatively few cell types. The environment of the explants can be carefully controlled and, furthermore, the culture procedure does not involve expensive serum-containing media. The latter is also important in lieu of the binding of hormones by serum proteins.

The technique as described by Wicks[4] and modified subsequently[5] has proven most useful in investigating the transcriptional and translational effects of hormones upon fetal liver enzymes as well as the normal ontogenic alterations that occur. This fetal liver explant technique used in our laboratory for the past three years gives predictable results with minimal expense. Its utility has also been extended to include studies on fetal rat skin and lung and on human skin.

Procedure

Materials

Pregnant rats, 18–20 days of gestation. Rats in an earlier stage of gestation can also be used.

Culture room—a small enclosed room with minimal traffic will suffice. No special precautions are necessary. The room should have electrical outlets and a supply of natural gas.

Tissue culture incubator.[6] A humidified temperature-controlled incubator is required which has adjustable air and CO_2 flow monitors. The gas mixture which has been used in this laboratory is 95% air/5% CO_2 with the temperature maintained at 37°. The incubator is located within the culture room.

Lamilar flow hood.[7] All preparations of the explants are conducted within a lamilar flow hood, which is located within the culture room.

O-Syl—a detergent-disinfectant germicidal material obtainable from Nationwide Paper Co., Atlanta, Georgia. A 3% solution in water (v/v) is used to scrub the hoods, table tops and culture room.

Sterile disposable plastic culture dishes. The Falcon No. 3005 plastic petri dishes, 35 mm in diameter, 10 mm in height, have been useful.

[4] W. D. Wicks, *J. Biol. Chem.* **243,** 900 (1968).

[5] K. Bürki, R. Seibert, and E. Bresnick, *Biochem. Pharmacol.* **20,** 2947 (1971).

[6] Napco Incubator obtainable from most distributors.

[7] We have employed a lamilar flow hood obtained from the Troemmer Co., Philadelphia, Pennsylvania.

Stainless steel grids. Stainless steel 16 wire diameter, type 316, 20 mesh, is cut into 2 cm squares. The four corners are bent to form legs 5 mm in height. The grids are then sterilized and placed individually in the plastic petri dishes under sterile conditions. These will serve as the platforms for the fetal liver tissue.

Fine curved forceps, for extirpation of liver from rat fetuses

Sterile scalpel blades, for preparation of cubes of fetal liver

Preparation of Culture Medium. Eagle's minimal essential medium with Hanks' balanced salt solution (without $NaHCO_3$) is prepared,[8] and glucose (1 g/liter), $NaHCO_3$ (0.75 g/liter), and tricine buffer (to 50 mM, pH 7.4) are added. In addition, penicillin–streptomycin, 100 units and 100 mg/ml, respectively, are required. The latter is obtainable from Grand Island Biological Co., Grand Island, New York. The medium is prepared fresh every 4 weeks and sterilized by passage through a Millipore filter.

Preparation of Fetal Liver Explants. A pregnant rat is decapitated, and the blood is allowed to drain freely. The decapitated rat is moistened profusely with 70% ethanol in order to prevent loose hairs from contaminating the field of operation. The skin is cut along the midline, and the two flaps are then pinned down on each side. The abdomen is washed with 70% ethanol, then opened up along the midline. With care not to puncture the intestine, the uterine horns containing the embryos are carefully dissected free and placed in a large petri dish, which is subsequently covered. The uterus is opened up and the embryos are placed in a sterile petri dish. All subsequent steps are to be conducted within the confines of the lamilar flow hood.

Each embryo is decapitated and an incision is placed along the midline; the fetal liver is easily removed by fine curved forceps and placed in a sterile petri dish (35 mm in diameter) containing 1 ml of sterile culture medium. When all the fetal livers have been placed in the petri dish, explants are prepared by dicing the livers with scalpel blades into cubes approximately 2 mm on each side. The cubes of liver are washed twice with culture fluid and randomized. Eight cubes of liver are then placed in the form of a circle upon each sterile stainless steel grid (approximately 2 mg of protein), and the grid is inserted into a sterile plastic petri dish. Sufficient culture medium is added (approximately 5.5 ml) to just wet the lower surfaces of the explants. The vapor-sealed petri dishes are then placed in the humidified incubator at 37° with circulating 95% air/5% CO_2. To prepare approximately 50 dishes (from fetal liver

[8] Obtainable as a dry powder from Grand Island Biological Co., Grand Island, New York, in amounts to make 1–10 liters of medium.

of two pregnant rats) would usually take 2 hours from the time of sacrifice of the rats to incubation. During the incubation, occasional adjustment of the pH to 7.4 may be required; sterile 0.1 *M* NaOH was utilized for this purpose.

Discussion

Fetal liver explants from either mice or rats can be maintained under the simple culture conditions described above for approximately 5 days as reported by Wicks[4] and confirmed in our laboratory. The addition of bovine calf serum, tryptose broth, or both, or increasing the O_2 tension does not have any further beneficial effect upon maintenance of the primary explants (at least during the 5-day period).

As reported by Wicks[4] and confirmed in our laboratory,[9] a substantial loss of blood cells and hematopoietic elements are noted in the liver explants. Under each grid in the petri dish may be noted a red pigmentation which is due to deposited erythrocytes.

Light and electron microscopic examination of the explants revealed the preservation of the morphological characteristics of the parenchymal cells in the intermediate and peripheral areas of the liver cubes for several days, although definite degeneration of the latter is apparent in the central zone of the explants. Some alterations have also been noted in the cytoplasm of the remaining hepatocytes, such as a decrease in mitochondria, diminution in the quantity of granular endoplasmic reticulum, free ribosomes, and glycogen, and a loss of microvilli on the sinusoidal and bile canalicular surfaces. These morphological changes may partially account for a decrease in RNA and protein of the explants[4] during incubation.

The hepatocytes in fetal rat liver are exclusively mononucleated and diploid,[10] and an increased number of binucleated and polynucleated cell types and an increase in the size of nuclei of the explants may be observed with increasing culture tissue. Whether the latter is related to ontogenetic changes in the ploidy has not been established.

Biochemically, the explants behave very well. Thus, specific functions e.g., synthesis of serum albumin and globulins,[11] the induction of α-ketoglutarate tyrosine transaminase by hydrocortisone,[4] the induction of arylhydrocarbon hydroxylase by 3-methylcholanthrene or flavone derivatives,[3,9] have been noted in explants of fetal mouse and rat liver.

[9] K. Cutroneo, R. Seibert, and E. Bresnick, *Biochem. Pharmacol.* **21**, 937 (1972).

[10] C. Nadal and F. Zajdela, *Exp. Cell Res.* **42,** 99 (1966).

[11] E. A. Luria, R. D. Bakirov, T. A. Yeliseyeva, G. I. Abelev, and A. Y. Friedenstein, *Exp. Cell Res.* **54,** 111 (1969).

[^{14}C]Orotic acid or [^{3}H]amino acid incorporation into RNA and protein, respectively, proceeds linearly for at least 12 hours. Similarly, labeled thymidine incorporation into DNA may be observed for at least 12 hours. It has also been noted that during a 5-day incubation, a few mitoses are apparent.

The extent of incorporation of precursors or transport of hormones is a function of the degree of permeability into the cubes and diffusion within the tissue. It was shown previously[4] and definitively established[5] that changes in permeability occur during the first 24–48 hours of incubation. Thus, for example, the net accumulation of polycyclic hydrocarbons in fresh explants was slower and more limited than in 24- or 44-hour preincubated cultures.[12] Consequently, explants should be routinely preincubated for 24 hours prior to addition of hormones or other investigative materials.

Acknowledgment

We wish to thank the Swiss Academy of Medical Sciences for a fellowship to one of us during most of the time these studies were being conducted and Dr. W. D. Wicks, who first acquainted us with the fetal rat liver explant techniques.

[12] K. Bürki, R. A. Seibert, and E. Bresnick, *Arch. Biochem. Biophys.* **162,** 574 (1972).

Section III

Heart

[6] Techniques for Perfusing Isolated Rat Hearts[1]

By J. R. Neely and M. J. Rovetto

The isolated, perfused rat heart has been used extensively in studies concerned with the regulation of muscle metabolism. This *in vitro* preparation affords several advantages over other heart and skeletal muscle preparations. (1) The muscle fibers are intact, and substrates and hormones are carried to the cells through the normal physiological channels by the coronary circulation. (2) The viability of the muscle cells can be monitored by measuring such physiological parameters as heart rate, ventricular pressure development, and cardiac output. (3) The preparation has a stable mechanical function for periods of 4–6 hours of *in vitro* perfusion. (4) The effects of hormones, pharmacological agents, and substrates on the mechanical function and metabolism of the tissue can be readily determined. (5) The levels of ventricular pressure development and cardiac output can be controlled, and the effects of these physiological functions on the metabolic activity of the tissue can be studied. (6) The rate of coronary flow can be regulated independently of ventricular pressure development and cardiac output, thus allowing the effects of various degrees of tissue ischemia on the mechanical function and metabolism of the heart to be studied in an *in vitro* system. (7) The tissue can be quickly frozen at the termination of an experiment with minimal changes in the intracellular levels of metabolites.

Three methods of perfusing the completely isolated rat heart will be described: (1) The classical Langendorff preparation,[2,3] in which the aorta is cannulated and the coronary vessels are perfused by introducing perfusate into the aorta. This preparation does not perform external mechanical work, but it does develop ventricular pressure.[4,5] (2) The working heart preparation,[4] in which the aorta and left atrial appendage are cannulated and perfusate is introduced into the left atrium, and left ventricular contraction pumps perfusate into the aorta. In this system, ventricular pressure development and cardiac output can be controlled by varying the left atrial filling pressure and/or aortic resistance. (3)

[1] This work was supported in part by NIH Contract No. 71-2499.

[2] O. Langendorff, *Arch. Gesamte Physiol.* **61**, 291 (1895).

[3] H. E. Morgan, M. J. Henderson, D. M. Regen, and C. R. Park, *J. Biol. Chem.* **236**, 253 (1961).

[4] J. R. Neely, H. Liebermeister, E. J. Battersby, and H. E. Morgan, *Amer. J. Physiol.* **212**, 804 (1967).

[5] J. R. Neely and H. E. Morgan, *Pa. Med.* **71**, 57 (1968).

The ischemic, working heart preparation,[6] in which coronary flow can be controlled.

Perfusion Medium. A modified Krebs-Henseleit[7] bicarbonate buffer, pH 7.4, equilibrated with $O_2:CO_2$ (95:5) at 37° was used. The final concentrations of salts in this buffer (m*M*) were: NaCl, 118; KCl, 4.7; $CaCl_2$, 2.5 plus 0.5 to balance EDTA; $MgSO_4$, 1.2; KH_2PO_4, 1.2; and $NaHCO_3$, 25. Na_2EDTA (0.5 m*M*) was added to chelate trace quantities of heavy metals that were present in the reagents. Substrates were included to give the final concentration desired. Use of long-chain free fatty acids as substrates required the addition of albumin, usually 3–4% bovine serum albumin. Washed red blood cells have been used to increase the oxygen-carrying capacity of the buffer but, owing to mechanical agitation, it was difficult to prevent hemolysis of the cells and denaturization of hemoglobin. As a result, mechanical activity of the heart usually deteriorated more rapidly when red blood cells were used. Adequate oxygen delivery was achieved without using red blood cells by gassing the perfusate with 95% O_2. At normal rates of coronary flow, the pO_2 of the coronary effluent never fell below 150 mm Hg.

Preparation of the Perfusion Medium. The perfusion medium as described above must be prepared each day and must be continuously equilibrated with $O_2:CO_2$ (95:5) throughout the day to prevent precipitation of calcium salts. The buffer was prepared each day from stock solutions of the individual components. Stock solutions of the following concentrations (grams/liter were prepared): NaCl, 9; KCl, 11.5; $CaCl_2$, 12.2; $MgSO_4 \cdot 7H_2O$, 38.2; KH_2PO_4, 21.4; $NaHCO_3$, 12.93; and Na_2EDTA, 40.7; the solutions were stored refrigerated for several weeks. On the day of use, the buffer was prepared by mixing these stock solutions in the ratio: NaCl (1000 ml); KCl (40 ml); $CaCl_2$ (30 ml plus 6 ml to balance EDTA); $MgSO_4$ (10 ml); KH_2PO_4 (10 ml); Na_2EDTA (6 ml); and $NaHCO_3$ (210 ml). If the $NaHCO_3$ solution was not fresh (usually made daily) it was important to mix the other components and to gas the mixture with $O_2:CO_2$ (95:5) before adding $NaHCO_3$. Gassing with 5% CO_2 lowers the pH and prevents precipitation of calcium salts when $NaHCO_2$ is added.

When long-chain free fatty acids were used as substrate, albumin (3%) was included in the perfusate as a carrier. A fatty acid–albumin complex was prepared by making the potassium salt of fatty acids and adding this salt to a warm 10% solution of albumin. The albumin solution was mixed constantly during addition of the fatty acid. The potassium

[6] J. R. Neely, M. J. Rovetto, J. T. Whitmer, and H. E. Morgan, *Amer. J. Physiol.* **225,** 651 (1973).

[7] H. A. Krebs and K. Henseleit, *Hoppe-Seylers Z. Physiol. Chem.* **210,** 33 (1932).

salt of fatty acids was prepared by dissolving the fatty acids in absolute ethanol (use the minimum volume required to dissolve the fatty acid with warming), adding a slight excess of K_2CO_3 and about 2 ml of H_2O. Ethanol was evaporated by carefully warming the mixture on a hot plate. The aqueous solution was then added to the warm albumin. The resulting fatty acid–albumin complex should be clear. If the amount of fatty acid exceeds the binding capacity of albumin, the solution will appear cloudy. The fatty acid:albumin molar ratio should not exceed 4. The albumin–fatty acid complex was dialyzed overnight against a large volume of Krebs-Henseleit buffer before use (10 volumes buffer:1 volume albumin–fatty acid complex). The dialyzed albumin was diluted with fresh buffer to the desired concentration and filtered through a Millipore filter (0.8 μm) to remove small particles of denatured protein.

Preparation of Heart. Rats weighing between 150 and 400 g have been used. For most purposes, however, 200–300 g rats were used. If smaller animals are used, it may be necessary to decrease the size of the cannulas that are described below. For routine perfusions, the animals were injected intraperitoneally with heparin sodium (2.5 mg) 1 hour before use. The rat was anesthetized with Nembutal (30 mg/animal, i.p.). The abdominal cavity was opened by making a transverse incision with the scissors just below the rib cage. The diaphragm was transected, and lateral incisions were made along both sides of the rib cage. The heart was exposed by folding back the cut section of the anterior chest wall. Prior to excising the heart, it was picked up with the fingers and the lungs were pushed toward the back. In order to retain the correct length of aorta and pulmonary veins for cannulation, the point at which the pulmonary veins join the left atrial appendage was identified and a single cut was made with scissors through this point and on through the other large vessels. This cut should leave about 5 mm of aorta on the heart and should transect the pulmonary veins just distal to their bifurcation. The heart was dropped into a beaker (50 ml) containing 0.9% sodium chloride chilled in ice water. Contractions stopped within a few seconds. After the diaphragm was ruptured the lungs stopped functioning and the heart was exposed to anoxic conditions. Therefore, speed was essential in removal of the heart from the animal. Before cannulation, the heart was cooled until contractions completely stopped (approximately 20 seconds).

Cannulation of the Heart. Using fine-tipped forceps, the heart was picked up by the aorta and any filamentous tissue, thymus, or lung that may have been removed with it were carefully pulled away without removing the heart from the cold saline. Care must be taken not to rupture the aorta during this process. The aorta was grasped on two sides with forceps and slipped about 3 mm onto the aortic cannula (Figs. 1–3). It was then held in place with a small (10 mm) serrefine. Perfusion of the coronaries was begun immediately by unclamping the tube from the washout perfusion reservoir. The heart resumed contraction within a few seconds. The aorta was then secured with a ligature and the serrefine was removed. Since the pulmonary artery was occasionally occluded in this ligation, a small incision was made near its base to ensure a free flow of coronary effluent. This cannulation was all that was required for the Langendorff technique (Fig. 1), and after 10 minutes of washout perfusion the heart was switched to recirculation perfusion as described below. For the working and ischemic techniques, the left atrial appendage was also cannulated.

After the aorta was secured with a ligature, the heart was rotated on the cannula to position the left atrial appendage to receive the atrial cannula (Fig. 2 or 3). The pulmonary veins may be difficult to identify at first. If the cut was made just distal to the bifurcation of the pulmonary veins, the lumina of the two veins can be located about midway between the right and left atria. A number of vessels may be identified in this area. The pulmonary veins can be distinguished from the other vessels by inserting the end of fine-tipped forceps into the lumen to see if it leads into the left atrium. One of the pulmonary veins was slipped about 3 mm onto the atrial cannula and secured with a ligature. This process also tied off the other pulmonary vein. The accuracy and competence of the cannulation were tested by unclamping the tube leading from the atrial bubble trap (Fig. 2 or 3) and observing the filling of the left atrium.

Cannulation of Pulmonary Artery. When oxygen consumption was measured, another cannula was placed in the rubber stopper (Langendorff preparation) or Teflon ball joint (working and ischemic preparations) for cannulation of the pulmonary artery. After the aorta was cannulated, the pulmonary artery was carefully freed from connective tissue down to its base. The artery was slipped onto the cannula and held in place with a serrefine while a ligature was secured. Since the openings into the right atrium were usually closed during cannulation of the left atrium, 90–100% of the coronary flow was ejected via the pulmonary artery. The pulmonary arterial cannula was connected to a short length of Tygon tubing for collection of coronary effluent. Samples of effluent for pO_2 measurements were collected in a 5-ml syringe by inserting a 20-gauge needle through the wall of the Tygon tubing into the lumen of the metal cannula.

Retrograde perfusion of the heart from the washout revervoir was routinely continued for 10 minutes. This perfusion period served to remove blood from the vascular bed, to allow time for cannulation of the

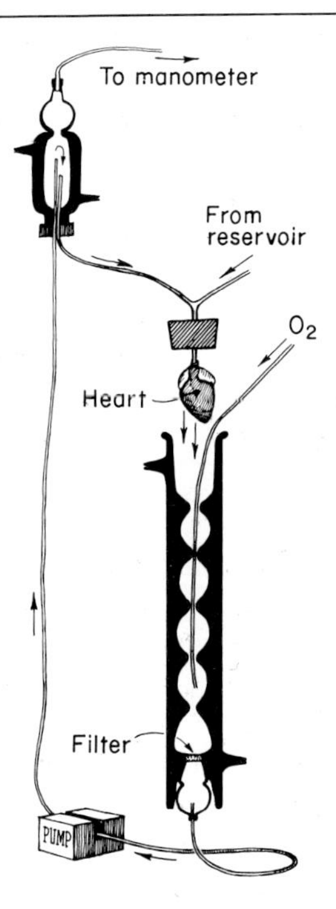

FIG. 1. Langendorff perfusion apparatus. See text for detailed description. Water jackets are indicated by black shading. Reproduced from J. R. Neely and H. E. Morgan [*Pa. Med.* **71,** 57 (1968)] by courtesy of *Pennsylvania Medicine.*

left atrium and pulmonary artery, and to allow the heart to recover from the period of anoxia associated with its removal from the animal. The coronary effluent from this perfusion was discarded.

Langendorff Perfusion Apparatus. The apparatus used for perfusing hearts by the Langendorff technique is shown in Fig. 1. This apparatus consisted of the following parts: (1) *Heart chamber:* A "ful-jak" allihn condensor (2.5 × 60 cm) with 29/42 standard taper ground joints served as a heart chamber, perfusate reservoir, and oxygenating chamber. The lower end of the condensor was fitted with a coarse-porosity, sintered-glass filter. (2) *Mechanical pump:* The Masterflex rotary pump (model 7014) equipped with a variable speed control is an inexpensive, reliable pump that can be used for this purpose. Viton tubing was used in the pump since Tygon tubing releases plasticizers into the perfusate. (3) *Combination bubble trap, pressure chamber:* This chamber consisted of the female portion of a water-jacketed, 10/30 standard taper glass joint which was connected to a mercury manometer by a short length of Tygon tubing

($\frac{3}{32}$ inch i.d.). The male plug was made of Teflon and fitted with a neoprene O ring to facilitate sealing. The Teflon plug held two short lengths of stainless steel tubing (0.058 inch o.d.) one of which was connected to a Tygon tube from the mechanical pump and the other to the aortic cannula via a short length of Tygon tubing (approximately 5 cm). It was important to place a short tube between the bubble trap and aortic cannula to prevent cooling of the perfusate. (4) *Aortic cannula:* This was made of 0.109 inch o.d. stainless steel tubing with a small groove located about 1 mm from the end to facilitate securing the aortic ligature. The cannula was held in a size 6 rubber stopper which was grooved to allow passage of a gas tube into the oxygenating reservoir and to allow gas to escape from the chamber. The aortic cannula was also connected to a second perfusate reservoir by Tygon tubing. This reservoir was positioned 75 cm above the heart and was used for the 10-minute preliminary washout perfusion (approximately 60 mm Hg hydrostatic perfusion pressure). The tubes carrying perfusate from the mechanical pump to the bubble trap and from the washout reservoir to the heart were water jacketed by enclosing them in larger (3/8 inch i.d.) Tygon tubes that were connected in series with the water-jacketed glass condensor and bubble trap. Water was recirculated through the jacket at 37° by a constant temperature recirculating system (Precision model 66600). It was important to equilibrate the gas mixture with water at 37° before passing it through the oxygenating reservoir to prevent evaporation of water from the perfusate.

Perfusion of Hearts by the Langendorff Technique. The apparatus was made ready for perfusion of the heart by placing the rubber stopper in the top of the heart chamber-oxygenating reservoir, which contained a measured volume of perfusate (minimum recirculating volume for this system was approximately 15 ml). The tube from the washout reservoir was clamped with a hemostat. Perfusate was pumped from the oxygenating reservoir through the aortic cannula. After all air bubbles were removed from the tube connecting the bubble trap and aortic cannula, this tube and the tube between the oxygenating reservoir and mechanical pump were clamped simultaneously with hemostats. The bubble trap contained about 3 ml of perfusate, and the pressure in the chamber was adjusted to about 60 mm Hg. The rubber stopper was then removed from the oxygenating chamber and held alongside the chamber by a castalloy clamp during cannulation of the aorta. The tube leading from the washout reservoir was filled with perfusate. The heart was removed from the animal and cannulated as described above. The preliminary perfusion was begun from the washout reservoir immediately after the aorta was cannulated. At the end of this preliminary perfusion (usually 10 minutes), the tube from the washout reservoir was clamped and the heart was quickly transferred to the heart chamber. Recirculation perfusion was begun by unclamping the tubes from the bubble trap to the aortic cannula and from the reservoir to the pump. Aortic perfusion pressure was adjusted by varing the speed of the Masterflex pump. For adequate coronary perfusion and oxygen delivery, perfusion pressure should not be less than 50 mm Hg. Perfusion under this condition could be continued for 3 hours without significant deterioration in mechanical performance. If it is not desirable to recirculate the perfusate, single flow-through perfusions can be conducted from the washout reservoir without the use of a pump and bubble trap. The hydrostatic perfusion pressure under this condition can be varied by adjusting the height of the reservoir relative to the heart.

Perfusion Apparatus for Working Hearts. Figure 2 illustrates the apparatus used to perfuse hearts that were performing external mechanical work. The complete apparatus is shown on the left and the cannula assembly, heart chamber and pressure chamber are shown in the enlargement on the right. This apparatus consisted of the following parts: (1) *Heart chamber.* The female portion of a 35/25 ground-glass ball joint was used. The chamber was held in place with a pinch clamp. The lower portion of the chamber was reduced in size and connected to a short length of Tygon tubing which served to carry coronary effluent back into the reoxygenating reservior. (2) *Cannula assembly.* This consisted of two stainless steel cannulas (0.134 inch o.d.), which were grooved about 1 mm from their tips to accommodate ligatures. The aortic cannula had a smaller tip made by soldering a short length of 0.109 inch o.d. tubing into the cannula. This cannula was also provided with a sidearm which was connected to the washout reservoir for preliminary perfusions and to a pressure transducer for measuring aortic pressure. The atrial cannula had a 0.5 cm tip-soldered on at a 90° angle to facilitate cannulation of the pulmonary vein. (3) *Aortic and atrial bubble traps and aortic pressure chamber.* These were made from female portions of 14/35 standard taper-water packeted points. The bubble traps were provided with sidearms for overflow of perfusate back into the oxygenating reservoir, and their tops were open. The sidearm for the atrial bubble trap was extended with Tygon tubing (3/8 inch o.d.) and connected to the oxygenating reservoir by an 18/9 ball joint. A sidearm for the aortic bubble trap was made from the male portion of a 28/15 ball joint which was connected to the oxygenating reservoir by 28/15 ball, 29/42 standard taper adapter. This design of the bubble traps eliminated pressure fluctuations due to the pump and ensured a constant hydrostatic pressure for filling the left atrium and perfusing the coronaries. The top of the pressure chamber

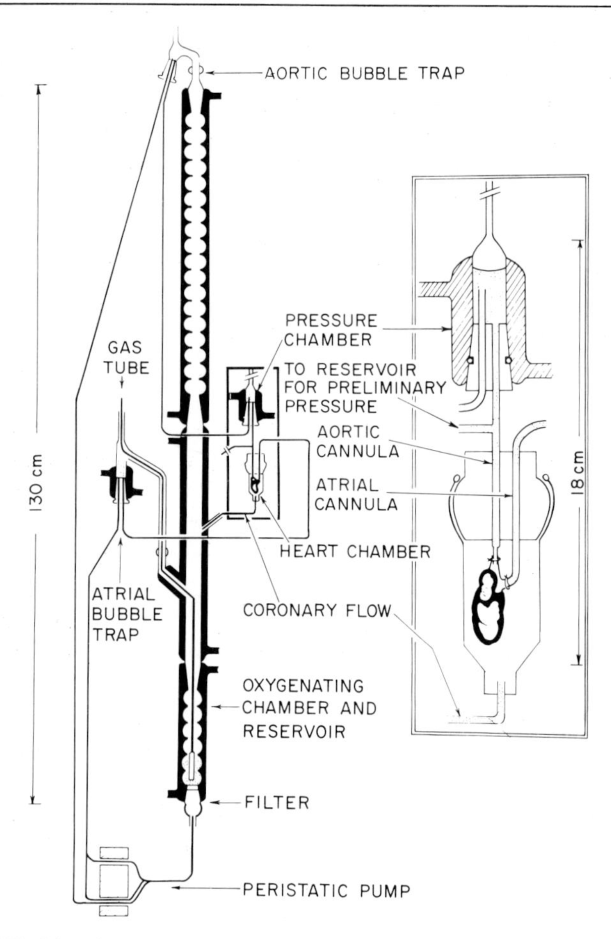

FIG. 2. Working heart perfusion apparatus. See text for detailed description. Water jackets are indicated by black shading. Fluid in bubble traps and pressure chamber is indicated by stippling. From J. R. Neely, *et al.*,[6] reproduced by courtesy of *American Journal of Physiology.*

was reduced in size and connected to a 5-ml syringe for adjusting the volume of air in the chamber. Male plugs for the bubble traps and pressure chamber were made of Teflon and were fitted with neoprene O rings.
(4) *Oxygenating reservoir.* This chamber was 130 cm in length and consisted of three water-jacketed condensors with 29/42 standard taper joints. "Ful-jak" allihn condensors, 60 cm and 20 cm long, were used for the top and bottom portions, respectively. The middle portion was a special condensor (H. S. Martin and Co., Evanston, Illinois) which was pro-

vided with sidearms for receiving overflow from the atrial bubble trap and coronary effluent from the heart chamber. The bottom condensor was fitted with a coarse porosity, sintered-glass filter. (5) *Mechanical pump.* Any pump capable of providing a flow of about 75 ml per minute is adequate. Both a peristaltic pump (model 505, Harvard Apparatus Co.) or a less expensive Masterflex rotary pump (Model 7014) with a variable speed control have been used. The Tygon tubes leading to the bubble traps and from bubble traps to the heart were water jacketed by inserting these tubes through larger (3/8 inch i.d.) Tygon tubes that were connected in series with the water jackets on the condensors. Water was pumped through the jacket at 37° by a constant temperature circulating system (Precision model 66600).

Perfusion by the Working Heart Technique. This apparatus was made ready for perfusing a heart by recirculating a measured volume of perfusate (minimum circulating volume was approximately 35 ml) until the tubes leading to the atrial and aortic cannulas were full of perfusate and free of air bubbles. These tubes were then clamped with hemostats, the heart chamber was removed and the tube leading from the washout reservoir to the aortic cannula was filled with perfusate. The heart was removed from the animal, the aorta was cannulated, and a 10-minute preliminary perfusion was begun as described above. During this time the left atrium was cannulated. Recirculation perfusion was begun by clamping the tube from the washout reservoir, replacing the heart chamber and starting retrograde perfusion from the aortic bubble trap. Heart work was begun by unclamping the tube from the left atrial bubble trap. The speed of the mechanical pump was adjusted to ensure an excess of flow to the atrial bubble trap. Cardiac output and ventricular pressure development was adjusted to the desired level by changing the height of the atrial bubble trap relative to the heart and/or by adjusting the compliance of the aortic outflow tract. Compliance in this system was dependent on the volume of air (routinely maintained at 2 ml) in the aortic pressure chamber.

After the left atrium was cannulated and ventricular filling was begun, cardiac output was more than sufficient for coronary flow. Excess cardiac output was pumped by the ventricle via the aortic pressure chamber and bubble trap to the top of the apparatus, where it overflowed into the oxygenating reservoir. The long, central oxygenating chamber was necessary to provide sufficient surface area for equilibration of O_2 and CO_2 between the gas phase and perfusate. Cardiac output could be measured by collecting the coronary effluent and the overflow from the aortic bubble trap. Since the right side of the heart was not cannulated, coronary effluent was returned to the right atrium and was pumped through the pulmonary artery by the right ventricle, where it collected in the heart chamber and was returned to the oxygenating reservoir by a short length of Tygon tubing. Aortic pressure was measured by connecting a pressure transducer to the sidearm on the aortic cannula.

Apparatus for Perfusing Ischemic Hearts. In the working heart preparation, the rate of coronary flow was a direct function of the aortic perfusion pressure. The cannula assembly of the working heart apparatus was modified to allow the rate of coronary flow to be controlled in hearts that were performing external mechanical work. Hearts were cannulated and control perfusions were conducted in the same way as described above for the working preparation. Ischemia was induced by two techniques. (1) Coronary flow was reduced by restricting cardiac output, while ventricular pressure development was maintained by increasing aortic resistance. (2) Coronary flow was reduced while maintaining cardiac output and ventricular pressure development by placing a one-way valve in the aortic outflow tract in such a way that it prevented retrograde perfusion of the coronary arteries during diastole.

MODIFICATION 1. The cannula assembly and arrangement of bubble traps that were used to induce ischemia by the low cardiac output, high aortic resistance technique are shown in panel A, Fig. 3. The arrows indicate direction of flow of the perfusate. This was the same cannula assembly used for the working heart preparation with the following modifications: (a) a water-jacketed heart chamber to help maintain myocardial temperature when the rate of coronary flow was reduced and (b) a bubble trap (left-hand side of panels A and B, Fig. 3) connected to the sidearm of the aortic cannula and used to maintain a minimum coronary perfusion pressure as ventricular failure occurred. This bubble trap was the same size as the atrial bubble trap used on the working heart apparatus, and its height relative to the heart could be adjusted. The tube leading from this bubble trap to the aortic cannula was fitted with a one-way ball valve. This valve allowed perfusate to flow into the aorta when the aortic pressure decreased below a preselected level and prevented flow of perfusate from the aorta into the bubble trap at normal levels of ventricular pressure development. Therefore, by adjusting the height of this bubble trap, it was possible to preselect a minimum aortic perfusion pressure and coronary flow during ventricular failure.

Perfusate was pumped from the main oxygenating chamber (see apparatus for working hearts) to the atrial bubble trap and to the minimal coronary flow bubble trap by a variable speed Masterflex rotary pump (model 7014). The overflow sidearm on the atrial bubble trap was set 10 cm above the left atrium. For control perfusions, the speed of the mechanical pump was adjusted so that perfusate overflowed from the atrial bubble trap at all times, thus providing a constant left atrial filling pressure of 10 cm H_2O.

After the 10-minute washout perfusions, hearts were electrically paced at approximately 300 beats/minute and perfusion as a working heart was continued for an additional 10 minutes before inducing ischemia. Ischemia was induced in this preparation by closing the aortic outflow tract and by reducing cardiac output to equal the desired rate of coronary flow. Cardiac output was reduced by adjusting the speed of the mechanical pump that supplied perfusate to the left atrial bubble trap. Decreasing the speed of this pump prevented overflow of perfusate from the sidearm of the atrial bubble trap and allowed the atrial filling pressure to decrease. Under this condition the adjusted flow from the pump was equal to left atrial input, to cardiac output and, therefore, to coronary flow as long as the left atrial pressure remained below 10 cm H_2O. A reasonable level of ventricular pressure development was maintained at the reduced cardiac output by clamping off the aortic outflow tube distal to the pressure chamber and by reducing the volume of air in the pressure chamber from 2.5 to 0.3 ml. In this arrangement, aortic resistance was equal to the resistance of the coronary vascular bed and ventricular pressure development was a function of left atrial filling pressure, coronary resistance and a small residual compliance in the aortic pressure chamber. With these manipulations, the rate of cardiac output and, therefore, coronary flow could be controlled by adjusting the speed of the mechanical pump. This perfusion technique allowed the rate of coronary flow to be controlled in hearts that were initially developing reasonable levels of ventricular pressure (about 90 mm Hg). It did not, however, allow for independent control of coronary flow and cardiac output. To achieve this goal, a second modification of the working heart apparatus was developed.

MODIFICATION 2. Panel B of Fig. 3 illustrates the placement of a one-way valve in the aortic cannula of the working heart apparatus to prevent perfusion of the coronary arteries during ventricular diastole. A detailed drawing of the one-way valve assembly is shown in Fig. 4. Since the largest fraction of coronary flow occurs during diastole, this one-way valve severely restricted coronary perfusion, but did not influence the initial rate of cardiac output or the ventricular afterload. The aortic cannula was provided with sidearms both above and below the one-way valve. These sidearms were connected by a short length of Tygon tubing, which provided a bypass around the one-way valve for control perfusions (similar to the working preparation). The bypass tube was also connected to a pressure transducer, to the preperfusion reservoir and to the minimum coronary flow assembly. The aortic cannula and sidearms were

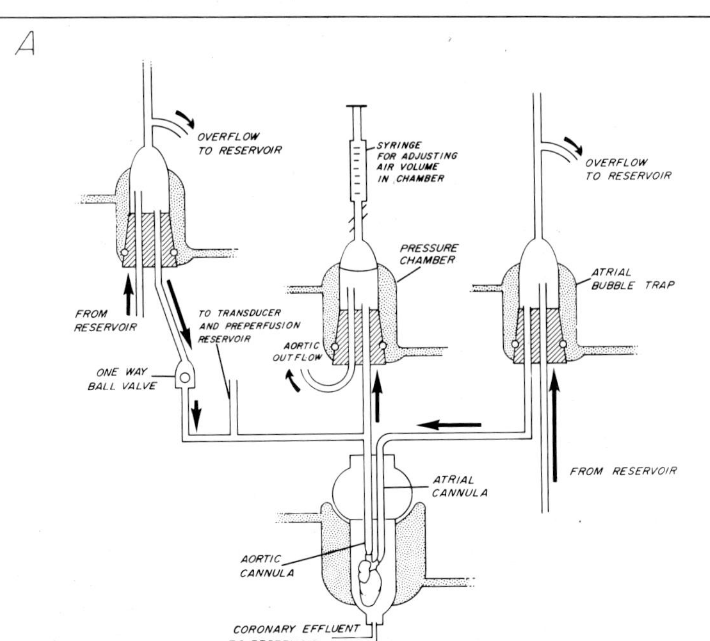
A
OVERFLOW TO RESERVOIR
SYRINGE FOR ADJUSTING AIR VOLUME IN CHAMBER
OVERFLOW TO RESERVOIR
PRESSURE CHAMBER
ATRIAL BUBBLE TRAP
FROM RESERVOIR
TO TRANSDUCER AND PREPERFUSION RESERVOIR
AORTIC OUTFLOW
ONE WAY BALL VALVE
FROM RESERVOIR
ATRIAL CANNULA
AORTIC CANNULA
CORONARY EFFLUENT TO RESERVOIR

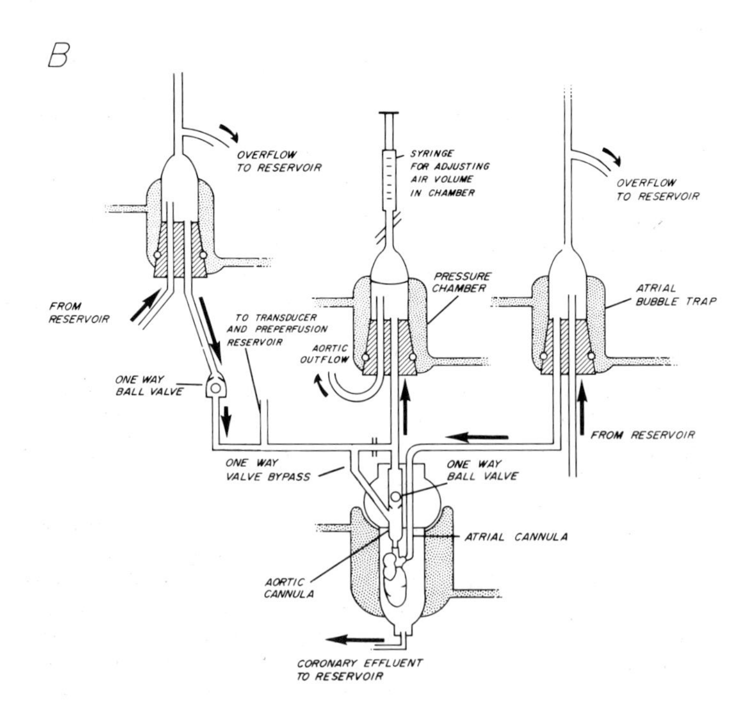
B
OVERFLOW TO RESERVOIR
SYRINGE FOR ADJUSTING AIR VOLUME IN CHAMBER
OVERFLOW TO RESERVOIR
PRESSURE CHAMBER
ATRIAL BUBBLE TRAP
FROM RESERVOIR
TO TRANSDUCER AND PREPERFUSION RESERVOIR
AORTIC OUTFLOW
ONE WAY BALL VALVE
FROM RESERVOIR
ONE WAY VALVE BYPASS
ONE WAY BALL VALVE
ATRIAL CANNULA
AORTIC CANNULA
CORONARY EFFLUENT TO RESERVOIR

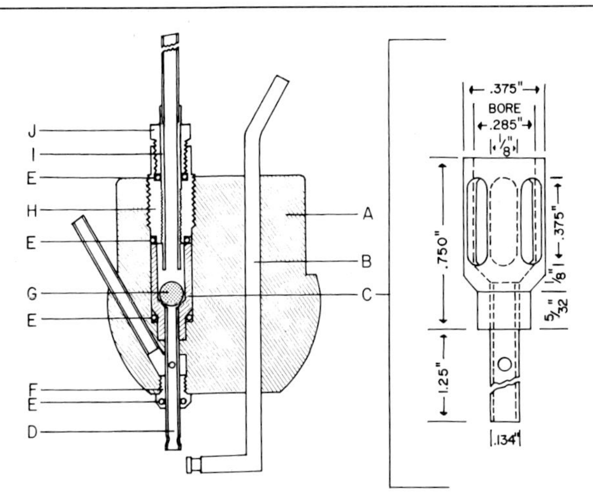

FIG. 4. Detailed drawing of one-way valve assembly. The valve seat is shown enclosed within the Teflon cannula holder in the drawing on the left. An enlargement of the valve seat is shown in the drawing on the right. The parts referred to by letters in the drawings are as follows: A, cannula holder—Teflon; B, atrial cannula—stainless steel; C, valve seat—stainless steel; D, aortic cannula—stainless steel; E, O ring; F, cannula seat-stainless steel; G, ball, 1/4-inch—stainless steel; H, valve retaining plug-stainless steel; I, outflow tube, 0.104 i.d.—stainless steel; J, packing nut—brass.

made of stainless steel tubing (0.109 inch o.d.). The one-way valve assembly consisted of 1/4-inch diameter stainless steel ball and a stainless steel ball seat which was contained in a Teflon plug. The walls of the valve seat were provided with 4 channels (1/4 inch long by 1/8 inch wide) to allow free flow of perfusate around the ball during ventricular ejection. The base of the valve seat was machined to a 45° angle so that the stainless steel ball made only a hairline contact with the seat when the valve

---

FIG. 3. Perfusion apparatus for ischemic hearts. Panel A shows the arrangement of bubble traps and cannula assembly used for the low cardiac output, high aortic resistance method of inducing ischemia. This apparatus was the same as that used for the isolated working rat heart except that a minimum coronary flow assembly was added (left-hand bubble trap). The one-way valve used in this assembly was a 1/8-inch Teflon ball with a ground-glass ball seat which was taken from an Oxford automatic pipettor. Panel B illustrates the placement of a one-way valve in the aortic outflow tract of the working heart apparatus. This valve allowed the ventricle to eject fluid but prevented retrograde perfusion of the coronaries during diastole. The valve assembly was constructed by Mr. Bailey Moore, Apparatus Shop, Vanderbilt University School of Medicine, Nashville, Tennessee. From J. R. Neely *et al.*,[6] reproduced by courtesy of *American Journal of Physiology*.

closed. The Teflon plug was machined to fit into the female portion of a 50/30 water-jacketed, ground glass joint which served to cover the heart and aided in maintaining the ambient temperature at 37°. In addition, coronary effluent was collected in this chamber and returned to the oxygenating reservoir.

Hearts were removed from the animals, the aorta and left atrium were cannulated, and a 10-minute preliminary perfusion was conducted as described for the working heart preparation. Cardiac work was begun by unclamping the tube from the left atrial bubble trap, the hearts were electrically paced at about 300 beats/minute, and perfusion with the bypass tube open was continued for 10 minutes. Ischemia was induced in this preparation by simply clamping the bypass tube at the point indicated in the figure. With this maneuver, the ventricle ejected perfusate via the one-way valve but against the same hydrostatic pressure as prior to clamping the bypass tube. The presence of the one-way valve prevented retrograde flow of perfusate into the aorta during diastole and allowed the diastolic aortic pressure to decrease. This decrease in coronary perfusion pressure resulted in a 60% reduction in coronary flow and in ventricular failure. The rate of flow continued to decrease as ventricular performance deteriorated. A minimum coronary flow could be maintained, however, by adjusting the height of the minimum coronary perfusion assembly (left-hand bubble trap in the figure) as described above.

Mechanical Performance of Isolated, Perfused Hearts. Pressure measurements were made using Statham pressure transducers (model No. P23Gb) connected to either a Sanborn (model 964) or Beckman dynograph (model RB) recorder. Aortic pressures were measured by connecting transducers to the sidearms on the aortic cannulas. In the ischemic heart preparations, aortic pressures were measured from the sidearm below the one-way valve. Intraventricular pressures were measured by inserting a 20-gauge needle through the apex of the heart into the left ventricle.

Figure 5 illustrates the effects of raising aortic perfusion pressure on ventricular pressure development in the Langendorff preparation. When perfusion pressure was raised from 40 to 120 mm Hg, peak systolic pressure increased from about 50 to 125 mm Hg. The rate of coronary flow increased from 35 to 144 ml per minute per gram dry tissue under these conditions. Although no provision was made for flow of fluid into the left ventricle in this preparation, ventricular pressure development was a function of the afterload. This increase in pressure development with increased afterload was explained by increased ventricular filling from thebesian veins and/or leakage through the aortic valve as perfusion pressure was raised.

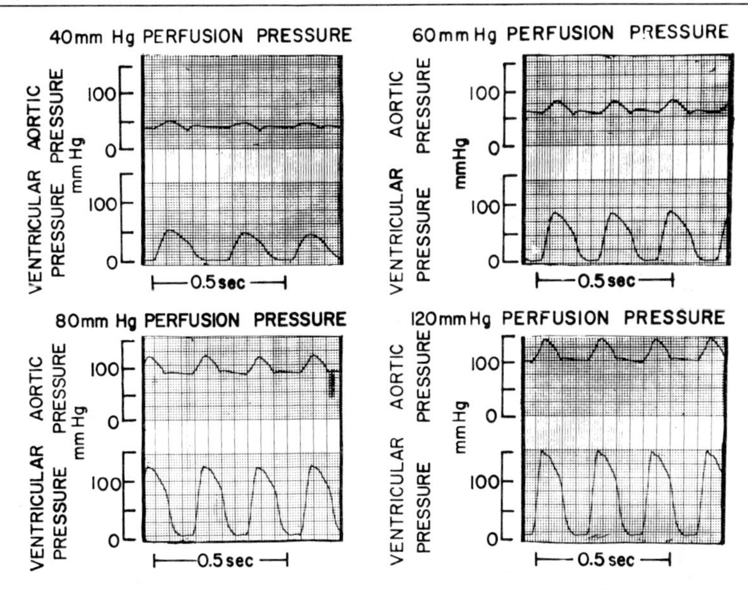

FIG. 5. Effects of perfusion pressure on aortic and ventricular pressures of the Langendorff preparation. Hearts were perfused as described in the text. Simultaneous recordings of aortic and ventricular pressures were made at a chart speed of 100 mm per second. Perfusate contained glucose (5.5 m*M*) as the only exogenous substrate. From J. R. Neely *et al.*,[6] reproduced by courtesy of *American Journal of Physiology*.

Figure 6 illustrates the effect of raising left atrial filling pressure on intraventricular and aortic pressures in the working heart preparation. Raising left atrial pressure from 0 to 20 cm H_2O increased peak systolic pressure in the ventricle from about 60 to 110 mm Hg. Cardiac output increased from about 60 to 300 and coronary flow from about 60 to 80 ml per minute per gram of dry tissue under these conditions.

Figure 7 illustrates typical aortic pressure tracings from ischemic hearts perfused by the low output, high aortic resistance method (panel A) and the one-way aortic valve method (panel B). During the 10-minute control perfusion, the heart in panel A developed 90 mm Hg peak systolic pressure with a diastolic pressure of about 35 mm Hg. Coronary flow was about 70 ml per minute per gram dry tissue. At zero time on the tracing, ischemia was induced by clamping the aortic outflow tract, reducing the volume of air in the pressure chamber from 2.5 to 0.3 ml and adjusting the rate of the atrial input pump to 7 ml per minute. As a result, cardiac output and coronary flow decreased to 7 ml per minute (41 ml per minute per gram dry tissue) and peak systolic pressure decreased from 90 to 60 mm Hg. At this reduced rate of coronary flow,

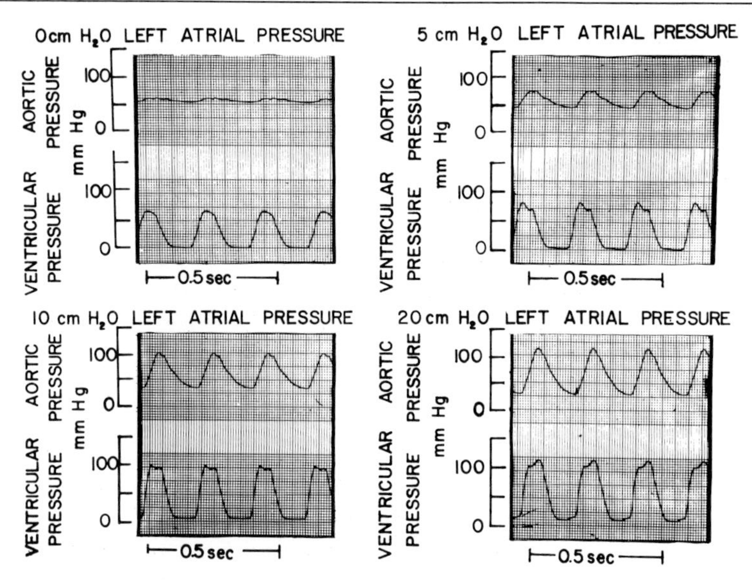

FIG. 6. Effects of left atrial filling pressure on aortic and ventricular pressures of the working heart preparation. Hearts were perfused as described in the text with buffer containing 5.5 m*M* glucose. Simultaneous recordings of aortic and ventricular pressures were made at a chart speed of 100 mm per second. From J. R. Neely *et al.*,[6] reproduced by courtesy of *American Journal of Physiology.*

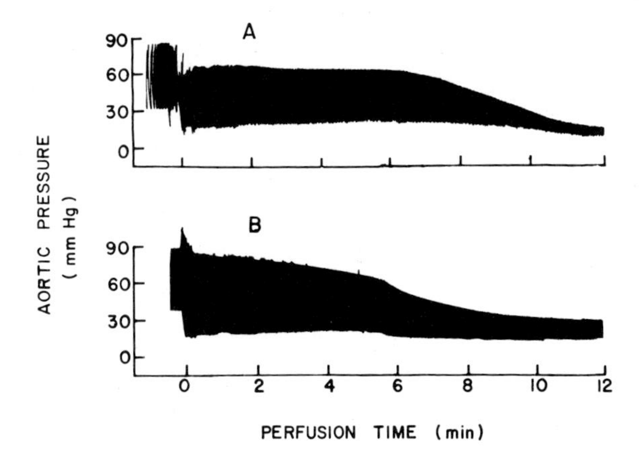

FIG. 7. Effect of myocardial ischemia on pressure development of the isolated heart. Ischemia was induced by the low output, high aortic resistance method (panel A) and with the one-way aortic valve (panel B). The hearts were electrically paced at 300 beats per minute. The perfusate contained 11 m*M* glucose as the only substrate. Ischemia was induced at zero time in each tracing. Chart speed was 10 mm per minute. From J. R. Neely *et al.*,[6] reproduced by courtesy of *American Journal of Physiology.*

ventricular pressure development was maintained for only about 7 minutes. Ventricular function decreased rapidly between 7 and 12 minutes of ischemic perfusion. The minimum coronary flow assembly was not used in this experiment, and aortic pressure continued to decrease as ventricular failure progressed. At the end of 12 minutes, peak systolic pressure had decreased to 15 mm Hg with a diastolic pressure of 4 mm Hg, and the rate of coronary flow was about 3 ml per minute per gram dry tissue.

At zero time in panel B (Fig. 3), ischemia was induced by clamping the bypass tube around the one-way aortic valve. Peak systolic aortic pressure was increased only slightly by diverting flow through the one-way valve. Diastolic pressure below the valve, however, decreased from 40 to 13 mm Hg. This low coronary perfusion pressure during diastole decreased the rate of coronary flow from about 70 to 30 ml per minute per gram dry tissue. Under this condition, ventricular pressure development was maintained for only about 5 minutes and decreased rapidly between 5 and 8 minutes. After 10 minutes, peak systolic pressure had decreased to 20 mm Hg, diastolic pressure decreased to 5 mm Hg, and coronary flow had fallen to about 10 ml per minute per gram dry tissue. In contrast to ischemic hearts, control hearts with unrestricted coronary flow continued to develop about 90 mm Hg peak systolic pressure throughout 30 minutes of perfusion under these conditions (data not shown).

Discussion. The isolated rat heart perfused by either the Langendorff or working heart techniques performed well over a wide range of ventricular pressures. These preparations were stable when well oxygenated for periods up to 3 hours with only slight deterioration occurring up to 5 hours. The rate of oxygen and substrate utilization increased in relation to increased ventricular pressure development in both preparations. Pressure development could be modified by adjusting the aortic perfusion pressure in the Langendorff and either left atrial filling pressure and/or ventricular afterload in the working heart preparation.

The two ischemic heart perfusion techniques that are described resulted in myocardial ischemia as indicated by (1) reduced rates of coronary flow and oxygen consumption, (2) reduced tissue levels of ATP and creatine P, (3) higher tissue levels of ADP, AMP, and lactate, and (4) ventricular failure.

A similar reduction in coronary flow resulted in essentially the same changes in mechanical performance and metabolism with either the low output, high resistance technique or the one-way aortic valve method of inducing ischemia. The first technique has the advantage of affording a more exact control of coronary flow in the initial phase of ischemia but has less flexibility in selecting initial levels of cardiac output and ventricular pressure development. The initial mechanical and metabolic changes that resulted from adjusting cardiac output and ventricular pressure development, however, could not be completely separated from the effects of reduced coronary flow. After these initial adjustments, further deterioration in mechanical performance and metabolism appeared to result directly from the lower rate of coronary flow.

The one-way valve procedure affords far more flexibility in selecting the initial levels of cardiac output and ventricular pressure development. Since inducing ischemia by this procedure does not change either the preload or the afterload, the initial rates of cardiac output and levels of pressure development were the same as the preischemic values. Thus the effects of ischemia on mechanical performance could be correlated with metabolic events throughout the ischemic period. With this technique, however, the rate of coronary flow was dependent on the aortic pressure during systole and the initial rate of coronary flow could not be controlled as accurately as in the low cardiac output procedure. In addition, it is difficult to maintain a competent one-way valve. The valves must be disassembled, cleaned and polished frequently to prevent backflow of perfusate during diastole.

These major advantages and disadvantages of the two methods applied only during the first few minutes of ischemia. After ventricular failure had occurred (5–10 minutes), the rate of coronary flow decreased to the same extent in both cases. Therefore, the long term effects of ischemia on metabolism, irreversible tissue damage and on the ability of various agents to improve cellular stability can be determined equally well with either procedure. With the minimum coronary flow assembly, it was possible to select various rates of coronary flow in either of the perfusion techniques.

[7] Frog Ventricular Muscle Suitable for Use in Measuring Rapid Metabolic Changes during the Contractile Cycle

By Gary Brooker

The isolated perfused frog heart ventricle strip preparation enables the study of metabolic changes occurring at various phases of a single myocardial contraction. Two strips are taken from the same ventricle, perfused with physiological salt solution and electrically stimulated out of phase from each other so that at any time the contractile state of each strip is in a different phase of the contraction cycle. The two strips are then fixed at the same time by releasing spring-loaded aluminum blocks precooled to the temperature of liquid nitrogen. The frozen tissue is then extracted by standard techniques, and the metabolite under investigation is measured.

Rana pipiens frogs have been used for this study. They are stored in plastic containers with 1–2 cm of water in the container at 4–6°. Several days before use they are allowed to equilibrate to room temperature, which varies between 22 and 24°. Physiological salt solution containing 91.3 m*M* NaCl, 25 m*M* $NaHCO_3$, 2.5 m*M* KCl, 1 m*M* $CaCl_2$, and 5 m*M* dextrose is made fresh each day. Concentrated stock solutions of each component are used to make the complete solution. Dextrose is weighed out each day. A stock solution of dextrose is not kept. The solution is vigorously oxygenated with 95% O_2–5% CO_2, and the $CaCl_2$ is added.

The ventricle strips are isolated and perfused in the following way. The frog is pithed and decapitated, and the chest cavity is opened. The pericardium is removed, and a small incision is made in the aorta. The heart is then perfused with physiological salt solution with a medicine dropper until free of blood. The ventricle is then removed by cutting at the atricular-ventricular groove and placed in a petri dish with gassed physiological salt solution. The ventricle is opened flat by two longitudinal cuts 180° from each other. Two similar strips are prepared and suspended between the strain gauge and binding post with 0.001-inch copper wire as shown in Fig. 1. Oxygenated physiological salt solution then is started at a flow rate of 1.5 ml per minute to bathe the muscles.

A Statham model No. UC3 strain gauge connected to a micrometer

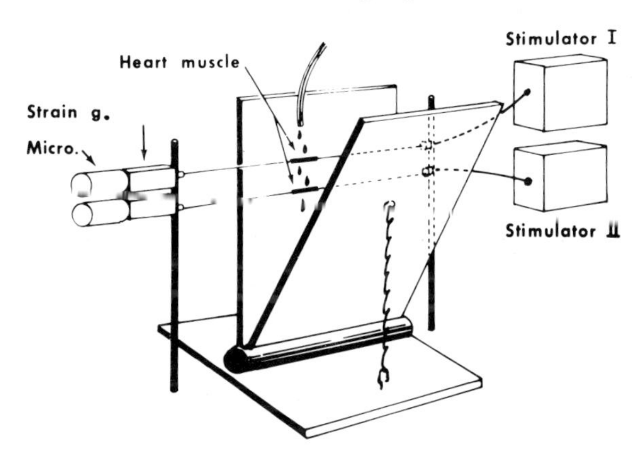

Fig. 1. Diagrammatic representation of the perfusion system and spring-loaded liquid nitrogen precooled clamps. Strain g = strain gauge; Micro. = micrometer movement.

movement was used to apply 500 mg of resting tension to each muscle strip. The copper wire was attached to a small eye hook soldered to a threaded nut, which was attached to the threaded nose of the transducer cell.

The strain gauges and respective stimuli were recorded on a 4-channel Beckman R-411 Dynograph. In addition, each contraction cycle was monitored on a four-channel Model No. 5103N Tektronix memory oscilloscope. To monitor the stimulus, 80,000 ohm resistors were placed in series between the two respective stimulators and the input circuit of the recorder. The strain gauge binding posts (onto which one copper wire from each muscle was connected) were connected to ground.

Because the variability in the absolute level of metabolic intermediates can be reduced by studying two or more experimental parameters using the same biological specimen, the approach taken here was chosen. One ventricle strip could be fixed in diastole while the other was in systole if they were paced by two electrical stimulators. As shown in Fig. 2, ventricle strip I was excited with a 4-V 10 msec square wave pulse by a Grass model S44 stimulator. In addition this stimulator has a delay function so that it can be set to trigger the second stimulator (Grass model S6) which was used to excite the second muscle. Thus, while strip I was in systole, strip II was in diastole. The relative position in time of the two strips with respect to each other is continuously variable simply by altering the delay time of stimulator 1. We have chosen to stimulate the strips at 12 per minute, but this parameter can also be varied.

The spring-loaded clamps are shown cocked open in position and ready to freeze the strips. The spring used is a door spring, and the plates are 1 cm thick highly polished aluminum.

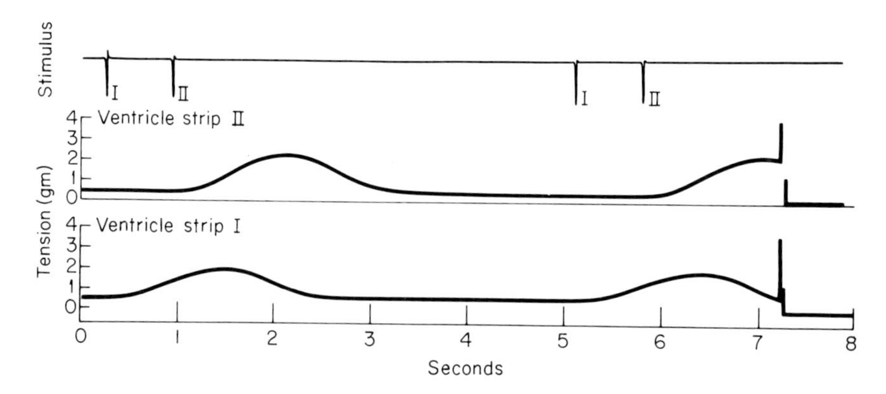

FIG. 2. High-speed recorder trace of each contractile record and respective stimuli one contraction cycle before fixation and during fixation.

The muscles are allowed to equilibrate 30 minutes after they are set up. During this time the clamps are removed and precooled in liquid nitrogen. Exactly 15 seconds before freezing the tissue, the clamps are cocked open and inserted as shown so that the muscle is between the opened jaws. In addition, perfusion is stopped. The apparatus is designed so that the jaws are at a sufficient distance from the muscle to avoid altering the muscle temperature before clamping. At the desired time the clamps are manually released and the muscles are clamped. The desired clamping point can be easily obtained by manually releasing the clamps. However, it would be easy to electronically trigger the release of the clamps.

The closed clamps are then immediately reimmersed in liquid nitrogen and once equilibrated both pieces of flattened tissue are quickly removed and placed in polypropylene tubes which have been precooled in liquid nitrogen. The samples are kept in liquid nitrogen until they are extracted and analyzed for metabolic intermediates.

Section IV

Skeletal and Smooth Muscle

[8] The Perfused *in Situ* Rabbit Skeletal Muscle

By JAMES T. STULL and STEVEN E. MAYER

The value of perfusion techniques in studying biochemical mechanisms in skeletal muscle has long been recognized.[1] Such techniques avoid artificial diffusion barriers that exist with isolated muscle resulting in adverse metabolic conditions, particularly with respect to oxygen supply, and delayed responses to drugs and hormones introduced in the medium. By injecting small amounts into the arterial vessel of the muscle *in situ,* hormones and drugs can be delivered directly to the muscle fibers with subsequent dilution in the general circulation of the whole animal. Significant physiological and biochemical responses to the agents are not elicited in other systems owing to this dilution, and hormonal, circulatory, and neurogenic influences on the muscle to be studied are not likely to be altered. Finally, the technique of autoperfusion avoids the needs of pumps, oxygenators, and a supply of extra blood. The preparation that will be described is a simple, easily controlled system developed for the study of glycogen metabolism in skeletal muscle.[2]

The rabbit was selected as the animal of choice for the following reasons: It is relatively inexpensive; it is easily handled because of its size and docility; and the enzymes participating in glycogenolysis in the rabbit have been more carefully characterized biochemically than the enzymes from other species.

The gracilis muscle of the rabbit was chosen for study because it is a superficial muscle and therefore easy to isolate surgically, the flatness of the muscle facilitates biopsy sampling, it is composed of essentially one fiber type (fast twitch-glycolytic fibers) with less than 1% slow twitch-oxidative and fast twitch-oxidative-glycolytic fibers,[3] and the mass of the muscle is sufficient to provide multiple biopsy samples. Usually one can perform experiments with each muscle as its own control and thereby minimize variation between muscles in statistical evaluation of data.

The primary disadvantage of the rabbit is the unpredictability of anesthesia. Not only is there great variation in the dose of anesthetic agent required, but the margin between surgical anesthesia and death is narrow. Each animal has to be treated individually in terms of proper dosage, and a means of measuring blood pressure is necessary to properly

[1] E. G. Martin, J. Field, II, and V. E. Hall, *Amer. J. Physiol.* **102,** 481 (1932).

[2] J. T. Stull and S. E. Mayer, *J. Biol. Chem.* **246,** 5716 (1971).

[3] C. R. Ashmore and J. T. Stull (unpublished observations).

evaluate the physiological state and anesthetic stage during the experiment.

Animal Preparation

Lift a 3.0–3.5 kg New Zealand white rabbit by the loose skin on the lower back above the pelvis and allow the head to dangle down. This method of handling usually results in little kicking of the hindlimbs, and thereby prevents scratches and forceful kicks that can break the rabbit's back. Place the animal on the floor and gently grasp the upper back from behind. After the rabbit has remained motionless for 5–10 seconds inject 1.2 g/kg ethyl carbamate by quickly plunging a 1.5-inch 22-gauge needle into the peritoneal cavity approximately 1–2 inches posterior to the ribs. Pull the plunger of the syringe back to be sure the needle is not in a blood vessel. The rabbit usually remains motionless during this injection period. Ethyl carbamate provides prolonged anesthesia after a delayed onset, but for complete surgical anesthesia supplementation with pentobarbital is necessary. After 30–60 minutes place the rabbit on a table and clip away the hair over the marginal ear vein with scissors; dilate the vein by swabbing with xylene. Insert a 1-inch 22-gauge scalp vein needle and inject a small amount of 0.9% NaCl to be certain the needle is in the vein. Hold the needle in place with a paper clip. The toe reflex is used as an index of the surgical stage of anesthesia and is elicited by squeezing between the toes of a hindfoot. Inject 5 mg/kg pentobarbital in the marginal ear vein, and after 2–5 minutes test for the toe reflex. If there is a response, repeat the pentobarbital injection. Caution and patience are needed at this time to prevent the administration of a fatal dose of pentobarbital. After the toe reflex can no longer be elicited, clip the hair from the throat and ventral right leg, and place the animal on its back on a flat operating board. A thermostatically controlled heating pad should be under the animal to maintain a rectal temperature of 37–38°. The rabbit is held in place on its back by means of ties around the distal portions of the legs.

Make a midline incision with scissors through the ventral skin of the neck from the caudal end of the larynx to the cranial end of the sternum. Cut the fascia as well. Use mosquito hemostatic forceps to clamp any small arteries or veins that were transected with the skin incision. Using blunt dissection separate the sternohyoid muscles along the midline to expose the trachea. Carefully separate the connective tissue from the trachea for about 1 inch. Slip two ligatures under the trachea and place aside until cannulation of the trachea.

Separate the muscles ventral to the trachea on the right side by blunt dissection. To expose the common carotid artery, carefully separate the sternomastoid from the sternothyroid muscle. The artery can be seen and felt pulsating within a heavy sheath lateral to the trachea. Separate the sheath from the fascia so it is free for about 2 inches. Penetrate the sheath by blunt dissection with full-curved eye dressing forceps and clear the sheath away from the artery for about 1.5 inches. Be careful not to tear the thin-walled internal jugular vein that is also in the sheath. Ligate the artery at its cranial end. Place another ligature loosely under the artery about 1.5 inches from the first ligature. Compress the artery below the loose ligature (cardiac end) with artery forceps (27 mm, Arthur H. Thomas Co.). Gently pull the cranial ligature to stretch the artery, and nick through the arterial wall with iridectomy scissors on the cardiac side of the ligature. Blood should momentarily flow from the nick. Insert a polyethylene cannula (outside diameter 0.075 inch) filled with 0.2% heparin in 0.9% NaCl into the cut, and direct to the artery forceps. The end of the cannula is cut at an angle to facilitate entry into the artery. Tie the cannula firmly with the loose ligature and reinforce with the ends of the cranial ligature. Remove the artery forceps and observe blood pulsating in the cannula. Flush with 0.2% heparin in 0.9% NaCl, and connect to a pressure transducer for measurement of blood pressure. The cannula may need occasional flushings during the experiment to prevent blockage by blood clots. The mean blood pressure should be 75 mm Hg or greater.

Lift the trachea by the posterior ligature and incise ¾ through and between two cartilage rings. Quickly insert a tracheal cannula (outside diameter 4.5 mm) pointing toward the lungs, and tie it in place with both ligatures. Connect the trachea to a small animal respirator with a tidal volume of 7–8 ml and a rate adjusted to body weight as determined from a ventilation nomogram (approximately 90 per minute).[4] These conditions maintained arterial pH between 7.32 and 7.47 for up to 6 hours as tested in eight animals. Cover the exposed areas of the neck with surgical gauze moistened with 0.9% NaCl. Inject 0.5 mg/kg *d*-tubocurarine (5 mg/ml in 0.9% NaCl) subcutaneously in the chest, and administer 0.5 mg/kg through the marginal ear vein. The intravenous injection of *d*-tubocurarine should be made slowly to minimize histamine release as demonstrated by a decrease in blood pressure.

If the mean blood pressure is not steady (an indication of light anesthesia), administer 5 mg/kg pentobarbital. Repeat as often as necessary during the course of the experiment to maintain anesthesia.

[4] L. I. Kleinman and E. P. Radford, *J. Appl. Physiol.* **19**, 360 (1964).

If contractions of the gracilis muscle will be elicited in the experiment, the origin of the muscle must be prevented from shifting; this procedure involves immobilization of the pelvis and leg.[5] The hair should be clipped over both ilia before the rabbit is placed on the operating board. Custom-made aluminum brackets are used for immobilization of the pelvis. The brackets are L-shaped with the upright 4.5 cm, the base 7.5 cm, and the width 4 cm. A slot in the base of the bracket allows adjustment of the position before the bracket is bolted to the operating board, one bracket on each side of the pelvis. A bolt with pointed tip is threaded through the upright 3.0 cm above the base. The brackets are secured snugly against the pelvis with the pointed ends of the threaded bolts opposite the crests of the ilia. The two bolts are screwed toward each other until the tips engage and firmly hold the ilia.

Muscle Preparation

Make a midline incision with scissors through the ventral skin of the right thigh from the lower abdominal musculature to the patella. Also cut through the fascia. Extend the incision parallel to the abdomen above the origin of the sartorius and gracilis muscles to properly expose the muscles with the skin forming a pocket. Usually hemostatic forceps are necessary to position the skin away from the gracilis muscle. Cover the muscles with gauze moistened in 0.9% NaCl.

By blunt dissection with full-curved eye dressing forceps, isolate the lateral circumflex artery and vein (Fig. 1). Ligate both vessels close to the femoral artery and vein so that immediate branches of the circumflex vessels are isolated from the femoral circulation. Repeat this process for the superior epigastric and deep femoral arteries and veins. Ligate the various small musculature branches around the deep femoral artery and vein that originate from the femoral vessels. Ligate the saphenous artery and vein medial to the gracilis muscle. Make a midline incision with blunt scissors in the tendon of the gracilis and sartorius muscles, and carefully extend the incision proximally into the muscle mass to separate the sartorius muscle from the gracilis muscle. Lift the tendon of the gracilis muscle as the incision is being made to prevent nicking the underlying saphenous and femoral vessels. The incision is continued to within 5 mm of the gracilis artery and vein. The gracilis and sartorius muscles are fused in the rabbit and therefore must be partially dissected in order to isolate the gracilis muscle. Ligate the exposed saphenous vessels proximal to their entry into the tendon. The femoral artery and vein are ligated distally

[5] E. G. Martin, J. Field, II, and V. E. Hall, *Amer. J. Physiol.* **102,** 476 (1932).

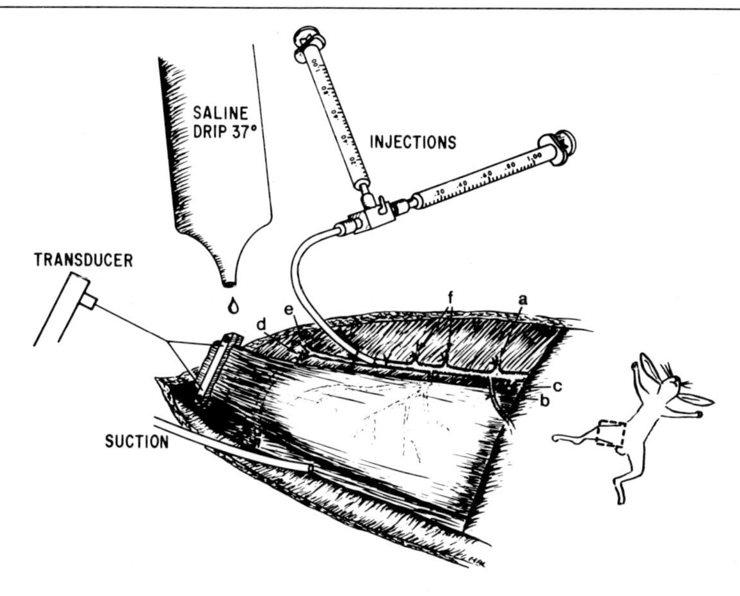

FIG. 1. Isolation of the right gracilis muscle. The femoral artery and its branches are emphasized in the drawing, but in reality are partially obscured by the medial edge of the sartorius muscle. Also, the respective veins that lie next to each artery are not shown. The ligated branches of the femoral artery are (a) lateral circumflex artery, (b) superior epigastric artery, (c) deep femoral artery, (d) saphenous artery, (e) the distal end of the femoral artery, and (f) small musculature branches.

from the saphenous branches. Now ligate small musculature branches of the femoral vessels that extend into the muscles beneath the sartorius muscle. Be particularly careful to ligate branches that extend deep into the muscle mass.

Using blunt dissection with full-curved eye dressing forceps, expose the femoral artery and vein in the sheath distal to the gracilis vessels. Cannulate the femoral artery using the same method described for the common carotid artery. After the cannula (outside diameter 0.038 inch) is inserted and extended to the artery forceps, tie a ligature snugly around the cannula in the artery (tight enough to prevent bleeding, but loose enough to slide the cannula farther into the artery). Release the artery forceps and observe blood pulsating in the cannula. Flush with 0.2% heparin in 0.9% NaCl. Now gently push the cannula into the artery until the tip is immediately distal to the origin of the gracilis artery. Tighten the ligature and bind the cannula in place. Inject 0.2 ml of the heparin solution and carefully observe all the muscles. Only the gracilis muscle should become pale with the injection. Repeat the injection as often as necessary to be certain the blood circulation is restricted to the gracilis muscle. Do not allow any air bubbles to be injected.

Make an incision across the tendon of the gracilis muscle about 1 inch from the distal end of the muscle fibers; extend the incision along the medial edge of the gracilis muscle after transecting the saphenous artery between the ligature and muscle. Hold the tendon firmly with forceps and gently lift the muscle as it is surgically isolated. Transect the few small arteries and veins that may penetrate the under surface of the gracilis muscle. Free the muscle to the origin and transect the obturator nerve at the underside of the origin of the gracilis muscle. During the isolation be careful not to damage the gracilis artery by stretching the muscle away from the femoral artery. Clamp the tendon between two $40 \times 6.5 \times 6.5$ mm methacrylate bars held in place by screws at each end. Position the muscle at a 30° angle to the leg and place 25 g tension at the methacrylate clamp. One may use a calibrated force displacement transducer (useful for measurements of mechanical responses) or a 25-g weight. A pulley is convenient to manipulate the proper angle of the muscle in relation to the source of tension. Adjust at 37°, 0.9% NaCl solution to drip on the tendon area immediately below the clamp. The saline should flow down over the surface of the muscle and should be removed by a suction hose in the pocket of the skin. Adjust flow to maintain a surface temperature between 34° and 36°. Occasionally, it is necessary to irrigate the under surface of the muscle with the saline solution to prevent drying. Again inject the heparinized saline into the femoral artery to test isolation of the circulation to the gracilis muscle. At the end of experiments India ink is injected to confirm the isolation of the circulation to the gracilis muscle. The rapid injection of 0.2 ml is sufficient to fill the femoral artery up to its entry into the leg before the solution flows through the gracilis artery. The volume of the femoral cannula is 0.18 ml. Drug concentrations are adjusted for injection in 0.2 ml and each injection is immediately followed by a rapid flush of 0.2 ml 0.2% heparin in 0.9% NaCl. In rabbits smaller than 3 kg this volume, 0.2 ml, usually exceeds the capacity of the isolated segment of the femoral artery and needs to be reduced accordingly.

If the muscle is to be stimulated to contract isometrically, then insert platinum electrodes (25 gauge) at the proximal and distal ends of the muscle. Check the immobilization of the leg and pelvis. For inducement of contraction, a square-wave pulse of 3 msec duration is applied as the stimulus with a supramaximal voltage determined for each muscle. The optimal resting tension for the maximal twitch tension is determined for each muscle.

Biopsy samples of the muscle (approximately 200 mg) are removed with scissors and forceps, and rapidly plunged into dichlorodifluoromethane held at its freezing point in liquid nitrogen (see Volume 38, [1] for details on rapid tissue fixation and extraction techniques). Analyses of the biopsy samples obtained from muscles in which careful attention was directed to maintaining an adequate physiological state (mean blood pressure $\gtrless$ 75 mm Hg, arterial pH between 7.32 and 7.47) and anesthesia (steady mean blood pressure) demonstrate that the constituents participating in glycogenolysis are in a nonstimulated state, i.e., only 7% of phosphorylase is in the *a* form, phosphorylase kinase is in the nonactivated form (ratio of activities pH 6.8/8.2 = 0.03), and cyclic AMP content is low (0.2 μmole/kg).[2] Injection of 0.9% NaCl or 4×10^{-9} mole of isoproterenol in a randomized sequence resulted in no change in these parameters at 60 seconds for the control injection, but the isoproterenol stimulated phosphorylase *a* formation (40% *a*), conversion of phosphorylase kinase to the activated form (ratio of activity pH 6.8/8.2 = 0.13) along with an increase in cyclic AMP content (0.7 μmole/kg). In a few cases there were no responses to isoproterenol in the seventh biopsy sample indicating that drug delivery was probably hampered; therefore, only six biopsy samples were routinely obtained from the muscle with 30–45 minutes between each sample.

The muscle preparation described here is the simplest model of a perfused tissue—autoperfusion *in situ*. Obviously more complicated techniques may be employed when modifications of experimental parameters are desired.[6]

[6] E. M. Renkin, *in* "Physical Techniques in Biological Research" (W. L. Nastuk, ed.), Vol. 4, p. 107. Academic Press, New York, 1961.

[9] A Technique for Perfusion of an Isolated Preparation of Rat Hemicorpus[1]

By L. S. JEFFERSON

A method for perfusion *in vitro* of a preparation of rat hemicorpus has been developed for studies of the metabolism of skeletal muscle. The hemicorpus is similar to preparations described by others for perfusion of rat hindlimbs,[2-6] but has been modified to include the psoas muscles.

[1] Supported by Grant No. AM 15658 from the National Institutes of Health.

[2] D. S. Robinson and P. M. Harris, *Quart. J. Exp. Physiol.* 44, 80 (1959).

[3] G. E. Mortimore, F. Tietze, and D. Stetten, *Diabetes* 8, 307 (1959).

Perfusion of psoas muscle is desirable for studies of protein metabolism, since this tissue can be readily homogenized by the gentle techniques required for isolation of polysomes and ribosomal subunits. The hemicorpus preparation offers a number of advantages for studying the control of skeletal muscle metabolism under a variety of well-defined situations. (1) The preparation consists mainly of muscle tissue. (2) Substrates and hormones are delivered to the cells by the normal capillary bed. (3) The preparation remains in a good physiological state during perfusion for periods up to 3 hours as judged by oxygen consumption, perfusion pressure, cell morphology, and hormone responsiveness. (4) Large samples of skeletal muscle and perfusate can be obtained rapidly and with ease for estimation of enzyme activities, metabolic intermediates, and substrate levels. The major disadvantage of the preparation is the inclusion of adipose tissue, connective tissue, skin, and bone. Effects on these tissues could affect the composition of the perfusate and thereby modify the metabolism of skeletal muscle. This limitation can be overcome in most instances by using a nonrecirculating perfusate.

Preparation of the Perfusion Medium. A modified Krebs-Henseleit[7] bicarbonate buffer containing bovine albumin and bovine erythrocytes constitutes the basic perfusion medium. Substrates and hormones are added to the basic medium as desired. The bicarbonate buffer contains the following salts in millimolar concentrations: NaCl, 118; KCl, 4.7; $CaCl_2$, 3.0 (includes 0.5 mM to balance Na_2 EDTA); $MgSO_4$, 1.2; KH_2PO_4, 1.2; and $NaHCO_3$, 25. Na_2EDTA (0.5 mM) is added to chelate trace quantities of any heavy metal ions that might be present in the reagents. An equal molar quantity of $CaCl_2$ is added to maintain the free Ca^{2+} concentration in the presence of EDTA. The buffer must be prepared fresh each day from stock solutions of the individual components. Stock solutions are made up in the following concentrations (grams/liter): NaCl, 9; KCl, 11.5; $CaCl_2$, 12.2; $MgSO_4 \cdot 7H_2O$, 38.2; KH_2PO_4, 21.1; $NaHCO_3$, 13.00; and Na_2TDTA, 40.7. These solutions can be stored in a refrigerator for several weeks. The buffer is prepared by mixing the following proportions (in milliliters) from the stock solutions: NaCl, 1000; KCl, 40; $CaCl_2$, 36; $MgSO_4$, 10; KH_2PO_4, 10; and Na_2EDTA, 6. This mixture is gassed with 95% O_2:5% CO_2 for 15 minutes at 0–4° prior to adding the $NaHCO_3$. This lowers the pH

[4] O. Forsander, N. Räikä, and H. Soumalainen, *Hoppe-Seyler's Z. Physiol. Chem.* **318,** 1 (1960).

[5] R. J. Mahler, O. Szabo, and J. C. Penhos, *Diabetes* **17,** 1 (1968).

[6] N. B. Ruderman, C. R. S. Houghton, and R. Hems, *Biochem. J.* **124,** 639 (1971).

[7] H. A. Krebs and K. Henseleit, *Hoppe-Seyler's Z. Physiol. Chem.* **210,** 33 (1932).

and prevents precipitation of calcium bicarbonate. After addition of $NaHCO_3$ (210 ml), the buffer is gassed for another 15 minutes. Fraction V bovine albumin (30 g/liter) is added and dissolved by mixing on a magnetic stirrer. The buffer containing albumin is passed through a Millipore filter (0.8 μm pore size) before the addition of erythrocytes. Bovine erythrocytes are prepared from blood obtained from the common carotid artery of stunned steers at a commercial slaughterhouse. The blood is collected into a heparinized container and kept at 0–4° until used. Blood is always used within 72 hours. Blood cells are collected by centrifugation (4°) at 3000 *g* for 15 minutes. They are washed 3 times with NaCl (9 g/liter) and twice with the Krebs buffer by repeated centrifugation and aspiration of the supernatant. The wash buffer is made up to contain the same concentration of substrates that are to be present in the final medium. The white cells form a fluffy layer on the erythrocyte pellet and are carefully removed by aspiration following the first wash step. The washed cells are put through a glass wool filter and added to the buffer containing albumin in the ratio of 1 volume of cells to 2 volumes of buffer. This gives a packed cell volume of 27–28%. At this stage, hormones or radioactive substrates are added to the perfusate. The final medium is placed in the rotating reservoirs of the perfusion apparatus, warmed to 37°, and equilibrated with 95% O_2:5% CO_2 prior to the start of perfusion.

The Perfusion Apparatus. The apparatus used for hemicorpus perfusion represents a modification of that described by Exton and Park[8] for liver perfusion. A photograph of a two-unit apparatus, which was constructed by The Apparatus Shop, Vanderbilt University, Nashville, Tennessee 37232, is shown in Fig. 1. Readers are referred to this volume [4] for a complete description of the apparatus. The perfusion system is identical with that described by Exton except for the following changes. (1) The variable speed peristaltic pump used to pump perfusion medium from the oxygenation chamber is fitted with a double section of Tygon tubing (1/8 inch i.d., 1/4 inch o.d., 1/16 inch wall, Formula S50HL, Norton Plastics and Synthetics Division, Akron, Ohio). This is shown in Fig. 1 and has been found to give satisfactory flow rates and to produce little hemolysis. (2) Vinyl tubing (size No. 22, Irvington Brand) is used to connect the bubble trap to an aortic cannula. (3) An aortic cannula replaces the portal cannula. This is 2 cm long and can be made from a No. 21 gauge hypodermic needle by blunting the sharp beveled tip on a small grindstone. (4) The animal tray consists of a stainless steel screen (length, 8 inches; width, 6 inches; 16 mesh) and a stainless steel trough (length,

[8] J. H. Exton and C. R. Park, *J. Biol. Chem.* **242**, 2622 (1967).

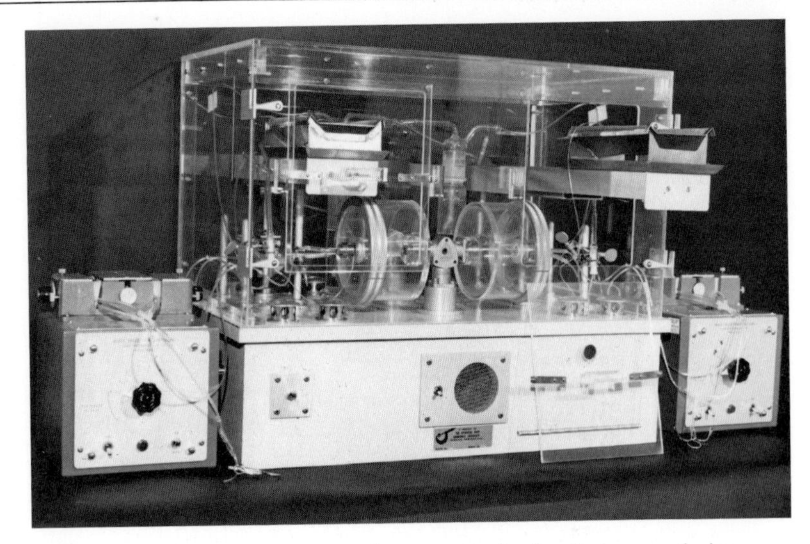

FIG. 1. Photograph of a two-unit apparatus for hemicorpus perfusion.

8 inches; width, 4½ inches; depth on deep end, 1½ inches). Bending the screen at right angles along each edge so that it fits snugly over the trough gives the unit stability. The screen and trough are mounted on a sliding platform which can be pulled through a door in the front of the Plexiglas box. This allows the surgical procedures to be done outside the box, while the temperature inside is maintained at 37°. The trough is formed so that it drains to one end, which has an outlet consisting of a short length of stainless steel tubing. Tygon tubing connected to this outlet 1/16 inch i.d., ¼ inch o.d., 1/32 inch wall) and passed through the peristaltic pump (double section, ⅛ inch i.d., ¼ inch o.d., 1/16 inch wall) allows the perfusate which collects in the trough to be pumped back to the oxygenation chamber. A three-way valve placed in the tubing between the pump and reservoir can be set to permit a fraction of the perfusate to be collected before recirculation is established. When a system with nonrecirculating medium is used, the valve is kept closed to the reservoir throughout the experiment, and the entire perfusate is collected as serial samples.

Prior to the start of perfusion, the components listed above are assembled as shown in Fig. 2. In experiments in which the perfusate is to be recirculated, an initial volume of 150 ml is added to the oxygenation chamber. Rotation of the chamber is started and the perfusate is warmed to 37° and equilibrated with humidified 95% O_2:5% CO_2. In experiments involving a nonrecirculating perfusate, an initial volume of up to 450

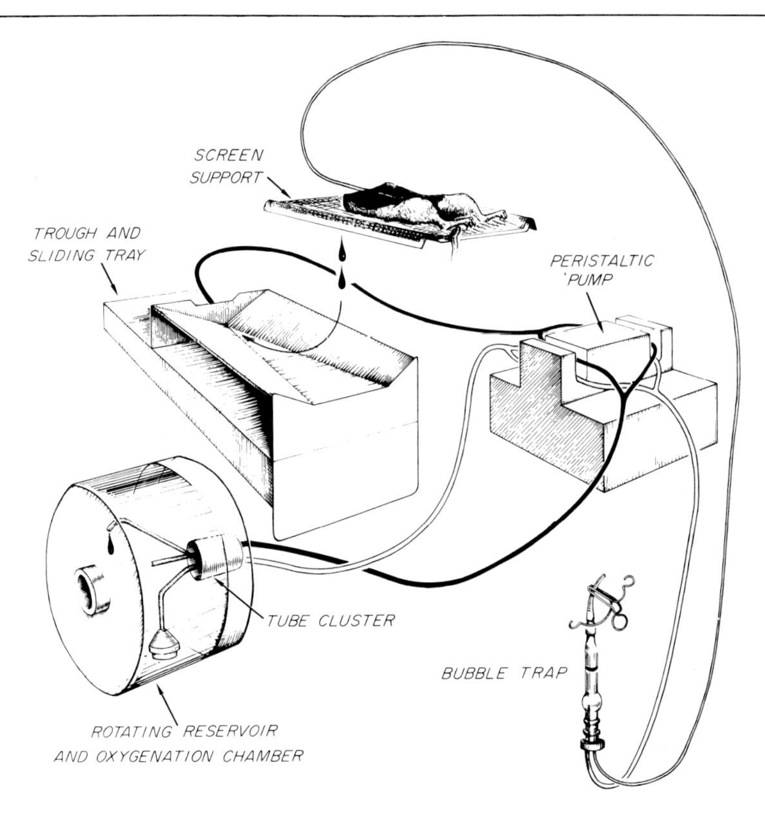

FIG. 2. Diagrammatic drawing of a perfusion setup. The tube cluster contains the following: (1) a tube fitted with a filter assembly for withdrawing the perfusate, (2) a tube for returning the perfusate, (3) a gas inlet tube, (4) a gas outlet tube, and (5) a tube for collecting samples of perfusate. The "arterial perfusate" and the "venous perfusate" are indicated by the unshaded and shaded tubing, respectively.

ml can be added to the oxygenation chamber. It is also possible for the perfusate to be pumped consecutively from two oxygenation chambers, or extra medium can be pumped into the system from an external reservoir. Prior to the start of the surgical procedures, perfusate is pumped from the oxygenation chamber to the aortic cannula. While the pump is running, approximately 2 ml of perfusate are allowed to collect in the bubble trap by opening the top clamp.

Perfusion of the Hemicorpus. Male rats weighing approximately 200–300 g are suitable for hemicorpus perfusion. If smaller animals are used, it may be necessary to modify the size of the aortic cannula and the flow rate. The animals are injected intraperitoneally with sodium heparin (10 mg/kg) 30 minutes before use. After induction of anesthesia by intraperitoneal injection of Nembutal (50 mg/kg), the rat is placed in a supine position on the screen support of the animal tray. The abdomen is opened by making midline and transverse incisions with scissors, and the following structures are ligated with surgical silk (size 00): coeliac artery, superior mesenteric artery, inferior mesenteric artery, renal and suprarenal arteries and veins, the spermatic cords, and rectum (Fig. 3). The chest is opened through vertical incisions along either side of the anterior thorax. Two loose ligatures are placed around the descending aorta, one immediately above the diaphragm and another approximately 1½ cm proximal to the diaphragm. While an assistant pulls the proximal ligature taut to prevent blood flow, the aorta is lifted by the lower ligature, and fine-tip scissors are used to make a small incision in the vessel wall between the ligatures. The aortic cannula is inserted into the vessel and tied into place. Perfusion is begun immediately at

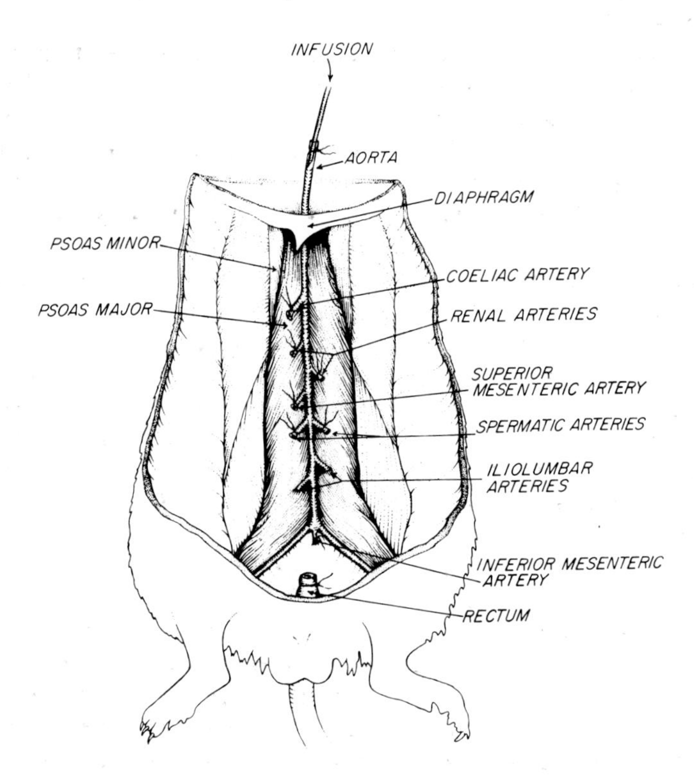

Fig. 3. The perfused hemicorpus. Reproduced from L. S. Jefferson, J. O. Koehler, and H. E. Morgan, [*Proc. Nat. Acad. Sci. U.S.* **69**, 816 (1972)] by courtesy of *Proceedings of the National Academy of Sciences.*

a flow rate of 7.0 ml per minute. Once the chest is opened it is essential that the cannulation procedures be done quickly since the animal is exposed to anoxic conditions. With practice, the procedures carried out between the time the chest is opened and perfusion is commenced can be completed within 60 seconds. The entire operative procedure requires approximately 5 minutes. After perfusion is begun, the animal is transected above the level of the aortic cannula; the stomach, large and small intestines, liver, and testes are removed; and the preparation is covered with polyvinyl film. The perfusate drips from the cut end of the inferior vena cava into the trough from which it is pumped through a pad of glass wool into the oxygenation chamber (in experiments where the perfusate is recirculated) or into collecting tubes placed in an ice-bath outside the Plexiglas box (in experiments involving a nonrecirculating perfusate). With a recirculating system, the first 50 ml of perfusate that drips from the vena cava are usually discarded, and 100 ml of perfusate are recirculated.

Samples of perfusate are obtained from the oxygenation chamber through a piece of transparent vinyl tubing (size No. 22, Irvington Brand) which is passed through the available inlet in the stainless steel tube cluster. The vinyl tubing is attached to a No. 21 gauge hypodermic needle and samples are drawn into a syringe of the desired size. Samples of "venous" perfusate are obtained from the vena cava by passing a No. 21 gauge hypodermic needle through the vessel wall and withdrawing the sample into a syringe. To obtain tissue samples for measurement of metabolic intermediates, the skin is removed from a hindlimb and a portion of thigh muscle is rapidly frozen. This can be done quickly by making an incision around the ankle and using a heavy, blunt-nosed hemostat to pull the skin forward to expose the thigh muscles. While still being perfused a sample of muscle, mainly gastrocnemius and posteroinferior thigh muscles, is rapidly frozen *in situ* with a Wollenberger clamp[9] cooled to the temperature of liquid nitrogen. To obtain tissue samples for preparation of polysomes and ribosomal subunits, the psoas muscles are rapidly removed, freed of connective tissue, and cooled to 0–4° prior to homogenization.

The weight of the hemicorpus is determined following perfusion by combining the final weight of the preparation with that of the tissue samples. The total weight of the hemicorpus has been found to be $43.0 \pm 0.004\%$ of the body weight. By weighing the hemicorpus after removal of the skin, feet, and tail and again after dissolution of the soft tissue with 30% (w/v) potassium hydroxide, it is possible to estimate

[9] A. Wollenberger, O. Ristau, and G. Schoffa, *Pfluegers Arch. Gesamte Physiol. Menschen Tiere* **270**, 399 (1960).

the relative percentages of skin (+ feet and tail), bone, and muscle tissue (muscle + fat) present in the preparation. These tissues have been found to represent 31.7 ± 0.8%, 6.3 ± 0.1% and 62.0 ± 0.8% of the total weight, respectively. Since skin, bone, and adipose tissue have low rates of oxidative metabolism, it has been possible to estimate that 90–95% of the total oxygen consumption by the hemicorpus preparation is due to the metabolism of skeletal muscle.

Properties of the Perfused Hemicorpus Preparation. Several parameters which have been measured indicate that the hemicorpus preparation remains in a good physiological state during a 3-hour perfusion (see the table). Water content of the psoas and gastrocnemius muscles did not

PROPERTIES OF THE PERFUSED HEMICORPUS PREPARATION[a]

Parameter	Unperfused	Perfused	
		90 minutes	180 minutes
Water content (ml/100 g)			
Psoas	74.0 ± 0.6[b]	73.6 ± 0.4	74.5 ± 0.5
Gastrocnemius	73.4 ± 0.6	73.6 ± 0.4	74.8 ± 0.4
Sorbitol space (ml/100 g)			
Psoas		20.3 ± 0.6	21.3 ± 0.3
Gastrocnemius		20.0 ± 0.9	21.8 ± 0.2
ATP content (μmoles/g)			
Gastrocnemius	6.02 ± 0.19	6.27 ± 0.30	6.14 ± 0.15
Creatine phosphate content (μmoles/g)			
Gastrocnemius	15.04 ± 0.29	14.29 ± 0.78	14.58 ± 0.58
Perfusion pressure (mm Hg)			
Initial		137 ± 4	137 ± 4
Final		136 ± 5	135 ± 4
Oxygen tension of perfusate (mm Hg)			
Arterial		599 ± 11	608 ± 9
Venous		39 ± 1	34 ± 3

[a] Preparations were perfused for either 90 or 180 minutes with perfusate containing 15 m*M* glucose and normal plasma amounts of all amino acids. [G. Nichols, Jr., N. Nichols, W. B. Weil, and W. M. Wallace, *J. Clin. Invest.* **32,** 1299 (1953)]. Water content, sorbitol space, ATP content, and creatine phosphate content are expressed per weight of wet muscle.

[b] Standard error of mean.

change during perfusion, and the value obtained for the sorbitol space of these tissues following 90 and 180 minutes of perfusion agreed with similar determinations in diaphragm[10,11] and with estimations *in vivo* of

[10] D. M. Kipnis and C. F. Cori, *J. Biol. Chem.* **244,** 681 (1957).

[11] F. C. Battaglia and P. J. Randle, *Biochem. J.* **75,** 408 (1960).

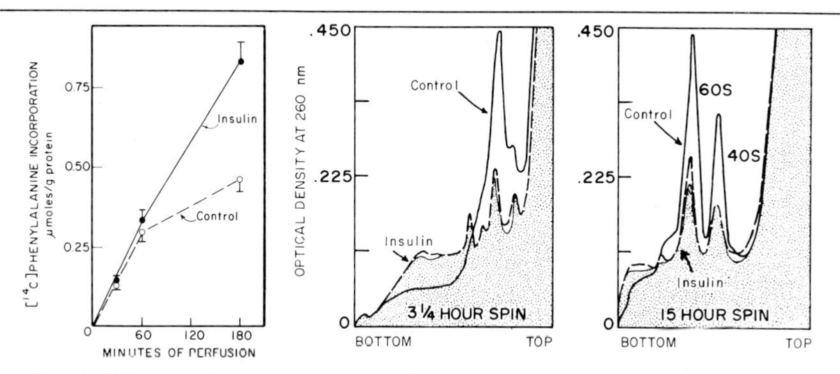

FIG. 4. Effects of time of perfusion and insulin on protein synthesis and ribosomal aggregation in psoas muscle. Preparations were perfused for periods up to 3 hours with perfusate containing 15 m*M* glucose, normal plasma concentrations of all amino acids, and [^{14}C]phenylalanine. The micromoles of phenylalanine incorporated per gram of protein were calculated using the mean specific activities of intracellular phenylalanine for each time period [H. E. Morgan, D. C. N. Earl, A. Broadus, E. B. Wolpert, K. E. Giger, and L. S. Jefferson, *J. Biol. Chem.* **246,** 2152 (1971)]. Ribosomal aggregation was assessed after 3 hours of perfusion. Following perfusion the psoas muscles were homogenized in 3 volumes of a solution containing 0.25 *M* KCl, 2 m*M* $MgCl_2$, and 10 m*M* Tris buffer, pH 7.4 [H. E. Morgan, L. S. Jefferson, E. B. Wolpert, and D. E. Rannels, *J. Biol. Chem.* **246,** 2163 (1971)]. Samples of a 10,000 *g* supernatant of the tissue homogenates were layered onto exponential sucrose gradients which had been formed in tubes of an SW 40 rotor using 15% and 2 *M* sucrose dissolved in homogenization buffer. After centrifugation at 40,000 rpm for either 3¼ or 15 hours, the optical density profiles of the gradients were monitored. Heavier ribosomal aggregates sedimented more rapidly and were seen more clearly in gradients that were spun for 3¼ hours (center panel). Large and small ribosomal subunits, labeled as 60 S and 40 S, respectively, were separated more clearly after 15 hours of centrifugation (right panel). The dashed and solid curves represent profiles obtained from tissues perfused in the presence and in the absence of insulin (25 mU/ml), respectively, and the stippled areas represent profiles obtained from unperfused psoas.

extracellular space in skeletal muscle.[12] These findings and the fact that the perfusion pressure was found to remain constant indicated that swelling did not occur during perfusion. A mean perfusion pressure of 136 mm Hg was found to be maintained when a flow rate of 7.0 ml per minute was used. The perfusion pressure is similar to the systolic pressure of the rat. The observed oxygen tension of the arterial and venous perfusate indicated that the perfusion conditions described above provide an adequate oxygen supply to the tissue of the hemicorpus preparation. In support of this contention, the ATP and creatine phosphate contents of

[12] G. Nichols, Jr., N. Nichols, W. B. Weil, and W. M. Wallace, *J. Clin. Invest.* **32,** 1299 (1953).

gastrocnemius muscle were found to be well maintained during a 180-minute perfusion.

An example of the hormone responsiveness of the hemicorpus preparation is illustrated by the data shown in Fig. 4. In these experiments the effects of time of perfusion and insulin on the rate of protein synthesis in psoas muscle were studied. The rate of protein synthesis was linear during the first hour of perfusion, but declined sharply during the second and third hours (left panel, Fig. 4). The decline in the rate of protein synthesis was associated with increased numbers of ribosomal subunits and decreased numbers of polysomes (center and right panels, Fig. 4). The number of polysomes and ribosomal subunits is influenced by rates of initiation and elongation of peptide chains. Increased numbers of subunits and decreased numbers of polysomes result when initiation is retarded relative to elongation.[13] These findings indicated that the decline in the rate of protein synthesis was due to a block in peptide-chain initiation which had developed during perfusion, perhaps as a result of depletion of hormones and substrates. Addition of insulin prevented development of the block in peptide-chain initiation as evidenced by maintenance of the initial rate of protein synthesis and *in vivo* levels of ribosomal aggregation. Other experiments indicated that glucose and fatty acid metabolism of the hemicorpus were also responsive to the *in vitro* addition of insulin.[14]

The maintenance of the tissues of the hemicorpus in a good physiological state for perfusion periods up to 3 hours and the responsiveness of metabolic pathways to *in vitro* addition of hormone indicate the suitability of the preparation for studies of the metabolism of skeletal muscle.

[13] H. E. Morgan, L. S. Jefferson, E. B. Wolpert, and D. E. Rannels, *J. Biol. Chem.* **246**, 2163 (1971).

[14] L. S. Jefferson, J. O. Koehler, and H. E. Morgan, *Proc. Nat. Acad. Sci. U.S.* **69**, 816 (1972).

[10] *In Vitro* Preparations of the Diaphragm and Other Skeletal Muscles

By Alfred L. Goldberg, Susan B. Martel, and Martin J. Kushmerick

Isolated mammalian muscles incubated *in vitro* offer numerous advantages for controlled investigations of muscle metabolism and neuromuscular physiology and pharmacology. This article will review certain preparations of skeletal muscle that have proved to be useful for biochemical and physiological studies *in vitro*. The major difficulty in maintaining tissues *in vitro* has been ensuring that adequate amounts of nutrients reach the individual cells by diffusion in the absence of a circulatory system. As a result, studies with incubated muscles have been restricted to especially thin muscles, obtained from small rodents.[1] The viability of such muscles *in vitro* generally exceeds that of slices from other mammalian tissues, such as liver, kidney, spleen, or brain. Perfusion of a specific muscle or whole hindlimbs has also been used with success by several laboratories[2]; however, the techniques involved in isolation and incubation of muscles *in vitro* are far simpler than those required for tissue perfusion.

In this discussion, special emphasis will be placed on the rat diaphragm which has been the most frequently studied muscle preparation since it was first introduced by Meyerhof and Himwich[3] nearly fifty years ago. The incubated diaphragm was popularized primarily by the work of Gemmill[4] and Kipnis and Cori,[5] who demonstrated that it was a convenient preparation for studies of hormonal control of carbohydrate metabolism and transport. The major advantages of the diaphragm are

[1] As a result of diffusion and metabolism, cells within an incubated muscle may be exposed to concentrations of nutrients different from those found in the incubation medium (A. V. Hill, "Trails and Trials in Physiology," p. 208, Williams & Wilkins, Baltimore, Maryland 1965). This difference will depend on the geometry of the tissue, the distance of a specific cell from the nearest surface, and the rate of utilization or release of substances from the cells. Thus in a flat tissue which consumes O_2 at a rate A (measured in units of cm^3 O_2 STP/cm^3 tissue per minute), the distance X from the surface at which the O_2 concentration falls to zero is given by $X = (2KY/A)^{1/2}$, where K is Krogh's diffusion constant and Y is the O_2 tension in the incubation medium measured as the partial pressure fraction. In muscle, K for O_2 is 1.66×10^{-5} cm^2 per minute at 37°. For the rat diaphragm, basal O_2 consumption rate at 37° is 0.5 μmole per gram per minute, and thus the critical depth into the muscle at which O_2 tension falls to zero is 0.54 mm for incubations in 100% O_2. For cylindrical muscles the appropriate formula is $R = 2(KY/A)^{1/2}$ Assuming similar rates of oxygen consumption for such muscles, the critical radius for *in vitro* incubations is about 0.77 mm. Since the Q_{10} for oxygen consumption is about 2.5 and for Krogh's constant K is about 1.1, it is clear that thicker muscles should be used only at temperatures below 37° to assure adequate oxygenation of all fibers; thus at 20°, the critical distances are approximately twice that at 37°.

[2] L. S. Jefferson, this volume [9]; J. T. Stull and S. E. Mayer, this volume [8].

[2a] N. B. Ruderman, C. R. S. Houghton, and R. Hems, *Biochem. J.* **124**, 639 (1971).

[3] O. Meyerhof and H. E. Himwich, *Pfluegers Arch. Gesamte Physiol.* **205**, 415 (1924).

[4] C. L. Gemmill, *Bull. Johns Hopkins Hosp.* **68**, 329 (1941).

[5] D. M. Kipnis and C. F. Cori, *J. Biol. Chem.* **224**, 681 (1957).

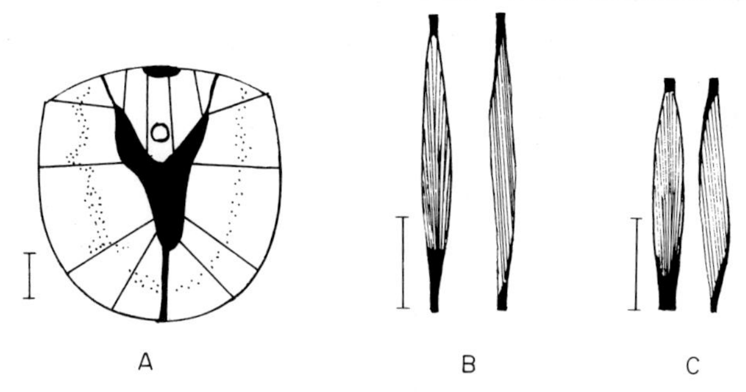

FIG. 1. Drawings of rat diaphragm (A), extensor digitorum longus (B), and soleus (C), after L. T. Potter [*J. Physiol.* (*London*) **206,** 145 (1970)] and R. A. Close [*Physiol. Rev.* **52,** 129 (1972)]. Darkened areas indicate tendon. For mature rats (200 g) the calibration line adjacent to each muscle indicates 1 cm; in younger rats routinely used in our labs (50–70 g) the calibration is approximately 0.5 cm. The diaphragm is drawn as viewed from the thoracic cavity with the dorsal segment on top. Lines are drawn radially to indicate the orientation of the muscle fibers from the central tendon. The ribs have been removed. Note the more diffuse arrangement of nerve terminals (dots) in the right hemidiaphragm. In drawings of the extensor digitorum longus (B) and soleus (C) the right member of the pair represents a lateral view of the muscle; on the left are anteroposterior views. The distal tendons are uppermost in each case. Note that the soleus is thicker than the extensor digitorum longus.

its rapid dissection and excellent viability *in vitro*. In addition, since the diaphragm can be readily dissected in combination with a considerable length of phrenic nerve, this preparation has been used widely by neurophysiologists and neurochemists.

Various other preparations of mammalian skeletal muscle have been introduced. One major motivation for identifying other muscles that are convenient for study *in vitro* is that mammalian muscles are composed of at least three distinguishable types of fibers which differ in their physiological, histochemical, and ultrastructural properties.[6] Different muscles can vary in the relative proportions of the three fiber types and therefore can differ appreciably in their metabolic, endocrinological, and pharmacological responses.

The Rat Diaphragm

Anatomy. In rats and mice, the diaphragm is an unusually thin sheet of muscle fibers that in young animals is even translucent. The muscle is composed of three distinct regions (Fig. 1): two larger lateral portions, which are generally referred to as "hemidiaphragms," and a smaller dor-

[6] R. A. Close, *Physiol. Rev.* **52,** 129 (1972).

sal segment adjacent to the spinal cord through which pass the esophagus, aorta, and vena cava. These subdivisions can be readily separated without damaging the fibers by cutting along the Y-shaped central tendon that delineates these three muscular segments. Within each hemidiaphragm, the muscle fibers run in a radial direction from the rib cage toward the central tendon. The neuromuscular junctions and the intradiaphragmatic portions of the phrenic nerve are localized in a well-defined thin band in the center of the muscle fibers that can be recognized with the naked eye (Fig. 1). In the diaphragm, red fibers predominate, but white and intermediate fibers each compose 20% of the total.

Most investigators have discarded the smaller triangular segment of muscle and compared the responses of the contralateral hemidiaphragms using one as the control. It is also possible to cut the hemidiaphragm parallel to the muscle fibers so as to produce several segments of tissue from the same diaphragm. For most metabolic investigations in this laboratory, we routinely bisect the hemidiaphragm and compare the four "quarter diaphragms" under different experimental conditions. Thus far, the behavior of the "quarter diaphragm" has proved indistinguishable from that of the hemidiaphragm. Comparisons between segments from the same muscle offer obvious statistical advantages in reducing biological variability. Such internally controlled data are more difficult to obtain with other muscle preparations and impossible in perfused muscles.

However, there are certain anatomical differences between the right and left hemidiaphragms. For example, the left hemidiaphragm with nerve attached has been used in most electrophysiological and biochemical studies of the neuromuscular junctions, because, on the left side, the region containing the nerve endings and neuromuscular junctions is more sharply delineated than on the right (Fig. 1).[7] In addition, the left phrenic nerve is simpler to dissect (or to section surgically) in the thorax than the right nerve. It thus seems advisable to check whether the behaviors of the contralateral hemidiaphragms are identical, at the outset of any investigation.

Dissection of the Diaphragm. In order to avoid problems of diffusion of nutrients into the incubated muscle,[1] it appears advisable to use rats weighing less than 100 g for most experiments. The best preparations of diaphragms from adult rats (280–320 g) lose most of their ATP and hexose phosphate intermediates upon incubation[8]; these findings contrast with those obtained with tissues from young rats (see below). For metabolic studies, we routinely use 60–80 g rats (3–4 weeks of age), in which the wet weight of the hemidiaphragms ranges between 40 and 50 mg.

[7] L. T. Potter, *J. Physiol.* (*London*) **206,** 145 (1970).
[8] K. A. Rookledge, *Biochem. J.* **125,** 93 (1971).

Such animals are rapidly growing, which complicates the planning of experiments. Obviously such rats are not appropriate for certain investigations (e.g., experiments on aging), and mouse diaphragm appears more appropriate for studies requiring adult tissues.

For rapid removal of the diaphragm from small rats or mice, cervical dislocation by pressure with a rod behind the neck has proved to be a simple and rapid method for sacrificing the animal. Many labs have successfully employed decapitation or stunning by a blow to the head. The latter methods are not advised when diaphragm and phrenic nerve are to be removed together, since they can injure the nerve or cause hemorrhage and blood clotting within the thoracic cavity. For such dissections, anesthesia with ether or barbiturates appears preferable (e.g., 40–50 mg of pentobarbital per kilogram of rat weight with supplementary 10 mg doses as necessary).

The diaphragm can be quickly excised along with a narrow ring of ribs to which the muscle is attached. After cutting through the skin on the ventral surface, the pectoral and abdominal muscles are trimmed from the rib cage. The lower edge of the diaphragm is next freed by cutting the several ligamentous attachments of the liver. The spinal column, esophagus, vena cava, and aorta are then severed below the diaphragm. The pleural cavity is entered by cutting through the sternum slightly above the xiphoid process, and the incision is extended on both sides to the spinal column. Finally the pericardial membranes and spinal column are severed above the diaphragm. The entire diaphragm with the supporting ribs and intercostal muscles is then transferred to a petri dish containing oxygenated buffer at room temperature. With practice, the time required from sacrifice to transfer to the petri dish can be shortened to less than 1 minute. The actual period of ischemia of the muscle is probably much shorter since the circulation remains intact until the aorta and vena cava are severed.

The preparation in the petri dish is trimmed by removing the xiphoid process, spine, and dorsal segment of diaphragm. The central tendon is then bisected to yield two intact muscle masses. If necessary, the intercostal muscles and ribs are further trimmed to minimize the amount of tissue extraneous to the diaphragm. Since these associated tissues can weigh 10–50 times the hemidiaphragm, they may affect the composition of the incubation medium either by leaking or utilizing substances.

The Diaphragm with and without Ribs

Unfortunately the diaphragm lacks a clearly defined tendon at the costal margins, and thus the investigator is forced either to incubate the muscle with a portion of the ribs attached or to risk damage to the diaphragm by attempting to trim away the attached remnants of the rib cage. Both preparations have been used successfully for different purposes.

For most studies it is advisable to incubate the hemidiaphragm with some ribs attached. As first shown by Kipnis and Cori,[5] such preparations more closely resemble the diaphragm *in vivo* and are more stable than hemidiaphragm or quarter diaphragm without ribs. The preparation with ribs is recommended for metabolic studies in which the products to be measured remain within the diaphragm, such as transport of substances into the intracellular space or incorporation of precursors into muscle protein, glycogen, fat, or nucleic acids. After incubation the diaphragm must then be cut away from the ribs, blotted, and weighed prior to chemical measurements.

Incubation with ribs, however, cannot be used for studies in which the experimenter monitors loss of substances from the medium (such as measurements of O_2 consumption) or evaluates metabolic products that might be released by the diaphragm into the surrounding media (such as measurements of transport out of the muscle, or investigation of CO_2, lactate, and amino acid production). For such experiments, the diaphragm must be cut away from its supporting structures prior to incubation, since the greater mass of muscle and bone attached will leak materials into the medium and thus contribute to the values measured. Trimming of ribs must be performed with fine scissors as close to the costal border as possible so as to minimize damage to diaphragm and yet avoid the inclusion of intercostal muscles in the incubation. Gentle blotting and weighing of the tissue is advisable prior to incubation because such preparations often swell during the course of incubation.

By a variety of criteria, the preparation with ribs more closely approximates the diaphragm within the organism than that with ribs removed, probably because the latter preparation contains appreciable numbers of injured fibers. (Kipnis and Cori[5] have even termed this latter preparation the "cut diaphragm" to distinguish it from the "intact diaphragm" preparation with ribs.) For example, even after 90 minutes of incubation in oxygenated buffer, quarter diaphragms with ribs attached contain amounts of ATP, inorganic phosphate, and phosphorylcreatine similar to muscles removed immediately from the rat (Table I). In such diaphragms, the extracellular space,[5] the permeability to pentoses,[5] and total intracellular amino acids (estimated in ninhydrin-reactive material or fluoresceamine-reactive material) also did not fall significantly even after incubation for 120 minutes. In addition, the preparations with ribs can maintain nearly normal resting potentials for 4 hours, are capable

TABLE I
COMPOSITION OF METABOLITES IN RAT DIAPHRAGM INCUBATED WITH AND WITHOUT RIBS[a]

Experimental conditions	Creatine	Phosphorylcreatine	Total creatine	ATP	Inorganic phosphate
With ribs					
Initial	9.1 ± 1	12.5 ± 1.2	21.6 ± 0.5	4.3 ± 0.5	5.7 ± 1.1
90-minute incubation at 37°	9.0 ± 0.2	10.8 ± 0.5	19.8 ± 0.5	4.3 ± 0.4	6.7 ± 0.2
Without ribs					
Initial	8.9 ± 0.5	12.2 ± 0.3	21.3 ± 0.4	4.0 ± 0.5	6.1 ± 0.9
90-minute incubation at 37°	5.6 ± 0.5	9.0 ± 0.9	14.7 ± 1.3	3.5 ± 0.5	6.9 ± 0.7

[a] Each value is the mean ± SE of four determinations, each of which involved quarter diaphragms from three animals. Values are expressed as micromoles per gram of blotted wet weight. After freezing of the tissues in isopentane at −150°C, the metabolites were extracted and analyzed by standard methods [M. J. Kushmerick, R. E. Larson, and R. E. Davies, *Proc. Roy. Soc. Ser. B* **174,** 293 (1969)].

of contraction for up to 12 hours, and incorporate labeled amino acids into protein at constant rates for 4 hours.

In contrast, in diaphragms without ribs attached, the levels of creatine and phosphorylcreatine fall during incubation for 90 minutes. The total creatine content (i.e., creatine and phosphorylcreatine combined) can be a useful measure of functional muscle mass,[9] and may even be a more reliable measure than wet weight, especially if some swelling or damage occurs. This fall in total creatine would suggest that some fibers in the muscle were damaged upon removal of the ribs and leaked metabolites into the medium. In support of this conclusion, creatine and phosphorylcreatine were recovered in the medium in amounts equal to those lost from the muscle (Table II). Glucose, insulin, and amino acids at 5 times plasma concentrations did not prevent this leakage of total creatine. Furthermore, the preparation leaks several percent of its protein into medium.

Although the diaphragm may leak intracellular components during the first half hour after the ribs are trimmed away, this preparation subsequently appears stable by several criteria. After the initial decrease, the concentrations of ATP and inorganic phosphate did not fall significantly during subsequent 90-minute incubation (Table II). The leakage of total creatine and proteins is about 10 times more rapid during the first half hour than afterward. The observation that the ratios of

[9] F. D. Carlson, D. H. Hardy, and D. R. Wilkie, *J. Physiol.* (*London*) **189,** 209 (1967).

TABLE II
EFFECT OF INCUBATION AT 37° ON VARIOUS METABOLITES IN RAT DIAPHRAGMS WITHOUT RIBS[a]

	Content of diaphragm (μmoles/g)		
	Initial	30 min incubation	120 min incubation
ATP	4.1	3.2	3.0
Phosphorylcreatine	9.3	9.8	7.7
Creatine	9.0	4.7	2.8
Inorganic phosphate	7.4	6.8	7.8
Tyrosine	0.17	0.13	0.13
Ninhydrin-reacting substances (leucine equivalents)	26	18	14
	Content of medium		
Total creatine (μmoles/g wet weight)	0	4.4	5.8
Protein (mg/mg dry weight)	0	0.034	0.044

[a] Quarter diaphragms from two or more animals were grouped for each experimental condition; means of two experiments are given. Methods of freezing the tissues, extraction, and analysis have been published elsewhere [R. Fulks, J. B. Li, and A. L. Goldberg, *J. Biol. Chem.* in press (1974); M. J. Kushmerick, R. E. Larson, and R. E. Davies, *Proc. Roy. Soc. Ser. B* **174,** 293 (1969)].

ATP:total creatine and phosphocreatine:total creatine remained constant throughout the incubation period suggests that the predominant number of cells maintain their energy reserves. In addition, intracellular concentrations of several amino acids (e.g., phenylalanine and tyrosine) remain constant during incubation.

Therefore, for studies of the diaphragm without ribs, in which measurements of the medium are to be made, a 30-minute preincubation of the tissue is recommended. Subsequently, the muscle is transferred to a second flask containing fresh medium, which is then closed and reequilibrated with the desired gas mixtures (see below). Following this preincubation, the rates of glucose oxidation, amino acid oxidation, protein synthesis, protein degradation, and nucleic acid synthesis are all linear for at least 2 hours. While membrane potentials appear generally reduced (ranging between −40 and −50 mV), vigorous contractions can still be demonstrated for at least 4 hours.

The actual physiological limitations of the diaphragms without ribs are unclear, since few studies have compared its properties with those of the intact preparation or the muscle *in vivo*. Diaphragms incubated without ribs synthesize proteins and actively transport amino acids and cations at normal or enhanced rates. It is, thus, incorrect to conclude that this preparation is generally leaky to small molecules or unstable as a consequence of injury to the fibers. The diaphragm without ribs also can respond to hormones, such as insulin, although the sensitivity of this preparation to endocrine factors may well be reduced. Nonspecific injury to muscle by itself has been shown to mimic certain actions of insulin.[10] Since other mammalian muscles can be studied without extraneous tissues attached and without cutting any fibers, comparative studies with such muscles may be performed to eliminate the possible effects of injury.

Other Muscle Preparations

Mammalian skeletal muscles have been reported to differ in important ways in their contractile and metabolic properties, apparently as a result of the differing types of muscle fibers of which they are composed. For example, in prolonged fasting or in response to large doses of glucocorticoids, the extensor digitorum longus and plantaris undergo marked atrophy, while the soleus shows little or no wasting.[11] Conclusions from studies of only the diaphragm are not necessarily valid for other muscles; for example, one unique property of the diaphragm is that upon denervation it undergoes a transient hypertrophy before the usual process of denervation atrophy. Similarly the levator anus muscle is uniquely dependent on the supply of androgens.[12] It, therefore, is important for many types of investigations to be able to compare several muscles of different physiological character.

The isolated soleus,[13,14] extensor digitorum longus,[15] and gracilis anticus[16] have all been used with success for metabolic or mechanical investigations. In addition, the rat peroneus longus,[17] rat epitrochlaris,[18] and mouse plantaris (unpublished) have been studied, and such preparations possibly merit further investigation. The anatomical locations of these muscles are found in Greene's atlas of rat anatomy.[19] Unfortunately the

[10] W. D. Peckham and E. Knobil, *Biochim. Biophys. Acta* **63**, 207 (1962).

[11] A. L. Goldberg, *in* "Muscle Biology" vol. 1, pp. 84–118. Dekker, New York, 1972.

[12] A. Arvill, S. Adolfsson, and K. Ahrén, this volume [11].

[13] I. H. Chaudry and M. K. Gould, *Biochim. Biophys. Acta* **177**, 527 (1969).

[14] A. L. Goldberg, C. M. Jablecki, and J. B. Li, *Ann. N.Y. Acad. Sci.* **228**, 190 (1974).

[15] V. M. Pain and K. L. Manchester, *Biochem. J.* **118**, 209 (1970).

[16] A. S. Bahler, J. T. Fales, and K. L. Zierler, *J. Gen. Physiol.* **51**, 369 (1968).

[17] K. L. Zierler, *Amer. J. Physiol.* **190**, 201 (1957).

[18] A. J. Garber, I. E. Karl, and D. M. Kipnis, *Amer. J. Clin. Invest.*, 65th Meeting, p. 31a (1973).

[19] E. C. Greene, "Anatomy of the Rat," The American Philosophical Society, Philadelphia, 1935; reprinted by Hafner, New York, 1963.

substrate requirements and viability of these preparations have not been studied in any detail with the exception of the extensor digitorum longus, whose properties have been carefully investigated by Pain and Manchester.[15]

The soleus and extensor digitorum longus (EDL) muscles are especially useful for comparative studies, since they are of relatively similar size and shape but contain substantially different proportions of fiber types (Fig. 1). Both are cylindrical fusiform structures, whose wet weights in a 50–60 g rat range between 20 and 30 mg. Since these muscles are significantly thicker than the diaphragm, use of rats larger than 60 g is not recommended.[1] In fact, upon incubation, EDL muscles exceeding 30 mg have been reported to show decreased ATP concentrations, poorer contractility, and increased extracellular fluid, unlike smaller muscles.[15] The problems of diffusion of precursors into such thick muscles may also explain certain published reports that the EDL contains multiple amino acid pools.[20] In composition the red soleus is one of the most uniform muscles of the body; approximately 80–90% of its fibers are of the slow twitch variety. Because these biochemical characteristics are adapted for continuous activity, the soleus may be a useful model for the behavior of cardiac muscle. On the other hand, the extensor digitorum longus is a pale muscle containing primarily fast twitch fibers and very few slow ones.

Both muscles can be easily dissected without damage to the fibers and with tendons intact at each end of the muscle. The soleus lies beneath the larger gastrocnemius and plantaris on the dorsal side of the lower limb. Together with the gastrocnemius, the soleus inserts into the Achilles tendon. The soleus can be induced to hypertrophy by cutting the contributions of the synergistic gastrocnemius to this tendon.[11] In addition, the soleus can be easily denervated by sectioning the sciatic nerve in the thigh. The EDL lies on the ventrolateral aspects of the lower leg beneath the tibialis anterior. It is innervated by the common peroneal nerve, which can be easily sectioned on the lateral aspect of the knee; compensatory hypertrophy of the EDL has been induced by tenotomy of the anterior tibialis.[21]

The gracilis anticus, located on the ventromedial aspect of the thigh, has also been used for studies of muscle mechanics. The gracilis is thinner than the soleus or EDL, but its outer surface is covered by thick fascia, which may *in vitro* constitute a barrier for diffusion of nutrients. In addition, this muscle is more difficult to dissect and its tendons are shorter than those of the soleus and EDL. In fact, in order to avoid cutting any

[20] J. B. Li, R. Fulks, and A. L. Goldberg, *J. Biol. Chem.* **248,** 7272 (1973).
[21] E. Mackova and P. Hnik, *Physiol. Bohemoslov.* **21,** 9 (1972).

muscle fibers it is advisable to remove portions of the pubis where the muscle originates, and the femur, where it inserts.

Conditions for Incubating Skeletal Muscles

These muscles have generally been incubated in Krebs-Ringer bicarbonate buffer[22] equilibrated with 95% O_2:5% CO_2 and contain 119 mM NaCl, 5 mM KCl, 3 mM $CaCl_2$, 1 mM $MgSO_4$, 1 mM KH_2PO_4, and 25 mM $NaHCO_3$ at pH 7.4.[22] Krebs-Ringer phosphate buffer equilibrated with 100% O_2 has also been widely used. Chloramphenicol (0.3 μg/ml) or other antibiotics are recommended for incubations lasting several hours, especially those employing labeled precursors. Addition of glucose (10 mM) to the incubation medium is advisable for incubations of several hours, although the diaphragm can oxidize its glycogen and triglyceride reserves for such periods in the absence of exogenous substrates. The addition of substrates is especially important for contracting muscle, whose metabolic requirements increase manyfold, and whose ability to contract may otherwise decrease rapidly.

Various types of vessels appear suitable for these incubations; we have found 25-ml Erlenmeyer flasks quite convenient. For incubation of hemidiaphragms without ribs, 3 ml of medium should provide a large excess of nutrients. Obviously for the preparation with ribs, larger volumes of medium may be required. Although smaller flasks may allow use of less medium, under such conditions release of materials (e.g., amino acids or potassium) from the incubated muscles can markedly affect the composition of the medium. It is useful if the flasks are fitted with resealable rubber serum stoppers through which a hypodermic needle can be inserted to introduce materials. Through such a needle, the air trapped in the flask prior to incubation can be flushed out of the flask with either O_2 or 95% O_2:5% CO_2, while a second hypodermic needle provides an exit for the gas. The 20 ml of gas contained within such flasks contains 1 mmole of oxygen, which is far greater than that required by the diaphragm which, at rest, consumes 0.5 μmole of O_2 per gram per minute at 37°.

Even though they are viable for 3–6 hours under these conditions, as evidenced by normal membrane potentials, ability to contract and to synthesize macromolecules, the diaphragm and other muscle preparations undergo net protein breakdown under these conditions.[23] In other words, protein degradation occurs several times more rapidly than protein syn-

[22] W. W. Umbreit, R. H. Burris, and J. F. Stauffer, "Manometric Techniques," 4th ed. p. 131. Burgess, Minneapolis, Minnesota, 1964.

[23] R. Fulks, J. B. Li, and A. L. Goldberg, *J. Biol. Chem.* (1974). In press.

thesis, and therefore these preparations leak amino acids into the medium. Addition of amino acids at the concentrations found in serum promotes protein synthesis, inhibits protein degradation, and helps maintain intracellular amino acid pools at normal levels.[23] When amino acids are added at 5 times normal concentrations, the muscle approaches normal protein balance. Leucine, isoleucine, and valine appear to be the critical amino acids affecting protein balance in muscle, possibly because this tissue rapidly degrades these compounds,[24] and they thus become rate limiting.

Addition of glucose (10 m*M*) and insulin (0.1 unit/ml) also promote protein synthesis and inhibit degradation in these incubated tissues. Other metabolizable substrates, such as fatty acids or ketone bodies, have not yet been found to influence viability of the diaphragm. Pain and Manchester[15] have reported that albumin (0.13%) decreased potassium loss and promoted contractility of the EDL. Albumin may specifically interact with the muscle or may simply bind noxious agents, such as heavy metals, in the incubation medium.[15] The effects of other proteins on muscle viability have not been studied.

One additional factor promoting muscle stability *in vitro* is the degree of tension (i.e., passive stretch) applied to the muscle.[14,15] Most investigators have incubated the diaphragm without any load, and such slack muscle tends to swell by taking up water. Passive tension of the diaphragm or extensor digitorum longus can retard this effect.[15] In addition, passive stretch of the diaphragm or soleus retards the net protein catabolism, characteristic of the incubated tissues.[14] Various groups have shown that stretch significantly enhances the long-term viability of incubated muscles from frog and rat.[14,15] For example, the diaphragm's ability to contract decay's rapidly 8 hours *in vitro* even in enriched medium, but if the contralateral muscle is incubated under slight tension, clear contractile responses can be demonstrated up to 20 hours. Pain and Manchester have described a useful method for passively stretching cylindrical muscle.[15] An alternative simple means of maintaining muscles at different lengths during incubation is to pin them to an inert support (e.g., Dow-Corning's Sylgard encapsulating resin). The mechanisms through which passive tension has these interesting effects *in vitro* are unknown, although it may be related to the marked ability of contractile activity to influence muscle size *in vivo*.[11,14]

Several methods have recently been introduced for maintaining the rat diaphragm in organ culture for up to 10 days.[25,26] These procedures

[24] A. L. Goldberg and R. Odessey, *Amer. J. Physiol.* **223**, 1384 (1972).

[25] D. Purves and B. Sakmann, *J. Physiol.* (*London*) **237**, 157 (1974).

[26] Z. Hall and D. Burg, *Science* **184**, 473 (1974).

involve use of tissue culture media, while maintaining the muscle in a stretched position with continuous oxygenation. Thus far, these methods have been used to study the development of denervation supersensitivity and the influence of repetitive stimulation[25,26] on the muscle membrane. In the future, these techniques should prove highly useful for investigations of the long-term consequences of muscular activity, drugs, and hormones.

Acknowledgments

We gratefully acknowledge research support from the National Institute of Neurological Disease and Stroke, the Air Force Office of Scientific Research, and the Muscular Dystrophy Associations of America (to ALG) and from the National Institute of Arthritis, Metabolic and Digestive Disease and the Massachusetts Heart Association (to MJK).

[11] The Use of the Levator Ani Muscle *in Vitro* in Evaluating Hormone Action

By A. Arvill, S. Adolfsson, and K. Ahrén

The levator ani (LA) muscle of the rat was early suggested as a possible endocrine test organ.[1] Existing only in the male rat, at least in the adult animal, its close relationship to the male sex organs initially made it interesting as an *in vivo* test organ of androgenic and anabolic substances.[2,3] In addition to its application *in vivo* the muscle has also been shown to be suitable as an *in vitro* preparation.[4]

Characteristics of the Muscle

The LA muscle develops from an originally uniform blastema as a bilateral and symmetrical structure. It is attached to the ventral bulbus cavernosus muscles and to the coccygeal fascia and is forming a loop around the caudal part of the rectum (Fig. 1).

The muscle develops similarly in both sexes until day 18 of embryonic

[1] P. Wainman and G. C. Shipounoff, *Endocrinology* **29,** 975 (1941).

[2] E. Eisenberg and G. S. Gordan, *J. Pharmacol. Exp. Ther.* **99,** 38 (1950).

[3] K. Ahrén, A. Arvill, and Å. Hjalmarson, *Acta Endocrinol.* (*Copenhagen*) **39,** 584 (1962); **42,** 601 (1963).

[4] A. Arvill and K. Ahrén, *Acta Endocrinol.* (*Copenhagen*) **52,** 325 (1966).

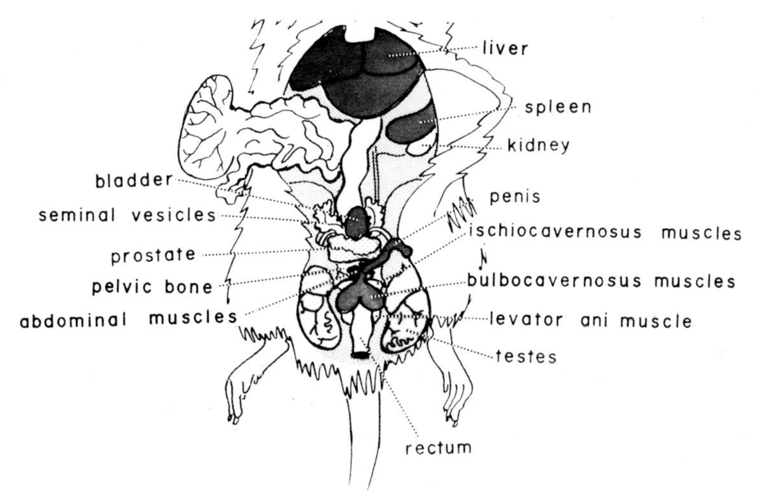

FIG. 1. Anatomical relations of the levator ani muscle.

life (21 days gestation). From this time it gradually retards in female rats and is completely involuted after 3 weeks.[5] In the newborn male rat the muscle is composed of myotubes without any cross-striation. Development is complete 14 days after birth and all ultrastructures commonly appearing in striated muscle can then be observed.[6,7] The LA muscle belongs to the "white" type of skeletal muscle and is composed of a uniform type of fibers with the myofibrils well packed together. The muscle is innervated bilaterally by the pudendal nerves.

The weight of the LA muscle in the prepubertal rats (body weight 50–60 g) is approximately 20 mg and in the adult rat about 200–250 mg. The muscle of the prepubertal rat is thin and flat (approximately $0.03 \times 2 \times 10$ mm). The extracellular space varies between 11 and 13% depending on the technique of determination and on the age of the rat.[4,8] The total water content is between 80 and 82% of the wet tissue weight.

In Vitro Preparations

Three different types of preparations are described. (1) The cut preparation: the LA muscle is dissected free from its attachments to the rest of the perineal complex before the incubation—analogous with the well

[5] R. Cihák, E. Gutmann, and V. Hanzlíková, *J. Anat.* **106,** 93 (1970).
[6] Z. Gori, C. Pellegrino, and M. Pollera, *Exp. Mol. Pathol.* **6,** 172 (1967).
[7] J. H. Venable, *Amer. J. Anat.* **119,** 271 (1966).
[8] T. M. Mills and E. Spaziani, *Biochim. Biophys. Acta* **150,** 435 (1968).

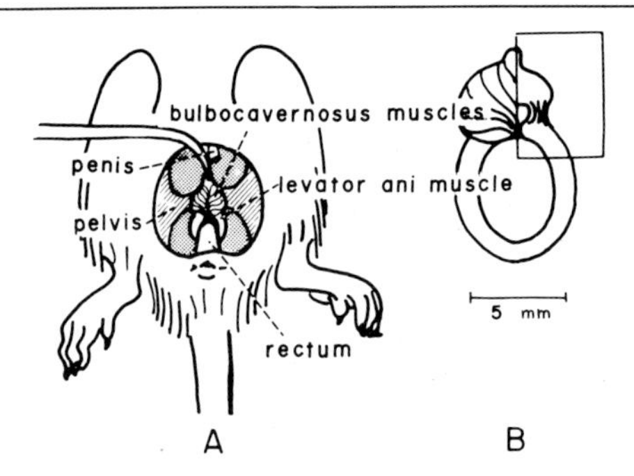

Fig. 2. Schematic drawings of the standard dissection procedure for the "intact" levator ani muscle preparation.

known "cut" diaphragm preparation.[9] (2) The "intact" preparation: the LA muscle is incubated together with the bulbocavernosus muscles as a ring-shaped complex—analogous with the diaphragm in the "intact" preparation of Kipnis and Cori.[10] (3) The "stretched" preparation: here the mid portion of the bulbocavernosus part of the "intact" LA preparation is cut off. Ligatures are applied to the remaining parts of the bulbocavernosus, and the LA muscle is mounted in a stretched position at a certain resting tension.[11]

Dissection Procedure

The killed rat is placed on its back, and a flap of the skin over the perineal region is cut away, leaving a circular opening into the testes and rectum region (Fig. 2A). The testes and the fat lying over the rectum are removed. The penis is cut at its connection to the bulbocavernosus muscles and these muscles are gently lifted with a forceps by pulling the root of the penis caudally. Bilateral incisions are made close to the puberal and ischiatic bones, cutting the ischiocavernosus part of the perineal complex. The rectum is cut cranially to the LA muscle and the ring of muscles is carefully dissected free from the rectum without touching or stretching the fibers of the LA muscle. The muscle ring is then placed in 0.9% NaCl and carefully trimmed free from fat and connective tissue under a stereomicroscope or other magnification. After blotting on filter paper the whole ring of muscles can then be incubated as an "intact" preparation. After some practice the total time of dissection will be about

[9] O. Meyerhof, K. Lohmann, and R. Meier, *Biochem. Z.* **157,** 459 (1925).
[10] D. M. Kipnis and C. F. Cori, *J. Biol. Chem.* **224,** 681 (1957).
[11] M. Buresová and E. Gutmann, *J. Endocrinol.* **50,** 643 (1971).

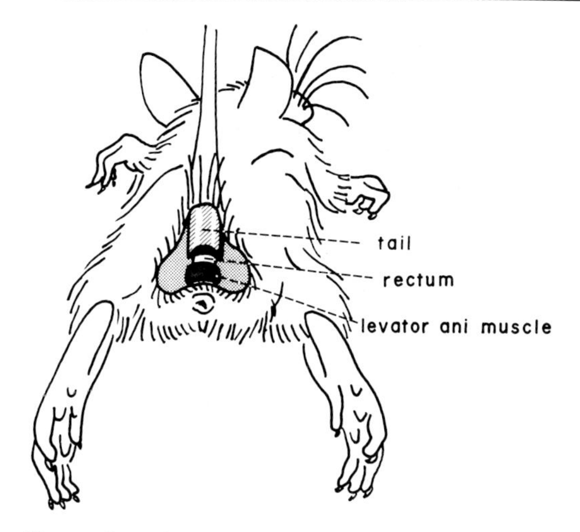

FIG. 3. Dorsal approach to the levator ani muscle.

1 minute, and it is to be noted that there is no need to touch the LA muscle with any instrument during the procedure, assuring fairly undamaged cell membranes. After the incubation period the attachments between the LA muscle and the bulbocavernosus muscles are cut (Fig. 2B), and the LA muscle is subjected to further analysis.

The LA muscle can also be used for acute *in vivo* experiments, e.g., for investigating the effects of induced muscle contraction. In this case the muscle is easily available through an incision made between the tail and anus (Fig. 3).

Incubation

The LA muscle, dissected as described above can be incubated by standard procedures. Ten milliliter Erlenmeyer flasks containing 1 ml of incubation medium is sufficient, owing to the small size of the preparation. The negligible amount of damaged cells included in the preparation minimizes the leakage of intracellular substances into the medium.[4] (Compare the intact diaphragm preparation.) When very long incubation times are used it is advisable to have continuous gassing or at least regas the flasks every other hour. The muscle can be stimulated to contractions *in vitro*[12,13] and contraction time and tensions be determined.[11]

[12] S. Adolfsson, *Acta Physiol. Scand.* **88,** 189 (1973).

[13] A. Arvill, Hormonal effects on the *in vitro* metabolism of a new intact mammalian muscle preparation. Thesis, Orstadius Ltd. Göteborg, Sweden (1967).

Basic Values for the Levator Ani Muscle *in Vitro*

The LA muscle of the prepubertal male rat is a fast phasic white-fiber muscle. The maximal twitch tension is about 3.5 g, maximal tetanus tension 9.0 g (fusion frequency 60–90 per second). Contraction time (time to peak) 46.0 msec and half relaxation time 33.0 msec.[5] The extracellular water content is about 20% after 15 minutes and about 27% after 2 hours of incubation (sucrose space and inulin space).[4] The total water content is about 82% even after long-time incubation (determined after drying as urea space). The muscle contains about 5.0 mg glycogen per gram wet weight, deprivation of food for 20 hours reduces the value to 3.5 mg/g.[12] Hormonal stimulation gives a maximal glycogen content of 9.0 mg/g wet weight.[14]

Up to now, there are basic values available for the "intact" LA preparation concerning the rate of transport of amino acids, including the model amino acids α-aminoisobutyric acid (AIB) and cycloleucine, incorporation of different precursors into protein and RNA as well as the rate of transport of monosaccharides (e.g., xylose and 2-deoxyglucose), uptake of glucose, glycogen synthesis, and the activity of some of the enzymes involved in the carbohydrate metabolism.[13,14]

Hormonal Effects

Testosterone. The LA muscle is highly dependent on normal androgen secretion for its development. After castration a rapid involution is obtained with a marked reduction of the total weight and, e.g., glycogen content. Fifteen days after castration of an adult rat, glycogen content is low and the LA muscle is reduced to less than 50% of its original weight.[15] In 45 days a steady state in function and structure is obtained.[7] Injections of testosterone accelerate the membrane transport of both amino acids and monosaccharides as well as the incorporation of amino acids into muscle protein and incorporation of precursors into muscle RNA as measured in the "intact" *in vitro* preparation.[8,16] Testosterone also stimulates glycogen formation by activating the glycogen synthetase enzyme.[14]

After the injections of testosterone a time lag of approximately 6 hours exists before any of the above-mentioned effects can be noticed. In the "standard" incubation medium (Krebs-Ringer bicarbonate buffer,

[14] S. Adolfsson, *Acta Physiol. Scand.* **88,** 234 (1973).

[15] E. Bergamini, G. Bombara, and C. Pellegrino, *Biochim. Biophys. Acta* **177,** 220 (1969).

[16] A. Arvill, *Acta Endocrinol.* (*Copenhagen*) **56,** Suppl. 122 (1967).

pH 7.4) with only glucose as substrate, no effects of testosterone are obtained when the hormone is added *in vitro*. *In vitro* effects of this hormone have been reported, however, by authors using "stretched" LA muscles incubated for 24 hours at 20°C in a tissue culture medium.[11]

Insulin. As for other striated muscles, also the LA muscle is sensitive to insulin. Insulin added *in vitro* to "intact" LA muscles from prepubertal rats markedly enhances the intracellular accumulation of several amino acids, including the model amino acid AIB[17] and increases the protein synthesis as well as the RNA synthesis.[18] Insulin *in vitro* also increases the uptake of different sugars[17] as, e.g., glucose, D-xylose, 2-deoxyglucose, and markedly stimulates the glycogen formation. It initiates an early increase of the active I-form of the glycogen synthetase enzyme, rate-limiting for glycogen synthesis.[14] The effects of insulin vary with the season of the year, with a higher effect in summertime and a lower in wintertime.[18]

Growth Hormone. With hypophysectomized rats an increase of the accumulation of monosaccharides, e.g., [^{14}C]D-xylose, is observed in isolated LA muscles as well as in diaphragm muscles, when growth hormone is added *in vitro*.[19] When injected *in vivo*, growth hormone makes the LA muscle insensitive to growth hormone later added *in vitro*,[19] this is also described for the diaphragm.[19] Under certain experimental conditions the LA muscle appears to react different to growth hormone than the diaphragm.[19]

Factors Influencing the Hormonal Effects

Denervation. As in the case of castration, denervation of the muscle is followed by a marked decrease of tissue mass, glycogen content, and inorganic phosphorus levels.[20] This similarity makes it interesting to compare the possible interaction of these two procedures on the metabolism of the LA muscle. It has been shown that denervation in the castrated adult rat completely abolishes the stimulatory effect of testosterone on the incorporation of precursors into nucleic acids and protein, while denervation in the immature rat does not have such an effect.[20] Thus, it appears that with sexual maturation, denervation abolished the protein-synthesis-stimulatory effect of testosterone and that this hormone in early stages of development can act as relatively nerve independent tropic agent.

[17] A. Arvill and K. Ahrén, *Acta Endocrinol.* (*Copenhagen*) **56,** 279 (1967).
[18] A. Arvill and K. Ahrén, *Acta Endocrinol.* (*Copenhagen*) **56,** 295 (1967).
[19] A. Arvill and Å. Hjalmarson, *Acta Endocrinol.* (*Copenhagen*) **56,** Suppl. 122 (1967).
[20] M. Buresová and E. Gutmann, *Life Sci.* **9,** 547 (1970).

Tension and Contraction. Besides denervation, other factors might influence hormonal effects on the metabolism of the LA muscle. It has been shown that optimal conditions for proteosynthesis are obtained when the muscle *in vitro* is set up at a certain resting tension (approximately 0.2 g).[21] There are also investigations dealing with the effect of contractions *in vitro*.[12,13] During a period of contraction LA muscle glycogen content diminishes, with no change of glycogen synthetase enzyme activity. There is an increased uptake of monosaccharides. There is also a decrease in the incorporation of precursors into muscle protein and nucleic acids. Some of these effects persist for at least 3 hours after contraction. Tension and contraction are of importance for the effects of hormones. Thus, an additive effect is found between insulin and contraction on the uptake of glucose and D-xylose.[14] As a contrast the stimulating effect of insulin on the accumulation of [^{14}C]AIB is decreased during the contraction period.[13]

Bioassay Aspects

Since the LA muscle is very sensitive to different hormonal influences, some of these effects seems to be suitable as biological test methods where other simple and reliable methods are not available. It has been shown that a correlation exists between insulin concentration and intracellular accumulation of [^{14}C]AIB in the "intact" LA preparation with a lowest effective insulin concentration of 1 μIU/ml and a straight-linear correlation in the range of 100–800 μIU/ml.[17] In comparison with other biological test methods this technique has proved to be quite reliable and accurate, with a simple laboratory procedure.[22] Another effect of insulin which can be adapted as a biological assay is the effect on the incorporation of [^{14}C]glucose into glycogen in the "intact" LA muscle *in vitro*.[23]

Advantages of the Preparation

1. Small and thin enough for *in vitro* diffusions of different substances
2. Possible to prepare undamaged with intact cell membrane functions
3. Simple preparation procedure
4. Small demands on the amount of incubation medium, radioactive precursors, and other substances which might be expensive or difficult to obtain

[21] M. Buresová, E. Gutmann, and M. Klicpera, *Experientia* **25,** 144 (1969).

[22] A. Arvill, G. Westberg, K. A. Jonsson, B. Hood, and K. Ahrén, *Diabetologia* **2,** 253 (1966).

[23] S. Adolfsson and O. Søvik, *Diabetologia* **4,** 309 (1968).

5. The irrelevant tissue necessary for the intact preparation is relatively small (cf. the intact diaphragm preparation)
6. Target organ for many hormones
7. Homogeneous in composition of myofibrils

Disadvantages of the Preparation

1. Preparations must be taken from immature or castrated rats
2. Each animal yields only one muscle
3. Low tissue weight, certain chemical analyses require pooling of muscles
4. As intact preparation, it contains some irrelevant tissue
5. In some aspects, not representative of the majority of the body muscles

[12] Systems for Evaluation of Uterine Responses *in Vitro*

By G. P. Talwar and Shail K. Sharma

In contrast to many organs, for example, liver, brain, and kidney, uterus of most experimental animals is not readily amenable to study in the form of tissue slices. It is also not easy to prepare cell suspensions from this tissue. The uterine responses can be studied *in vitro* by three types of approaches. These are: (1) homogenates and subcellular preparations; (2) intact uterus; (3) uterine explants in organ culture.

Homogenates and Subcellular Preparations

Binding of the steroid hormones to tissue "receptors" can be studied in homogenate preparations. The tissue is minced with scissors in a small beaker immersed in an ice bucket. Finely minced material is homogenized either with a Dounce type of tight-fitting glass homogenizer or still better with a Polytron homogenizer (type PT 10 OD, supplied by Kinematica G.m.b.H., Lucerne, Switzerland). The latter is more effective for an organ like uterus, which is homogenized with difficulty by the conventional methods. Standard procedures based on centrifugation through sucrose gradients are applicable to the uterus for preparation of nuclei, mitochondria, microsomes, and cell membrane fractions. Responsiveness to hormones and other affector agents is invariably lost in homogenate preparations owing to the derangement of intercellular and intracellular organization. However, these preparations are serviceable for various types of studies, such as the binding of hormones and assay for enzyme activities.

Intact Uterus

Earlier studies have mostly been done with uterine segments. There are, however, certain limitations of this method. We describe here a procedure by which isolated uterine horns can be studied conveniently and which gives fairly repeatable results. This experimental approach was evolved as a consequence of the observation that uterine horns when incubated *in vitro* in physiological media (such as Eagle's balanced salt solution containing vitamins and amino acids) contracted in size. Freshly excised uterine horns from ovariectomized rats weighing 100–150 g measure about 3.5 ± 0.3 cm. These shrink to about 2.0 ± 0.2 cm during incubation. The contraction of the tissue, presumably because of the smooth muscle component, would result in a reduction of about 40% of the exposed surface: this in turn would lead to slow diffusion of oxygen and metabolites in and out of the tissue. These difficulties are circumvented in the present method by tying the uterine horns with a surgical thread to hooks on a glass mount, so that the original length of the uterus is maintained. The mount serves also as a useful support for transfer of the horns to various media, and thus avoids direct handling and damage to the tissue. The detailed procedure is described below.

Animals. Uterine horns from normal cycling rats, as well as from ovariectomized animals can be used for study. The method is applicable to uteri from other animals with suitable modifications in the dimensions of the mount.

Removal of Tissue. Uteri are dissected and transferred to the incubation medium in a petri dish. The adhering adipose tissue is stripped off while the tissue is bathing in the medium, and the two horns are separated.

Temperature for Receiving the Isolated Uteri. The biological responsiveness of *intact* uteri is better maintained if they are handled at room temperature. Chilling and processing at 0–4° should be avoided, a procedure commonly employed for most biochemical experiments.

Mounting of the Uterine Horns. Uterine horns from a given animal are mounted with a surgical thread on two different mounts, one serving as control and the other as experimental. This approach permits comparative studies of the effect of an agent on contralateral horns of the same animal. Care is taken to stretch the tissue lightly so that it measures to approximately the same length as *in situ*. Overstretching of the tissue is likely to cause damage.

Description of the Mount. For rat uteri, Pyrex glass rods of length 5.5 cm and diameter 0.2 cm are used (Fig. 1). Two glass hooks are made at each end of the rod, with a knoblike projection on one side. As has been mentioned above, the contralateral uterine horns from a given animal are taken for experimental and control studies. These are therefore mounted on two different mounts, which will be incubated in separate tubes ± the affector agent. The orientation of the knob not only serves to mark the horns from a given animal, but also provides a socket for picking up of the mount with the forceps.

Incubation Media. While the studies can be conducted in any medium demanded by the needs of the experiment, it is recommended that an enriched type of medium suitable for tissue culture may be used. This may either be medium **199** or Dulbecco's modified Eagle's medium, with L-glutamine. Sodium bicarbonate (0.35 g per liter of medium) is added to adjust the pH to 7.4. The medium is saturated with **95%** oxygen and 5% CO_2. It is also advisable to include in the medium phosphatidylserine and phosphatidylinositol which have been found to restore the sensitivity

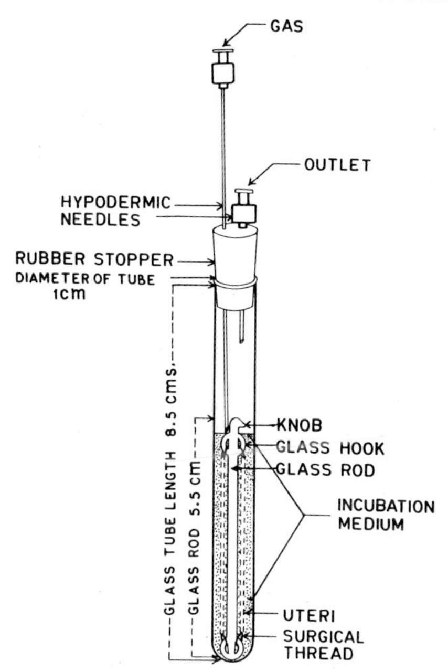

FIG. 1. Section diagram of the mount and tube used for studies *in vitro* on the uterus.

of adenyl cyclase to stimulation by hormones in solubilized myocardial and liver plasma membrane preparations.

Need for Preincubation. The tissues should be incubated in an atmosphere of 95% oxygen, 5% CO_2 at 37° in enriched medium for a minimum period of 15 minutes before experimental studies in order to restore the basal state. The ionic balance, ATP content, and excitability of the cells have been observed to alter in the course of the removal of the tissue from the body and subsequent handling.

Incubation. The mounts carrying the uterine horns are transferred to glass tubes containing incubation medium pregassed with 95% oxygen and 5% CO_2. Usually tubes 8.5 cm in length and 1.0 cm in diameter are used. If one mount is loaded in each tube, 2.0 ml incubation medium are sufficient. However, tubes of bigger size could be used, if more than one mount has to be incubated in each tube. The tubes are fitted with a rubber stopper containing two hypodermic needles. The long needle which touches the surface of the incubation medium is attached to the gas source, the other needle is the outlet. At the end of incubation, the mounts are removed by forceps and rinsed successively in four beakers containing isotonic saline at 0–4°. The washing procedure takes about 5 seconds per rinse per mount. The tissues are then processed for extraction and measurements according to the needs of the experiment.

Expression of Results. When the incubation period is longer than an hour, results should be expressed on the basis of DNA or dry weight, as increase in weight due to water imhibition may give misleading results.

Typical Results. Repeatability of response of contralateral horns of rat uterus is shown in Table I. Although there are variations in the uptake and incorporation of [5-^{3}H]uridine from one animal to another results show that contralateral horns from the same rat behave similarly if used on the glass mount by the above procedure. On the other hand, there is variation even within the same animal if the tissues are studied as segments. Uteri studied by the glass mount method show better metabolic response than contralateral horns examined as segments. It is interesting to note that the glass mount uterine preparations are responsive to affector agents like 3′,5′-cyclic adenosine monophosphate, while uterine segment preparations do not show a consistent response to this cyclic nucleotide (Table II).[1]

Utility and Limitations of in Vitro System for Studies on Uterine Responsiveness. Early actions of estrogens, cyclic nucleotides, and other agents can be obtained *in vitro* using this system. The system is also useful for study of metabolic responses of uterine horns *in vitro* after

[1] S. K. Sharma and G. P. Talwar, *J. Biol. Chem.* **245,** 1513 (1970).

TABLE I

REPEATABILITY OF RESPONSE OF CONTRALATERAL UTERINE HORNS STUDIED AS SEGMENTS AND BY THE METHOD RECOMMENDED

	Rat	Horn	Tissue preparation	dpm [5-^{3}H]uridine per mg wet weight in 30 min incubation
Set I	1	L	Glass mount	550
	1	R	Glass mount	520
	2	L	Glass mount	1600
	2	R	Glass mount	1400
	3	L	Glass mount	1670
	3	R	Glass mount	1580
Set II	4	L	Segments	558
	4	R	Segments	260
	5	L	Segments	715
	5	R	Segments	460
	5	L	Segments	730
	5	R	Segments	304

treatment with hormones and affector agents *in vivo*. The later effects of hormones and other agents, however, are not obtained in a manner identical to their action *in vivo*. This is perhaps due to the fact that uterine horns from ovariectomized rats are prone to structural damage during *in vitro* incubation. Besides other changes, the mitochondria show

TABLE II

EFFECT OF 3′,5′-CYCLIC ADENOSINE MONOPHOSPHATE (cAMP) ON UPTAKE AND INCORPORATION OF [5-^{3}H]URIDINE INTO RNA

Rat	Horn	Tissue preparation	Addition (0.1 mM)	dpm in acid-soluble fraction per mg DNA × 10^3 in 2 hours	dpm in RNA per mg DNA × 10^3 in 2 hours
1	L	Glass mount	—	560	164
1	R	Glass mount	cAMP	800	280
2	L	Glass mount	—	980	376
2	R	Glass mount	cAMP	1500	640
3	L	Segments	—	1140	420
3	R	Segments	cAMP	940	460
4	L	Segments	—	820	292
4	R	Segments	cAMP	792	252

swelling, a lesion that can be prevented to some extent by prior administration of estrogens to the animal.[2,3]

Organ Culture

The above-mentioned methods are suitable for all studies conducted over a limited period (2–4 hours) after removal of the tissue. For studies of longer duration, uteri cultivated in organ culture systems are recommended. A typical procedure is given below.

Preparation of Explants. Explants are made from the fetal uterus by cutting it into segments 2–4 mm long and from prepubertal and adult animals by dissecting endometrium 3 mm square and 0.5–1 mm thick.

Method. Uterine explants can be cultivated by using a watch glass method.[4] The culture medium used consists of 600 mg of agar, 10 ml of egg albumin, 10 ml of chick embryo extract (diluted 1:3 in Tyrode's BSS) and 80 ml of Tyrode's BSS. Hormones such as progesterone (2 μg/ml) and estrogens (1 μg/ml) dissolved in ethanol are added to the medium. Petri dishes containing cultures are incubated at 36° in a McIntosh and Filds anaerobic jar through which a mixture of 95% O_2 and 5% CO_2 is passed at 1 liter per minute for 40 minutes before sealing. Tissues may be cultured in a liquid medium with the use of stainless steel grid and lens paper.[5,6] The cultures are incubated at 37° under a gas phase of 95% O_2 and 5% CO_2. Cultures are fed with fresh medium and regassed every third day.

[2] I. Ljungkvist and T. Terenius, *Steroids* **15,** 413 (1969).
[3] K. N. Rao, S. K. Malhotra, and G. P. Talwar, *Cytobios* **5,** 201 (1972).
[4] J. Everett, *J. Endocrinol.* **24,** 491 (1962).
[5] O. A. Trowell, *Exp. Cell Res.* **16,** 118 (1959).
[6] H. M. Maurer, D. E. Rounds and C. W. Raiborn, *Nature (London)* **213,** 182 (1967).

Section V

Endocrine and Reproductive Tissue

[13] Isolation, Cloning, and Hybridization of Endocrine Cell Lines

By PETER O. KOHLER

The basic methods for the successful culture of living cells are amazingly simple, and the successful accomplishment of these techniques depends mainly on the care with which they are applied. It is impossible to give a complete summary of all the methodology used in cell culture. I will therefore describe some procedures that have been utilized to culture hormone-producing and hormone-responsive cell lines and to form somatic cell hybrids.

Although many valuable studies may be performed with primary cultures containing mixed cell populations, the use of cloned cells offers several advantages to the investigator interested in a specific function of one cell-type. In effect, a cloned cell line represents an artificial organ *in vitro* composed of cells which are all the progeny of a single precursor. Cooperative effects between different cell types can thus be eliminated or controlled. The use of cloned cell culture facilitates experiments in which specific populations of cells are exposed to known concentrations of drug or hormone for a defined length of time. Many studies can be performed in cloned cell culture which would be impossible in primary culture, organ culture or whole animals.

Several generalizations should be kept in mind regarding cloned established cell lines. Permanent cell lines appear to be "immortal" in that they will continue to grow and function as long as the proper culture environment is maintained. These cells may gradually undergo genetic changes as different stem lines appear. However, established cell lines provide a relatively stable and uniform population for repeated studies. The novice beginning to work with endocrine cell culture should be aware of the following points:

1. Normal somatic cells usually gradually dedifferentiate, i.e., lose their specialized function, in culture unless "transformation" to a tumor cell occurs. This loss of function may take weeks or years and is frequently accelerated by subculturing.

2. Normal nontumorous endocrine cells have not yet been adapted to long-term suspension culture.

3. In contrast to normal tissue, tumor cells which retain their appropriate organ-specific function *in vivo* frequently will retain the specialized function in culture. These cells usually can be cloned with reasonable ease and often are adaptable to suspension culture.

For these reasons, tumors of hormone-secreting or hormone-responsive tissue frequently make appropriate starting material for the development of cloned, differentiated cell lines.[1,2] Tumors may be generated in animals by chemical carcinogens or endocrine manipulation[3,4] with the eventual goal of adapting the tumor to culture. The remainder of this chapter will include a few specific points about equipment and a reasonably detailed description of some methods which can be used to develop cell lines and hybridize cells.

Equipment

Hood. The use of a laminar flow hood is not absolutly essential for work in tissue culture. However, if there are large amounts of dust or mold, air movement, or animal danders in the area where the cultures are handled, a hood may be highly desirable to avoid infection of the cultures. Cells from primates and other animals as well as human cells may harbor potentially infectious agents. Therefore, hoods with a vertical air flow are preferable because any potentially infectious material cannot be directed at the operator as in the case of hoods with a horizontal air flow.

Incubator. The incubator should have a stable temperature control with less than a 1.5° fluctuation in temperature. Currently, water-jacketed incubators are believed to be unnecessary with improvements in the temperature control devices. Water-jacketed incubators are more stable in the event of power failures, etc. For cloning and many studies where culture dishes and bicarbonate-buffered media are used, the incubator should be humidified and gassed. The usual gas mixture is 95% air–5% CO_2, although higher or lower CO_2 or lower O_2 concentration may be desirable for certain cells.[5] Small sealed plastic boxes which can be gassed and humidified are available for use in incubators without gas and humidity controls, but these are often cumbersome to handle. A recording thermometer on the incubator is recommended. This will give a permanent record of any possible temperature variations during procedures or experiments.

Glassware. Unless the laboratory is prepared to carry out time-consuming special washing of glassware, disposable plastic dishes, flasks, and

[1] G. H. Sato and Y. Yasumura, *Trans. N.Y. Acad. Sci.* **28,** 1063 (1966).

[2] P. O. Kohler, W. E. Bridson, J. M. Hammond, B. Weintraub, M. A. Kirschner, and D. H. Van Thiel. *Acta Endocrinol.* (*Copenhagen*) Suppl. 153, 137 (1971).

[3] J. N. Dent, E. L. Gadsden, and J. Furth, *Cancer Res.* **16,** 171 (1956).

[4] H. Kirkman and F. T. Algard, *Cancer Res.* **25,** 141 (1965).

[5] A. Richter, K. K. Sanford, and V. J. Evans, *J. Nat. Cancer Inst.* **49,** 1705 (1972).

pipettes should be used. These are available from commercial sources in a variety of sizes. Plastic dishes and flasks for cell culture have been specially treated, so the investigator should be aware that not any sterile plastic petri dish will suffice for successful culture. If special washing facilities are available, reusable glass pipettes and culture bottles may be more economical. However, great care must be used in the washing to avoid contamination with detergents, heavy metals, etc., which may damage delicate cells. A recommended glass-washing protocol has been described by Levintow and Eagle.[6]

Media. Although many investigators use one medium formulation for a specific type of cell, it is frequently advantageous to use several different media when attempting to isolate new cell strains. Medium which is optimal for growth of fibroblasts or established cell lines may be inappropriate for isolation of slow growing epithelioid cells. The classical media generally contained a bicarbonate buffer system. This has frequently resulted in pH fluctuation as CO_2 escaped from the culture vessel or as CO_2 was produced by rapidly metabolizing cells. New organic acid buffer systems such as HEPES[7,8] appear to provide better pH control. These may be preferable to bicarbonate buffer systems for some cells[9] and in some instances when a reliable CO_2 source is not available.

The serum used in any media is one of the critical determinants of success or failure. It is unfortunately true that there is a wide variation in growth-promoting properties in batches of horse or fetal bovine serum even from the same supply company. In fact, some lots of serum may be toxic to cells. In addition, serum is frequently a source of mycoplasma infection. For these reasons, it is important to check a batch of serum for possible mycoplasma contamination[10] and to determine its efficiency in supporting cell growth. Perhaps the best method for checking serum lots is to compare the cloning efficiency (see below) using standard cell lines. The serum lots selected are those which most effectively promote the growth of single cells into clones. A relatively large amount of a desirable serum can be ordered and frozen for use over a period of several months.

Ideally, antibiotics should be avoided in the medium. However, antibiotics are used by most tissue culture laboratories. Penicillin (50 U/ml) and streptomycin (50 μg/ml) are routinely added to the media in our laboratory without apparent deleterious effects. Development of L forms of bacteria rarely becomes a problem. Many investigators add ampho-

[6] L. Levintow and H. Eagle, this series, Vol. 5, p. **77**.

[7] C. Shipman, *Proc. Soc. Exp. Biol. Med.* **130,** 305 (1969).

[8] H. Eagle, *Science* **174,** 500 (1971).

[9] H. R. Massie, H. V. Samis, and M. B. Baird, *In Vitro* **7,** 191 (1972).

[10] L. Hayflick, *Tex. Rep. Biol. Med.* **23,** 285 (1965).

tericin B (10–25 μg/ml) to the medium if fungal contamination is frequent. Mycoplasma infections can be treated with specific antisera if the type is known or with tylosine or tetracyclines. However, once cultures have become infected with bacteria, mycoplasma, or fungi, successful treatment is the exception to the rule and the cultures should usually be discarded.

Initiation of Primary Cultures

For optimal cell growth, viable tissue should be obtained surgically and placed directly in culture. Endocrine tissue should be removed from the animal during anesthesia or immediately after death. A long delay in excision of tissue permits autolysis and cells do poorly in culture. This is particularly true of human endocrine tissue, which is seldom viable when obtained at necropsy. If circumstances prevent immediate culture of material, the tissue may be maintained one of two ways. If the period prior to culture is relatively short, less than 48 hours, the tissue may be cut into small pieces a few millimeters in diameter to allow diffusion of gas and nutrients and placed in a balanced salt solution containing both serum and antibiotics. Excess alkalinity such as might occur in a bicarbonate-buffered medium exposed to air so that the CO_2 escapes should be avoided. Exposure to only saline for more than a few minutes may also decrease the yield of viable cells. Pieces of tissue in HEPES-buffered medium or a balanced salt solution with serum may be placed in an ice bucket, but the tissue *should not be frozen* without special treatment. For the purpose of shipping specimens, ambient temperatures between 4 and 37° are satisfactory for short periods of time. The other method of holding cells for culture is to freeze them as described below.

When one is ready to initiate the primary cultures, the tissue is optimally divided into 3 portions, which are handled according to 3 different procedures. Any of these may be omitted, but for the greatest liklihood of success in establishing the appropriate cell strains, we use the following procedure.

1. The first portion of the tissue is placed in simple explant culture. This is accomplished by placing the tissue in a very small amount of medium to prevent drying, and mincing with sharp curved scissors or scalpels. A tearing action should be avoided and the cutting made as cleanly as possible into explants measuring 0.5–1.0 mm in diameter. The explants are placed in plastic culture dishes or flasks. Sterile Pasteur pipettes with curved tips are used to spread the explants evenly over the surface of the vessel. A selection of different media–sera combinations is frequently helpful. Although several media are adequate, we have used

media such as McCoy's 5a (Gibco), Ham's F-12 (Gibco) with 13.5% horse serum and 3.2% fetal bovine serum, and Medium 199 with 20% fetal bovine serum during the original explant growth. Attachment of explants to the culture surface is enhanced by using a very small volume of medium for the first 4–6 hours of culture. Care must be taken to avoid drying of the explants or floating them off the surface with a rapid movement of medium. We generally use 1.0 to 1.5 ml of medium in the 60-mm plastic dish for the first 4–6 hours, at which time an additional 1 ml is added. Over the next few days 1 ml of fresh or conditioned medium (see below) may be added at intervals of 24–48 hours.

We obtain conditioned medium by placing fresh medium in cultures of normal cells 3–5 days after subculture. The medium is removed after 24 hours and filtered in a Millipore HA filter to assure that no cells are contaminating the medium. This conditioned medium appears to have a beneficial effect on newly cultured cells. Once the explants are firmly attached and migration of cells from the explants occurs, complete changes of a full 5-ml volume of medium may be made.

One important feature of some explant cultures is finding that the cells of interest may attach poorly and float in the medium. To avoid losing floating cells or explants, it is frequently valuable to save the first complete change of medium, centrifuge at 100 *g*, and resuspend the sedimented cells in fresh medium and culture in a new dish. Continued reculturing of floating viable cells may yield a strain of poorly adherent but functional cells, which might also be easily adapted to suspension culture.

A word of caution is in order in regard to attempting to identify appropriate cells according to their morphology in culture. The appearance of cells in culture does not necessarily correlate well with the histologic appearance *in vivo*. However, many hormone-secreting cells have an epithelioid appearance in culture, while hormone-responsive cells may appear either epithelioid or fibroblastlike.

After cells have begun to migrate out from explants, a frequent problem is overgrowth of the endocrine cells by stromal fibroblastlike cells. Although some tumor cells will aggressively displace fibroblasts, it is advisable to identify areas in the culture that appear to contain the appropriate cell type. These areas may be mechanically transferred to another culture dish by using a scalpel blade to separate the cells from undesired cell types. A Pasteur pipette pulled into a fine glass loop may be used to gently tease the cells free from the surface of the dish. If the colony is large and contracts into a tight roll of cells, it should be again divided into small pieces or enzymatically dispersed (see below) prior to reculture.

An alternate method for transferring cells from explant culture is to place a stainless steel or plastic cloning cylinder directly over the desired area. The bottom of the cylinder is first coated with sterile silicone grease to form a seal isolating the cells within the cylinder from the rest of the culture. The cells within the cylinder may then be mechanically or enzymatically removed and recultured. If the cell density is very low in the secondary culture, conditioned medium or feeder layers may be used. This is also an appropriate time to attempt to clone the cells. These techniques are described below.

2. The second major portion of the tissue to be cultured may be prepared for complete dispersion and culture as individual cells (monodisperse suspension) rather than clumps. Again the tissue is minced with scissors or scalpels into small pieces about 1 mm in diameter. To remove blood, the fragments should be washed two or three times in HEPES-buffered media or a phosphate-buffered saline without calcium and magnesium.

After the tissue has been washed, it is ready for dissociation with enzyme or sodium ethylenediaminetetraacetate (EDTA). The exact solution and conditions depend on the type of cells to be disaggregated. When very large numbers of cells (grams) are needed, dispersion can be carried out in sterile screw-cap "trypsinization flasks" (Bellco). We have used sterile screw-cap polystyrene tubes (Falcon) for smaller amounts of tissue. Commercial trypsin solutions contain other enzymes such as elastase which may be important in dispersing the cells.[11] These "trypsin" preparations successfully dissociate most endocrine tissue. The 0.25% trypsin solution or the 0.05% trypsin 0.02% EDTA solutions from Gibco have been generally efficacious in dispersing endocrine cells by the following procedure:

The tissue fragments are washed in a large excess of the trypsin solution, centrifuged at 100 *g* and resuspended in the plastic tubes in 10 ml of the solution. The tubes are placed in a 37° incubator for 20 minutes. At this time the tubes are inspected and gently agitated. If the fragments are poorly dissociated, they may be agitated briefly in a Vortex mixer and replaced. When the cells are dissociated, the cell suspension should be gently agitated and poured through a double layer of sterile cheesecloth to remove clumps. We usually limit the total enzyme digestion to a maximum of 1 hour because these cells appear to plate well and further exposure is not necessary. However, when a high yield of cells is needed from a relatively small amount of tissue, the remaining fragments of tissue may be retained in the original tube by allowing them to settle by gravity and then the tissue may be resubjected to enzyme treatment.

[11] H. J. Phillips, *In Vitro* **8**, 101 (1972).

The filtered cells should be kept at room temperature or even in an ice bucket as temperatures over 20° may cause clumping. An aliquot of the cells should be diluted and counted immediately in a hemacytometer, and whole medium containing serum should be added to the remainder of the suspension to stop the action of the trypsin. The use of several different media as described previously may increase the yield of endocrine cells.

The major problems encountered with culture of monodisperse cell populations from endocrine tissues are a low plating efficiency and overgrowth of the slow growing endocrine cells by rapidly growing fibroblasts. The plating efficiency indicates the number of dispersed cells that will form viable colonies. A number of manuevers may be used to attempt to circumvent these problems and increase the liklihood of endocrine growth. These include: (1) Primary cloning, using methods similar to those described by Steinberger,[12] may permit immediate selection of the appropriate cell. (2) The addition of an appropriate hormone to the medium may be used to select for responsive cells. Norris *et al.*[13] have used testosterone in the medium to select for growth of a testosterone-induced hamster ductus deferens tumor. Sato and co-workers have used a variety of tropic hormones to help select for endocrine target cells.[14] (3) A drug such as bromodeoxyuridine may be used to kill rapidly proliferating cells.[15] Bromodeoxyuridine may be alternated with a tropic hormone in an attempt to alternately kill fibroblasts and stimulate the hormone-responsive cells. (4) Alternate culture to animal passage of a mixed tumor may be utilized in some instances to remove fibroblasts (see below).[16] (5) Finally, we have used feeder layers after the method of Puck *et al.*[17] to isolate endocrine cells.

Our feeder layers[18] are made up of normal human fibroblasts which are morphologically distinct from the cell we are trying to isolate. The feeder layers are irradiated with 10,000 rads which will prevent cell growth and division but allow the cells to continue to remain metabolically active for about 10 days. These cells in effect "condition" the medium for the more delicate endocrine cells. The feeder layers are plated at about 0.5 to 1.0 $\times$ 10^6 cells per 60 mm dish. This allows areas between the feeder cells for the dispersed cells to attach to the dish. The dispersed

[12] A. Steinberger, this volume, [14].

[13] J. M. Norris, J. Gorski, and P. O. Kohler, unpublished observations.

[14] G. H. Sato, J. L. Clark, M. Posner, H. Leffert, D. Paul, M. Morgan, and C. Colby, *Acta Endocrinol.* (*Copenhagen*), Suppl. **153**, 126 (1971).

[15] T. T. Puck and F. T. Kao, *Proc. Nat. Acad. Sci. U.S.* **58**, 1227 (1967).

[16] V. Buonassisi, G. H. Sato, and A. I. Cohen, *Proc. Nat. Acad. Sci. U.S.* **48**, 1184 (1962).

[17] T. T. Puck, P. I. Marcus, and S. J. Cieciura, *J. Exp. Med.* **103**, 273 (1956).

[18] P. O. Kohler and W. E. Bridson, *J. Clin. Endocrinol. Metab.* **32**, 683 (1971).

cells are plated at concentrations of 50, 500, and 5000 per dish to allow colonies of cells to develop even if plating efficiency is quite low. Plating is carried out at a cell density such that colonies of growing cells can achieve a satisfactory size without being overgrown by similar colonies of stromal cells. When the colonies attain a satisfactory cell number, usually at the time they measure about 0.5–1.0 cm in diameter, they are transferred by methods previously described to another dish with a fresh feeder layer. Endocrine function can frequently be tested at this passage to assure that the cells are appropriate prior to the time-consuming process of cloning.

3. The third portion of tissue may be used for animal passage if there is no constant supply of tumor tissue. If the tumor has been developed and serially passed in an animal species, this may be continued. However, if the tumor is of human origin or from an exotic animal species not kept in the laboratory, we use the hamster cheek pouch for *in vivo* growth after the method of Hertz.[19] The hamster cheek pouch is apparently a partially immunologically privileged site which will allow growth of many xenografts for periods up to 20 days.

Female Golden Syrian hamsters 4–6 weeks in age are anesthetized with intraperitoneal pentobarbital. The cheek pouch is everted with forceps and cleaned of debris with a dry gauze pad or with 70% ethanol. A 2- to 4-mm slit is made in the distal part of the pouch and a 1-mm fragment of tissue introduced with great care not to touch the outside of the everted pouch. The tissue is moved with external manipulation by curved forceps to about 1 cm from the incision. The area is blotted and, without suturing the cheek pouch, is gently tucked back into the hamster's mouth. However, no effort is made to completely return the pouch to its usual location. This will be done spontaneously by the hamster after awakening. After 4–5 days, the hamsters are reanesthetized and examined for tumor growth. Tumors will appear as blue nodules through the pouch wall. White nodules usually indicate infection, although some tumors derived from culture cells do not vascularize.[18] Tumor tissue should be removed after 12–14 days, prior to the onset of rejection. Animals obviously cannot be reused after tumor growth. Tissue taken from the animal may be placed in culture and reimplanted in a new hamster host. This technique is also useful as a method of removing stromal fibroblasts from a mixed cell culture which is difficult to clone. The fibroblasts generally grow rapidly in culture, but will not grow in the hamster unless transformation to a tumor has occurred. Endocrine tumor cells will often grow in the hamster.

[19] R. Hertz, *Proc. Soc. Exp. Biol. Med.* **102,** 77 (1959).

Cloning

The requirements for cloning include the ability for the tissue or cells in culture to be completely dispersed, plated as single cells, and multiply under the specific culture conditions. In order to clone cells effectively, all conditions should be optimal for cell growth. This includes the selection of the growth medium, temperature, and pH which will permit the highest plating efficiency of similar cells. Excessive exposure to enzymes or EDTA after the cells have been completely dispersed and wide temperature fluctuation which may occur when cultures are repeatedly taken out of an incubator for microscopic inspection at room temperature should be avoided during the cloning procedure.

Various maneuvers such as the use of "conditioned medium" or chick embryo extract have often been successful in increasing the yield of clones. Coon has described a primary cloning technique[20] in which he adds 2% chicken serum to a 0.1% trypsin solution to dissociate the cells. The chicken serum at this concentration apparently does not interfere with the enzymatic action of trypsin as do other types of serum.

In our laboratory we have again employed irradiated feeder layers of fibroblasts to support growth of the clones. We usually have used normal human fibroblasts irradiated with 10,000 rads for this procedure. The feeder layers are plated three days prior to the cloning attempt and irradiated 2 days before cloning. The medium is changed on the day prior to cloning so that this medium is "conditioned " by the irradiated cells before the dispersed cells are added. The cells to be cloned should be in an optimal growth stage, i.e., not from sparse, unhealthy looking cultures or cultures which have been confluent for an excessive length of time. The cloning procedure is carried out as follows:

1. Feeder layers are prepared in advance. Cells with a high plating efficiency will not require feeder layers.

2. The cultures to be cloned should be disaggregated with a minimum of enzyme treatment. After washing once with (4 ml for a 60 mm dish) 0.05% trypsin and 0.02% EDTA, 2 ml of the solution are layered on the cells and the culture is left at room temperature for 10 minutes. Gentle pipetting of the trypsin solution can be used to wash off and disaggregate the cells. A small aliquot may be examined on a slide or in a hemacytometer to determine when the cells are monodispersed.

3. As soon as the cells are dissociated in the trypsin solution, medium containing serum should be added to the suspension. If the dispersion can successfully be carried out in a small volume of trypsin solution,

[20] R. D. Cahn, H. G. Coon, and M. B. Cahn, *in* "Methods in Developmental Biology" (F. H. Wilt and N. K. Wessells, eds.), p. 493. Crowell-Collier, New York, 1967.

the washing procedure requiring centrifugation may be avoided. This is preferable since it reduces the manipulations required prior to plating and it avoids possible clumping of cells in the cell pellet that may require further pipetting to disperse. The serum in the medium stops the action of enzyme and dilutes the EDTA to a concentration where it has no apparent effect on cells.

4. Cells are immediately plated on the feeder layer at 50, 500, and 1000 cells per 60-mm dish. The dishes are swirled gently to provide uniform distribution of the cells in the dish and placed into a highly humidified incubator with a correct CO_2 tension.

5. The plates are examined the next day for attached single cells. This should preferably be done at 37° to avoid temperature fluctuations. The scanning should be performed rapidly to avoid major pH changes caused by CO_2 escape from the medium. Single endocrine cells can easily be differentiated from the feeder cells. These should be circled with a marking pencil. To assure that the developing colony is a true clone, a cloning cylinder may be placed over the cell and sealed to the dish with silicone grease. This will prevent other cells from being dislodged and adhering to a previously marked cell. The entire marking procedure should take less than 10 minutes. The medium in the cultures is changed less frequently than with vigorously growing established cells. We frequently change one-half of the medium every 2 days until the cells start to grow well.

6. Once a colony develops within the cloning cylinder, this may be transferred to another dish or feeder plate.

A variation in the technique is the use of microtiter test plates[21] with individual wells for the cloning procedure. Feeder layers are not possible with these plates, but conditioned medium may be used. The cells are plated into the wells. The plates are examined, and the wells containing a single cell are marked. This type of procedure has been described by Steinberger.[12]

In using the microtiter plates to develop permanent cell lines, one must be careful to maintain a high humidity in the incubator to avoid partial dehydration in the small wells leading to harmful hypertonicity of the medium. After a colony has developed within a well, it may be transferred to a regular culture dish for continuous culture.

Suspension Culture

Many biochemical studies can be accomplished more efficiently with suspension culture. Harvesting of cells requires only the removal of

[21] J. E. K. Cooper, *Tex. Rep. Biol. Med.* **28**, 29 (1970).

aliquots of medium followed by centrifugation. This obviates the need for enzymatic treatment or scraping of cells to remove them from the culture vessel; and quantification of cells is simplified.

The ease with which different cells can be adapted to suspension culture varies considerably. Other than hematopoietic cells, normal non-tumorous cells are almost impossible to grow in suspension. Usually, cells are first grown in monolayer culture and then placed in suspension culture. However, a few cell types adhere poorly to a culture vessel and may be rapidly adapted to suspension. Recultivation of floating cells or clumps of cells from a monolayer culture may yield cells that adapt easily to suspension.

To prevent settling of cells by gravity and possible attachment to the floor of the culture vessel, constant agitation is usually required. Some form of spinner apparatus is used in most laboratories, although other methods such as shakers and tumble tubes are available.[22,23] Special spinner flasks are effective (Bellco), but a simple stoppered Erlenmeyer flask with a magnetic stirring bar may suffice. The magnetic stirring mechanism should be used at the slowest possible speed which will prevent cell aggregation. However, some cell lines grow in suspension as small clumps rather than individual cells.

Elimination or reduction of calcium in the medium and additional phosphate buffer favors the growth of cells in suspension. Again, unless the cells of interest have special specific nutritional requirements and the investigator has the time and interest to construct special media, the use of commercial sources for suspension media is advisable. A large number of media which are specially formulated for suspension culture are available through the biological supply companies. Special additional factors such as serum have been necessary for growth of endocrine cells in suspension in the past. However, hormone-responsive cells can survive for variable periods in serum-free medium for special studies.[24]

The cells to be adapted to suspension culture should be in a state of vigorous growth, if possible. The cells may be dispersed with trypsin and added directly to special medium containing 5–20% serum without centrifugation and washing. The serum in the medium will inactivate the trypsin. It is critical that the final cell count be at least 1×10^5 viable cells per milliliter of medium. At concentrations lower than this, most cells

[22] O. V. H. Owens, M. K. Gey, and G. O. Gey, *Ann. N. Y. Acad. Sci.* **58,** 1039 (1954).

[23] W. R. Earle, E. L. Schilling, J. C. Bryant, and V. J. Evans, *J. Nat. Cancer Inst.* **14,** 1159 (1954).

[24] E. B. Thompson, G. M. Tomkins, and J. F. Curran, *Proc. Nat. Acad. Sci. U.S.* **56,** 296 (1966).

will not survive and multiply. The reason for this is not completely understood, but presumably relates to metabolic products in the medium from the living cells which promote growth. As in the case with stationary cultures, it is advantageous to try a variety of serum concentrations using horse serum and fetal bovine serum and a mixture of the two. Once adapted to suspension, a few very vigorous cell types will multiply in serum-free medium containing bovine serum albumin or lactalbumin hydrolyzate.

At times, cells which resist adaptation to growth in suspension culture can be passed through alternate periods of suspension and stationary culture.[25] Although the technique is not completely reliable, the number of viable cells in suspension can be approximated daily using a 0.3% trypan blue dye exclusion technique.[26] Viable cells will exclude the dye while dead cells will be stained. When the viable cells plateau or begin to decrease, the cells may be centrifuged and plated in stationary culture. As the cells again begin log phase growth, they can again be dispensed and placed in suspension. Presumably, this sequence selects for cells that can survive in suspension. The progeny of these cells may later multiply continuously in suspension culture.

Preservation and Storage of Cells by Freezing

Cell culture like many other techniques is subject to a variety of both mechanical and human errors. However, in few other endeavors can a mistake prove as costly. Infection with viruses, mycoplasma, bacteria, or fungi can alter or kill a cell line. Overheating due to a faulty thermostat, loss of air-conditions, etc., can destroy an entire incubator full of cultures in a short time. In addition, cell lines can undergo functional changes as the result of mutation, chromosome loss or the development of aggressive stem lines during prolonged periods of continuous culture. For these reasons, it is advantageous to preserve cells at many intervals during the development of cell lines so that cells can be retrieved to reestablish the line or examine the changes which have occurred. Fortunately, methods are available for preserving cells by freezing without detectable changes (after thawing and recultivation) in function as a result of the storage.[26]

The temperature at which the cells are stored is critical. In general, the lower the temperature the better. Storage at any temperature above —70° is inadequate for any length of time.[26] An ultra-low temperature chest-type freezer at —90° has been adequate for periods of two to three

[25] J. S. Hougham, personal communication.

[26] W. D. Peterson, Jr., and C. S. Stulberg, *Cryobiology* **1**, 80 (1964).

years. However, fluctuations of temperature up to —40° to —50°, which can occur when the freezer is opened frequently or for prolonged periods, are harmful to the stored cells. The use of a liquid nitrogen container is probably the optimal method for storing cells. Immersion of the storage vials in the liquid nitrogen leads to the risk of a small leak of liquid nitrogen into the vial, which could vaporize and cause explosion of the vial on warming. Storage of the vials in the vapor phase above the liquid nitrogen avoids this possibility and is at a sufficiently low temperature (—150° to —175°) to provide long-term storage with good recovery of viable cells.

Freezing of cells without special solvents such as glycerol or dimethyl sulfoxide causes lethal changes in the cells as the result of ice crystal formation and alteration in permeability, osmolality, and pH. Glycerol at 5–20% or dimethyl sulfoxide at 5–15% in the medium lowers the freezing point and prevents the deleterious effects of freezing. However, freezing should be done slowly to protect the cells. We have used both reagent-grade glycerol and dimethyl sulfoxide and found both to be effective, although many investigators now prefer dimethyl sulfoxide because it permeates the cell more rapidly and provides higher recoveries of some cell types.[26]

Frozen cells have usually been stored in 1-ml thick-walled glass ampules sealed at the tip with a flame. However, new screw-cap polymer vials are available which are easier to use and have been generally satisfactory.

The procedure we have used for freezing the cells is:

1. Cells in the rapid phase of growth are dispersed with 0.05% trypsin 0.02% EDTA, centrifuged and resuspended in special sterile freeze medium consisting of growth medium with 15% serum and 10% dimethyl sulfoxide or glycerol.

2. Cell concentration is adjusted to 2×10^6 cells/ml.

3. Cells are transferred to the special vials in 1-ml aliquots.

4. To allow gradual cooling, which is critical, the vials are left at room temperature for 15 minutes, then transferred to a 4° refrigerator for 2 hours. The cells are then placed in a —25° freezer for 2 hours, and finally placed in the ultra-cold freezer or vapor phase liquid nitrogen storage. Glass ampules are preferred for immersion in liquid nitrogen. Sophisticated devices are available for the gradual cooling of cells but are generally not necessary.

5. Thawing of the cells after storage should be done as rapidly as possible, in contrast to the cooling. The vial or ampule should be thawed in a 40° water bath in less than 60 seconds. As soon as the ice melts, the vial is placed in 70% ethanol at room temperature, and then dried

with a sterile towel and opened. (Ampules which have been immersed in liquid nitrogen must be warmed and opened by nicking with a file using gloves and preferably a welder's mask because of the possibility of explosion.) The contents of the vial are diluted with fresh growth medium containing serum so that the final concentration of glycerol or dimethyl sulfoxide is less than 1%. The cells are then plated in a culture dish for continuous culture.

Somatic Cell Hybridization

One of the major advances in methodology for studying control of gene function in cloned mammalian cells has been the development of techniques for somatic cell hybridization. Eukaryotic cells from two distinct parental lines may be fused to form heterokaryons. These are single cells that contain two or more nuclei from different parent cells. Under appropriate growth conditions, the two nuclei of binucleate heterokaryons undergo simultaneous mitosis on a single spindle apparatus producing mononucleate hybrid daughter cells. Immediately after fusion the nuclei of these daughter cells should contain the total number of chromosomes from both parent nuclei.

Hybrid strains can be derived from parent lines of different species.[27] Most of these interspecific hybrid cells have the advantageous property of preferentially discarding the chromosomes from one parent as growth in culture continues. This allows the investigator to derive a hybrid strain with as few as one chromosome from one of the parent lines.

The first observation of a new hybrid cell with a single nucleus containing chromosomes from two different parent lines was made by Barski *et al.*[28] in 1960. These hybrid cells were thought to have occurred spontaneously during co-culture of two mouse lines. The frequency of formation of this type of hybrid was extremely low. Littlefield[29] made a major contribution by using the HAT selective medium[30] (see below) which killed enzyme-deficient parent cells and greatly facilitated selection of hybrids. Finally, the use of inactivated Sendai virus to fuse cells[31-34] has greatly

[27] B. Ephrussi and M. C. Weiss, *Proc. Nat. Acad. Sci. U.S.,* **53,** 1040 (1965).

[28] G. Barski, S. Sorieul, and F. Cornefert, *C. R. Acad. Sci.* **251,** 1825 (1960).

[29] J. W. Littlefield, *Science* **145,** 709 (1964).

[30] W. Szybalski, E. H. Szybalska, and G. Ragni, *Nat. Cancer Inst. Monogr.* **7,** 75 (1962).

[31] Y. Okada, *Exp. Cell Res.* **26,** 98 (1962).

[32] G. Yerganian and M. B. Nell, *Proc. Nat. Acad. Sci. U.S.* **55,** 1066 (1966).

[33] H. Harris and J. F. Watkins, *Nature (London)* **205,** 660 (1965).

[34] H. G. Coon and M. C. Weiss, *Proc. Nat. Acad. Sci. U.S.* **62,** 852 (1969).

increased the yield of hybrid strains. Somatic cell hybridization has proved to be a valuable method for studying mammalian genetics and gene function, including examples of both negative and positive control.[35] The selective loss or segregation of chromosomes from one species in an interspecific hybrid is of great importance in studies directed at linkage analysis and assignment of genes to chromosomes.[36,37]

The process used to establish hybrid strains usually has two steps. First, the two types of cells must be placed in close contact or fused. Although not always necessary, fusion is usually accomplished with Sendai virus which has been inactivated with ultraviolet light[38] or β-propiolactone.[39] Lysolecithin has also been used to promote fusion between cells.[40] After heterokaryon formation, a variable number of mononucleate hybrid strains will develop. At this stage, some selection system is necessary to distinguish and select the hybrid strains from parent cells which might overgrow the hybrids.

Although other selection systems are being developed,[41] the HAT selection system of Littlefield[29] has been used widely. This system depends (shown schematically in Fig. 1) on the use of mutant parent cells, one line lacking thymidine kinase (TK) and the other lacking hypoxanthine guanine phosphoribosyltransferase (HGPRT). Cells deficient in either of these enzymes can form nucleotides for DNA synthesis only through synthesis de novo from amino acids and sugars. When this pathway is blocked by the folic acid antagonist aminopterin, these enzyme-deficient cells eventually die. However, hybrid cells which have corrected the enzyme deficiencies by intergenomic complementation may survive and grow in HAT medium which contains aminopterin, hypoxanthine, thymidine, and glycine with no folic acid.

Enzyme-deficient mutants may be produced by exposure to 8-azaguanine or thioguanine for HGPRT and to bromodeoxyuridine for TK. However, this is a laborious process with several pitfalls.[42,43] Therefore, the half-selection technique of Davidson and Ephrussi[44] has been useful for endocrine cells which have no enzyme-deficient, drug-resistant

[35] J. A. Peterson and M. C. Weiss, *Proc. Nat. Acad. Sci. U.S.* **69**, 571 (1972).
[36] M. C. Weiss and H. Green, *Proc. Nat. Acad. Sci. U.S.* **58**, 1104 (1967).
[37] F. H. Ruddle, *Nature (London)* **242**, 165 (1973).
[38] Y. Okada and J. Tadokoro, *Exp. Cell Res.* **26**, 108 (1962).
[39] J. M. Neff and J. F. Enders, *Proc. Soc. Exp. Biol. Med.* **127**, 260 (1968).
[40] J. A. Lucy, *Nature (London)* **227**, 815 (1970).
[41] T. T. Puck, *In Vitro* **7**, 115 (1971).
[42] F. D. Gillin, D. J. Roufa, A. L. Beaudet, and C. T. Caskey, *Genetics* **72**, 239 (1972).
[43] D. J. Roufa, B. N. Sadow, and C. T. Caskey, *Genetics* **75**, 515 (1973).
[44] R. L. Davidson and B. Ephrussi, *Nature (London)* **205**, 1170 (1965).

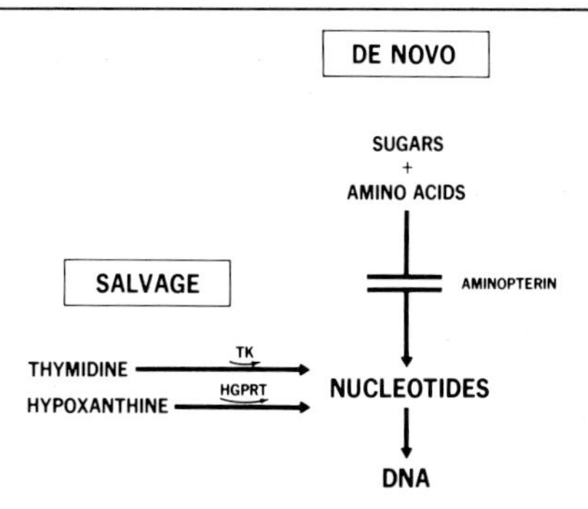

FIG. 1. Schematic representation of HAT selection system. Medium containing aminopterin blocks the *de novo* synthesis of nucleotides from sugars and amino acids. Therefore, the only way for cells to synthesize DNA is through the "salvage" pathways requiring both thymidine kinase (TK) and hypoxanthine guanine phosphoribosyltransferase (HGPRT). Cells deficient in either of these enzymes will die in HAT medium. However, hybrids which have both enzymes will grow in the medium, which contains hypoxanthine and thymidine as well as aminopterin. Modified after B. Ephrussi and M. C. Weiss [*Sci. Amer.* **220,** 26 (1969)].

markers. This technique which will be described, requires the identification of the hybrid colonies by gross morphology because the endocrine parent cell line will also grow in HAT medium. The hybrid nature of the strain must be proved by karyology or other markers.

A method for isolating hybrid strains of endocrine cells by the modified HAT selection technique requires: (1) endocrine cells which may be cloned tumor line or freshly isolated normal cells; (2) TK or HGPRT deficient cell lines which may be developed by drug treatment (some mouse and hamster lines are now available commercially); (3) a preparation of inactivated Sendai virus; (4) trypsin solution for cell dispersion; (5) regular medium and HAT medium without folic acid and containing aminopterin (4×10^{-7} M), hypoxanthine (1×10^{-4} M), and thymidine (1.6×10^{-5} M).[34] These media should contain optimal concentrations of fetal bovine serum and may also contain antibiotics; (6) plastic culture dishes; (7) a low power scanning microscope; and (8) plastic or stainless steel cloning cylinders.

Procedure for Half-Selection HAT System

1. Fusion of cells: The use of a virus such as inactivated Sendai to fuse cells will often increase the yield of hybrid strains as much as 100-fold.[34] There are no detectable differences in the properties of hybrids

formed with or without inactivated virus.[34] There are several methods for adding virus to the cells.[45] One technique is to disperse one group of cells, treat these with virus, and then add these virus-coated cells to an untreated monolayer.[34] A second approach is to cocultivate the parental lines and then add inactivated Sendai virus or lysolecithin.[37] The final method of dispersing both parent cell lines and treating with virus has been used effectively by many investigators.[31,33,34] Both lines of cells are dispersed as described previously with trypsin, centrifuged and resuspended in F-12 medium. The cells are mixed in plastic tubes in ratios of about 10–1000 TK- or HGPRT-deficient cells to 1 endocrine cell. The higher number of enzyme-deficient cells are used because these can be killed later with HAT medium and large numbers of nonhybridized endocrine cells may cause a subsequent problem with hybrid isolation. The total number of cells per tube should be 10^7 cells. The mixed cells should be centrifuged with removal of the supernatant and resuspension in 0.3 ml of cooled inactivated Sendai virus preparation containing 100–1000 HA units of virus[34] in F-12 medium without serum. After agitation for 10 minutes at 4° to permit virus adsorption, the tubes are transferred to a metabolic shaker at 37° to promote cell fusion. All of these steps obviously are performed using precautions to avoid infection with bacteria or fungi. After the hour of agitation at 37°, the cells may be plated in F-12 medium containing serum.

2. Selective medium: Twenty-four hours after plating, the whole F-12 medium is removed and replaced by the selective F-12 HAT medium containing serum. The hybrids should be maintained in selective medium to kill the enzyme-deficient parent cells.

3. Identification and isolation of hybrid colonies: After several days of growth, the plates should be scanned with a microscope for colonies of cells which appear morphologically distinct from the endocrine parent. These may appear a few days or even weeks after plating. The key determinant to successful isolation of the hybrids is identification of colonies before they are overgrown by the nonhybrid endocrine cells. Colonies may be circled with a grease pencil for observation until they are large enough to transfer. The colony may also be isolated within a cloning cylinder as described previously in the cloning section. Dispersion for cloning should be performed at the time of transfer. The hybrid nature of the selected clones must be confirmed by specific markers or karyology.

The disadvantage of the half-selection system is that hybrids which are morphologically similar to the endocrine cell may not be selected.

[45] F. J. Wiebel, H. V. Gelboin, and H. G. Coon, *Proc. Nat. Acad. Sci. U.S.* **69**, 3580 (1972).

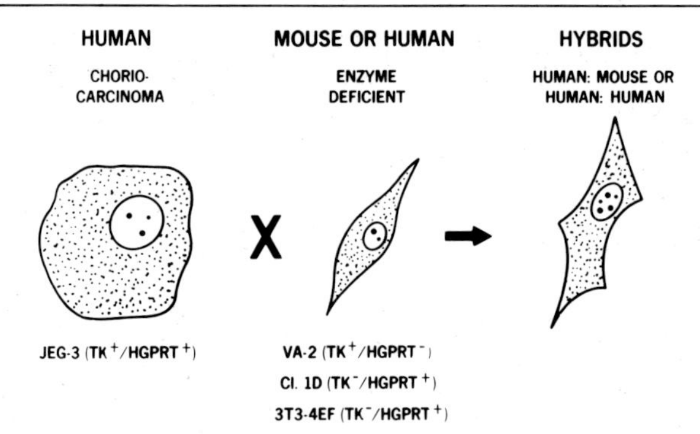

FIG. 2. The half-selection HAT system employs the use of one parent line which is deficient in either hypoxanthine guanine phosphoribosyltransferase ($HGPRT^-$), such as the human VA-2 line, or deficient in thymidine kinase (TK^-), such as the mouse lines clone 1 D or 3T3-4EF. These cells are fused with a cell, such as the JEG-3 human choriocarcinoma line, which contains both TK and HGPRT. The ratio of cells during the fusion is such that the TK^- or $HGPRT^-$ cells are in a 10- to 100-fold excess. The TK^- or $HGPRT^-$ parent cells which do not subsequently fuse and develop into hybrid strains will be killed by the HAT selection medium. However, the $TK^+/HGPRT^+$ human choriocarcinoma cells will continue to grow. Therefore, hybrids are selected on the basis of cell morphology. Cells that are morphologically different from the enzyme-competent parent line are mechanically isolated and recultured. Their hybrid nature must be confirmed by karyology or other specific markets.

Many endocrine cells appear epithelioid in culture while most of the TK- or HGPRT-deficient cells are fibroblastlike. Therefore, most of the hybrids selected by visual inspection will have a fibroblastlike appearance (Figs. 2 and 3). Reversion of a TK- or HGPRT-negative parent line to positive must also be kept in mind as a possible cause of fibroblasts surviving in HAT medium. Fortunately, reversion of parent cells to TK or HGPRT positive can be screened for by treatment of the isolated parent cells with medium containing 30 μg of bromodeoxyuridine or 8-azaguanine per milliliter, respectively. Specific enzyme assays may be desirable in some instances since drug-resistant cells often contain TK or HGPRT.[42,43]

Once hybrid strains have been isolated and cloned, specific endocrine or other functions can be tested. For example, earlier studies had indicated that hormone synthesis was extinguished when growth hormone-producing cells were hybridized with fibroblasts.[46] However, we have isolated hybrids between a gonadotropin-producing human trophoblast line

[46] C. Sonnenschein, U. I. Richardson, and A. H. Tashjian, Jr., *Exp. Cell. Res.* **69**, 336 (1971).

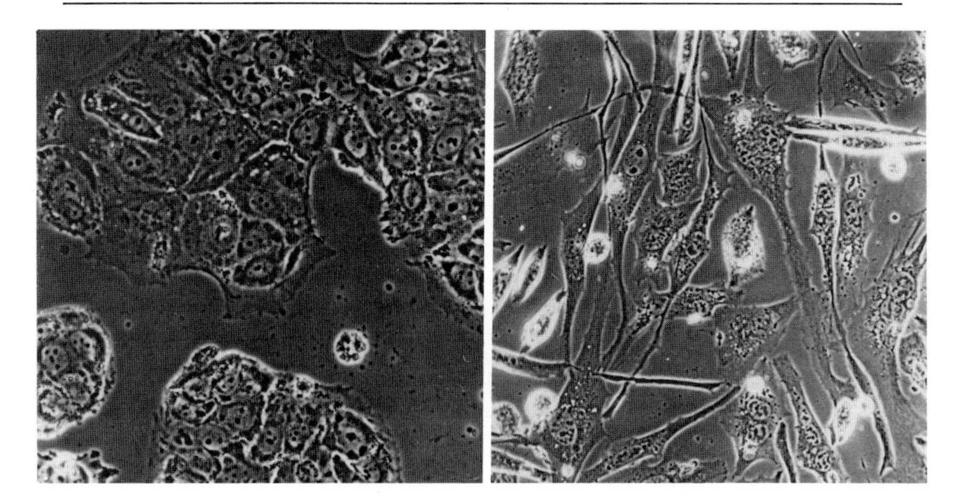

FIG. 3. Example of a hybrid strain initially selected on the basis of gross morphology in culture. The JEG-3 human trophoblast parent line is shown in the left panel. The human:mouse hybrid which survived in HAT medium is on the right. The hybrid was initially selected because of its morphologic appearance. The hybrid nature was confirmed by karyology.

and both mouse and human fibroblasts which have continued to synthesize hormones.[47] In the unstable interspecific hybrids, function can be examined longitudinally and compared with changes in chromosomal constitution. We found that a number of human:mouse hybrids between the gonadotropin-producing JEG-3 trophoblast line and two mouse fibroblast lines initially synthesized immunoreactive chorionic gonadotropin. However, as the human chromosomes were lost, the hybrids ceased producing detectable levels of hormone.[47] Hybridization can be used in the study of control mechanisms. Croce *et al.* have used hybridization studies to assign the gene for the repressor of corticosteroid-induced tyrosine aminotransferase to the X chromosome.[48] Other selection systems based on the correction by complementation of enzymatic defects have been developed,[42] and more will be devised in the future. For the present, the half-selection HAT system can easily be applied to endocrine cells. However, the full-selection HAT system is superior if the investigator can afford the time to develop TK or HGPRT-deficient mutant lines of endocrine cells.

[47] P. O. Kohler, H. D. Ruder, and H. G. Coon (in preparation).

[48] C. M. Croce, G. Litwack, and H. Koprowski, *Proc. Nat. Acad. Sci. U.S.* **70**, 1268 (1973).

Somatic cell hybridization has tremendous potential for examining the genetic basis for both the control and process of hormone synthesis as well as hormone action. By studying the responses of hybrids containing small numbers of chromosomes from one parental species our understanding of factors responsible for positive and negative control of hormone synthesis or action should be greatly increased.

Acknowledgment

I would like to thank Drs. Donald Roufa, Mary Bordelon, and Katherine Stefos for helpful comments in the preparation of this manuscript. This work supported in part by Grant AM17307 from the National Institutes of Health.

[14] Initiation of Clone Cultures of Rat Anterior Pituitary Cells Which Secrete a Single Gonadotropic Hormone

By ANNA STEINBERGER

Numerous problems related to the physiology of pituitary hormones have been successfully investigated *in vitro* utilizing incubations of whole or parts of pituitary gland or mixed cell cultures. The desirability of homogeneous cell populations over whole organs or mixed cell cultures is apparent when biochemical studies or studies of the direct effects of hormones or other factors on a specific cell type is the objective. However, homogeneous populations of specific cell types are difficult to isolate by using density gradients or velocity sedimentation techniques, unless the desired cells differ significantly in size or density from other cell types present in the tissue of origin. In addition, great limitations are imposed on the conditions that can be used for the separation, if cell viability is to be retained.[1]

Cell cloning, on the other hand, assures homogeneous populations of viable cells since each clone represents the progeny of a single cell. There are two major requirements in the application of cell cloning techniques for the isolation of specific cell types: (1) the desired cells must be able to proliferate actively *in vitro* and (2) must possess some unique morphologic or biochemical characteristics by which they can be readily identified.

The cloning method, described here, has been utilized primarily for initiating cultures of luteinizing hormone (LH) or follicle-stimulating

[1] K. Shortman, *Annu. Rev. Biophys. Bioeng.* **1**, 93 (1972).

hormone (FSH) secreting cells from rat adenohypophysis.[2] A similar method should, in principle, be applicable in isolation of other hormone-secreting cells from the rat pituitary or pituitaries of other species. This method, like other tissue culture procedures, requires sterile techniques, meticulously clean glassware, and good quality media and equipment. Conditions for single-cell growth are much more critical than necessary for mass cultures.

Cells for cloning are obtained from short-term primary cultures of the anterior pituitary.[3] Attempts to clone cells directly from cell suspensions obtained from fresh tissue gave very low cloning efficiencies (5–11%) compared to cloning efficiencies obtained with cells from cultures several days old (40–65%). The primary cultures seem to serve the purpose of either selecting cells which are more capable of growing *in vitro* or allowing the cells to become adapted to the culture conditions.

Primary Culture

1. Animals are sacrificed by either cervical dislocation or an overdose of chloroform. The pituitary gland is exposed through a skull incision by removing most of the brain tissue. When cutting the skull, care should be taken to avoid excessive bleeding which would obscure the pituitary. The whole pituitary gland is lifted from the sella turcica with forceps and placed into a small petri dish containing Hanks' balanced salt solution (BSS) pH 7.2. The anterior portion of the gland is separated with small forceps from the neurohypophysis and placed into a second dish containing several drops of Hanks' BSS.

2. This tissue is minced with iridectomy scissors and transferred to an Erlenmeyer flask containing 0.25% trypsin (Difco Laboratories) in Ca^{2+}- and Mg^{2+}-free Hanks' BSS. The trypsin solution should be prewarmed to 37° and adjusted to pH 7.5–7.6 with 7.5% $NaHCO_3$. One to 2 ml of the trypsin solution are used for each anterior pituitary. The flask is incubated at 37° in an atmosphere of 5% CO_2/95% air for 15 minutes and is agitated manually every 3–4 minutes.

3. After incubation, the contents are gently triturated with a 10-ml pipette. The undigested tissue fragments are allowed to settle to the bottom and the supernatant, containing predominantly single cells, is transferred to a centrifuge tube. To stop the action of trypsin, calf serum in a final concentration of 2% or lima bean trypsin inhibitor 0.2% (w/v) (LBI, Worthington Biochemical Corp.) is added to the cell suspension, and the tubes are spun at 900 *g* for 7 minutes in a refrigerated centrifuge

[2] A. Steinberger, M. Chowdhury, and E. Steinberger, *Endocrinology* **92**, 18 (1973).

[3] A. Steinberger, M. Chowdhury, and E. Steinberger, *Endocrinology* **93**, 12 (1973).

(4°). The supernatant is discarded, and the cells are resuspended in several milliliters of culture medium and kept at 4°.

4. After the cells released by the first trypsin treatment have been removed, fresh trypsin solution is added to the incubation flask and the incubation, centrifugation, etc. are repeated two or three times until the tissue fragments become completely dispersed.

5. The cells recovered after each centrifugation are combined. One hundred microliters of the cell suspension are mixed with 50 μl of 1% trypan blue dye in isotonic saline, and both stained and unstained cells are counted in a hemacytometer. Cells which take up the trypan blue dye are considered to be nonviable.[4] Usually 6 to 9×10^5 cells, of which 92–98% are viable, are obtained from a single anterior pituitary of an adult (200–250 *g*) rat.

6. Plastic, disposable 25 cm^2 culture flasks (Falcon Plastics, Bio Quest) are seeded with 5×10^5 cells in 5 ml of culture medium and incubated at 37° in an atmosphere of 5% CO_2/95% air. The culture medium consists of Eagle's minimum essential medium[5] supplemented with m*M* glutamine, nonessential amino acids, sodium pyruvate and 10–20% calf serum (Baltimore Biological Laboratories). It should be pointed out that while this medium consistently yielded satisfactory results, extensive comparison with other commercially available media was not carried out. However, fetal calf serum was found to be superior to calf serum, or sera from other animal species. Unfortunately different lots of serum vary considerably in their ability to support growth of the pituitary cells.

7. Within 24 hours after seeding, most cells become attached to the culture vessel, flatten out, and begin to divide. After 4–6 days of incubation, the dividing cells form a monolayer which seldom reaches complete confluency. These primary cultures, which represent a mixed population of various anterior pituitary cells (as evidenced by presence of LH, FSH, ACTH, TSH, GH, and prolactin in the medium) are used for cloning.

Cloning Procedure

1. The medium is decanted, and the culture flask is rinsed twice with 5 ml of Hanks' BSS free of Ca^{2+} and Mg^{2+}. Five milliliters of trypsin (0.25% in Ca^{2+}- and Mg^{2+}-free Hanks' BSS, pH 7.4) are added to the culture and incubated at 37° for about 5 minutes or until most of the cells become detached from the surface. The cell suspension is pipetted gently but thoroughly, and spun in a centrifuge at 900 *g* for 7 minutes.

[4] H. J. Phillips and J. E. Terryberry, *Exp. Cell Res.* **13**, 341 (1957).
[5] H. Eagle, *Science* **130**, 432 (1959).

Serum (2%) or trypsin inhibitor (0.2%) is added prior to spinning to inhibit the action of trypsin. The cell pellet is thoroughly resuspended in several milliliters of culture medium, and the suspension is strained through gauze to eliminate remaining cell clusters.

2. A viable cell count is performed in triplicate, using the trypan blue exclusion method[4] and a hemacytometer. If the 3 counts do not fall within 10% of the mean they should be repeated. An accurate determination of the cell concentration and absence of cell clusters are very crucial in the cloning procedure.

3. The cell suspension is diluted to a concentration of 1 cell per 100 μl of culture medium, and 100-μl aliquots of this suspension are dispensed into individual wells of Microtest II Tissue Culture Plates (Falcon Plastics, Bio Quest). Each plate contains 96 microwells of approximately 300-μl capacity. After seeding, the plates are covered with Microtest II lids (Falcon Plastics, Bio Quest) and incubated at 37° in an atmosphere of 5% CO_2 and 95% air.

4. Sixteen to 20 hours after seeding, the plates are scanned with an inverted microscope using 100–150× magnification, and microwells containing a single cell are recorded. The microwells which are found to contain 2 or more cells are considered unsuitable for clone cultures. According to the Poisson[6] distribution, seeding of wells with multiple cells as well as with cell-free aliquots would be expected to occur. After scanning, the plates are returned to the 37° incubator.

5. After 6–8 days of additional incubation the wells are rescanned microscopically to determine presence of clones. Cloning efficiencies up to 70% have been achieved when 20% fetal calf serum was used in the medium. However, there is considerable variation in the quality of different lots of fetal calf serum. Cloning efficiency can vary by as much as 30% with different serum lots. The cloning efficiency decreases by approximately 15% when the serum concentration is reduced to 10%.

6. Clones composed of at least 4 cells are treated with 100 μl of culture medium containing hypothalamic releasing factor for LH and FSH (LRF) in order to increase hormone concentration in the medium. In case a synthetic preparation of LRF is not available, an acid extract of rat hypothalami, containing LRF can be used.[2]

7. Four to 24 hours after the addition of LRF, the medium from *each* microwell is removed and assayed for LH and FSH using a double antibody radioimmunoassay method.[2,3] Fresh culture medium is added to the wells, and the plates are reincubated at 37° to allow further cell growth. Media can be collected from the same well after several 24-hour

[6] R. G. D. Steel and J. H. Torrie, "Principles and Procedures of Statistics." McGraw-Hill, New York, 1960.

intervals, in order to accumulate larger volumes if needed for assay of several hormones.

Pituitaries from normal adult male rats yield approximately 5% of clones which secrete only LH and about 40% of clones which secrete only FSH (these ratios may be different for clones derived from pituitaries of female rats). Occasional wells are found to contain both LH and FSH.[2] Presence of both gonadotropins in the same well, most likely, results from a chance seeding of the well with multiple cells, which escape detection during the initial scanning or may be due to the capacity of some pituitary cells to secrete both hormones.

8. The LH or FSH secreting clones can be subcultured to larger culture vessels using 0.1 ml of 0.1% trypsin per well. The subcultured cells continue to proliferate and to secrete hormones over a period of several weeks. The maximum period during which the hormones remain detectable in the culture medium depends on a number of factors such as the quality of the culture medium, proper pH control, treatment with LRF, etc.

The LH- and FSH-secreting clones cannot be identified on the basis of morphology alone when examined with either bright light or phase optics. However, these clones exhibit differential tinctorial properties following staining with periodic acid Schiff–methyl blue–orange G (PAS-MG-OG) as described by Rennels[7] for staining histologic sections of rat pituitary. The clones which secrete LH stain PAS-red and those which secrete FSH stain PAS-purple. Nevertheless, not all PAS-red clones secrete LH. This may be due to the fact that in histologic sections of rat pituitary gland stained with PAS-MG-OG both the thyrotrophs and LH-secreting gonadotrophs appear PAS-red.[7]

The main disadvantage of the above described method is that it is time consuming. Approximately 3 weeks are required for cell growth and identification of the gonadotropin-secreting clones. The double antibody radioimmunoassay for LH and FSH itself requires 7–9 days. Another serious limitation of the method is the small yield of cloned cells. This can be partly remedied by pooling clones that secrete the same hormone and extending their growth period after subculture to larger culture vessels. In this case the useful growth period will be determined by the ability of cells to secrete the hormone.

[7] E. G. Rennels, *Z. Zellforsch. Mikrosk. Anat.* **45,** 464 (1957).

[15] Basic Electron Microscopy Techniques for Endocrine Tissue

By ROBERTO VITALE

The combined use of morphological, histochemical and biochemical methods in the analysis of endocrine tissues at the cellular level has become, in recent years, a fruitful approach to the study of endocrine function. The electron microscope, an elegant morphological tool, has been successfully applied for over two decades to the endocrine cells; its contributions to our understanding of their detailed structural organization are many. Among these, the electron microscope has shown the degree of involvement of different organelles in the synthesis, storage, and release of hormones. Significant contributions have also been made in studying the nature of the barriers that hormones must penetrate in order to reach the blood, lymph, and thereafter the target cells.

It is beyond the scope of this chapter to consider the details of every aspect in the preparation and interpretation of various electron microscopy techniques and its relevance to endocrine tissues. Our attention rather will be concentrated on general examples of the different types of techniques used in the preparation of tissues for electron microscopic studies. We shall mention also the basic principles of the various microscopes and briefly discuss the information they provide.

The specialist will surely be dissatisfied with the superficial treatment of his field, but it is hoped that the nonspecialist particularly will profit from this overview and have a useful point of departure in the literature to pursue any of the topics in greater depth.

Transmission Electron Microscopy (TEM)

A. Fixation

Comprehensive reviews are available on various procedures for the fixation of tissues for TEM (Parson[1]; Hayat[2]; Dawes[3]). We shall

[1] D. F. Parson (ed.), "Some Biological Techniques in Electron Microscopy." Academic Press, New York, 1970.

[2] M. Hayat, ed., "Principles and Techniques of Electron Microscopy," Vols. 1–3. Van Nostrand-Reinhold, Princeton, New Jersey, 1970.

[3] C. J. Dawes, "Biological Techniques in Electron Microscopy." Barnes & Noble, New York, 1971.

describe briefly the routine methods commonly in use and consider some recent developments that show potential improvements over the current techniques. For proper fixation, the fixative vehicle seems to be particularly important, and several buffer systems have been extensively used as vehicles. *sym*-Collidine, cacodylate, and phosphate are among the general purpose buffers which seem to be preferred by the majority of laboratories.

It has been recognized for a long time that the osmolarity of the fixative has some effect on the appearance of the fixed tissue. Isotonic fixatives affect the appearance of the fixed tissue and usually do not prevent swelling of the cells; hypertonic fixatives, on the other hand, give a more satisfactory preservation. With aldehyde fixation some degree of hypertonicity is achieved at the usual concentrations for routine work without the addition of other solutes. Even the extreme hypertonicity found in some aldehyde fixatives, such as Karnovsky paraformaldehyde–glutaraldehyde mixture, can give satisfactory preservation. For example, there is some evidence that in the nervous system an initial perfusion with diluted fixative (slightly hypertonic) followed by a full straight fixative improves the preservation.[4] Different tissues apparently require a specific degree of hypertonicity for optimal results, and several concentrations and osmolarities should be tried to find the one most suitable.

The temperature of the fixatives is another aspect of fixation that has not been completely explored. Most authors seem to favor the use of aldehydes at room temperatures and postfixation in osmium at low temperature.

Conclusive evidence is available showing that the best way to expose the cells to the fixative is through perfusion of the organ, and several techniques have been devised based on the original and comprehensive paper of Palay *et al.*[5] The perfusion method easily overcomes the usual disadvantages of immersion fixation, such as uneven penetration of fixatives, mechanical distortion during slicing, and alterations due to sudden changes in blood pressure. A general perfusion schedule would include the following steps:

1. Two 500-ml intravenous solution bottles for hospital use (one for fixative and one for physiological saline) are connected via intravenous administration sets to a three-way stopcock. An additional piece of short tubing is added (8 inches) with a hypodermic needle at the tip.

2. The bottles should be leveled at about 150 cm over the animal, depending on the arterial pressure of the species used. With this system,

[4] T. S. Reese and M. J. Karnovsky, *J. Cell Biol.* **34**, 207 (1967).

[5] S. L. Palay, S. M. McGee-Russell, S. Gordon, and M. A. Grillo, *J. Cell Biol.* **12**, 385 (1962).

it is possible to ensure that the fixative is preceded by a few milliliters of saline.

3. Under anesthesia the selected vessels are surgically exposed and cannulated, and the venous side is severed to permit the egress of the solutions. After a brief washing with a few milliliters of saline, the stopcock is turned to allow the flow of the fixatives. Individual preference and experience will determine the amount of fixative to perfuse and the temperature.

1. Buffers

Cacodylate and collidine buffers are considered general purpose buffers and give particularly good preservation of the ground cytoplasm. Phosphate buffers usually give excellent preservation of membranes.

Cacodylate Buffer (0.2 M; pH 7.4)

Add 4.28 g of Na cacodylate trihydrate (cacodylic acid, Na salt) to 100 ml of distilled water. Adjust the pH to 7.4 with 0.1 *N* HCl (3–6 ml). Use with various fixatives.

sym-Collidine Buffer (0.2 M; pH 7.4)

Add 2.67 ml of *sym*-collidine (Merck-TM Laboratories) to 50 ml of distilled water. Shake and add 8 ml of 1 *N* HCl. Dilute to 100 ml and adjust pH to 7.4. Use with various fixatives.

Phosphate Buffer (0.1 M; pH 7.7)

Stock solutions
- Solution A: Dissolve 27.8 g of monobasic sodium phosphate ($NaH_2PO_4 \cdot H_2O$) in 1000 ml of distilled water (0.2 *M*).
- Solution B: Dissolve 53.65 g of dibasic sodium phosphate ($Na_2HPO_4 \cdot 7H_2O$) in 1000 ml of distilled water (0.2 *M*).

Working solution: 10 ml of solution A, 90 ml of solution B; dilute to 200 ml

Phosphate-Buffered Saline (pH 7.2)

Dissolve the following in distilled water: 1.20 g of $NaH_2PO_4 \cdot H_2O$; 5.70 g of $Na_2HPO_4 \cdot 7H_2O$ or 3.02 g of anhydrous dibasic sodium phosphate; 24 g of NaCl. Bring volume to 3 liters; refrigerate.

The following two buffers are used in some electron-opaque tracer techniques (see section on tracers).

Citric Acid-Sodium Citrate Buffer (0.05 M, pH 3.9)

Stock solutions
- Solution A: Dissolve 2.101 g of citric acid in distilled water and dilute to 100 ml
- Solution B: Dissolve 2.940 g of sodium citrate · $2H_2O$ in distilled water and dilute to 100 ml.

Working solution: 61.2 ml of solution A, 38.7 ml of solution B, dilute to 280 ml; refrigerate.

Tris-HCl (0.05 M, pH 7.6)

Stock solutions
- Solution A: Dissolve 2.423 g of Tris(hydroxymethyl)aminoethane (THAM) in 100 ml of distilled water.
- Solution B: 0.1 *N* HCl

Working solution: 25 ml of THAM solution, 38.5 ml of HCl; dilute to 100 ml.

Maleate buffer (0.2 M)

This buffer is used with block staining techniques.

Stock solutions
- Solution A: Dissolve 23.2 g of maleic acid, add 200 ml of 1 *N* NaOH and dilute to 1000 ml with distilled water
- Solution B: 0.2 *N* NaOH

Working solutions: (1) pH 5.15: 50 ml solution A + 7.2 ml solution B; (2) pH 6.00: 50 ml solution A + 26.9 ml solution B

2. Fixatives

A large variety of aldehyde and buffer combinations have been proposed as primary fixatives. The following are most frequently used.

Traditional 5% Glutaraldehyde

This is the most commonly used aldehyde at concentrations of 2.5–6% and in combination with different buffers. It usually gives a good preservation of most cellular components including microtubules and ground cytoplasm.

Working solutions: (1) 10 ml: 25% glutaraldehyde (TAAB laboratories, Reading, England); (2) 40 ml: collidine or cacodylate buffer (0.2 *M*; pH 7.4); (3) 1 ml: 1% $CaCl_2$ (should be omitted if phosphate buffer used)

Karnovsky[6] *High Osmolarity Fixative*

This is an excellent general purpose and widely used combination of glutaraldehyde and paraformaldehyde. The fixation time with this mixture depends upon the size and type of tissue, and individual preference. It varies from 10 minutes to overnight at room temperature or 4°. For example, cultured cells or embryonic material usually are well preserved after 15 minutes at room temperature, while nervous tissue requires longer fixation times and low temperature for better results. In general prolonged exposure to aldehydes will increase the amount of myelin figures at any temperature. In some cases, better results are obtained with this fixative by diluting it 1:2 or 1:4 with buffer.

Working solution: Add 2 g of paraformaldehyde (Matheson, Coleman and Bell) to 20 ml of distilled water, and warm to 80°; add 2–3 drops of 0.2 *N* NaOH to clear the solution and cool.

To the cooled solution, add: 9.8 ml of 25% glutaraldehyde (TAAB laboratories, Reading, England), 18 ml of cacodylate or collidine buffer, and 2 ml of 0.2% of $CaCl_2$.

Diluted Paraformaldehyde-Glutaraldehyde (Karnovsky,[7] Dvorak *et al.*[8])

This is a diluted version of the original Karnovsky method; it is particularly well suited for cell cultures and monolayers. It is recommended also for cell suspensions.

Working solution: 50 ml of a 2% paraformaldehyde solution, 5 ml of a 25% glutaraldehyde solution, 45 ml of 0.025% $CaCl_2$ in 0.1 *M* cacodylate buffer.

Trinitro Compounds (Ito and Karnovsky[9])

In this fixative, a trinitro compound is added to the traditional aldehyde. Membranous components of the cells are generally well preserved

[6] M. J. Karnovsky, *J. Cell Biol.* **27**, 137 A (1965).

[7] M. J. Karnovsky, *J. Cell Biol.* **35**, 213 (1967).

[8] A. M. Dvorak, M. E. Hammond, H. F. Dvorak, and M. J. Karnovsky, *Lab. Invest.* **27**, 561 (1972).

[9] S. Ito and M. J. Karnovsky, *J. Cell. Biol.* **39**, 168a (1968).

by this method. *It should be remembered that nitro compounds are potentially explosive.*

Stock solution: Add 4 g of paraformaldehyde to 100 ml of distilled water; warm to 80°, and clear with several drops of 0.1 *N* NaOH. Dilute paraformaldehyde solution to a half with 100 ml of appropriate buffer (phosphate or cacodylate). Add 40 mg of trinitrocresol. This solution is stable for 1 week at 4°.

Working solution: Dilute 50 ml of stock solution to a half with 50 ml buffer. Add 0.8 ml of 25% glutaraldehyde to 9.2 ml of stock fixative. If cacodylate buffer is used, 0.4 ml of 0.2% $CaCl_2$ should be added. When perfusion is used, 1:1 dilutions with buffer seem to give better results.

Acrolein Mixtures (Aoki *et al.*[10])

The addition of small concentrations of distilled acrolein (hydroquinone free) to glutaraldehyde and paraformaldehyde mixtures gives very good preservation with embryonic material and vacuolated structures.

Working solution: 4 ml of a 25% glutaraldehyde solution, 50 ml of a 2% paraformaldehyde solution, 1 ml of freshly distilled acrolein (caution: acrolein is very toxic and should be handled with extreme care under the fume hood). Dilute to 100 ml with 0.2 *M* cacodylate or collidine buffer.

B. Postfixation, Dehydration, and Embedding

After the initial aldehyde fixation, the tissue must be postfixed in a solution containing osmium tetroxide. The following general procedure may be used: Wash in cold buffer (same as used in aldehyde fixation) 15 minutes to overnight. If cacodylate is used, it should be diluted to a half with a 2% solution of $CaCl_2$. Postfix in a solution containing equal parts of 2% OsO_4 and the buffer used previously. Time and temperature can be changed depending upon condition and individual preferences. Small volumes should be used, just enough to cover the pieces.

Palade's variation of postfixation with osmium tetroxide is sometimes used to advantage in membrane preservation.

Stock solutions

Solution A: Add 1.12 g of sodium acetate (0.14 *M*) and 2.89 g sodium Veronal (sodium barbital) (0.14 *M*) to 100 ml of water.

[10] A. Aoki, R. Vitale, and A. Pisano, *Z. Zellforsch. Mikrosk. Anat.* **98**, 9 (1969).

Solution B: 0.1 *N* HCl

Solution C: 2% OsO_4 in distilled water

Working solution: 5 ml of solution A, 5 ml of solution B, 12.5 ml of solution C. Adjust volume to 25 ml with distilled water. The pH should be 7.2–7.5. Fix at 4°, 2–4 hours.

After postfixation the tissue is dehydrated in increasing grades of cold alcohols or acetones, with about 1 minute in each up to 100%. While in absolute alcohol, the tissue should warm to room temperature; make 3 additional changes in 100% alcohol or acetone at 20-minute intervals. If alcohol is used, 5–15 minutes in propylene oxide will be necessary after the last change of alcohol. Infiltrate in half acetone or propylene oxide and half plastic with accelerator from 2 hours to overnight. Infiltrate for 2 hours in pure plastic, and embed overnight to 3 days at 60°.

In humid weather, be sure to heat plastic solutions above the dew point before opening tightly closed caps. All the times specified are average, and the ideal schedule in relation to type of tissue and size of the pieces should be found.

Embedding Plastics

Several types of embedding plastics are available according to the anticipated use and particular tissue; among the most commonly used are the following.

Araldite (Luft[11])

This plastic offers some difficulties in trimming the blocks and thick sectioning, but thin sectioning is easier. This plastic is also less damaging to the diamond knife than Epon. Compresion in the sections can be erased by whiffing xylene over them on a stick. Another problem with Araldite is that it stains with more difficulty than Epon.

Stock solution: 100 g Araldite 6005, 75 g of DDSA. The mixing can be done more easily if the components are heated 10–15 minutes in a 60° oven.

Working solution: Before use, add 1–2% of DMP-30, heat a few minutes to 60°, and stir very well or the embedding will be unsatisfactory. A vacuum oven can be used to remove air bubbles formed during stirring.

Embedding can be done in flat aluminum pans, plastic capsules, dried gelatin capsules, or Silastic embedding plates.

[11] J. Luft, *J. Biophys. Biochem. Cytol.* **9**, 409 (1961).

Epon (Luft[12])

This is a very popular plastic that stains easily but is somewhat more difficult to section than the others. Sections can be stretched with trichloroethylene if necessary. Stock solutions: The proportions that follow have an A:E (anhydride:epoxy) ratio of 0.7 and an epoxide equivalent in the Epon 812 of 150 ± 5.

Epon 812/DDSA/NMA

Mixture A: 66 ml Epon 812, 100 ml DDSA, no NMA
Mixture B: 100 ml Epon 812, no DDSA, 84 ml NMA

The solutions can be stored in the refrigerator for several weeks.

Working solutions: for routine work, mix equal parts of mixture A and B, plus 2% of accelerator DMP-30; stir for 15 minutes avoiding air bubbles.

Epon-Araldite (Newcomb[13])

This mixture is supposed to overcome the disadvantages of the previous plastics.

Stock solutions

Epon 812, 21.5 g
Araldite 6005, 21.5 g
DDSA, 57 g

Shake vigorously after warming in a 60° oven. The solution can be stored in the refrigerator for several weeks. Before use, add 2% of accelerator DMP 30 and mix well. Blocks must be cured at least overnight at 60°; longer periods will give harder plastic. This combination is easy to section with good infiltration and staining properties.

A number of cytochemical procedures in electron microscopy requires the embedding of the tissues in water-miscible plastics. With this technique, the problems created by the exposure of the tissues to strong dehydrating agents are somewhat minimized. The following technique is one of several available for this purpose.

Water-Miscible Glycol Methacrylate Embedding

Stock solutions

(1) Embedding mixture of glycol methacrylate (GMA); 7 parts of 97% glycol methacrylate with 3% water and 3 parts of nondestabilized butyl methacrylate with 2% Luperco

[12] J. Luft, *in* "Advanced Techniques in Biological Electron Microscopy" (J. Koehler, ed.). Springer-Verlag, Berlin and New York, 1973.

[13] E. H. Newcomb, Polyscience data sheet. Paul Valley, Warrington, Pennsylvania.

(2) Prepolymerized mixture (to reduce swelling artifacts): Take a small amount of embedding mixture in a large Erlenmeyer flask less than 1 cm thick. Heat until boiling just begins with rapid swirling. With vigorous agitation immerse the flask in a large ice bath. Repeat if necessary until a thick syrup is obtained. This can be stored in deep freeze.

Procedure. This should be carried on in a cold room at 4°.

1. Fix in diluted Karnovsky's in cacodylate buffer.
2. Rinse in buffer.
3. Dehydrate in 80% of GMA monomer and 20% water for 20 minutes; 97% of GMA monomer and 3% water for 20 minutes; unprepolymerized embedding mixture 20 minutes.
4. Place in prepolymerized mixture overnight.
5. Embed in gelatin capsules filled with fresh prepolymerized mixture. Wait 30 minutes with uncapped capsules to eliminate air bubbles.
6. Expose the capsules in a wire support to UV light for 1–3 days (3150 Å wavelength).
7. Mount sections on coated grids and stain 30 minutes to 2 hours in uranyl and 1–5 minutes in lead.

C. Staining

In electron microscopy structures can be identified only by their morphology and electron opacity. Heavy metals are extremely valuable in increasing the density of certain cellular components, and therefore the contrast. Staining can be achieved after fixation, during dehydration, or after embedding. We shall describe some typical examples that usually give consistent results.

Uranyl Acetate–Maleate Buffer Block Stain

This is an example of postfixation staining of tissues. The treatment of blocks before dehydration with uranyl salts greatly improves the preservation and staining properties of membranes.

1. After osmication, wash blocks of tissue in maleate buffer (5.15 pH) three times for 5 minutes each. If phosphate buffer has been used, extend the washing to 30 minutes.
2. Stain 1 hour at room temperature in 1–2% uranyl acetate in maleate buffer at pH 6.0.
3. Dehydrate as usual.

Uranyl Acetate-Collidine Buffer Block Stain

1. After osmication, wash three times 5 minutes each in acid water. (1 drop of 0.2 *N* HCl in 200 ml of water).

2. Stain 1 hour at room temperature in 1% uranyl acetate in collidine buffer pH 6.0.

3. Wash in acid water three times, 5 minutes each.

4. Dehydrate and embed as usual.

1. Grid Staining

Heavy metals can be used after thin sections are made and mounted. For routine purposes these are the best techniques for increasing the contrast of the cell components.

Uranyl Acetate (Method 1)

Stock solution: saturated uranyl acetate in distilled water

Working solution: Mix stock solution and acetone 1:1 in a disposable glass syringe. Refrigerated solution is stable for a few weeks.

1. Cover the bottom of a petri dish with a piece of Parafilm.

2. Put drops of the stain on the Parafilm and place the grids on the drops with the sections facing the stain. Usual staining times vary from 15 seconds to 1 minute.

3. Wash grids carefully with a gentle jet of distilled water.

Uranyl Acetate (Method 2)

This is similar to the previous method, but uses 3% uranyl acetate in distilled water.

Lead Citrate

This can be used alone when only membranes need to be stained, or in combination with uranyl acetate for general staining of all structures.

Procedure

1. Wash graduated centrifuge tube with concentrated NaOH, followed by careful wash with soap and generous rinsing.

2. Place 20 mg of lead citrate in the tube and add 10 ml of fresh double glass distilled water.

3. Add 4 drops of 5 *N* NaOH and shake well to dissolve lead citrate. Transfer to a glass disposable syringe.

4. Use in a similar way to the uranyl acetate with staining times of 30 seconds to 1 minute. Keep in the refrigerator and avoid contact with air, since CO_2 will easily form lead carbonate precipitates.

2. Negative Staining

In the study of particulate specimens (whole mounts) with the electron microscope it is necessary to increase the density of the background in order to visualize the particles (negative staining). One of the most widely used stains for this purpose is phosphotungstic acid.

Phosphotungstic Acid

Staining solution: 1% of phosphotungstic acid with 0.4% of sucrose

Procedure

1. Place a small drop of this solution on a piece of wax.
2. Touch the surface of the drop with the coated grid containing the sample.
3. Draw off the liquid with filter paper and dry. During observation, use the anticontamination device of the microscope since the rate of contamination of the column with this material is high. More details and applications of this procedure will be found in Horne.[14]

3. Extracellular Surface (Glycocalyx) Staining

In recent years, several techniques have been developed to stain the sugar containing extracellular coat. These techniques are useful in delineating clearly the cell contour, pynocytic vesicles, and other sugar-rich structures.

Karnovsky's OsO_4-Potassium Ferrocyanide (Dvorak *et al.*[15])

This method gives good staining of cell membranes, extramembranous material, and cellular glycogen. Nucleus, chromatin, and ribosomes are less well stained.

[14] R. W. Horne, *in* "Techniques for Electron Microscopy" (D. Kay, ed.), p. 328. Davis, Philadelphia, Pennsylvania, 1965.

[15] A. M. Dvorak, M. E. Hammond, H. F. Dvorak, and M. J. Karnovsky, *Amer. J. Pathol.* **66,** 72a (1972).

Working solution: 1 part 2% OsO_4; 1 part 1.5% potassium ferrocyanide in 0.1 *M* sodium phosphate buffer at pH 6.0

Procedure

1. After aldehyde fixation and rinsing, fix for 2 hours at 4°C in working solution.
2. Embed as usual.

Luft's Ruthenium Red[16]

Ruthenium Red stains only extracellular materials with fine grain, high resolution, and good contrast.

Working Solutions and Procedure

1. Fix 1 hour at any temperature in 0.5 ml of 3.5% glutaraldehyde or 2% acrolein in water; 0.5 ml of 0.2 *M* cacodylate buffer pH 7.3.
2. Rinse 10 minutes in cacodylate (3 changes).
3. Fix 3 hours at room temperature in: 0.5 ml: 5% OsO_4 in water, 0.5 ml: 0.2 *M* cacodylate buffer, pH 7.3; 0.5 ml: Ruthenium Red stock solution (1500 ppm in water).
4. Rinse briefly with buffer.
5. Dehydrate and embed as usual.

Shea's Lanthanum[17]

Extracellular coats with this method are stained dark.

Working Solutions and Procedure

1. Fix 2 hours at room temperature in diluted Karnovsky to which 0.5% cetylpyridinium hydrochloride has been added.
2. Postfix in 1% OsO_4 and 1% lanthanum nitrate in 0.1 *M* collidine buffer at pH 8.0 for 2 hours at room temperature.
3. Dehydrate and embed in Epon as usual.

Alkaline Bismuth (Ainsworth *et al.*[18])

This method stains periodate reactive mucosubstance and polysaccharides containing 1,2-glycols.

[16] J. Luft, *Anat. Rec.* **171,** 347 (1971).
[17] S. M. Shea, *J. Cell Biol.* **51,** 611 (1971).
[18] S. K. Ainsworth, S. Ito, and M. J. Karnovsky, *J. Histochem. Cytochem.* **20,** 995 (1972).

Solution and Procedure

1. Routine thin sections are mounted in 200 mesh grids.

2. Float grids 10–15 minutes at room temperature in 0.8 g of H_5IO_6 in 70 ml ethanol reagent.

3. Add 10 ml of 5 *M* sodium acetate and 20 ml of distilled water. This solution in the dark and at 18° is stable for a month.

4. Wash grids thoroughly (10 minutes in distilled water with 3 changes).

5. Float grids in 400 mg of sodium tartrate in 10 ml of 2*N* NaOH; add this solution dropwise to 200 mg of bismuth subnitrate while stirring. Solution should be clear. Prior to use, dilute 1:50 with distilled water. This is stable for a month at 4°.

D. Electron Opaque Tracers

1. Enzymatic Tracers

A variety of enzymatic electron-opaque techniques have been recently introduced. With these methods, it is possible to study aspects of cell permeability, pinocytosis, phagocytosis, immunology, and molecular localization at the ultrastructural level.

Peroxidase Technique (Graham and Karnovsky[19])

A general technique will be described with this tracer (MW 40,000 particle size 40–60 Å).

Peroxidase by Injection

1. Inject horseradish peroxidase (RZ approximately 3) (4–8 mg/0.5 ml saline) into the tail vein of a mouse.

2. After 10–20 minutes, fix the tissues in Karnovsky buffered with cacodylate for 4–6 hours.

3. Rinse overnight in buffer.

4. Prepare 40 μm sections with a Sorvall TC-2 tissue sectioner and store them in 0.05 *M* Tris·HCl buffer, pH 7.6.

5. Incubate the sections at room temperature at 37° for 15 minutes to an hour in the DAB solution described below.

6. Wash three times in 0.05 *M* Tris buffer.

7. Postfix osmium; dehydrate, and embed as usual. Sections should be observed in the microscope unstained or slightly stained with lead.

[19] R. C. Graham and M. J. Karnovsky, *J. Histochem. Cytochem* **14,** 291 (1966).

Incubation medium: Add 5 mg of diaminobenzidine tetrachloride (DAB) to 10 ml of 0.05 *N* Tris buffer, pH 7.6. To this, add 0.1 ml of 1% H_2O_2 before use. An excess of H_2O_2 will damage the tissues and inhibit the reaction.

Peroxidase by Incubation

1. Place small pieces of tissues in 1 ml of saline containing 50 mg of horseradish peroxidase for 10–20 minutes at room temperature or 37°.
2. Fix and rinse as described above.
3. Incubate in DAB media 30 minutes to 2 hours. If the tissue does not turn brown quickly, add another 0.05 ml of H_2O_2.
4. Wash, osmicate, dehydrate, and embed as usual.

In both cases, sites of enzymatic activity will show a dense precipitate.

Cytochrome c Technique (Karnovsky and Rice[20])

This enzyme is also used as a tracer (MW 12,000, particle size 20 Å) in a similar way to peroxidase.

Stock solutions

Phosphate-buffered saline (PBS)

Cytochrome (40 mg) in 0.5 ml PBS; adjust pH to 7.2 with 0.2 *N* NaOH just before use.

Citric acid-sodium citrate buffer (50 m*M* pH 3.9)

Tris·HCl buffer, 0.05 *M*, pH 7.6

Prepare 0.3 ml of 30% H_2O_2 in 10 ml of citric acid-sodium citrate buffer, 40 m*M*, pH 3.9.

Preincubation medium: Prepare 5 mg of DAB in 10 ml of this buffer (0.05 *M* pH 7.6) just before use.

Incubation medium: Prepare 5 mg of DAB in 9.5 ml of citric acid-sodium citrate buffer (50 m*M*, pH 3.9) and add 0.5 ml for H_2O_2 solution.

General Procedure

1. Surgically clamp the kidneys of a mouse and inject cytochrome solution into the tail vein: wait the desired time (10 minutes) and fix in Karnovsky solution (2–4 hours).
2. Rinse overnight in buffer.
3. Place in preincubation media for 1–3 hours.

[20] M. J. Karnovsky and D. F. Rice, *J. Histochem. Cytochem.* **17**, 751 (1969).

4. Place in incubation media for 15–30 minutes at 37°.
5. Osmicate, dehydrate, and embed as usual.

Sites of enzymatic activity will show a dense precipitate.

Catalase Technique (Venkatachalam and Fahimi[21])

Catalase is a larger molecule (MW 240,000, particle size 100 Å) and can be used the following way:

1. Twice crystallized beef liver catalase (suspension with 26 mg/ml and 0.1% thymol) is dissolved before injection in a sonifier for 15–30 minutes at a temperature below 40°.
2. Inject 0.5 to 2 ml of the solution slowly into the tail vein of a mouse. At 50–30 minutes after injection fix small pieces of tissue in Karnovsky's for 2–4 hours and rinse overnight in cold cacodylate buffer.
3. Prepare 40 μm sections with the Sorvall TC-2 tissue sectioner and collect them in 50 m*M* Tris buffer.
4. Incubate at 37° for 1–4 hours in a medium containing 10 mg of DAB in 10 ml of 50 m*M* Tris buffer, pH 7.5 and 0.1 ml of 1% H_2O_2.
5. Wash sections in Tris buffer, postfix in osmium, dehydrate and embed as usual.

Sites of enzymatic activity will show a dense precipitate.

Myoglobin Technique (Anderson[22])

This technique uses the peroxidatic activity of the myoglobin in a similar way to the previous tracers. It has an MW of 18,000 and a particle size of 20 Å.

Procedure

1. Inject 10–30 mg of myoglobin in 0.2–0.5 ml of saline per 100 g of body weight.
2. After 10 minutes to 1 hour, fix small pieces of tissue in Karnovsky's solution for 2–4 hours. Rinse overnight in buffer and prepare 40 μm sections in the tissue sectioner.
3. Incubate sections 30 minutes to an hour in a medium containing 9.5 ml of 50 m*M* citrate buffer pH 3.9, 0.5 ml 1% H_2O_2, and 20 mg of DAB.
4. Wash sections at 4° in 0.1 *M* phosphate buffer, pH 6.8; three changes, 5 minutes each.
5. Postfix in osmium, dehydrate, and embed as usual.

[21] M. A. Venkatachalam and H. D. Fahimi, *J. Cell Biol.*, **42**, 480 (1969).
[22] W. Anderson, *J. Histochem. Cytochem.* **20**, 672 (1972).

Sites of enzymatic activity will show a very dense precipitate. Tissue fixed in 2% formaldehyde should give dense precipitates.

2. Nonenzymatic Tracers

Lanthanum is a heavy metal that can be used for a variety of situations, including the outlining of cells and junction studies. It is not always possible to obtain good precipitates, and the chemistry of the reaction is not known.

Lanthanum should be included in the fixative, buffer, and osmium solutions.

Stock Lanthanum (Revel and Karnovsky[23])

Slowly adjust to a pH of 7.0 with 0.1 N NaOH, 4% $La(NO_3)_3$ solution.

Fixation: Karnovsky fixative with collidine buffer and lanthanum solution 1:1. If perfusion is used the washing solution should be NaCl saline. Fix 2–4 hours. Rinsing buffer: *sym*-collidine 0.2 M and lanthanum solution 1:1.

Osmium fixation: use 2% OsO_4, 2 parts; collidine buffer, 1 part; lanthanum solution, 1 part. Dehydrate and embed as usual. Grids should be studied under the microscope without staining

Thorotrast Tracer

Commercially available 1% colloidal thorium dioxide (Thorotrast; Kapco, Inc., Kalamazoo, Michigan) is used. The tissues are exposed to this solution by local injection or incubation, fixed, and processed as usual. The particle size of this tracer is 60–80 Å and will show as very dense dots in the preparation.

Iron Dextran-Complex Tracers

Available commercially as Inferon (Lakeside Laboratories, Milwaukee, Wisconsin), this tracer is used in a similar way to the previous one (MW 70,000, particle size 70 Å). Process tissues routinely and examine sections without staining. The tracer will appear as particles of moderate electron density.

Ferritin Tracer

This is another commonly used electron-opaque tracer of high molecular weight (500,000) and 110 Å particle size.

[23] J. P. Revel and M. J. Karnovsky, *J. Cell Biol.* 33, C7 (1967).

Tissues are exposed by injection or by immersion to twice crystallized cadmium-free ferritin and, after a suitable period of time, processed as usual. Sections can be observed without staining, and ferritin will appear as particles of moderate electron density. Ferritin will show an increase in the particle size and density when observed under the microscope. To increase the electron opacity, a ferritin staining procedure can be used (Ainsworth and Karnovsky[24]).

1. Dissolve 400 mg of sodium tartrate in 10 ml of 2 *N* sodium hydroxide.
2. Add this solution dropwise to 200 mg of bismuth subnitrate, stirring with a magnetic bar. The final solution should be clear and is diluted 1:50 with distilled water. This is stable for 1 month when refrigerated.
3. Float grids at room temperature on drops of bismuth solution for 30–60 minutes.
4. Wash them carefully and dry.

Dextran and Glycogen Tracers (Simionescu and Palade[25])

Commercially available dextrans and glycogens can also be used in permeability studies. These particles, adequately stained, provide a relatively large set of probes (diameter 45–300 Å).

Glycogen solution: 15% of shellfish or rabbit liver glycogen are ultrasonically dissolved in 0.9% saline before use.

Dextran solution: 10% (w/v) of dextran; choose molecular weight dextran according to the type of probe needed.

Procedure

1. Inject animals resistant to histamine release by dextrans with 1.5 ml/100 g body weight of one of the previous solutions.
2. Fix in Karnovsky's 0.1 *M* phosphate buffer pH 7.4.
3. Wash in buffer.
4. Postfix in OsO_4 phosphate buffer 1:1.
5. Dehydrate and embed as usual in Epon.
6. Stain thin sections in 7.5% magnesium uranyl acetate for 1 hour at 60° followed by lead citrate.

E. Cytochemistry

Significant progress has been made in the last few years on the ultracellular localization of enzymes and chemical components (Shnitka and

[24] S. K. Ainsworth and M. J. Karnovsky, *J. Histochem. Cytochem.* **20**, 225 (1972).

[25] N. Simionescu and G. Palade, *J. Cell Biol.* **50**, 616 (1971).

Seligman[26]). The exact requirements for preservation of tissue fine structure, retention of enzyme activities, and the localization of precipitated reaction products are known for a number of cytochemical systems. When these techniques are used, morphological findings can also be interpreted to some extent in biochemical terms.

The number of methods available is rather extensive, and we shall describe some typical procedures that exemplify the application of these techniques to problems in enzyme cytology, physiology, and immunology.

Acid Phosphatase

Incubation medium: Prepare 50 m*M* sodium acetate buffer, pH 5.0 by mixing 50 m*M* sodium acetate, 70 ml, and 50 m*M* acetic acid, 40 ml. Add 0.12 mg of lead nitrate to 70 ml of sodium acetate buffer. Shake to dissolve and add 0.3 g of glycerophosphate. Adjust to pH 5.0 with 50 mM acetic acid. Bring to 100 ml with buffer. Leave overnight at 37°; adjust again to pH 5.0 (if necessary) and add 100 ml of distilled water.

Procedure for Light Microscopy

1. Dry unfixed cryostat section onto slides and incubate in the above solution 1 hour at 37°.
2. Rinse in 2 changes of distilled water, 1 minute each.
3. Rinse for a few seconds in 1% acetic acid.
4. Rinse in 2 changes of distilled water, 1 minute each.
5. Rinse in 1% ammonium sulfide for 2 minutes.
6. Wash in running water for 5 minutes.
7. Rinse in 80% ethanol for 1 minute.
8. Can be counterstained in Mayer's paracarmine for 30 seconds.

Procedure for Electron Microscopy

1. Place dry unfixed cryostat sections on coverglasses.
2. Follow the light microscopy procedure.
3. Fix 1 hour in OsO_4, 1 part; Gordon's buffer, 1 part (Gordon *et al.*[27]).
4. Dehydrate and run through propylene oxide-plastic 1:1 mixture.
5. Clean the bottom of the coverslip with 96% ethanol. With the coverslip on a flat surface, cover with plastic, avoiding overflow under coverslip. Polymerize.

[26] T. Shnitka and A. Seligman, *Annu. Rev. Biochem.* p. 375 (1971).
[27] G. B. Gordon, L. R. Miller, and K. G. Bensch, *Exp. Cell Res.* **31**, 440 (1963).

6. After polymerization, the glass can be removed by rapid freezing and thawing, then peeling away with a razor blade.

Alkaline Phosphatase

1. Prepare 40 μm sections from glutaraldehyde-fixed tissues.
2. Rinse sections in water.
3. Incubate at 37° for 5–10 minutes in Tris maleate buffer 0.2 *M*, 1 ml; sodium β-glycerophosphate (1.25%), 2 ml; distilled water, 5.7 ml; lead nitrate (1%), 1.3 ml; $MgCl_2$ as activator may be added.
4. Rinse.
5. Postfix in osmium, dehydrate, and embed.

Digestion of RNA (Hay and Gurdon Technique[28])

1. Fix 30 minutes in 2.5% glutaraldehyde in 0.2 *M* cacodylate buffer, pH 7.2.
2. Rinse several hours in buffer and prepare (handmade) very thin slices (1 mm).
3. Incubate 6 hours at 37° in 0.1% solution of RNase in 0.1 *M* phosphate buffer, pH 7.7. Controls are incubated in buffer without the enzyme.
4. Rinse in buffer for 1 hour.
5. Postfix in OsO_4-collidine buffer 2:1 for 1 hour.
6. Dehydrate and embed as usual.

Glycogen Staining (Revel and Karnovsky, unpublished)

Mount thin slices on nickel or tungsten grid. Insert them in dents made with a razor blade on a wax plate; this allows one to perform the procedure without manipulating the grids. Other methods can also be used.

Solution: Add 0.5 g of $NaClO_2$ to 19 ml of distilled water. Before use, add 2 ml of 10 *N* acetic acid, saturated uranyl acetate in distilled water, and freshly prepared 3% H_2O_2.

Procedure

1. Place tissue in 3% H_2O_2 for 10 minutes.
2. Rinse with distilled water.
3. Place in 1% periodic acid for 30 minutes.
4. Rinse carefully.
5. Chlorite solution for 15 minutes in the hood and in the dark.
6. Rinse carefully.
7. Uranyl solution 5–10 minutes.

[28] E. Hay and J. B. Gurdon, *J. Cell Sci.* **2**, 151 (1967).

Controls

1. Omit periodate oxidation.
2. Block aldehyde groups produced by periodate in the following way: after step 4 above, place sections for 30 minutes in 11% *m*-aminophenol in glacial acetic acid.
3. Rinse well and proceed with chlorite stain.

Glycogen Staining with Thiosemicarbazide (TSC) (Vye and Fischman[29])

Proceed in a similar way to the previous method.

1. Place in 1% periodic acid for 30 minutes.
2. Rinse carefully.
3. Place for 30–60 minutes in 0.2% TSC in 20% acetic acid.
4. Rinse carefully.
5. Place for 30 minutes in the dark in 1% aqueous solution of silver protein.
6. Rinse.

Controls: Omit TSC

Colloidal Iron and Colloidal Thorium Staining of Acidic Groups (Revel[30])

Colloidal Iron

Stock solutions
- Acetic acid, 30%
- Colloidal iron stock, 5% in 30% acetic acid

Procedure. Transfer methacrylate section with a platinum wire loop through the following steps:

1. Acetic acid, 30%, 5 minutes.
2. Colloidal iron, 5 minutes.
3. Acetic acid, 30%, 5 minutes.
4. Acetic acid, 30%, 5 minutes.
5. Distilled water, 5 minutes.
6. Mount sections on coated grids.

Colloidal Thorium

Stock solution: In this technique, Thorotrast (colloidal thorium dioxide) is used in a similar way to the colloidal iron.

[29] M. V. Vye and D. A. Fischman, *J. Cell Sci.* **9**, 727 (1971).
[30] J. P. Revel, *J. Microsc.* **3**, 535 (1954).

Procedure. Methacrylate sections are treated as follows.

1. Acetic acid, 30%, 2–5 minutes.
2. Thorotrast, 1% in 30% acetic acid, 5 minutes.
3. Acetic acid, 30%, 2–5 minutes.
4. Acetic acid, 30% 2–5 minutes.
5. Distilled water, 5 minutes.
6. Mount sections on coated grids.

Lead citrate can be used to lightly stain the membranes, and a thin carbon coat will prevent extreme contamination of the microscope column.

Ultrastructural Localization of Hormones Using Immunochemical Methods

Several techniques have been improved recently to localize hormones at the ultrastructural levels. Immunoglobulins labeled with peroxidase is one of the most commonly used; as an example, we shall describe the procedure developed by Moriarty and Halmi,[31] for localization of adrenocorticotropin-producing cells using unlabeled antibody and the soluble peroxidase-antiperoxidase complex.

1. Fix in 2.5% glutaraldehyde in 0.1 *M* phosphate buffer, pH 7.4, for approximately 2–4 hours. Another alternative is fixation in formaldehyde-picric acid.
2. Wash, dehydrate, and embed in methacrylate or Araldite as usual.
3. Mount light gold sections on Formvar-coated nickel grids.

If Araldite is used, etch sections by 20 minute flotation in a drop of 10% H_2O_2. Float grids for 3 minutes in each of the following:

1. Normal goat serum diluted 1:50.
2. Rabbit anti-ACTH antiserum diluted 1:20.
3. Goat anti-rabbit IgG diluted 1:10.
4. Peroxidase anti-peroxidase complex diluted 1:50 (rabbit anti-peroxidase).

For dilutions of sera, use Tris-buffered saline with 8% normal goat or sheep serum.

1. Stain 3–4 minutes in the following solution: 92 mg diaminobenzidine; 175 ml of Tris buffer, 0.5 *M*, pH 7.6; 1.5 ml of 0.3% H_2O_2.
2. Filter the solution before use.
3. During the incubation, the solution must be stirred with a magnetic bar.

[31] G. C. Moriarty and N. S. Halmi, *J. Histochem. Cytochem.* **20,** 590 (1972).

4. Wash three times in distilled water.
5. Stain for 20 minutes at room temperature in aqueous 4% OsO_4.
6. Stain in uranyl acetate for 20–30 minutes and lead citrate for 3–5 minutes.

For dilutions of sera, use Tris-buffered saline with 8% normal goat or sheep serum.

Scanning Electron Microscopy

The scanning electron microscopy (SEM) is a recent introduction in the morphological armamentarium but has already proved to be a very useful tool in biological studies to study topographical or surface features (Hollenberg and Erickson[32]).

The instrument is quite different in principle and application to the conventional transmission electron microscope (TEM). In the SEM, the sample is bombarded with a fine probe of electrons 100 Å in diameter or less and operates usually in voltages of up to 50 kV. A system of electromagnetic deflection coils in synchronization with a suitable detector, and display tubes are used to scan the sample. Typical operating values are 500–1000 lines of scan and from a few seconds to a few minutes to accumulate the picture. The image is recorded photographically from the cathode ray tube display. The result is a magnified image of the surface scanned with a resolution of about 100 Å in commercial microscopes.

Further technical details on this type of microscope can be found in Kimoto and Russ,[33] Hayes and Pease,[34] Nixon,[35] Echlin,[36] Crewe,[37] Hearle *et al.*,[38] Everhart and Hayes.[39]

In order to study the topography of tissues with the SEM, the tissues must be placed in the specimen chamber of the instrument completely dry. Since the tissues are not embedded in any type of material the dehydration technique becomes the most important step in the preparation.

Early studies used air-dried aldehyde-fixed osmicated tissues. Obviously, the liquid surface tension during the evaporation created forces which produced gross artifacts and distortions that were easily seen with the microscope.

[32] M. J. Hollenberg and A. M. Erickson, *J. Histochem. Cytochem.* **21**, 109 (1973).
[33] S. Kimoto and J. C. Russ, *Amer. Sci.* **57**, 1 (1969).
[34] T. L. Hayes and R. F. Pease, *Advan. Biol. Med. Phys.* **12**, 85 (1968).
[35] W. C. Nixon, *Phil. Trans. Roy. Soc. London Ser. B* **261**, 45 (1971).
[36] P. Echlin, *Phil. Trans. Roy. Soc. London Ser. B* **261**, 51 (1971).
[37] A. V. Crewe, *Phil. Trans. Roy. Soc. London Ser. B* **261**, 61 (1971).
[38] J. W. S. Hearle, J. T. Sparrow, P. M. Gross, *in* "The Use of the Scanning Microscope," (J. Hearle *et al.*, eds.). Pergamon, Oxford, 1971.
[39] T. E. Everhart and T. L. Hayes, *Sci. Amer.* **226**, 54 (1972).

To overcome this problem, the freeze-drying technique was introduced, thereby minimizing to some extent the artifacts of drying. Theoretical discussions of freeze-drying have been given by Rowe[40] and Stephenson,[41] and these papers should be consulted for design characteristics of equipment used in the technique. Basically, aldehyde-fixed tissues with or without osmication are washed in saline containing a cryoprotective agent to avoid the ice crystal formation. Tissues are then frozen and placed in a low-temperature vacuum chamber for a variable period of time to sublimate the water (Rebhun[42]). With this procedure, evaporation artifacts are greatly minimized (Boyde and Wood[43]).

Further improvement in the drying procedure came with the introduction of the Anderson's critical point method originally developed for the preservation of bacteria for the TEM.[44] In short, this technique uses the principle discovered in 1822 by Cagniard de la Tour and defined by Ramsey in 1880 as "the point at which the liquid, owing the expansion, and the gas, owing the compression, acquire the same specific gravity and consequently mix with each other."

When a liquid is at its critical point, the liquid and the gas phases are in equilibrium, and there are no phase boundaries. A tissue in that liquid can be dried with a minimum of artifact due to surface tension. This has probably become the most popular technique of preparing tissues for the scanning microscope. The critical point of water is 374°, which is too high for practical use. In Anderson's technique, fixed tissues are dehydrated in graded alcohols and finally placed in amyl acetate. Under pressure, amyl acetate is replaced with liquid CO_2 (critical temperature and pressure 36.5° and 1080 psi, respectively). At that temperature and pressure or above, liquid CO_2 will become a gas; after slow release of the gas, the specimens are dry and ready for examination.

To prevent some artifacts of image formation and to improve the quality of the pictures, a conducting layer of carbon is applied to the specimens under vacuum. Carbon coating can also be used in combination with metals like gold or silver (Cross[45]). Details on all steps of these

[40] T. Rowe, *Ann. N.Y. Acad. Sci.* **58**, 641 (1960).

[41] J. Stephenson, *in* "Recent Research in Freezing and Drying" (A. S. Parkes and A. U. Smith, eds.), p. 122. Blackwell, Oxford, 1960.

[42] L. I. Rebhun, *in* "Principles and Techniques of Electron Microscopy" (M. Hayat, ed.), Vol 2. Van Nostrand-Reinhold, Princeton, New Jersey, 1972.

[43] A. Boyde and C. Wood, *J. Microsc.* **90**, 221 (1969).

[44] T. F. Anderson, *in* "Physical Techniques in Biological Research" (A. Pollister, ed.), Vol. III. Academic Press, New York. 1966.

[45] P. M. Cross, *in* "The Use of the Scanning Microscope" (J. Hearle *et al.*, eds.), p. 87. Pergamon, Oxford, 1972.

procedures and some variations will be found in Cohen *et al.*,[46] Koller and Bernard,[47] Smith and Finke,[48] and Kanagawa *et al.*[49].

In standard TEM, the information obtainable represents an extremely small area of the surface of the cell, and if the specimens are not wholly homogeneous as to the distribution of the structure studied, very laborious and sometimes impossible serial sectioning is needed before one can draw any conclusion. The SEM is thus very useful to complement TEM studies with surface information otherwise difficult to obtain.

High Voltage Electron Microscopy

Unlike light microscopy, where systems such as phase contrast or Nomarski differential interference contrast provides an image of the living cell to compare with fixed and embedded cells, electron microscopy lacks that possibility. At the present time, the alternative is to use a variety of methods and to establish whether the information from one method to the other correlates well and does not conflict with results obtained with light microscopy, diffraction studies, and biochemistry. However, recent advances in electron microscopy technology (Parsons *et al.*[50]; Fernández-Morán[51]; Cosslett[52]) indicate that eventually it may be possible to examine cells under the high voltage electron microscope in the living hydrated state. Parsons and co-workers, using a differentially pumped, aperature limited hydration chamber in a high voltage electron microscope, have shown pictures of several types of wet cells including bovine spermatozoa.

The high voltage electron microscope operates at 2 mV, allowing the possibility of observing specimens of greater thickness. The chief limitation lies in the decrease of image contrast, which is the concomitant of increased penetration. It also gives clear images with much improved resolution of thick plastic sections (1–10 μm) (Favard *et al.*[53]; Chan-Palay and Palay[54]). Owing to the great depth of focus provided by the instru-

[46] A. L. Cohen, D. P. Marlow, and G. F. Gardner, *J. Microsc.* **7**, 331 (1968).

[47] T. Koller and W. Bernard, *J. Microsc.* **3**, 589 (1964).

[48] M. S. Smith and E. H. Finke, *Invest. Ophthalmol.* **11**, 127 (1972).

[49] H. Kanagawa, E. Hafez, C. A. Baechler, W. C. Pitchford, and M. I. Barnhart, *Int. J. Fert.* **17**, 75 (1972).

[50] D. F. Parsons, V. R. Matricardi, J. Subject, I. Uydess, and G. Wray, *Biochim. Biophys. Acta* **290**, 110 (1972).

[51] H. Fernández-Morán, *Ann. N.Y. Acad. Sci.* **195**, 376 (1972).

[52] V. E. Cosslett, *Phil. Trans. Roy. Soc. London Ser. B* **261**, 35 (1971).

[53] P. Favard, L. Ovtrach, and N. Carasso, *J. Microsc.* **12**, 301 (1971).

[54] V. Chan-Palay and S. L. Palay, *Z. Anat. Entwicklungsgesch.* **137**, 125 (1972).

ment, details throughout the thickness of a section are in focus in the final image.

It is usually desirable to record stereopairs of micrographs by tilting the specimen through a small angle (10°) between two exposures. The prints, when mounted in a stereoviewer, will show the tridimensional structure of the section clearly.

Yet another possibility of high voltage electron microscopy being explored in some laboratories is selected area diffraction and localized chemical characterization (Chan-Palay[55]).

The biological significance of these techniques have only recently been appreciated, and severe technical problems such as limitation in contrast and poor resolution call for further developments and make its routine applicaton as yet impractical. Eventually, high voltage electron microscopy might prove to be an important technique to complement light microscopy and conventional electron microscopy in the studies of the structure and organization of cells.

Acknowledgment

The author thanks Dr. Claire Huckins for her critical reading of this manuscript.

[55] V. Chan-Palay, *Z. Anat. Entwicklungsgesch.* **139,** 115 (1973).

[16] An Ultrastructural Approach to the Study of Endocrine Cells *in Vitro*

By A. JOSEPHINE MILNER[1]

The highly specialized nature of endocrine glands renders them uniquely difficult to study experimentally. By definition, their function is to synthesize and secrete into the circulation hormones that affect the metabolism of other cell types. This activity is modulated by the blood levels of other compounds, including hormones, impinging upon the secretory cells within the gland. As might be anticipated, the vasculature of an endocrine gland plays a critical role in its functional activity and is often extremely complex, limiting the possibilities for studying the gland *in situ*. However, where specialized perfusion techniques have been developed to study a particular gland *in vivo*, as described in sections of this volume, the results have proved highly informative (see, for

[1] Recipient of a research grant from the Wellcome Trust.

example, McCracken *et al.*[2]). Nonetheless, the endocrine cells in the intact animal remain susceptible to the influence of other tissues and factors which are, as yet, largely undefined. This can be countered by studying the endocrine gland *in vitro*. In this approach the methodology is relatively straightforward and the conditions to which the cells are subjected can be controlled. The interpretation of the results is now, however, open to the question: How does the functioning of the tissue under such conditions relate to the situation *in vivo?* Clearly, the crux of the problem lies in our lack of understanding of the fundamental relationships that exist between cells in the intact animal. The elucidation of such relationships will doubtless progress through the assimilation and integration of data obtained from experimental approaches both *in vivo* and *in vitro*.

When an endocrine gland is studied *in vitro* with the aim of relating the results to its functioning *in vivo,* basic precautions must be taken to maintain the tissue under conditions approaching those in the intact animal. For example, the pH of the incubation medium should be adjusted and maintained at a level equivalent to that normally maintained by homeostasis. If the experimental conditions are such that the tissue is adversely affected during incubation, it may undergo morphological changes detectable by light microscopy. For example, slices of tissue that have been incubated for prolonged periods will undergo substantial necrosis when the diffusion of substrates and metabolites through the cell mass is incomplete. Any such degradation of the tissue will affect its biochemical activity and should be assessed by light microscopy, so that, if necessary, the experimental procedure may be modified accordingly. More detailed information can be obtained by electron microscopy. However, electron microscopy is time consuming, and therefore in contemplating this approach the following question must be considered: Does the ultrastructural appearance of cells provide sufficient additional information about their functional state to justify the use of electron microscopy?

The functional state of a cell is dependent upon its environment, as implied above, and also upon the integration of metabolic events within that cell. Clearly, when cells are subjected to experimental procedures *in vitro* their environment is altered; if this is not carefully controlled there is the possibility that structural damage may occur below the resolution of the light microscope. Since most individual metabolic processes within a cell are associated with specific intracellular loci, it is highly likely that progressive loss of structural integrity will affect the functional integrity of the cell. It follows that the ultrastructural examination of cells will provide a more exact indication of their functional integrity than that obtainable by the use of light microscopy.

[2] J. A. McCracken, J. C. Carlson, M. E. Glew, J. R. Goding, D. T. Baird, K. Green, and B. Samuelsson, *Nature (London) New Biol.* **238**, 129 (1972).

At this point an important distinction may be drawn between the terms "functional integrity" and "functional state" of a cell. The term integrity in the present context is taken to refer to the integration of those metabolic processes common to all cells. The "functional state" of a cell reflects the specialized functions of a given cell type; it is, of course, dependent upon the integrity of that cell. The functional state of many endocrine cells is reflected in the morphological appearance of subcellular structures which are frequently sites associated with the synthesis, storage, or release of a hormonal product. A general consideration of this topic has been presented by Fawcett *et al.*,[3] and it will suffice here to emphasize the fact that the ultrastructural appearance of many endocrine cells provide a very useful indication of their functional state, whether *in vivo* or *in vitro*. The reliability of this approach for assessing the functional state of a cell should, however, first be established by combined morphological and biochemical studies to determine the extent of correlation between structure and function in the cell type under investigation. The way in which such a correlation has been established for adrenal cortical cells grown in tissue culture will now be outlined, giving details of the methodology used to examine the ultrastructure of the cells.

The cortical cells are cultured from fetal rat adrenals. The adrenals are collected in balanced salt solution, chopped into small pieces, and transferred into plastic culture dishes. Culture medium is then added to each dish and the cultures are maintained in a humidified atmosphere of 5% carbon dioxide in air at 37°. After 2 days most pieces of adrenal tissue are attached to the surface of the dishes by a layer of outgrowing cells. The culture medium is changed every 3 or 4 days and contains 50% Hanks' lactalbumin, 25% Eagle's minimum essential medium, and 25% calf serum. The pH of the medium is maintained between 7.4 and 7.6 by adjusting the gas flow into the incubator. Stringent precautions are taken to maintain sterility throughout the culture period, and no antibiotic cover against bacterial contamination is used.

The steroidogenic activity of the cultures is assessed by measuring the conversion of radioactive precursors to radioactive steroidal products as described in Milner and Villee[4] and Milner.[5] Radioactive steroidal products are extracted from the medium, characterised and quantitated as described by Milner and Mills[6] and in.[4,5] The cultured cells are left intact and prepared for electron microscopy *in situ*.

[3] D. W. Fawcett, J. A. Long, and A. L. Jones, *Recent Progr. Horm. Res.* **25**, 315 (1969).

[4] A. J. Milner and D. B. Villee, *Endocrinology* **87**, 596 (1970).

[5] A. J. Milner, *Endocrinology* **88**, 64 (1971).

[6] A. J. Milner and I. H. Mills, *J. Endocrinol.* **47**, 369 (1970).

Prior to fixation the cultures are examined by phase contrast microscopy, and colonies showing a good outgrowth of cortical cells are selected and marked for ultrastructural examination. This preliminary selection is necessary for two reasons. First, there are some colonies that appear to contain no cortical cells. These have presumably arisen from pieces of connective tissue removed with the adrenals during dissection. Second, the colonies contain different cell types, the cortical cells forming a shoulder, several cells deep, immediately surrounding the original piece of explanted tissue. This shoulder of cells is in turn surrounded by a monolayer of other cell types, mainly fibroblastic, occupying a relatively large surface area of the colony (see Fig. 1B). Once the cells have been fixed and embedded the whole colony appears as a dark area within which visual discrimination between the different cell types is impossible. A clear mental image of the layout of the cell types within the colony is therefore essential in order to obtain sections of the cortical cell region. The general principles of electron microscopy have been detailed elsewhere (see, for example, Hayat[7]), and the following description will be restricted to a consideration of the difficulties associated with the ultrastructural examination of cells grown in tissue culture. It is hoped that the modifications which have been developed to minimize these difficulties will serve as a general indication of how the techniques involved in electron microscopy may be adapted to suit a particular experimental system.

Electron Microscopy

The cultures are fixed and embedded *in situ* as detailed in Table I. Since the shoulder of cortical cells is only a few cells deep, the penetration of fixatives is rapid and the fixation times are correspondingly short. Similarly dehydration of the cells can be effected rapidly, and, at this stage, prolonged exposure to ethanol is undesirable because it would lead to excessive extraction of lipids from the cortical cells. When ethanol is employed for dehydration a transitional solvent is normally used prior to embedding in resin since most resins are not readily miscible with ethanol. Propylene oxide is commonly used as a transitional solvent to facilitate the penetration of the resin into the cells. However, the plastic of the culture dishes is soluble in propylene oxide and therefore this step is omitted from the present procedure. Instead, the cells are left in a mixture of resin and ethanol (1:1) for 1 hour and then in resin overnight to allow complete infiltration. Finally, the cells are embedded in fresh

[7] M. Hayat, ed., "Principles and Techniques of Electron Microscopy: Biological Applications." Van Nostrand-Reinhold, Princeton, New Jersey, 1970.

TABLE I
FIXATION AND EMBEDDING PROCEDURE USED TO PREPARE TISSUE CULTURES OF ADRENAL CORTICAL CELLS FOR ELECTRON MICROSCOPY

Step	Fixative	Time	
Fixation	5% Glutaraldehyde[a] in BSS[b] at pH 7.4	30 min	at room temperature
Wash	BSS followed by phosphate buffer	20 min	
Postfix	1% Osmium tetroxide[a] in phosphate buffer	20 min	
Dehydrate	50% Ethanol	2 min	
	70% Ethanol	2 min	
	90% Ethanol	2 min	
	100% Ethanol	2 × 3 min	
Embedding	100% Ethanol + resin(1:1)	1 hr	
	Resin (Taab)	Overnight	
	Fresh resin	48 hr	at 60°C

[a] Solutions made up freshly.
[b] BSS = Hanks' balanced salt solution (Flow Laboratories Inc.). The resin is made up and stored in sealed 20-ml plastic syringes at −20°; it is allowed to thaw at room temperature before use.

resin which is polymerized at 60° for 48 hours. For ease in sectioning the colonies which were selected by phase contrast microscopy are embedded in a cylinder of resin, as illustrated in Fig. 1A.

For sectioning, the resin cylinders are simply broken away from the culture dish. The procedure for trimming and sectioning is straightforward, and the major problem is to obtain sections which include some cortical cells. In the author's experience this is best achieved by trimming away most of the original explant and most of the monolayer region at the top and bottom of the section face, as indicated in Fig. 1B.

The combined use of the biochemical and morphological techniques described above has produced evidence for a close correlation between the ultrastructure and functional activity of adrenal cortical cells. The rationale behind this approach is given in Milner and Hamilton[8] and rests upon the fact that particular steroidogenic enzymes appear to be associated with specific intracellular organelles (see Christensen and Gillim[9] for a review of this topic). The two major sites of steroidogenic activity are the mitochondria and the smooth endoplasmic reticulum. In

[8] A. J. Milner and D. W. Hamilton, *in* "*In Vitro* Methods in Reproductive Cell Biology" (Karolinska Symp. Res. Methods Reprod. Endocrinol., 3rd) (E. Diczfalusy, ed.), p. **62**, Bogtrykkeriet Forum, Copenhagen, 1971.

[9] A. K. Christensen and S. W. Gillim, *in* "The Gonads" (K. W. McKerns, ed.), p. **415**, Appleton, New York, 1969.

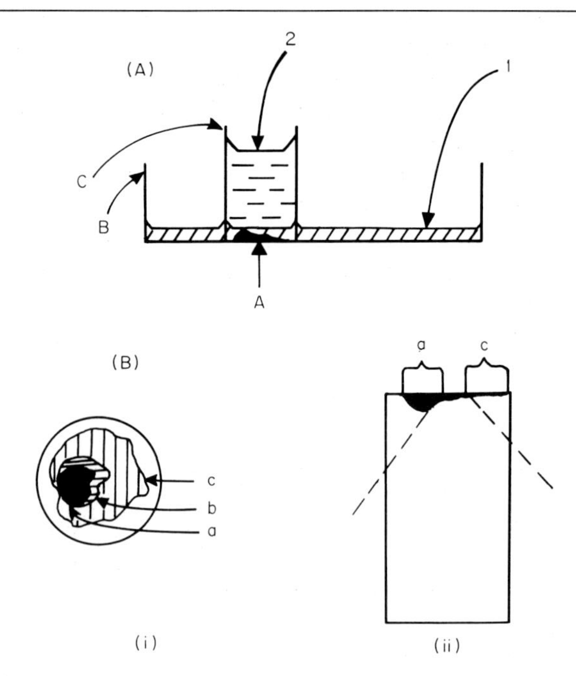

FIG. 1. A, Diagram to illustrate the method for embedding colonies of adrenal cells for electron microscopy. A = original explant of tissue surrounded by outgrowing cells. C = plastic embedding capsule cut to produce a cylinder and placed over the colony of cells. The first addition of resin (1) is partially polymerized at 60° to form a seal between the capsule and the culture dish (B) before the capsule is filled with additional resin (2). B, Adrenal cell colony embedded in resin. A surface view of the block which has been broken away from the culture dish is represented in diagram (i) to show the distribution of cell types within the colony. a = original tissue explant, b = cortical cells, c = surrounding monolayer of different cell types. In diagram (ii) a profile of the block is shown. The dashed lines indicate where the block would be trimmed to obtain sections of the cortical cell region.

the absence of adrenocorticotropin (ACTH) these two organelles are poorly developed in cortical cells grown in culture.[10] ACTH induces ultrastructural changes in the mitochondria and the smooth endoplasmic reticulum[10] which are accompanied by a progressive increase in the steroidogenic capacity of the cells.[4] Three days after exposure to ACTH, the ultrastructure of the cells resembles that observed *in vivo*.

Because the morphology of cortical cells cultured in the absence of ACTH differs from that observed *in vivo* there is the problem of how to identify the cells in the electron microscope. In this situation it is es-

[10] A. I. Kahri, *Acta Endocrinol.* (*Copenhangen*) **52,** Suppl. 108 (1966).

sential to have a morphological marker to enable cell type identification, and the most reliable marker for the cortical cells is the appearance of the mitochondria. When cultured in the absence of ACTH the cortical cells contain mitochondria which are small with vesiculolamellate cristae (Fig. 2). Similar mitochondria have not been observed in any of the other cell types present in the tissue cultures. Following exposure to ACTH the mitochondria in the cortical cells enlarge and transform to the vesicular type which are characteristic of adrenal cortical cells in the rat (Fig. 3). It is interesting to note that this vesicular conformation of the inner membrane appears to be correlated with the activity of the mitochondrial 11β-hydroxylating system, responsible for the conversion of deoxycorticosterone to corticosterone in the rat.

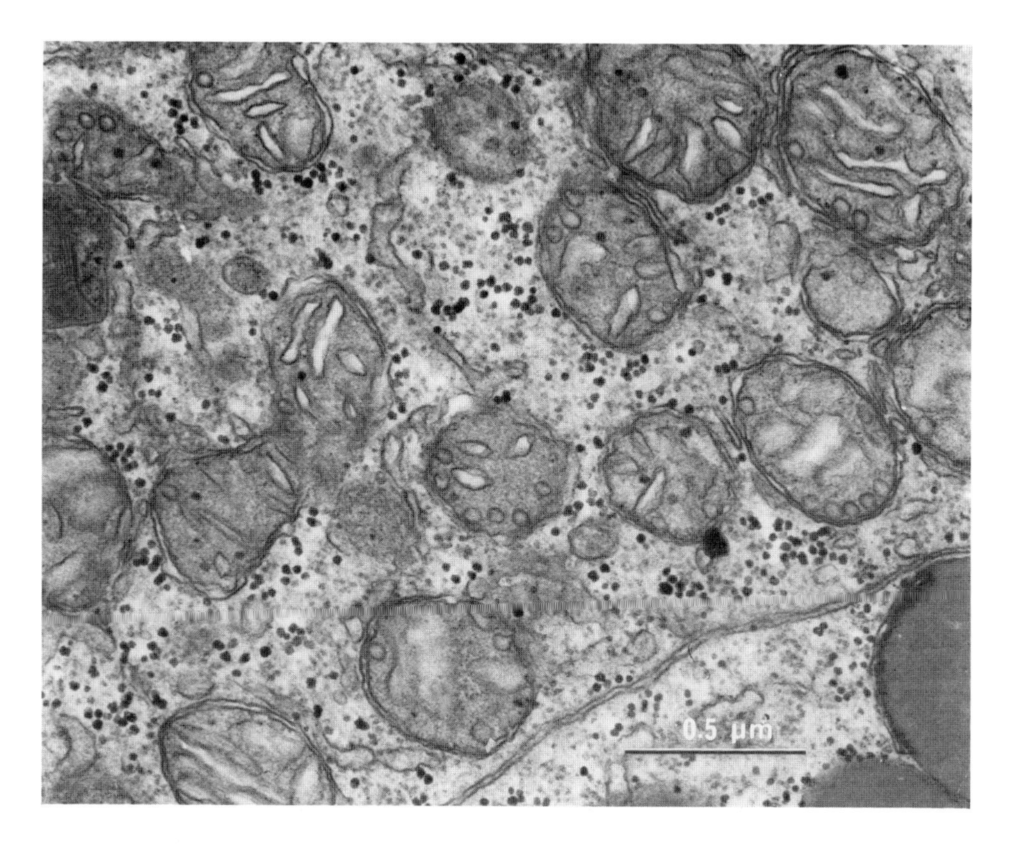

FIG. 2. Electron micrograph showing the appearance of the mitochondria in an adrenal cortical cell which has been cultured in the absence of adrenocorticotropin. Note the vesiculolamellate cristae and the small size of these mitochondria in comparison with those in Fig. 3.

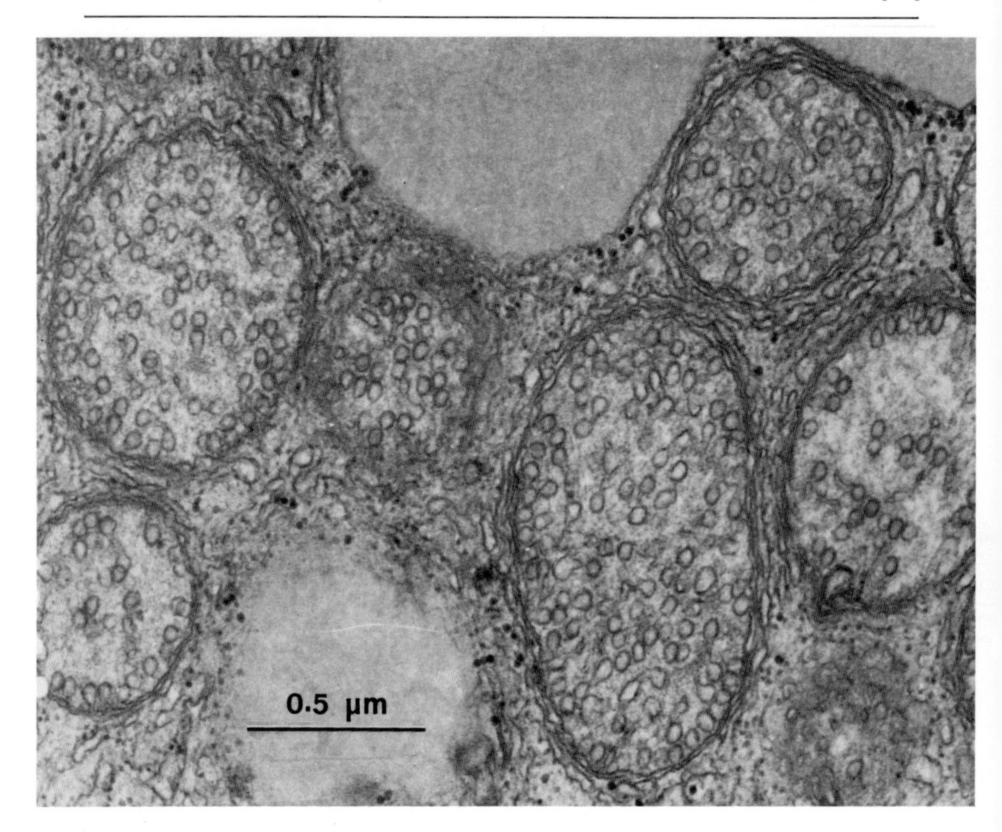

FIG. 3. Three days after exposure to adrenocorticotropin, the mitochondria in the cortical cells transform, as illustrated in this electon micrograph. The mitochondria increase in size and the inner membrane forms vesicular cristae; such mitochondria are characteristic of the corticosteroid-secreting cells of the rat adrenal.

In summary, the normal functioning of endocrine cells is in part dependent upon their structural integrity. Structural damage may occur when the cells are subjected to experimentation *in vitro,* and any gross damage should be assessed by morphological examination of the preparation by light microscopy at the end of the experimental period. More detailed information can be obtained by electron microscopy, although it should be borne in mind that the functioning of a cell may be affected through structural changes below the present resolving power of the electron microscope. As mentioned at the beginning of this chapter the nature of the information obtained by electron microscopy falls into two classes, one relating to the *functional integrity* of the cell, and the second to its *functional state.* On this basis a distinction between the systems which

TABLE II
EXPERIMENTAL SYSTEMS USED TO STUDY THE FUNCTIONING OF ENDOCRINE CELLS *in Vitro*[a]

System	Comments	
Superfusion	Pieces of tissue, e.g., adrenal quarters, superfused with medium which is subsequently analyzed for hormonal products	Group 1
Incubations	(i) Tissue slices, etc., incubated in medium which is subsequently analyzed for hormonal products. Suitability of incubation conditions should be checked by light microscopy	Group 1
	(ii) Isolated cells, prepared by tissue digestion with proteolytic enzymes and incubated as in (i). Cells friable, should be checked before and after incubation period by light microscopy. Permeability of cells to nigrosin should also be tested to check for membrane damage.	Group 1
Organ culture	Usually short term, often equivalent to *maintenance* rather than *culture* of piece of organ. Cells should be examined by light microscopy	Group 1
Perfusion	May be carried out with the whole gland *in situ*, transplanted *in vivo*, or *in vitro*	Group 2
Tissue and cell culture	Long term. Suitable for examination by light and electron microscopy	Group 2

[a] For explanation of Groups 1 and 2 see text.

are currently used to study the functioning of intact endocrine cells *in vitro* may also be drawn (Table II). The first group includes those systems which, while aimed at maintaining the cells in a state approximating to their functional state *in vivo*, are *not* designed to support cell growth and division. Ultrastructural examination of cells in this group provides a means of assessing the structural, and hence the functional, integrity of the cells. The second group includes those systems designed to support the growth and division of endocrine cells, that is, to maintain the cells under conditions in which their functional integrity may be assumed. Under these circumstances ultrastructural examination provides an extremely useful indication of the functional state of the endocrine cells.

[17] Methods for Studying Pituitary-Hypothalamic Axis *in situ*[1]

By JOHN C. PORTER

Successful experimental manipulation of the hypothalamus and pituitary *in situ* is facilitated by exposure of the ventral diencephalon to enable one to see clearly this particular portion of the brain; but this task is not easy. Indeed, achievement of satisfactory exposure of the diencephalon without compromising the viability of the animal is tedious and requires an element of proficiency. In order to acquire such proficiency, some may find it necessary to seek assistance and to practice diligently before learning to recognize pertinent anatomical landmarks and developing the necessary coordination. The outline which follows includes a description not only of the step-by-step operative procedure but also a description of mechanical aids which make the task easier—at least we believe they do so.

The hypothalamus and pituitary are difficult to study *in situ* in mammals primarily because both structures lie deep within the cranium. The complex, i.e., the hypothalamic–pituitary system, is protected superiorly by the skull and brain and inferiorly by bones of the skull and jaw and by an extensive array of soft structures involved in head movement, eating, breathing, and transport of blood. Therefore, an early task of the investigator is to circumvent these barriers or reduce them to a level of insignificance. The procedure for doing so, which is described here, is applicable specifically to the rat, but with appropriate modification may be adapted to certain other species.

Exposure of the Ventral Diencephalon of the Rat

Anesthesia. Although many anesthetic agents have been tried in order to achieve a level of anesthesia in the rat suitable for surgery, we now use sodium pentobarbital. The agent is given slowly via a tail vein during a 4- to 5-minute period so as not to kill the animal by overdosage. The infusion continues until the animal ceases to respond to pain but is stopped before respiration is suppressed. These physiological indices are more reliable than a fixed, predetermined dose; however, it is worth

[1] Supported by a grant from the National Institute of Arthritis and Metabolic Diseases (AM01237) and by a grant from The Population Council, New York, New York.

noting that anesthesia is usually achieved after 35–45 mg of sodium pentobarbital per kilogram of body weight has been administered. Because a perfect anesthetic agent probably does not exist, the investigator must be prepared to accept the fact that the state of anesthesia may have its own characteristics which may not be the same in detail as those of the conscious condition.

Head Holder. After anesthesia has been induced, the rat's head is placed in a special holder to anchor the head and to secure the animal in a supine position. A metal holder used by our laboratory for several years for this purpose is illustrated in Fig. 1. Its principal features are short posts with adjustable ear bars and a plate upon which the rat lies. (Although this holder is not available commercially, it can be made easily by a machinist.) At first, it may be helpful to cut the ear lobes before inserting the bars into the auricular canals. (However, with practice, the procedure becomes routine; and it is unnecessary to make such an incision.) Care should be taken to ensure that the ear bars are inserted into the auricular canals, not into the soft tissue surrounding the ears. When

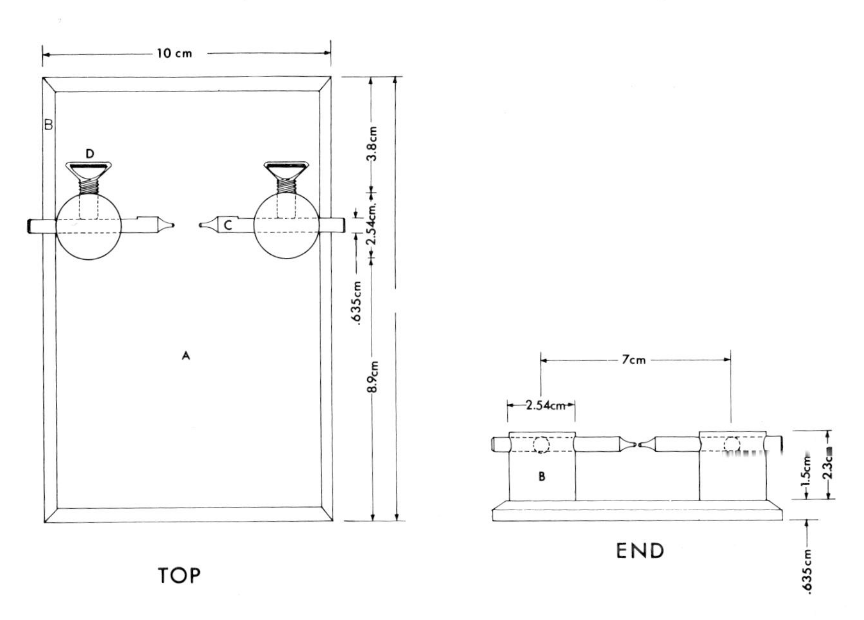

FIG. 1. Holder to fix the rat's head. The animal is placed on its back on the base plate (A), and the head is held rigidly by means of the ear bars (C). The bars are held by vertical supports (B) and secured with screws (D). The dimensions of the holder are indicated in the drawing. Reproduced from J. C. Porter, R. S. Mical, I. A. Kamberi, and Y. R. Grazia, *Endocrinology* **87**, 197 (1970).

the bars are placed correctly, one cannot move the rat's head from side to side. It is firmly fixed. As you move the muzzle right and left, if the head moves in a "rubbery" fashion, the ear bars are not positioned correctly.

Tissue Retractors and Operating Boards. Although the role of the head holder is to fix the skull, the purpose of the operating board is to provide points of attachment for the retractors and to provide a base for the head holder. The retractors may be obtained from Brookline Surgical Specialties, Westford, Massachusetts. Four holders for securing the legs of the rat are needed, and one maxillary incisors holder is used to hold the muzzle by means of the upper teeth firmly to the board. Five single retractors, H ½ inch, and two double spring retractors, R-6, are needed to hold the soft tissues around the operating area out of the field of view.

Two types of operating boards are used in our laboratory. They are made from plywood (1 inch thick) and covered on the top and sides with formica for ease of cleaning. The octagonally shaped board shown in Fig. 2 is used for most purposes. Fasteners for fixing the ballchains of the retractors and leg holders are made from electrical solderless terminals

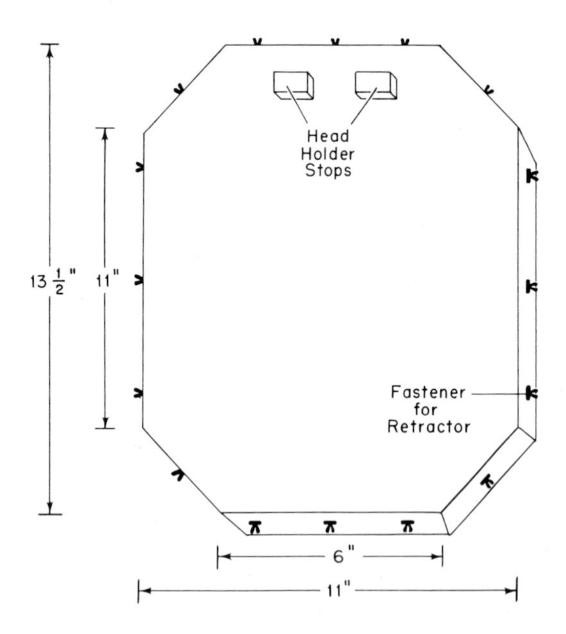

FIG. 2. General purpose operating board. The fasteners are used to secure the ballchain of the retractors. The dimensions of the board are indicated in the drawing.

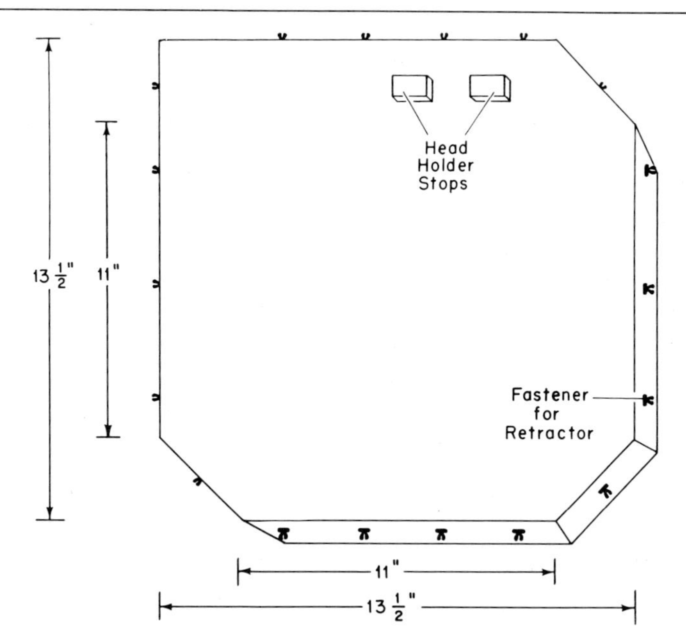

FIG. 3. Operating board used during the cannulation of a hypophyseal portal vessel. The dimensions of the board are indicated in the drawing.

(spade lugs) for 16–14 gauge wire. The shanks of the lugs are flattened and attached to the board by means of screws. The prongs of the lugs are bent to form a V-shaped slot to hold the chain. This arrangement works well and makes it easy to attach and detach the ballchains.

The other operating board used in our laboratory is wider (Fig. 3). This board is used when one is cannulating a hypophyseal portal vessel. The extra width and enlarged corner make it possible to place the micromanipulator, which is used in the cannulation, on the board beside the head of the rat.

Two bolts with large, square heads are fixed to the anterior end of each board (Figs. 2 and 3). The purpose of these bolts is to provide a stop for the head holder.

Exposure of the Ventral Diencephalon. The animal, which is in the head holder, is placed on the operating board with the nose of the rat directed toward the anterior end of the board. The incisors holder is placed over the teeth, and the animal and head holder are pulled tightly against the bolt stops. Next, the legs of the rat are secured to the board by leg holders. (Do not stretch the forelegs too tightly. If you do so, inspiration will be impaired.) The animal is now arranged for the surgical

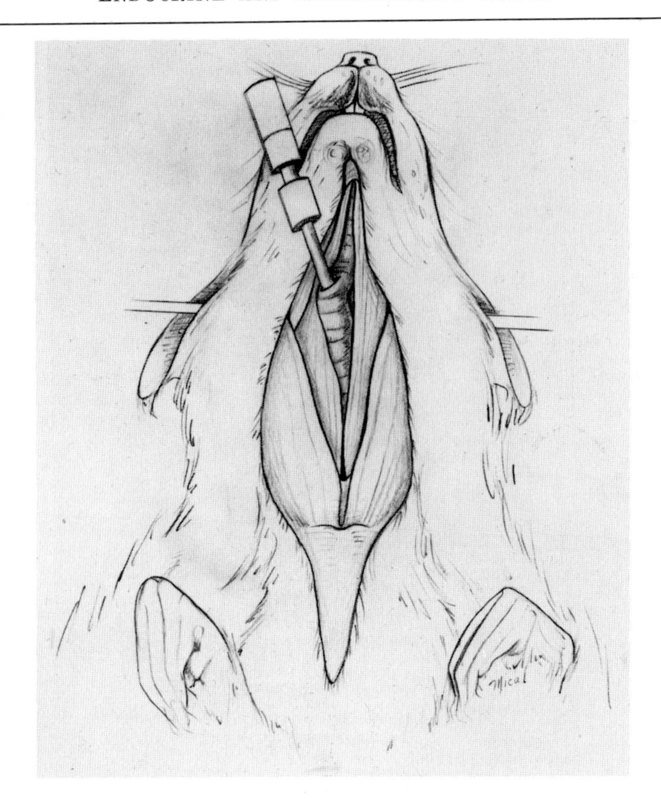

FIG. 4. The initial incision showing the sternohyoid and omohyoid muscles and trachea. Reproduced from J. C. Porter, R. S. Mical, J. G. Ondo, and I. A. Kamberi, *in* "Karolinska Symposia on Research Methods in Reproductive Endocrinology, 4th Symposium: Perfusion Techniques" (E. Diczfalusy, ed.), p. 249. Karolinska Institutet, Stockholm, 1971.

procedures and experimental manipulations. The trachea is exposed through an incision in the midline of the neck, and a polyethylene cannula is inserted into the trachea and secured with a ligature. The arrangement of the head of the animal at this point is illustrated in Fig. 4. The rats can be supplied with oxygen or air by means of a rodent respirator (Harvard Apparatus Co., Millis, Massachusetts) to ensure adequate respiratory ventilation.

Subsequent manipulations are performed with the aid of a binocular operational microscope equipped with an objective which provides a working distance of 200 mm. The scope should be equipped with coaxial light of sufficient intensity to light the operating field well. The stereo microscope, MTX-002/V-T-F-200, supplied by Olympus Corporation of

America (New Hyde Park, New York), has proved satisfactory in our hands.

The incision in the midline of the neck, which was made during the tracheotomy, is extended anteriorly to the level of the interramal vibrissae. The sternohyoid and omohyoid muscles are retracted laterally. It is usually convenient, but not necessary, to transect the thyreohyoid muscles at a point about 0.5 cm below the thyroid. The hyoid bone is elevated by means of curved, dissecting forceps (Clay-Adams, Catalog No. B-631/C) to expose the muscles overlying the occipital bone, the posterior margin of which can be identified by the foramen magnum (Fig. 5). Seven retractors, five single retractors, and two double spring retractors,

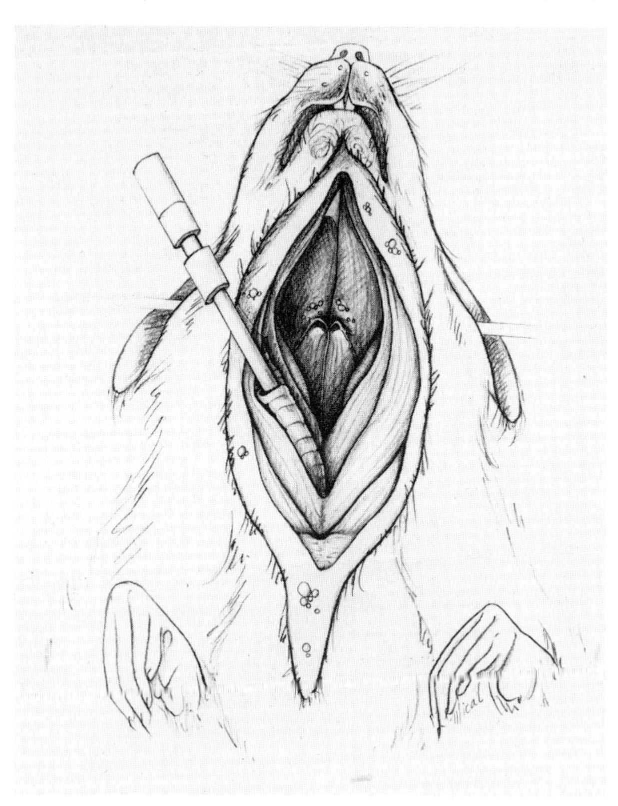

FIG. 5. The trachea, sternohyoid, and omohyoid muscles are retracted laterally, and the hyoid bone is pulled anteriorly to expose the muscles covering the first vertebra and base of the skull. Reproduced from J. C. Porter, R. S. Mical, J. G. Ondo, and I. A. Kamberi, *in* "Karolinska Symposia on Research Methods in Reproductive Endocrinology, 4th Symposium: Perfusion Techniques" (E. Diczfalusy, ed.), p. 249. Karolinska Institutet, Stockholm, 1971.

are next placed around the edges of the opening and drawn tightly to give maximal exposure (Fig. 6). The double spring retractors are used at the posterior edges of the opening. Caution is appropriate when fixing the two posterior retractors. If these two retractors are drawn too tightly, the flow of blood through the internal carotid and pterygopalatine arteries is diminished. Obstruction of these vessels will result in a slow death which is often preceded by rigor in one or both hind legs. The remaining five retractors can be drawn as tightly as desired, however.

The base of the skull is dissected free of overlying tissue. A 3- to 4-inch long metal tube (made of 16- to 18-gauge hypodermic tubing) with a blunt tip connected to a mild vacuum source is excellent for the dissection. By scraping the bone, beginning in the midline and proceeding laterally, one can remove all tissue covering the occipital, basisphenoid, and presphenoid bones. The occipital bone should be sufficiently free of

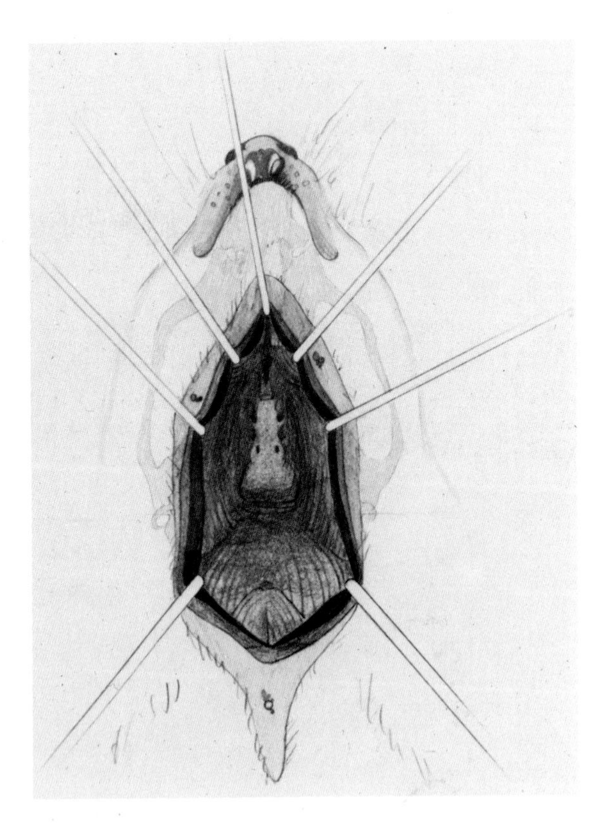

Fig. 6. The position of the retractors around the opening are indicated, enabling one to see the occipital, basisphenoid, and part of the presphenoid bones.

tissue to enable one to see the internal carotid and ptergopalatine arteries and much of the tympanic bullae. The retractor holding the hyoid bone should be kept tight at all times. If it becomes loose as a result of the dissection, tighten it again. The basisphenoid is dissected free of tissue until one can see the pterygopalatine foramina, pterygoid processes, middle lacerated foramina, and the anteromedial portion of each tympanic bulla. At this point, tighten again the most lateral retractors. But, be certain that these retractors are placed near the anterior edges of the tympanic bullae. If so, the flow of blood in the external jugular veins and carotid and pterygopalatine arteries will not be obstructed. Obstruction is apt to occur, however, if these retractors are placed near the posterior margins of the tympanic bullae. The dissection is continued anteriorly until much of the presphenoid bone is exposed. At this time, it is well to check all retractors for proper tension and position (Fig. 6).

In order to expose the anterior pituitary, pituitary stalk, and ventral hypothalamus, it is necessary to remove part of the occipital bone, all of the basisphenoid bone, and part of the presphenoid bone. This region is illustrated by the light portion of the skull, which is shown in Fig. 7. Begin by fracturing the ptergoid processes with forceps and removing them. The bones of the skull are removed by grinding. For this purpose, we have found it convenient to use a drill consisting of an engine, a 7-D handpiece (Foredom Electric Co., New York), and a No. 8 spherical dental burr. For fine control, the drill is operated at maximal speed, approximately 5000 rpm. The drilling should be interrupted frequently to prevent accumulation of excess heat and subsequent damage of the underlying tissue. The bone dust that is formed during drilling is excellent for stopping bleeding. Thus, a skilled operator accumulates the dust near the working area and uses it as needed. In addition to bone dust, a soft bone wax (W-31G wax, Ethicon, Inc., Somerville, New York) is useful in plugging broken vessels; however, bone wax should be used sparingly since excessive quantities become hindrances in the operating field.

There are three sets of blood vessels which should be approached carefully. These are the internal carotid arteries as they pass by the anterior pituitary, the sinus of the basisphenoid bone, and the cavernous sinuses. With the exercise of care, the internal carotids and the cavernous sinuses will not be broken. However, it is necessary to destroy the sinus of the basisphenoid; and some bleeding will inevitably occur when one does so. But bleeding can be minimized by the following procedure. By means of the drill, remove the bone covering the sinus as much as practicable. Then, while holding a Q-tip (Johnson & Johnson Cotton Buds) with a small piece of bone wax pressed into the cotton, obliterate the sinus with the drill and press the Q-tip into the opening that once held the sinus.

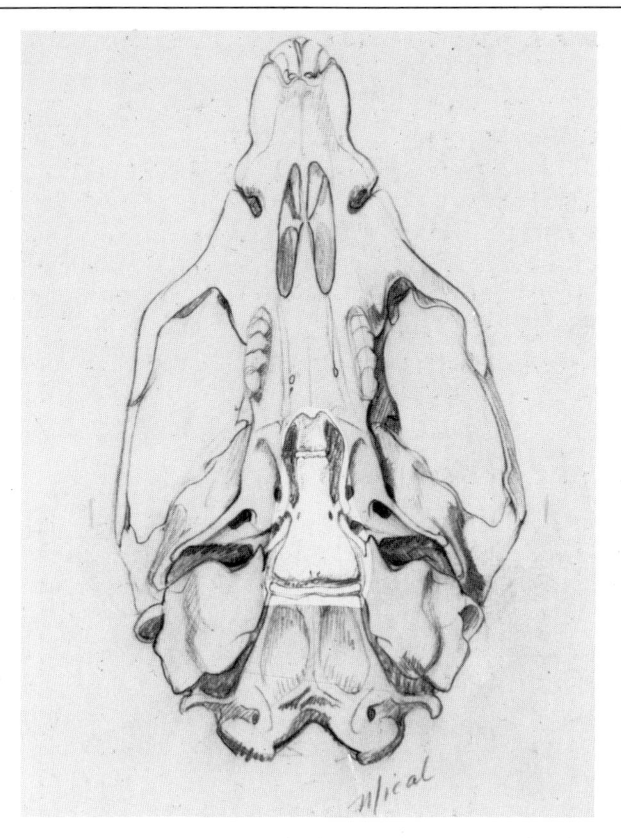

FIG. 7. The posterior margin of the presphenoid, the presphenoid–basisphenoid suture, the basisphenoid, the basisphenoid–occipital suture, and the anterior margin of the occipital bone are illustrated in the lightly shaded region. The relationship of these structures to other important landmarks such as the tympanic bullae, pterygoid processes, and lacerated foramina are shown in the drawing. Reproduced from J. C. Porter, R. S. Mical, J. G. Ondo, and I. A. Kamberi, *in* "Karolinska Symposia on Research Methods in Reproductive Endocrinology, 4th Symposium: Perfusion Techniques" (E. Diczfalusy, ed.), p. 249. Karolinska Institutet, Stockholm, 1971.

Give the Q-tip a sharp twist to press wax into the bone. This procedure stops the bleeding completely. Since this operation takes only a few seconds, very little blood is lost. The drilling continues until all bone in the opening is removed, with the exception of a thin residual sheet of the basisphenoid. At this point, substitute a No. 4 dental burr for the larger No. 8 burr and use the smaller burr to make the final enlargement of the opening. The anteromedial corners of the tympanic bullae should also be removed. But, when doing so, beware of the internal carotid arteries.

These surgical procedures take 45–60 minutes to complete when done by a skilled operator, and little blood is lost during the operation. However, beginners should be prepared to spend 3–4 hours or more to complete the same procedure, and more often than not they will bleed their first few animals to death. However, this should not discourage one. The beginner is urged to persevere. When success comes, as it will, the ends justify any anguish which may have been experienced initially. One is rewarded with an experimental animal preparation which enables him to discern and manipulate experimentally a precise portion of the hypothalamus and/or pituitary.

Collection of Pituitary Stalk Blood

The rat is a good species in which to collect pituitary stalk blood for several reasons. These include the existence of a complete diaphragma sellae—which isolates the pituitary in a fossa, the absence of a sella turcica, and a relatively long pituitary stalk which enters the pituitary fossa through the diaphragma sellae. Consequently, one can remove the pituitary and collect blood from the pituitary stalk with relative ease. To do so, one needs only to open the pituitary fossa through the dura mater while leaving the diaphragma sellae fully intact as follows (Fig. 8a–g). Using a Ziegler needle knife (Storz Instrument Co., St. Louis, Missouri, Catalog No. E-153), make a transverse incision in the dura covering the ventral surface of the anterior pituitary. The incision should be near the posterior edge of the pituitary and extend as far laterally as practicable without cutting the internal carotid arteries. [Note: Do not cut the anterior pituitary when performing this operation.] In making the incision, we find the following procedure helpful. The tip of the knife blade is used to cut a small hole in the dura. Then, using an O'Brien foreign body spud (Storz Instrument Co., Catalog No. E-858), pull the dura away from the anterior pituitary and proceed to enlarge the incision with the knife. With Vannas capsulotomy scissors (Stotz Instrument Co., Catalog No. E-3386), enlarge the opening by cutting a V-shaped slot in the dura as illustrated in Fig. 8b, d. The major structures attached to the pituitary are (1) the stalk, (2) the inferior hypophyseal arteries, and (3) the veins draining the pituitary. The internal carotids pass over the corners of the anterior pituitary. Using a small, blunt probe (e.g., an O'Brien foreign body spud with a blunt point), break the pituitary veins and inferior hypophyseal arteries and lift the corners of the gland above the internal carotid arteries. It may be helpful to press small pieces

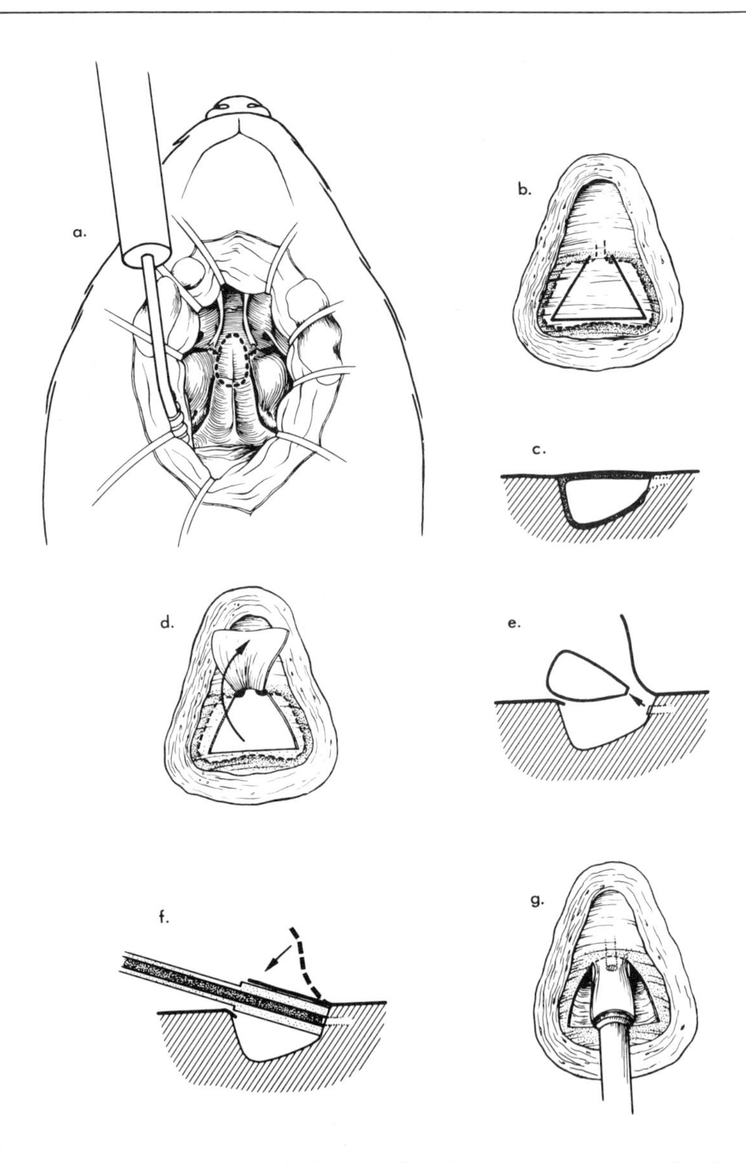

FIG. 8. Cannulation of the pituitary stalk. The region of the skull which is removed is indicated in drawing (a) by the broken line. The extent of the triangular incision made in the dural covering of the anterior pituitary is illustrated in (b). Insets (c), (d), and (e) show the various steps involved in removing the pituitary. Insets (f) and (g) illustrate the position of the collecting cannula. Reproduced from J. C. Porter, B. D. Goldman, and J. F. Wilber, *in* "Hypophysiotropic Hormones of the Hypothalamus: Assay and Chemistry" (J. Meites, ed.), p. 282. Williams & Wilkins, Baltimore, Maryland, 1970.

of gel foam against the cavernous sinuses to slow the retrograde flow of blood from these vessels. After a few minutes when bleeding slows, inject 2 mg of heparin intravenously and begin a slow infusion of lactated Ringer containing sufficient heparin (38 μg per minute) to prevent clotting of the blood. Next, transect the pituitary stalk and remove the pituitary. After transection, a small stump of the stalk protrudes through the diaphragma sellae (Fig. 8e); but one should take care not to cut the stalk too short or too long. After experience, the operator can usually decide for himself what is the optimal length. We prefer to cut the stalk with angled capsulotomy scissors of the Vannas type. However, this particular operation is delicate, and people tend to vary in their preference of surgical tools. Some prefer straight scissors. When cutting the stalk, one should not puncture the diaphragma sellae nor tear the stalk free from its attachment to the diaphragma sellae. After the pituitary has been removed from the fossa, one can see blood flowing from the cut ends of the long hypophyseal portal vessels which lie on the ventral surface of the stalk. These vessels receive blood from the capillary plexuses of the stalk and anterior median eminence. One can also see one or two vessels near the dorsal surface of the stalk. These vessels are believed to arise from the peduncular arteries and to drain capillary plexuses of the posterior median eminence.

The blood from the hypophyseal portal vessels is collected in a cannula which is placed over the stump of the stalk (Fig. 8f,g). The cannula consists of a short segment of polyethylene tubing (Clay-Adams, PE-90) into one end of which is inserted a longer segment of PE-20 tubing. The opposite end of the PE-90 tubing is flared slightly by holding it near a small flame. This segment of the cannula is illustrated in Fig. 9.

The flared end of the cannula is placed over the stalk and held against the diaphragma sellae by means of a special micromanipulator to form a liquid-tight seal (Fig. 8f,g). To confirm that such a seal has indeed been effected—which is essential if one is to collect pure pituitary stalk blood uncontaminated with extraneous fluids—we routinely fill the pituitary fossa and cover the end of the cannula with a solution containing lissamine green. The cannula is considered correctly positioned when, and only when, pure blood with no dye passes through the cannula. Stalk blood is usually collected one of two ways. One, the blood is allowed to flow into a tube placed below the level of the rat. Or, two, the blood is withdrawn into a long segment of tubing by means of a pump. The average flow from the stalk is approximately 7 μl per minute, and one can easily collect blood from an animal for 2–4 hours, and sometimes longer.

The type of manipulator used to hold the cannula over the stump

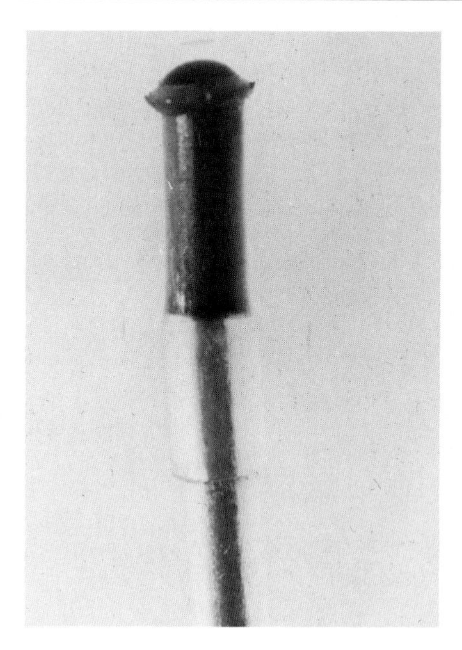

FIG. 9. A photograph of the end of the cannula which is placed over the stalk. Reproduced from J. C. Porter, I. A. Kamberi, and Y. R. Grazia, *in* "Frontiers in Neuroendocrinology, 1971" (L. Martini and W. F. Ganong, eds.), p. 145. Oxford Univ. Press, London and New York, 1971.

of the stalk and against the diaphragma sellae is important. The instrument designed by us and which has been used continuously for several years in our laboratory is illustrated in Fig. 10. This manipulator and others like it were constructed in the Laboratory for Bio-Engineering at our institution. In addition to the usual controls for three-dimensional movement, the manipulator has one control for moving the cannula holder in a circular plane and another control for moving the cannula holder back and forth along a chord of this circle. These two movements are especially useful in the placement of the cannula over the stalk, where fine control is needed to hold the cannula against the diaphragma sellae. It is essential that no pressure be applied laterally on the portal vessels, since the blood pressure in these vessels is low and can be easily obstructed.

Several advantages accrue from this method of collecting stalk blood. First, the brain itself remains enclosed completely by the meninges throughout the collection. Second, since the portion of the pituitary stalk which protrudes through the diaphragma sellae is covered by the can-

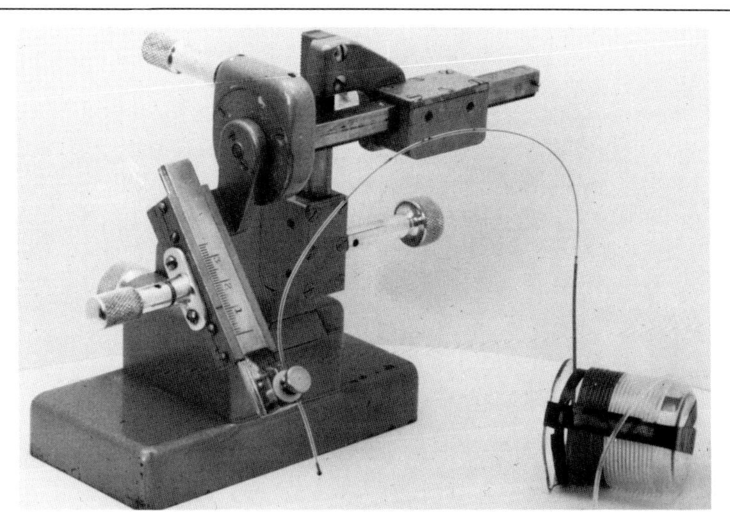

FIG. 10. Micromanipulator used to hold the cannula over the stalk. Polyethylene tubing serves as the collecting reservoir for the stalk blood. Reproduced from J. C. Porter, I. A. Kamberi, and Y. R. Grazia, *in* "Frontiers in Neuroendocrinology, 1971" (L. Martini and W. F. Ganong, eds.), p. 145, Oxford Univ. Press, London and New York, 1971.

nula which is filled with blood, the stalk is not exposed to air and its desiccating effects. Third, pure stalk blood is obtained that is uncontaminated with cerebrospinal fluid and extraneous blood. And, fourth, all the blood that passes through the long portal vessels is collected, which makes it possible, at least theoretically, to determine secretory rates of hypothalamic releasing factors.

Cannulation of Portal Vessels and Perfusion of the Anterior Pituitary

The rat is well-suited for the cannulation of a hypophyseal portal vessel (Fig. 11). The pituitary stalk in this animal passes horizontally from the brain to the anterior pituitary, and most of the long portal vessels lie on the ventral surface of the stalk. Thus, when the animal is on its back, these vessels face the operator. For the cannulation procedure, the operating board illustrated in Fig. 3 is used since a relatively large micromanipulator is placed on the upper, left-hand corner of the board adjacent to the head holder and next to the animal's head.

In order to expose the median eminence and pituitary stalk, one makes a long incision in the meninges which is parallel to the anterior median eminence and is about 1 mm lateral to the midline of the brain.

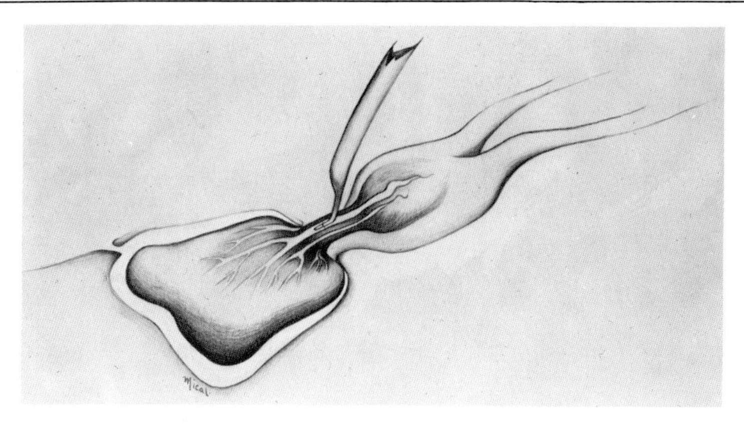

FIG. 11. Drawing indicating the approximate position of the cannula in a portal vessel. Reproduced from J. C. Porter, R. S. Mical, I. A. Kamberi, and Y. R. Grazia, *Endocrinology* **87,** 197 (1970).

A Ziegler needle knife is useful for this purpose. When making the incision, one first cuts through the dura mater. The arachnoid cannot actually be seen, but its location is easily ascertained by the fact that after transection of the dura, cerebrospinal fluid appears to bulge through the incision but does not run out the opening. A second incision opens the arachnoid, allowing the cerebrospinal fluid to escape and the dura to collapse onto the surface of the brain. Lift the meninges using an O'Brien spud, and enlarge the incision in the meningeal membranes in such a way as to expose completely the median eminence and pituitary stalk to the level of the anterior margin of the pituitary. After exposing the portal vessels, either cover the brain with an isotonic salt solution or proceed immediately to cannulate a vessel. As long as the vessels are moist, it is easy to insert a glass cannula into a portal vein. However, if the tissue dries, the task becomes difficult.

Microcannulas are made from capillary tubing having nominal dimensions of 1 mm o.d. and 100 mm length (Curtin Scientific Co., Catalog No. V48302-265033). The tubing is drawn to a fine tip by means of a micropipette puller. We have found the apparatus supplied by Industrial Science Associates, Inc. (Ridgewood, New York) to be suitable for this purpose. The fine tip of the pipette is bent about 60° by holding it above a hot wire. As the glass heats, its weight is sufficient to blend the tube. With minimal practice, one can make quite satisfactory cannulas. The final adjustment of the tip size is not made until the vessel to be cannulated has been selected. A typical cannula is illustrated in Fig. 12.

Because the micropipette is fragile and unwieldy, it is necessary to

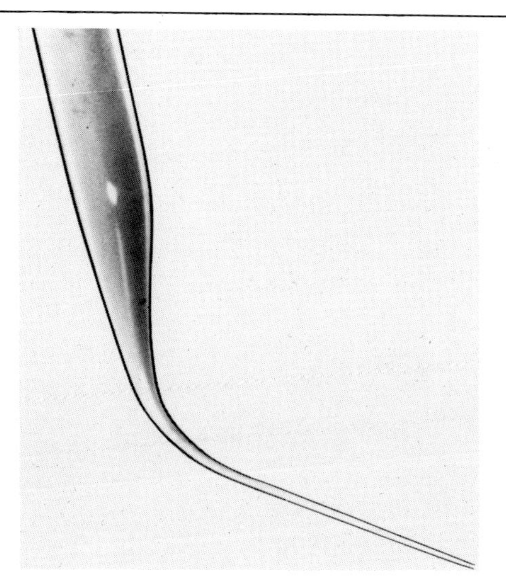

FIG. 12. Microcannula suitable for the cannulation of a pituitary stalk portal vessel. Reproduced from J. C. Porter, R. S. Mical, I. A. Kamberi, and Y. R. Grazia, *Endocrinology* **87,** 197 (1970).

have some means by which the pipette can be connected to a micromanipulator. In addition, small particles, when present in the solution which is passed through the cannula, will plug the tip. Therefore, a pipette holder was made to contain a filter to trap all but the smallest particles. A pipette holder is illustrated in Fig. 13. Such a holder can be made in most machine shops. Its essential features consist of a polyethylene tube, a, with a flared tip. The end of the polyethylene tube is held tightly against the end of a steel tube, c, by means of a screwcap, b. The opposite

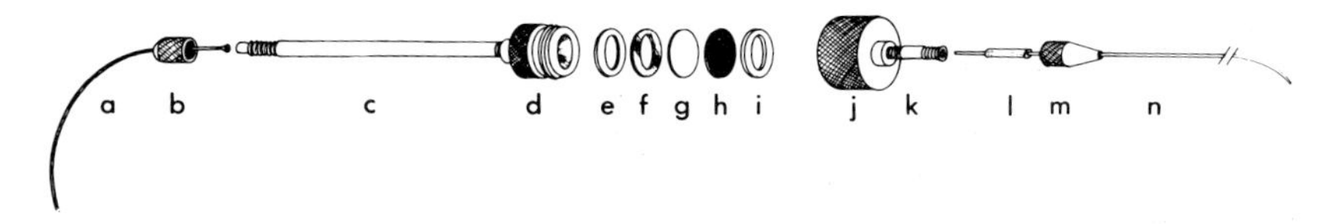

FIG. 13. Details of the cannula holder: a = polyethylene tubing (0.96 mm o.d.) with a flanged tip; b = female coupling (11 mm long, 8 mm in diameter) used to hold the polyethylene tube tightly to the male conduit c (75 mm long, 4.8 mm in diameter); d and j = male and female parts of a Swinny syringe filter holder (Millipore Corp., Bedford, Massachusetts); e, f, g, h, and i = Teflon gasket, steel ring, filter, steel grid, and Teflon gasket, respectively, for the Swinny holder; m = female coupling (7 mm long, 6 mm in diameter) used to attach the cannula n (Fig. 12) to the male conduit k (15 mm long, 4.8 mm in diameter); l = silicone rubber gasket (7 mm long, 3 mm in diameter). Reproduced from J. C. Porter, R. S. Mical, I. A. Kamberi, and Y. R. Grazia, *Endocrinology* **87,** 197 (1970).

end of tube c is joined to a stainless steel Swinny syringe filter holder, d–j (Millipore Corp., Bedford, Massachusetts). Parts d and j are the male and female parts of the filter holder, whereas e, f, g, h, and i represent a Teflon gasket, steel ring, filter, steel grid, and Teflon gasket, respectively. Part k is another steel conduit, and l is a silicone rubber gasket (7 mm in length, 3 mm o.d.). Part m is a female coupling (7 mm in length, 6 mm in diameter) used to attach the glass cannula n to conduit k.

The cannula is attached by means of its holder to a micromanipulator. We have found the Leitz Wetzlar micromanipulator to be highly satisfactory. The complete setup is shown in Fig. 14. The polyethylene tube a is connected to a syringe containing the solution to be infused. The fluid is forced from the syringe by means of a suitable infusion pump.

Before use, each solution is passed through a succession of Millipore filters having nominal pore diameters of 8, 3, 0.8, and 0.3 μm. A filter having a pore size of 0.47 μm is used in the cannula holder to effect a final filtration immediately before the solution enters the cannula.

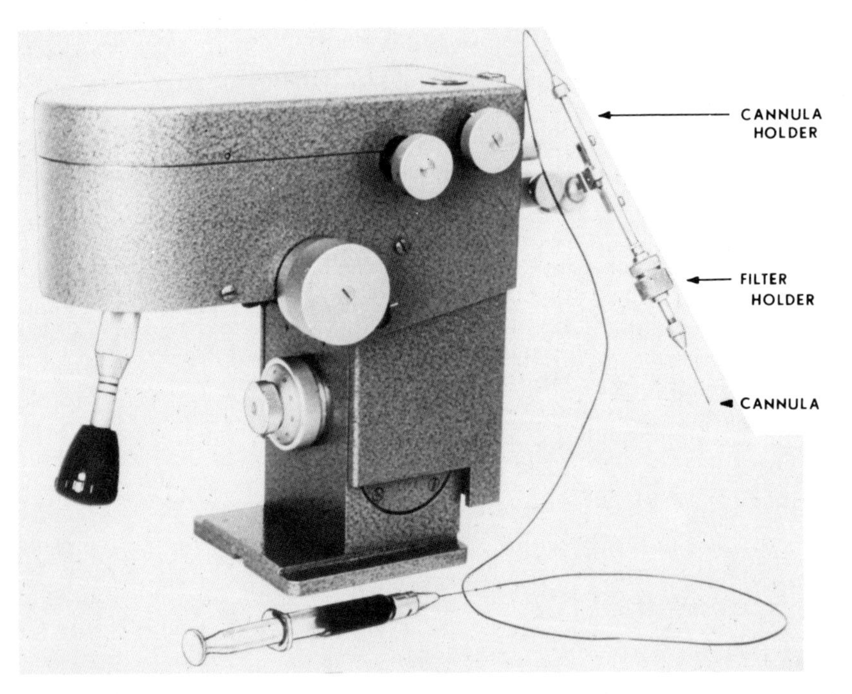

FIG. 14. Cannula holder including an inline filter. The cannula was attached to a Leitz Wetzlar micromanipulator, which was used to insert the cannula into a portal vessel. Reproduced from J. C. Porter, R. S. Mical, I. A. Kamberi, and Y. R. Grazia, *Endocrinology* **87**, 197 (1970).

The micromanipulator is placed on the left anterior corner of the operating board adjacent to the animal holder. The cannula holder is affixed to the manipulator, and the microcannula is lowered to the level of the pituitary stalk. As portal vessels in the rat tend to vary in size even within a single animal, one does not adjust the size of the cannula tip until the vessel to be cannulated has been selected. Place the microcannula close to the selected portal vessel, and, using fine scissors, cut the tip of the cannula so that the outside diameter of the tip is slightly smaller than the diameter of the vessel. The final outside diameter of most cannula tips after cutting is usually 20–40 μm. As a result of cutting, the tip of the cannula is uneven and has sharply jagged points. These sharp points are desirable and make it easier to cannulate a vessel. Now, with the microscope on a suitable magnification, e.g., 40×, place the tip of the cannula on top of a vessel near the junction of the stalk and brain. Lower the cannula so as to indent the vessel. Then, by means of the universal control of the micromanipulator, pull the tip of the cannula into the lumen of the vessel. It is sometimes useful to have fluid flowing from the cannula during the cannulation process. Thus, when the cannula enters the vessel, one sees immediately the flow of fluid in the vessel toward the anterior pituitary. We routinely infuse at 2 μl per minute. On the basis of the blood flow of the anterior pituitary (Porter *et al.*[2]), one can estimate that 1–2 μl per minute is the probable flow in a single vessel. We usually continue an infusion for 30 minutes.

[2] J. C. Porter, M. F. M. Hines, K. R. Smith, R. L. Repass, and A. J. K. Smith, *Endocrinology* **80**, 583 (1967).

[18] Methods for Assessing Hormone-Mediated Differentiation of Ovarian Cells in Culture and in Short-Term Incubations[1]

By Cornelia P. Channing and Florence Ledwitz-Rigby

In order to examine control of hormone-mediated differentiation of ovarian cells a suitable model system is required. Granulosa cells have been chosen since they are a homogeneous cell population which are easy to isolate from a variety of mammalian ovaries and can differentiate into luteal cells in culture. Furthermore, they are already in suspension

[1] Supported by research grants from the National Institutes of Child Health and Human Development, USPHS (HD-03315), and from the Population Council of New York (M72-024C).

and do not require trypsinization or other enzymatic treatment to disaggregate them. Granulosa cells are located within the ovarian follicle and are separated from the blood supply by a basement membrane as shown diagrammatically in Fig. 1. Normally granulosa cells do not luteinize until after ovulation. Luteinization can be hastened, however, if cells harvested from a large preovulatory follicle are cultured. Luteinization is defined as the morphological transformation from a small cell, with a low cytoplasmic–nuclear ratio and which secretes a small amount of progestins, to a large cell containing a large number of lipid droplets and granules which secretes elevated amounts of progesterone (Fig. 2). The morphological aspects of luteinization in the rhesus monkey have been defined at the ultrastructural level both *in situ* and in culture.[1a] Descriptions of the steroidogenic events which occur during luteinization of granulosa cells of the mare,[2-5] rhesus monkey,[6,7] human,[8] and pig[9-12] have been made.

Luteinization of the granulosa cell clearly is an example of hormonally controlled differentiation. Evidence for this is obtained in the observation that cells obtained from large growing follicles of the rhesus monkey during the follicular phase luteinize in culture only if they are harvested during or immediately after the preovulatory LH surge.[13] Furthermore, one can take granulosa cells from medium or small follicles, which do not luteinize spontaneously in culture, and get them to luteinize by addition of LH or FSH or a mixture of LH and FSH (monkey,[6,7] pig,[10] mare[5]). The hormonal control of luteinization is mediated by cyclic 3′,5′-AMP which can cause luteinization of monkey,[7] pig[14] and bovine[15] granu-

[1a] T. M. Crisp and C. P. Channing, *Biol. Reprod.* **7,** 55 (1972).
[2] C. P. Channing, *J. Endocrinol.* **43,** 381 (1969).
[3] C. P. Channing and S. A. Grieves, *J. Endocrinol.* **43,** 391 (1969).
[4] C. P. Channing, *J. Endocrinol.* **43,** 403 (1969).
[5] C. P. Channing, *J. Endocrinol.* **43,** 415 (1969).
[6] C. P. Channing, *Endocrinology* **87,** 49 (1970).
[7] C. P. Channing, *Recent Progr. Horm. Res.* **26,** 589 (1970).
[8] C. P. Channing, *J. Endocrinol.* **45,** 297 (1969).
[9] D. W. Schomberg, *in* "The Gonads" (K. W. McKerns, ed.), p. 383. Appleton, New York, 1969.
[10] C. P. Channing, *Endocrinology* **87,** 156 (1970).
[11] D. H. Van Thiel, W. E. Bridson, and P. O. Kohler, *Endocrinology* **89,** 622 (1971).
[12] Bjersing, L., *Acta Pathol. Microbiol. Scand.* **55,** 127 (1962).
[13] C. P. Channing, *in* "Regulation of Mammalian Reproduction" (S. J. Segal, R. Crozier, P. A. Corfman, and P. G. Condliffe, ed.), p. 505. Thomas, Springfield, Illinois, 1972.
[14] C. P. Channing and J. F. Seymour, *Endocrinology* **87,** 165 (1970).
[15] V. J. Cirillo, O. F. Andersen, C. A. Ham, and R. B. Gwatkin, *Exp. Cell. Res.* **57,** 139 (1969).

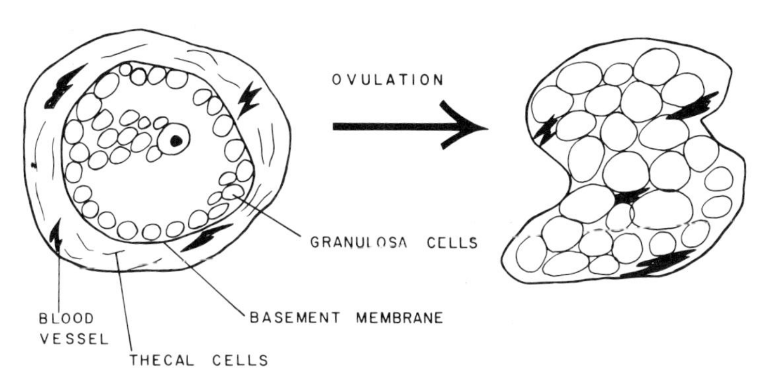

FIG. 1. Transformation of a follicle to a corpus luteum at ovulation. The process is controlled by pituitary gonadotropins, LH and FSH. Notice that the granulosa cells in the follicle (left) are separated from the blood supply by a basement membrane, whereas in the corpus luteum (right) the cells are intimately associated with blood vessels.

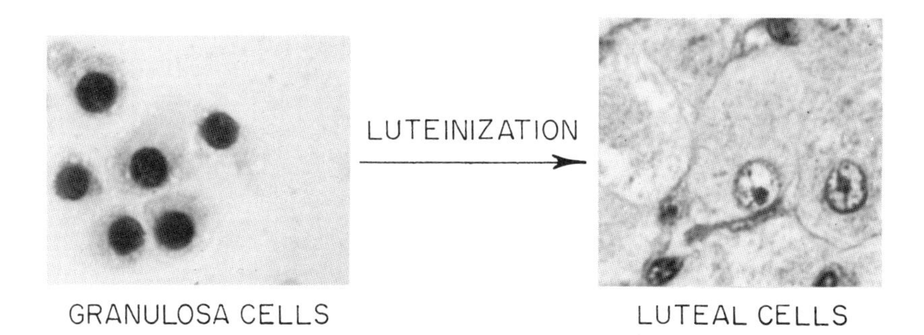

FIG. 2. Appearance of granulosa cells (left) and luteal cells (right) of the mare. These cells resemble those of most other species. In most instances the granulosa cells form most of the corpus luteum. Hematoxylin and eosin, ×800.

losa cells. Furthermore LH and FSH can stimulate cyclic 3′,5′-AMP formation from adenine[16] and *de novo* from endogenous precursors[17] in porcine granulosa cells.

In the maturation of the granulosa cell the differentiation process leading up to luteinization can be divided conveniently into three arbitrary stages according to the size of the follicle *in situ* from which the

[16] J. Kolena and C. P. Channing, *Biochim. Biophys. Acta* **252**, 601 (1971).
[17] J. Kolena and C. P. Channing, *Endocrinology* **90**, 1543 (1972).

TABLE I
FOLLICLE DIAMETERS OF VARIOUS SPECIES

Follicle category	Size of follicle (mm)			
	Monkey	Pig	Mare	Human
Small	1	1–2	<10	1
Medium	2–3	3–5	10–30	2–3
Large	3–10	6–12	30–60	3–12

granulosa cells are obtained. The stage of nonatretic follicular growth for a given species is very easy to estimate; one merely measures the diameter of the follicle one wants to use; in most species the follicles are near the surface of the ovary and can be measured with a ruler. The three stages of differentiation leading up to granulosa cell luteinization occur at specific times *in situ* but also can be studied conveniently *in vitro* in a controlled hormonal environment.

Stage 1. This stage occurs when a follicle grows from a small follicle to a medium-sized follicle under the influence of low levels of LH and FSH. Granulosa cells obtained from small follicles are completely undifferentiated; they do not luteinize in culture[7]; they contain very few LH-hCG receptors[18,19]; they cannot respond to LH with an elevation in cyclic 3′,5′-AMP levels[20]; and they cannot respond to LH or FSH alone with resultant luteinization in culture.[6,7] Exogenous dibutyryl cyclic 3′,5′-AMP can, however, induce luteinization of such cells in culture.[7] "Small" follicles differ in diameter according to the species used. The description is a functional rather than a literal definition of a follicle size. A summary of small and other follicle sizes is found in Table I. Robertson has described follicular growth in relation to the stage of the estrous cycle for the pig, cow, and other domestic animals.[21]

Medium follicle cells come from more mature follicles and are slightly more differentiated. The stage of differentiation is not evident morphologically *in situ,* and these cells still fail to luteinize spontaneously when cultured. They will, however, luteinize in culture in the presence of LH or FSH *alone*.[6,7,10] Furthermore, addition of LH or FSH to such cells exerts some stimulatory influence upon cyclic 3′,5′-AMP levels.[17,20]

[18] S. Kammerman, R. E. Canfield, J. Kolena, and C. P. Channing, *Endocrinology* **91,** 65 (1972).

[19] C. P. Channing and S. Kammerman, *Endocrinology* **92,** 531 (1973).

[20] C. P. Channing, *Symp. Int. Congr. Endocrinol. 4th 1972, Excerpta Med. Found. Int. Congr. Ser.* **273,** 914 (1973).

[21] H. A. Robertson, *Vitam. Horm.* **27,** 91 (1969).

Stage 2. This stage occurs when a follicle grows from a medium- to a large-sized preovulatory follicle under the influence of a surge of LH and FSH. Morphologically, granulosa cells obtained from a large preovulatory follicle are small and do not differ markedly from those obtained from smaller follicles. In contrast, functionally they have been primed to luteinize fully since if they are removed from the follicle and cultured they will luteinize spontaneously. They are differentiated, but the expression of the differentiation is held in check within the follicle because of some inhibitor present in follicular fluid. Characteristics of such cells are that they have manyfold more LH-hCG receptors compared to small or medium follicle cells.[19] Furthermore they respond to LH with a greater elevation in cyclic 3′,5′-AMP levels compared to cells obtained from smaller follicles,[20] and when incubated for a 20-minute period they have higher levels of cyclic 3′,5′-AMP than smaller follicle cells.[20]

Stage 3. This stage occurs when a large follicle ovulates and luteinization itself occurs. The cells increase in size from 12–15 μm to 35–40 μm and undergo extensive ultrastructural changes,[1] which initially include an accumulation of lipid droplets, a proliferation of the Golgi complex and agranular endoplasmic reticulum, and the development of fine filaments. Mitochondrial changes are striking, with the transformation of lamelliform (platelike) cristae to velliform (tubular) cristae and an increase in density of the mitochondrial matrix. During the final luteinization process, the cells secrete elevated amounts of progestins, primarily progesterone.

Each stage of the differentiation process can be examined for endocrine regulatory mechanisms by taking the cells which are at the start of the given stage (i.e., small follicle cells would be used to examine stage 1 of differentiation) and examining the process either in long-term cultures or in short-term *in vitro* incubations. Both stimulation of luteinization and inhibition of any stage of the luteinization process can be examined. Schomberg[22] has demonstrated an inhibitory action of a uterine luteolytic agent in porcine granulosa cell cultures. Maintenance of the luteinized state once it has been achieved, *a stage 4*, should also be considered and has been included in the culture studies by Kohler and his colleagues[23] and by Channing.[24]

The choice of long-term cultures or short-term incubations for these studies will depend upon what is to be examined. Usually a long-term culture will be used for any process which is to be examined over a 2-,

[22] D. W. Schomberg, *J. Endocrinol.* **38**, 359 (1967).

[23] R. L. Goldenberg, W. E. Bridson, and P. O. Kohler, *Biochem. Biophys. Res. Cummun.* **48**, 101 (1972).

[24] C. P. Channing, *Endocrinology* **94**, 1215 (1974).

4-, 6-, to 20-day period, and a short-term incubation will be used when close timing of the reaction to be examined is required, i.e., 5 minutes up to 48 hours. Since a monolayer culture takes about 2 days to get started one cannot study short-term changes in it which occur in less than about 2 days. Both monolayer culture and short-term incubation methods are described in detail below.

Emphasis will be made in describing procedures using porcine cells since they are more readily available in most laboratories, as large quantities of pig ovaries can be obtained from a local slaughterhouse. Furthermore the *in vivo* status of the ovarian material of the pig can be reasonably ascertained from gross examination of the ovary itself. A reasonable knowledge of the *in vivo* hormonal status of the animal before granulosa harvest is critical in studies of this sort and should be assessed whenever possible. Details of methodological description outlined below will include (1) methods for granulosa cell harvest in the pig, monkey, human, and rat; (2) methods for tissue culture of granulosa cells including a critical review of media available for this; (3) methods for short-term incubations of granulosa cells; and (4) methods for assessing luteinization in granulosa cell cultures and suspensions which include morphology, steroidogenesis, cyclic 3′,5′-AMP levels, and binding of cells to iodinated hCG. Methods to stimulate and inhibit differentiation will be included followed by a brief resume of methods for assessing hormonally mediated differentiation in other ovarian cell preparations.

Isolation of Granulosa Cells from Ovarian Follicles

Pig

Ovaries are obtained from 4- to 6-month-old or older nonpregnant pigs at a local slaughterhouse within about 20 minutes of death. The ovaries are dissected free from the uterus and most of the fallopian tube after the pig has been in the scalding tank and has been slit, and prior to or after removal of the viscera. The ovaries are placed on ice in 0.9% NaCl containing (per milliliter) 100 IU of penicillin, 100 μg of streptomycin, and 2.5–25 μg of fungizone. One to 6 hours can be allowed to elapse between collection of the ovaries and later granulosa cell harvest, provided the ovaries are kept on ice during this time. Three types of follicle sizes can be used as a source of granulosa cells: small (1–2 mm); medium (3–5 mm), and large (6–12 mm) depending upon what stage of differentiation and maintenance is to be investigated (stages 1–4). The appearance of pig ovaries at various stages of follicular development is shown in Fig. 3. A detailed

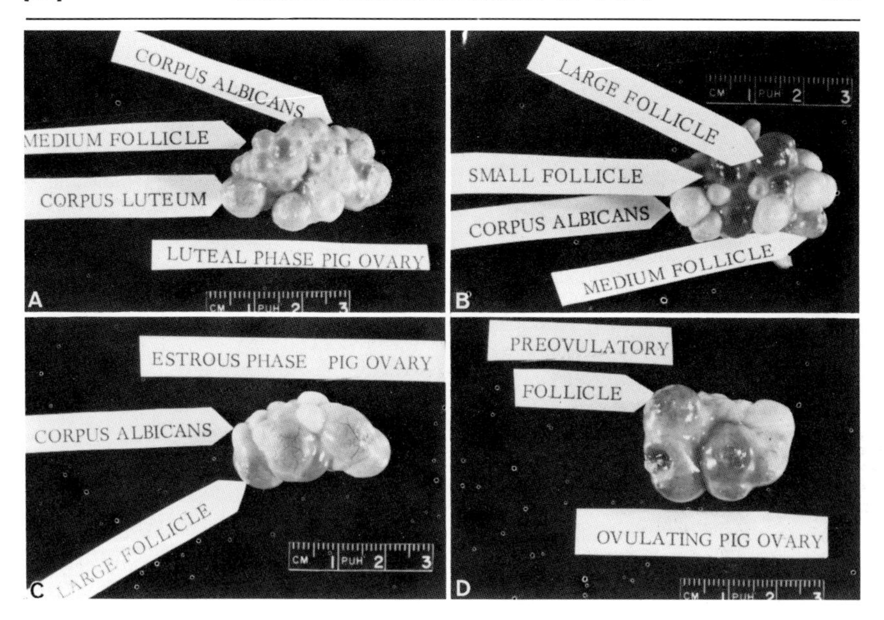

FIG. 3. Appearance of porcine ovaries at different stages of the estrous cycle. (A) Midluteal phase pig ovary showing pink corpus luteum, regressing corpora albicantia of the previous estrous cycle, and medium-sized follicles (3–5 mm). (B) Late luteal phase pig ovary showing large follicles, 7 mm in diameter, regressing corpora lutea (now corpora albicantia), and small 1–2 mm follicles. (C) Estrous phase pig ovary showing large vesicular follicles and corpora albicantia of the previous cycle. (D) Ovulating pig ovary showing 2 ovulated follicles and 1 unovulated, preovulatory follicle.

colored pictorial description of pig ovaries obtained at various stages of the estrous cycle has been made[25] and may be referred to. Small 1–2 mm follicles are present in all types of ovaries throughout the estrous cycle, and granulosa cells obtained from small follicles in all types of ovaries behave indistinguishably and may be pooled together. The first day of estrus is defined as day 1 of the pig estrous cycle. After ovulation, which occurs 2–3 days later, a corpus luteum is formed which is pink and persists throughout most of the 21-day estrous cycle (Fig. 3A). Medium follicles are present from days 3–15. After day 15 or so the corpus luteum regresses; at the same time medium follicles grow into large ones. As the corpora lutea further regress they become white (corpus albicans) and large follicles can be found (Fig. 3B). Follicles greater than 5 mm are found from

[25] E. L. Akins and M. C. Morrisett, *Amer. J. Vet. Res.* **29**, 1953 (1968).

day 15 of the cycle until ovulation. Rapid growth occurs between day 15 and day 18 going from a diameter of 5 mm to 7.8 mm. By the onset of estrus at day 21, follicles are 8.2 mm, and by the time ovulation is reached they are 9 mm in diameter, according to Robertson.[21] The large follicles mature and secrete estrogen to initiate another estrous cycle. An ovary containing corpora albicantia in the presence of large follicles (6–12 mm in diameter) (Fig. 3C) is most likely to come from a pig in estrus or 2 days before. It is also possible to have large follicles present without adjacent corpora albicantia if that pig is in its first estrous cycle. More consistent results can be obtained from large-follicle granulosa cells if they are obtained from large follicles that are in the presence of adjacent corpora albicantia. Large follicles that are not adjacent to corpora albicantia may also be used as a source of large follicle granulosa cells, but, for reasons not clear to us, these cells have consistently fewer binding sites and a lower responsiveness to LH in terms of cyclic 3′,5′-AMP levels compared to cells obtained from animals containing large follicles adjacent to corpora albicantia. If cells obtained from both types of large follicles are used, they should be kept separate. If, for a specified experiment, an inadequate number of cells are obtained from either type of follicle the cells can be pooled and a notation made of how many of the ovaries contained corpora albicantia and how many did not. Such a notation has been very useful in later assessment of data. If greater than 40% of the ovaries in a given batch used as a source of large follicles have corpora albicantia, a good responsiveness of the cells in terms of LH stimulation of cyclic 3′,5′-AMP can be expected. Follicles greater than 12 mm should not be used, as they most likely are cystic. The method for harvesting of granulosa cells outlined below depends upon the size of follicle used as a source of granulosa cells.

Medium and Small Follicle Cells. Cells from these follicles are harvested by aspiration of the cells with a 20–23 gauge 1-inch needle and 5-ml syringe. Prior to cell harvest the ovaries are rinsed in two changes of sterile 0.9% NaCl. The ovaries are transferred to a disposable petri dish one at a time and grasped with forceps while the cells are aspirated. Small and medium follicle cells are kept separate, and the aspirate in the syringe which contains cells plus follicular fluid and ova is periodically transferred to a sterile graduated 12-ml empty centrifuge tube. The number of follicles used is counted with a hand tally. The cells are collected at room temperature, and as soon as a centrifuge tube is filled with cells and follicular fluid it is refrigerated. The cells are separated from the follicular fluid by centrifugation at 600 *g* for 5 minutes. The packed cell volume is estimated using the graduation in the graduated centrifuge tube and the follicular fluid (about 10–12 ml) is decanted and recentrifuged at 2000 *g* for 20 minutes to remove cellular debris and frozen for later use. The initial cell pellet is resuspended in 10 ml of medium Y consisting of 15% lamb or pig serum plus tissue culture medium 199 containing Earle's salts and 25 m*M* HEPES buffer plus (per milliliter) 50 U of penicillin, 50 μg of streptomycin, 2.5 μg of fungizone, and also 2 m*M* L-glutamine added immediately prior to use. Fifty micrograms per milliliter of gentamycin may be used instead of the other antibiotics. Tissue culture medium 199 and other culture reagents, sera, and antibiotics may be purchased from either Grand Island Biological Co., Grand Island, New York, Flow Laboratories, Rockville, Maryland, or other reputable suppliers. After resuspension by gentle agitation in 10 ml of medium Y, the cells are recentrifuged for 5 minutes at 600 g followed by removal of the supernatant medium and resuspension of the cells in an additional 10 ml of medium Y. The tube containing the cell suspension is stoppered with a rubber stopper and refrigerated until used. The cells may be used immediately thereafter or a period up to 24 hours may elapse before use in either culture or short-term incubations in which binding to gonadotropins or responsiveness of the cells to gonadotropins in terms of cyclic 3′,5′-AMP levels is estimated, as outlined below.

Large Follicle Cells. Since granulosa cells residing in large follicles adhere to the follicular wall they must be gently scraped out from the follicle. Simple aspiration with a needle and syringe does not adequately remove cells from such follicles. The procedure for harvesting follicular fluid and cells from large follicles has therefore been modified from that for small and medium follicles as follows. An ovary with large follicles is chosen and the follicular fluid and some loose granulosa cells are aspirated with a needle and syringe and transferred to a 12-ml graduated centrifuge tube. The remaining "deflated" follicle is then slit in a V formation with fine scissors and the granulosa cells within removed by gently rinsing of the inner wall of the follicle with a Pasteur pipette filled with medium Y. The apex of the V incision in the follicle is held up with fine forceps while the inside of the follicle is rinsed 3–5 times with medium Y. The rinsings are pooled in 10 ml of medium Y in a 12-ml graduated centrifuge tube. The cells from about 10 ovaries (about 50 follicles) are pooled in the same 10 ml of medium Y. Subsequently the cell suspensions in medium Y and in follicular fluid are centrifuged at 600 *g*, and the packed cell volume is estimated for the cells suspended in the medium Y and in the follicular fluid. The number of large follicles used and the number of ovaries containing corpora albicantia are counted. After centrifugation the follicular fluid is decanted into another 12-ml centrifuge tube, centrifuged at 2000 *g* for 20 minutes, decanted and frozen for later use. The packed cells obtained from both the cell suspensions in medium Y and those separated from the follicular fluid are pooled and resuspended in an additional 10 ml of medium Y and recentrifuged for 5 minutes at 600 *g*. The supernatant fluid is discarded, and the cells are resuspended in 10 ml of medium Y and refrigerated until used. Large follicle cells (cells from large follicles) must be washed at least twice before use to remove any traces of follicular fluid, which contains large amounts of estrogens and progestins that may interfere with subsequent studies.

Cell Counting. After thorough mixing of the cell suspension with a Pasteur pipette, a small sample (0.1 ml out of 10) is transferred to a small test tube (12×75 mm) and mixed with 0.9 ml of medium Y (1:10 dilution). Depending upon the packed cell volume of the original cell suspension, a sample of the 1:10 diluted cell suspension is further diluted in 0.2% lissamine green (naphthalene green) in Hanks' solution. In most instances a 1:10 dilution of the diluted cell suspension in the dye solution suffices. The cells are mixed thoroughly with the dye solution in a small tube (12×75 mm), and a drop of cell suspension is added to a hemacytometer. The cells in the four corner areas are enumerated (one area measures $1 \times 1 \times 0.1$ mm^3). The dilution of the initial cell suspension in medium Y, and dye solution should be made so that 20–100 cells can be counted in one corner area, measuring 1 mm^2, of the hemacytometer. Since granulosa cells are similar in size to white blood cells (12–16 μm), similar principles can be used for counting both types of cells. Adequate mixing of the cells is important to avoid clumping of the cells in the hemacytometer. Only cells that have clearly defined nuclei and cytoplasm are counted. Granulosa cells may also be counted in a Coulter counter. Cells may also be quantitated by a Lowry protein determination,[26,27] but cellular debris is also included which limits the reliability of a protein determination as an estimate of cell mass. There are about 10^7 cells per milligram of protein. Figure 4 illustrates the typical appearance of granulosa cells harvested from small, medium, and large pig follicles.

The yield of granulosa cells will depend on how many ovaries are available and how many cells are needed for an experiment. From 1 to 5×10^6 granulosa cells are required to inoculate 1 Leighton tube culture, and 1 to 10×10^7 cells are required for each vial of a short-term incubation in 1 ml of medium in which responsiveness of the cells to LH in terms of cyclic 3′,5′-AMP levels are measured. From 5×10^6 to 2×10^7 cells are required per incubation tube for studies involving binding of

[26] O. H. Lowry, N. J. Rosebrough, A. L. Farr, and R. J. Randall, *J. Biol. Chem.* **193,** 265 (1951).

[27] V. I. Oyama and H. Eagle, *Proc. Soc. Exp. Biol. Med.* **91,** 305 (1956).

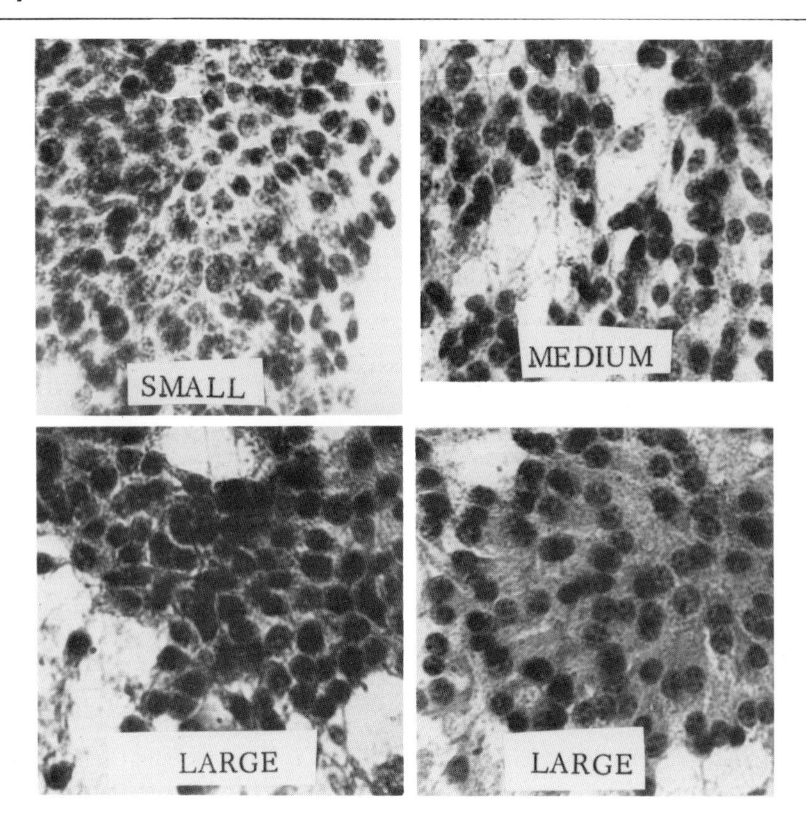

FIG. 4. Appearance of hematoxylin and eosin stained smears of granulosa cells harvested from small (1–2 mm), medium (3–5 mm), and large (6–12 mm) porcine follicles. $\times 275$.

the cells to iodinated hCG. In our experience a day's collection at the slaughterhouse yields 100–250 ovaries. In 19 separate experiments the average number of small follicles harvested from was 3045 ± 217 (mean $\pm$SE), which yielded $99 \pm 16 \times 10^7$ granulosa cells with a packed cell volume averaging 1.1 ± 0.11 ml. The average number of cells per small follicle (averaged out in each experiment) was $3.3 \pm 0.5 \times 10^5$. An average of 657 ± 46 medium follicles were obtained yielding $(16 \pm 2.2) \times 10^7$ cells with a packed cell volume of 0.21 ± 0.02 ml. The cells per medium follicle obtained were $(7.2 \pm 4.3) \times 10^5$. An average of 27 ovaries containing about 5 large follicles each yielded $(27 \pm 4) \times 10^5$ cells per follicle and a total packed cell volume of 0.53 ± 0.10 ml. These data are summarized along with the calculations of the follicles per packed cell volume and the cells per packed cell volume in Table II.

TABLE II
RELATION BETWEEN FOLLICLE SIZE, FOLLICLE NUMBER, CELL NUMBER, AND PACKED CELL VOLUME FOR PORCINE GRANULOSA CELLS

Parameter	Follicles per 1.0 ml packed cell volume	Cells per follicle (× 10^5)	Cells per 1.0 ml packed cell volume (× 10^7)	Total No. follicles	Total cell No. (× 10^7)	Total packed cell volume (ml)
Small follicles						
Average	3021	3.3	109	3045	99	1.1
SE	143	0.5	23	217	16	0.11
n	19	19	19	21	19	20
Medium follicles						
Average	3775	7.2	112	657	16	0.21
SE	485	4.3	22	46	2.2	0.02
n	19	18	18	20	19	19
Large follicles						
Average	285	27	68.9	—	26.9	0.53
SE	27	4	10.6	—	4.1	0.10
n	17	17	17	—	20	18

Parameter	Number of ovaries (total)	Number of ovaries with corpora albicantia
Average	27	11
SE	3.9	2.1
n	17	17

Such a table has a practical value in assessing the number of follicles required to yield an adequate number of cells for a given series of experiments.

Rhesus Monkey

Methods for harvesting granulosa cells from *normally cycling rhesus monkey follicles* have been published in detail previously[6] and will be summarized here, annotated with recent modifications. The ovary is exposed at laparotomy, and the ovarian arterial blood supply is clamped off with a hemostat. Ovaries containing 3–11-mm follicles of monkeys from day 7 to day 15 of the menstrual cycle are used (Fig. 5 shows the typical appearance of a monkey ovary *in situ* containing a medium-sized follicle). The size of the follicle and the state of maturation of the granu-

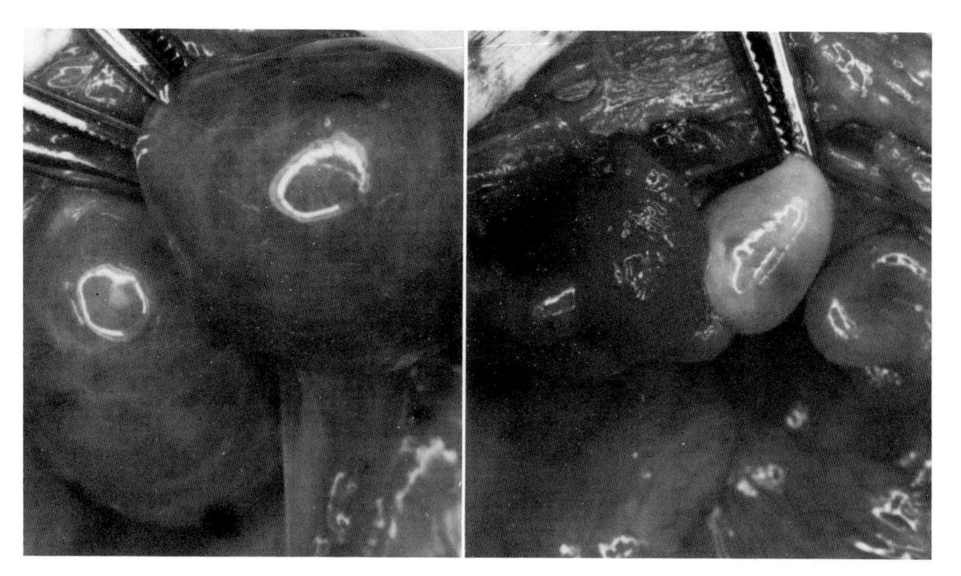

FIG. 5. Appearance of a normal monkey ovary containing a growing medium-sized follicle (right) and a pair of ovaries in a monkey who was stimulated with pregnant mare's serum gonadotropin. About ×3.

losa cells contained in the follicle will depend upon what stage of the menstrual cycle the monkey is at (see footnote 6 and below). The follicle wall of the Graafian follicle is slit, and the contents of the follicle, including granulosa cells, oocyte, and follicular fluid, are aspirated with a Pasteur pipette. Subsequently the follicle is rinsed with 3–4 washes of about 0.2 ml of a suitable culture medium, such as medium W containing 80% medium 199, and 20% normal male monkey serum plus 50 IU of penicillin per milliliter and 50 μg of streptomycin per milliliter. One drop of heparin (1000 IU/ml) is added per milliliter of washing medium to prevent clotting if blood is inadvertently mixed with the granulosa cells. Red blood cells do not appear to interfere with the cultures since most of them are poured off along with the first medium change after 2 days in culture. The washings and original aspirate are pooled in a small vial and kept on ice until cultured 1–12 hours later.

Methods for Evaluation of the in Vivo Hormonal Status of the Rhesus Monkey Prior to Granulosa Cell Harvest. Since the original description of monkey granulosa cell harvest was written,[6] better methods than measuring peripheral plasma progestins have been devised to determine the endocrine events leading up to and including ovulation. Perhaps the best method is to measure blood LH levels as outlined by Knobil and his associates.[28,29] The radioimmunoassay for monkey LH is quite laborious to set up and requires a good antiserum which may not be easy to obtain. Hotchkiss and her colleagues made the fortuitous observation that blood estrogen levels measured by radioimmunoassay in the monkey closely overlap LH levels, with peak levels of the two hormones occurring immediately prior to ovulation.[30] In many instances the peak of the two hormones occurs on the same day. Therefore, the radioimmunoassay for estrogen may be used as a rough indicator of blood LH levels. This method has an advantage in the fact that it may be performed rapidly, within half a day, and therefore may be used for a day to day assessment of follicular development throughout the follicular phase of the menstrual cycle since the growing follicle secretes most of the estrogen. This is what is used currently in our laboratory; 200 μl of plasma are extracted with 2.5 ml ether, and the ether extract is assayed for total estrogen.[30] Monkeys are sampled for blood daily from day 7 of a normal menstrual cycle until a laparotomy is performed. If a medium follicle is desired, one which yields granulosa cells which do not luteinize spontaneously in culture, the cells are harvested before the estrogen levels start to surge (i.e., before they go above about 100–150 pg/ml, depending upon what the base-line level of estrogen is for that animal at day 7). If granulosa cells from a large preovulatory follicle are desired, the animal is laparotomized on the first day after the estrogen starts markedly to go above the baseline, i.e., from about 100–150 to about 200–300 pg/ml or more (Table III). In most monkeys with a normal 28 day cycle this increase occurs between day 11 and 13 of the cycle but occasionally it may occur earlier, sometimes as early as day 9, or it may occur later. Serum samples from such sequential samplings can be saved for LH analysis at a later date.

The peripheral plasma progestin levels, used earlier as an indicator of the preovulatory events in the monkey as outlined by Johansson *et al.*,[31] were useful but suffered the disadvantage that if a monkey was stressed the plasma progestin values would fluctuate markedly, making the levels difficult to interpret. In our experience animals new to the colony, as were many of our animals, are more easily stressed than those used by the Johansson, Neill, and Knobil group, thus making the plasma progestin estimations unsatisfactory as indicators of when ovulation would occur. In the past, before use as a source of granulosa cells, mon-

[28] S. E. Monroe, W. D. Peckham, J. D. Neill, and E. Knobil, *Endocrinology* **86**, 1012 (1970).

[29] S. E. Monroe, L. E. Atkinson, and E. Knobil, *Endocrinology* **87**, 453 (1970).

[30] J. Hotchkiss, L. E. Atkinson, and E. Knobil, *Endocrinology* **89**, 177 (1971).

[31] E. D. B. Johansson, J. D. Neill, and E. Knobil, *Endocrinology* **82**, 143 (1968).

TABLE III
PERIPHERAL PLASMA LEVELS OF PROGESTINS, PROGESTERONE, ESTROGEN, AND LH IN A NORMALLY CYCLING RHESUS MONKEY PRIOR TO LAPAROTOMY AND GRANULOSA CELL HARVEST (MONKEY 737)[a]

Menses (day)	Progestins (ng/ml)[b]	Progesterone (ng/ml)[c]	Estrogen (pg/ml)[d]	LH (ng/ml)[e]
7	<1	0.25	178	2.7
8	1.8	0.2	103	2.6
9	<1	0.27	74	2.4
10	<1	0.17	140	2.5
11	1.5	1.1	186	5.4
12 (AM sample)	8.4	0.69	318	20.2
12 (PM sample)	27.1	2.4	220	—
Left ovarian vein	43.3	7.4	3665	—

[a] This monkey previously had a 27-day menstrual cycle. The laparotomy was started at 12:30 PM on day 12 of the menstrual cycle and the ovarian vein and peripheral plasma samples were taken at 1:00 PM (designated PM sample). The left ovary contained a 7-mm follicle. Granulosa cells harvested from that follicle luteinized spontaneously in culture.

[b] Progestins were determined by a protein binding assay [E. D. B. Johansson, J. D. Neill, and E. Knobil, *Endocrinology* **82,** 143 (1968)].

[c] Progesterone was determined by radioimmunoassay without chromatography [I. H. Thorneycroft and S. C. Stone, *Contraception* **5,** 129 (1972)].

[d] Estrogen was determined by radioimmunoassay [J. Hotchkiss, L. E. Atkinson, and E. Knobil, *Endocrinology* **89,** 177 (1971)].

[e] LH assays were done in the laboratory of Dr. Ernst Knobil using a radioimmunoassay method [S. E. Monroe, W. D. Peckham, J. D. Neill, and E. Knobil, *Endocrinology* **86,** 1012 (1970); and S. E. Monroe, L. E. Atkinson, and E. Knobil, *Endocrinology* **87,** 453 (1970)].

keys were routinely sampled for one complete menstrual cycle and the samples were analyzed for progestin to determine whether an animal was ovulatory and when ovulation occurred. This is still done except that the plasma is assayed for progesterone by radioimmunoassay[32] rather than by the protein-binding assay. The antiserum kindly donated by Dr. Ian Thorneycroft is highly specific for progesterone and, in our laboratory, it cross-reacts less than 5% with other known progestins. This is advantageous since it becomes unnecessary to chromatograph the crude ether extract of the plasma. If other biological fluids such as follicular fluid are analyzed, it is necessary to chromatograph them since they contain large amounts of progestins other than progesterone. For the progesterone radioimmunoassay, 50 μl of plasma are mixed with 200 μl of water and the mixture extracted with 2.5 ml of ether followed by radioimmunoassay

[32] I. H. Thorneycroft and S. C. Stone, *Contraception* **5,** 129 (1972).

as outlined elsewhere.[32] In our experience slightly better results are obtained when the ether extract is incubated with the antiserum overnight than after a shorter 1–2-hour incubation. Progesterone and progestin levels during a normal follicular phase of a monkey are shown in Table III. An example of a stress value for progestins is the value obtained at 1 PM on day 12 of the cycle, which is a half hour after initiation of the laparotomy. The progesterone value is reasonable, 2.4 ng/ml, whereas the progestins are 27.1 ng/ml.

As a means of increasing yields of granulosa cells in anovulatory, young, or irregularly cycling rhesus monkeys *follicular growth can be stimulated by exogenous human menopausal gonadotropin (HMG) or by pregnant mare's serum gonadotropin (PMSG)*. Methods of administration of the gonadotropins have been presented in detail elsewhere.[33,34] PMSG is the preferable means of treatment since it is available free of cost from the National Pituitary Agency and produces reliable stimulation of follicular growth. A subcutaneous (s.c.) dose of 1000 IU on day 1 and 2 followed by 500 IU on days 3, 4, and 5 followed by 250 IU on days 6, 7, and 8, followed by ovariectomy on day 10, produces a satisfactory response in most animals (Fig. 5). The multifollicular ovaries are removed surgically, and the granulosa cells are harvested by slitting the follicle walls and aspirating the granulosa cells and follicular fluid with a Pasteur pipette. The interior of each follicle is rinsed out with medium W until most of the granulosa cells are removed. Cell harvesting is best done under 10–40$\times$ dissecting microscope. A pair of ovaries such as those seen in Fig. 5 will yield about 1×10^9 cells with a packed cell volume of 0.1–0.2 ml. This is enough cells for about 100 Leighton tube cultures. These cells fail to luteinize spontaneously but will do so in the presence of physiological amounts of pituitary hormones in the culture medium.[24,34]

Rat

Rat granulosa cells have been successfully harvested by puncture and aspiration of follicles with a fine needle and syringe[35] under the dissecting microscope. Because of its small size the rat is not desirable for obtaining granulosa cells in large quantities. An increase in the yield of granulosa cells from rat follicles is achieved by s.c. administration of 5 IU PMSG to 26- to 28-day-old rats. Sixty-five hours later the ovaries contain numerous 1-mm follicles which yield about 250,000 granulosa cells per follicle

[33] C. P. Channing, *Prostaglandins* **2,** 331 (1972).

[34] C. P. Channing and S. Kammerman, *Endocrinology* **93,** 1035 (1973).

[35] W. C. Redmond, *Anat. Record* **166,** 366 (1970) (Abstract).

according to Dr. Thomas Crisp (personal communication). Each ovary is dissected free of adhering bursa and fat and rinsed several times in culture medium before harvesting granulosa cells. Each follicle is punctured with a needle, and with another needle the wall is depressed to force the granulosa cells and ovum out of the follicle. About 1×10^6 cells can be harvested from one ovary, which is enough for 1 Leighton tube culture. A suitable medium for culturing rat granulosa cells consists of 80% medium 199 plus 20% fetal calf serum containing per milliliter 50 IU of penicillin and 50 μg of streptomycin. According to Thomas Crisp (personal communication) granulosa cells harvested from PMSG-treated rats fail to luteinize spontaneously, but will luteinize in the presence of a pituitary explant or purified bovine pituitary LH. In this regard they resemble granulosa cells obtained from PMSG-treated monkeys.[34] Estrous-phase large-follicle rat granulosa cells luteinize spontaneously, as has been demonstrated histochemically[35,36]

Other Species

Ovine and bovine granulosa cells can be isolated with a needle and syringe in a similar manner to that used to harvest porcine granulosa cells.

Methods for Tissue Culture of Granulosa Cells

General Culture Procedures including Culture Inoculation and Quantitation of Cells

Regardless of the species of animal or follicle size used as a source for granulosa cells, tissue culture methods for examining hormonal control of luteinization *in vitro* are basically similar and will be discussed in general terms with digressions upon different species when warranted. Complete up to date descriptions of general tissue culture techniques have been prepared by others[37-40] and should be referred to for details and

[36] T. V. Fischer and R. H. Kahn, *In Vitro* 7, (1972).

[37] G. H. Rothblat and V. J. Cristofalo, "Growth, Nutrition, and Metabolism of Cells in Culture," Vols. 1 and 2. Academic Press, New York, 1972.

[38] E. N. Willmer, "Cells and Tissues in Culture," Vols. 1–3. Academic Press, New York; Vols. 1, 2, 1965; Vol. 3, 1966.

[39] J. Paul, "Cell and Tissue Culture," 4th ed. Livingstone, Edinburgh, 1970.

[40] Third Karolinska Symposium on Research Methods in Reproductive Endocrinology entitled "*In Vitro* Methods in Reproductive Cell Biology," *Acta Endocrinol.*, Suppl. 153 (1971). Especially pertinent is the chapter by L. Hamberger, A. Hamberger and H. Herlitz, pp. 41–61, which covers methods for metabolic studies on isolated granulosa and theca cells.

TABLE IV
EQUIPMENT NEEDED TO CULTURE GRANULOSA CELLS

1. Bacteriological dry-type incubator or water-jacketed incubator
2. Water bath
3. Plastic hood with glass bottom or small booth or room with UV light source and gas outlet or laminar flow hood
4. Inverted ocular microscope with or without Polaroid camera attachment
5. Binocular dissecting microscope
6. Bausch and Lomb Simplex viewer
7. Hot air oven
8. Autoclave

appropriate background material. Culture techniques are used when cells are to be studied beyond a period of 2 days and up to at least a month. Methods for examination of cells for periods less than 2 days are outlined under methods in short-term incubation below. Detailed descriptions of culturing monkey[6] and porcine[10] cells have appeared previously and will be summarized with up to date modifications included in detail. Equipment required for culturing granulosa cells is minimal and is summarized in Table IV.

Leighton Tube Cultures. Routinely, at a concentration of about 1×10^6 cells/culture, granulosa cells are added to Leighton tubes (Bellco Glass Co., Vineland, New Jersey) which contain 10.5×35 mm glass coverslips; 1.5 ml of culture medium are added with 5% CO_2 in air as the gas phase, and the culture tube is stoppered and incubated at 37° in a bacteriological type or other dry-type incubator. If electron microscopic observations are desired, the cells are grown upon carbon-coated coverslips and fixed after various days in culture 2–4 hours in Karnovsky's paraformaldehyde–glutaraldehyde fixative[41] followed by storage in 0.2 *M* phosphate buffer, pH 7.4, containing 3% sucrose.[1a] Later the cultures are postfixed in 1–2% OsO_4 solution followed by dehydration in ethanol and acetone according to the method of Brenner[42] and are embedded in Epon 812 using the technique of LaVail.[43] Prior to fixation of the coverslip culture, the coverslip is rinsed in warm Hanks' solution to remove adhering cellular debris and serum protein present in the culture medium. If light microscopic observations are to be made, the cells are grown on a plain glass coverslip, and at the end of the experiment the coverslip culture is rinsed in warm Hanks' solution and fixed for 10 minutes in Bouin's solution followed by hematoxalin and eosin staining.

[41] S. Ito and M. J. Karnovsky, *J. Cell. Biol.* **39,** 168A (1968).
[42] R. M. Brenner, *Amer. J. Anat.* **119,** 429 (1966).
[43] M. M. LaVail *Tex. Rep. Biol. Med.* **26,** 215 (1968).

On alternate days throughout the culture period the culture medium is changed (1.5 ml per culture) and the medium is frozen for subsequent steroid analysis. After each change of medium the cultures are examined with phase contrast microscopy. The cell numbers can be approximated after 2, 4, 6, . . . , days by one of two methods. The first method, which is more accurate but also more laborious, consists of projecting the culture upon a piece of paper using a Bausch and Lomb Tri-simplex viewer, ×48.[6] The areas covered with colonies of cells are traced on the paper, cut out and weighed, and expressed as a percentage of the total area projected. At the end of the culture period the cells per unit area are estimated using the fixed stained culture. This method is satisfactory for a small series of experiments but becomes cumbersome and time consuming for experiments larger than 50 cultures. For the larger experiments an alternative method of approximation of cell growth is used. On alternate days the culture is projected with the Tri-simplex viewer, and the percent area covered with cells is approximated. It is important that the same investigator estimate all the cultures in a given experimental series. At the end of the culture the coverslips are fixed and stained with hematoxylin and eosin and the cells per unit area are estimated and used to calculate the cell numbers at previous culture times. Both of these methods have the advantage that the culture does not have to be fixed to estimate cell numbers after 2, 4, and 6, etc., days in culture. The fixed coverslip at the end of the culture serves as a permanent record of the culture and can be referred to at any time.

Porcine, equine, and monkey granulosa cells may be used up to 2 days after cell harvest provided that the cells are stored in the refrigerator. It is preferable to isolate the granulosa cells from the biopsy specimen on the same day as the biopsy and to store the cells in suspension in an appropriate culture medium, such as medium Y or W. The plating efficiency, which is 20–30% for porcine granulosa cells[11] (C. P. Channing, unpublished observation) decreases with time delay between obtaining the biopsy and culturing, but decreases less if the cells are stored in culture media at refrigerator temperature (4°) than if the cells are stored in the ovarian tissue itself. Dr. Peter Kohler and his associates can deep freeze porcine granulosa cells under liquid nitrogen and use them at any time.[44]

Small Vessel Cultures. If the investigator has access to only a limited number of cells, as is the case with human biopsy specimens, but still wants to examine a large number of variables, such as the effects of various hormones and culture conditions upon the cells, there are available

[44] P. O. Kohler, personal communication.

miniature disposable culture chambers (Miles Laboratories, Wesmont, Illinois, Lab-Tek Culture Chambers, Nos. 4808 and 4804), which can accommodate 8 small cultures each of 0.4 ml of culture medium (Fig. 6). About 10^5 rather than 10^6 cells can be used as an inoculum. A similar type of culture vessel is available divided into four compartments. Since the construction is not gas-tight we find that if we gas the cultures and place the entire group of 4 or 8 cultures into a large Leighton tube (Bellco Glass Co., Vineland, New Jersey), and gas the large Leighton tube and stopper it, there is no need for a CO_2 incubator. Alternatively the cultures can be placed inside a gas-tight container, such as a small desiccator, and the interior of the container gassed with the 5% CO_2 in air. At the end of the culture period the medium is decanted, the partitions of the Lab-Tek chambers peeled off and the slide containing the 8 cultures fixed and stained with hematoxylin and eosin. Media are changed on alternate days using Pasteur pipettes to suck off the used medium and to add fresh medium.

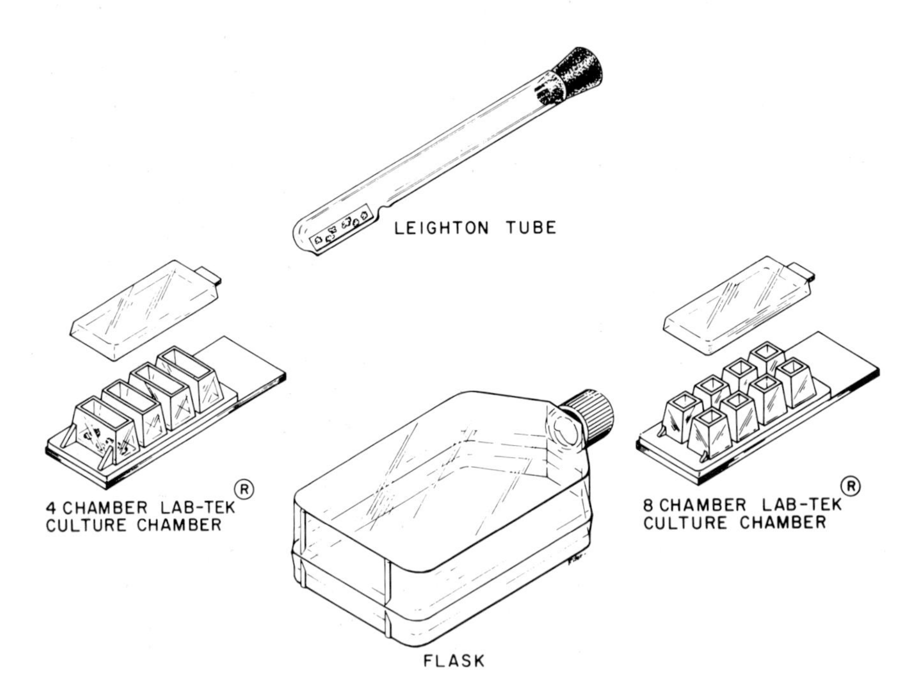

FIG. 6. Four types of culture vessels of various sizes. Leighton tube (top), Lab-Tek culture chamber with 8 chambers (right), Lab-Tek culture chamber with 4 chambers (left) and 250-ml flask (bottom). The drawings are not necessarily to scale.

Larger Culture Vessels. Various types of disposable plastic culture vessels are available commercially from a number of suppliers, e.g., Falcon Plastic vessels, and they are supplied already sterile. The choice of the vessel depends upon how many cells are to be cultured and whether or not a closed gas-tight system is desired. If the investigator has a CO_2 incubator available, petri dishes may be used. The sizes vary. If larger culture vessels are desired a 250-ml plastic flask with screw cap is recommended (Fig. 6) (e.g., Falcon No. 3024). Flasks have the advantage of being a closed system and not requiring a CO_2 incubator. About 4 to 10×10^7 granulosa cells are used in the inoculum and 15–20 ml of medium are used in the 250-ml flask.

Smaller flasks are available and the size chosen will depend upon the experiment. Kohler and his colleagues used a 30-ml plastic flask and inoculated it with 2.5×10^6 porcine granulosa cells.[23] Borzynski and colleagues[45] cultured porcine granulosa cells in Earle T-30 flasks and used 4×10^6 cells in the inoculum along with 5 ml of culture medium. Cells grown in the larger plastic vessels are not suitable for fixing and staining at the end of the culture period unless coverslips are placed on the bottom of the vessel prior to inoculating them with cells. Day-to-day observations of cells grown in this manner can be made with an inverted ocular phase contrast microscope. Cells can be quantitated at the end of the culture period by detaching them from the plastic surface with a mixture of trypsin and EDTA (0.05% crude trypsin and 0.02% sodium EDTA). The cells are first washed with Ca^{2+}-free Ringers or 0.9% NaCl followed by incubation at room temperature with the trypsin-EDTA mixture for 10 minutes. The cells are then easily dislodged from the surface with agitation using a Pasteur pipette. The cell suspension is then transferred to a centrifuge tube containing 1 ml of lamb serum and centrifuged at 600 g for 5 minutes. The cells are then counted in 0.2% lissamine green or the protein content is measured by the Lowry method.[26,27] If protein determinations are to be done, the cells are washed twice in Hanks' solution to remove traces of serum and cellular debris. Only a small aliquot of the total cell suspension is required for counting or protein analysis; the rest of the cells may be used for further investigation as outlined below. If trypsin will interfere with subsequent experiments with the cells they may be treated with 0.02% EDTA in 0.9% NaCl or may simply be scraped off the surface with a rubber policeman. If the trypsin is omitted the cells come off the surface in clumps which cannot be dispersed adequately for quantitation by hemacytometer. In these instances the

[45] L. J. Borzynski, W. J. McDougall, R. L. Gist, M. D. Vogel, and D. A. Norton, *Comp. Biochem. Physiol.* **40B**, 575 (1971).

quantitation must be done by Lowry protein determination. After scraping, the surface of the vessel should be checked for completeness of cell removal under the microscope. The Bausch and Lomb Simplex viewer is suitable for this purpose since if any cells are left behind they can be located and scraped off while viewing.

Culture Media

One of the biggest problems in this field is that there are as many different culture media as there are investigators in the field. This creates problems since duplication of data from laboratory to laboratory is hindered because of individual differences in culture media. To exacerbate this, most media contain serum which is undefined from an endocrinological viewpoint and contains hormones and other unknown agents present in unknown quantities. It is difficult to study hormonally mediated differentiation in a medium that contains unknown amounts of hormones. Traditional culture media contain 15–20% fetal calf serum since, historically, investigators found that it promoted rapid growth of cells in culture. Caution must be exercised, since rapid cell growth may not be compatible with complete differentiation in nontumorous cells. Furthermore fetal calf serum may contain high amounts of placental hormones, which may act like pituitary hormones. Therefore, in order to study regulation of endocrine differentiation in culture it is recommended not to use serum in the medium. This is fraught with problems, since serum contains numerous growth factors helpful for promoting good growth in culture; these factors may be nonhormonal but are lacking in serumless media. We are attempting to find a good serumless medium for culturing granulosa cells and have made some progress in this field. The media we have used are by no means as good as serum, but they do support growth and progesterone secretion of monkey and pig cells up to about 8 days. Data on these defined media are not complete and are the subject of a long-term study. (Details of these media will be reported in detail elsewhere.) If serum is to be used in media we recommend using a more endocrinologically defined type of serum, such as adult male human serum for human cultures, which can be assayed for LH and FSH. A small amount of fetal calf serum can be added, i.e., in such small amounts as not to contain an appreciable amount of pituitary or steroid hormone, but which adds to the growth-promoting properties of the other serum used. Table V summarized recommended media to use in culturing granulosa cells from various species. Notes on how to obtain "gonadotropin controlled" sera are included below. Detailed general considerations on

TABLE V
CULTURE MEDIA SUITABLE FOR GRANULOSA CELL CULTURES[a]

Medium designation	Base medium[b]	Serum	Serum substitutes	Species of cells to use with medium
Y	85% 199	15% Porcine	—	Porcine
S	85% 199	15% Lamb	—	Porcine or sheep or other
W	80% 199	20% Monkey[c]	—	Monkey or human
H	80% 199	20% Human[d]	—	Human
199D	199	—	0.4% BSA[e] 0.2% Lactalbumin hydrolyzate[f]	Any species
SMD	Serumless medium of Neumann and Tytell[g]	—	0.4% BSA	Any species

[a] All media contain per milliliter 50 μg gentamycin (Schering) and also supplementary 2 m*M* L-glutamine added immediately prior to use.

[b] Medium 199 contains 25 m*M* HEPES buffer and Earle's salts and is available from the Grand Island Biological Co., Grand Island, New York, or from Flow Labs, Rockville, Maryland. Formulation of medium 199 was made by J. F. Morgan, H. J. Morton, and R. C. Parker [*Proc. Soc. Exp. Biol. Med.* **73**, 1 (1950)].

[c] Monkey serum is obtained from normal immature male monkeys or mature female monkeys from days 1–6 or 21–20 of the menstrual cycle (nonmidcycle female monkeys), amenorrheic or prepubertal female monkeys.

[d] Human serum is obtained from normal male adult humans and should have a low gonadotropin content.

[e] BSA—bovine serum albumin fraction V can be obtained from the Metrix Corp. or the Armour Corp. It is made up as 10% stock solution in medium 199 and the pH is adjusted to 7.4 with 1 *N* NaOH; it is sterilized by filtration through a 0.45 μm filter followed by a 0.22 μm filter. It is stored frozen. The final concentration, whether 0.4% or 1%, is made by mixing the appropriate amount of stock solution with medium 199.

[f] Lactalbumin hydrolyzate can be obtained from Grand Island Biological Co. as a 10% solution. It is added to give a final concentration of 0.2%.

[g] The serumless medium of Neuman and Tytell can be obtained from Grand Island Biological Co.; the original mixture originated in the laboratory of R. E. Neuman and A. A. Tytell [*Proc. Soc. Exp. Biol. Med.* **104**, 252 (1960)]. The medium contains 0.2% lactalbumin hydrolyzate. It is stated in the Grand Island Biological Co. catalog that the medium contains 1 μg of insulin per milliliter, but we find that activity is absent from commercial preparation, therefore, insulin should be added separately by the investigator if desired. Grand Island will no longer add the insulin to their medium, thus making the medium an endocrinologically defined one.

construction of media are presented by Waymouth.[46,47] A summary of the role of serum in control of multiplication of cells in culture has also been presented.[48]

Sera. If serum is to be included in culture media it is essential that measurement of LH and FSH as well as progestins, estrogens, androgens, and adrenal steroids be made on the serum so that proper interpretation of the role of the serum per se in promoting luteinization be adequately assessed. Progestins and adrenal corticoids can be measured by the corticosterone-binding globulin assay of Johansson *et al.*[31]; estrogens can be measured by radioimmunoassay according to Hotchkiss *et al.*[30] or other suitable assay methods. LH and FSH measurements can be done by a radioimmunoassay which will depend upon the species of serum employed.[49] Sera should be obtained from individuals with a known endocrine status, the ideal one being the hypophysectomized state. Since hypophysectomized animals and humans are not always readily available, a practical method for obtaining sera at instances where gonadotrophins and gonadal hormone levels are known to be low can be used and is outlined below. Hormone levels in such sera have been determined in our laboratory and in those of others, and the results support the concept that sera obtained in such a manner have low LH and FSH levels which are the primary known hormones involved in controlling luteinization. Gonadal estrogens and progestins are also low in such sera. Except for the lamb serum all sera should be obtained fresh by the investigator under controlled circumstances. Sera should be obtained from individuals or animals which have fasted overnight. Sera should be kept cold, either on ice or in the refrigerator immediately after collection, a precaution which protects against destruction of labile hormones, such as insulin and other serum components which may be unstable. In each instance, whenever possible the blood is collected aseptically and allowed to clot overnight in the refrigerator. The next day it is centrifuged at 2000 g and the serum decanted and frozen until used. If the blood cannot be collected aseptically the serum is filtered through a 0.45 then a 0.22 μm filter prior to use. Sera are *not* heat inactivated. An aliquot of each batch of serum is tested in a broth tube for sterility. Another aliquot is tested with a Leighton tube of large-follicle porcine granulosa cells, or other type of

[46] P. A. Kitos, R. Sinclair, and C. Waymouth, *Expr. Cell Res.* **27**, 307 (1962).

[47] L. Whittington Gorham and C. Waymouth, *Proc. Soc. Exp. Biol. Med.* **119**, 287, (1965).

[48] H. M. Temin, R. W. Pierson, Jr., and N. C. Dulak, *in* "Growth, Nutrition, and Metabolism of Cells in Culture" (G. H. Rothblat and V. J. Cristofalo, eds.), Vol. 1, p. 49. Academic Press, New York, 1972.

[49] A. R. Midgley, Jr., *Endocrinology* **79**, 10 (1966).

granulosa cells, to see if it supports spontaneous luteinization and is not toxic to cell growth. If a batch of serum is toxic to cell growth it is discarded.

HUMAN SERUM. Serum is collected from normal adult males at mid morning when both adrenal steroid hormones and gonadotropins are low. The subject should be as unstressed as possible and have fasted overnight and eaten no breakfast. A needle and syringe or other device are used to collect blood from the antecubital vein. Sera from several individuals should not be pooled until they are tested for hormone content and toxicity. Later on they may be pooled once all are known to have low gonadotropin levels and an absence of toxicity to cells. Once a given individual has been bled and tested the same individual may be used repeatedly.

MONKEY SERUM. Immature male monkeys are the best and most consistent source of low gonadotropin and low gonadal steroid hormone sera.[50] If they are not available, mature male monkeys also have low gondotropins and may be used, or females not at midcycle, i.e., from days 1–8 or 20–30 of the menstrual cycle, may be used.[28-30] Irregularly cycling monkeys or immature female monkeys less than 2.5 years old may also be used.[50] Monkeys are fasted overnight and bled at midmorning. Once a given monkey is known to provide good low gonadotropin serum, it can be repeatedly bled; 20 ml and 10 ml of blood are taken per week from males and females, respectively. Normal monkey sera behaves like hypophysectomized monkey serum in the fact that granulosa cells harvested from medium-sized monkey follicles fail to luteinize and secrete low levels of progestins in medium W containing these sera. These cells did, however, respond to exogenous gonadotropins added to the serum-supplemented culture medium (medium W).[51] Mature male monkeys may be tranquilized before blood is collected.

PIG SERUM. Immature male or female pig serum can be used. A pig may either be bled in the laboratory or at the slaughterhouse. A pig less than 4–6 months old is kept 1 week or overnight and bled the next day via a catheter inserted into the jugular vein passing into the right atrium. After the pig has been exsanguinated, the ovaries or testes are examined to be sure that they are in an immature state. Immature pigs can be bled at the slaughterhouse by slitting the throat, collecting the blood and later examining the ovaries or testes to be sure they are in the infantile state. A prepubertal ovary has no corpora albicantia or corpora lutea present and may have numerous small follicles present. Porcine sera from such animals consistently have low LH levels,

[50] D. J. Dierschke, G. Weiss, and E. Knobil, *Endocrinology* **94**, 198 (1974).

[51] C. P. Channing, unpublished observations.

<1 ng/ml in terms of NIH-LH-Sl units (Gay and Channing, unpublished). Since it is difficult to know whether a pig in the slaughterhouse has been rigidly fasted prior to sampling a useful way to tell is, if the serum obtained is clear rather than cloudy with fat, the chances are good that the pig has not eaten in the last day or so. Cloudy lipidemic sera are discarded since they are difficult to filter and difficult to work with, and indicate that the animal has eaten recently. Commercial pig sera are unsatisfactory since they are often cloudy and are not collected under endocrinologically defined conditions. Pig sera from a number of suppliers have been examined in our laboratory, and none was found to be consistently satisfactory.

LAMB SERUM. Serum can be collected from lambs in the laboratory or at the slaughterhouse, or it can be obtained commercially. Commercially available lamb serum has LH levels of 6–7 ng/ml using NIH-LH-S1 as standard (M. Nekola, personal communication). Studies in our laboratory using 15% lamb serum in porcine granulosa cell cultures have demonstrated a loss in responsiveness of porcine granulosa cells to exogenous LH as a function of culture time as well as less hCG receptors, a phenomenon which occurs to a lesser degree using 15% pig serum or serumless media. Lamb serum does however promote good growth in culture (Channing and Ledwitz-Rigby, unpublished).

Antibiotics. All culture media have the same antibiotics added per milliter, which are as follows: 50 U of penicillin, 50 μg of streptomycin, 2.5 μg of fungizone, 50 μg of tylosine and sometimes 50 μg of kanamycin. The inclusion of a spectrum of antibiotics obviates the need for rigid aseptic techniques and is especially helpful in the case of porcine material obtained from the slaughterhouse. In the case of monkey granulosa cell cultures the kanamycin is omitted because they are collected under rigid sterile surgical conditions. Since many of the hormone preparations are not filtered prior to use the inclusion of the antibiotics in the culture media retards possible contamination. Most antibiotics are purchased as a 100 × solution stored frozen and added to the culture medium immediately before use. Fifty micrograms per milliliter of gentamycin (Schering) may be used instead of the other antibiotics.

Balance of the Culture Medium. The basic part of the culture media is customarily medium 199 [52] containing Earle's salts and 25 m*M* HEPES buffer. Medium 199 is purchased commercially with the HEPES buffer already added. (Flow Laboratories or Grand Island Biological Co). If medium 199 is not available with HEPES buffer, it can be added by the investigator. A maximum of a 10 m*M* concentration of HEPES buffer can be tolerated in medium 199 without alteration of the salt concentra-

[52] J. F. Morgan, H. J. Morton, and R. C. Parker, *Proc. Exp. Biol. Med.* **73**, 1 (1950).

tion of the medium and with maintenance of physiologically satisfactory ionic strength. The NaCl concentration of the medium is reduced proportionately to accommodate a greater than 10 m*M* concentration of a HEPES buffer. In preliminary experiments using porcine granulosa cells, medium 199 supplemented with 15% porcine serum promoted better growth and progestin secretion than Eagle's medium supplemented with 15% porcine serum. A mixture of HEPES buffer and Earle's salts provides an excellent buffer and is gassed with 5% CO_2 in air. If the investigator does not want to gas with 5% CO_2 in air, medium 199 with Hanks' salts and HEPES buffer may be used.

Defined Media (Media Containing No Serum). Media containing no serum are preferable since they are defined from an endocrinological point of view and permit an exact definition of the hormonal requirements for luteinization. Two such media have been devised from existing commercially available media (Table V). Medium 199D (199 defined) has been shown to support cell growth and maintenance up until about 8 days in culture. Likewise medium SMD (serumless medium defined) supports growth up until about 10 days. A number of other media have been tested, namely Ham's medium NF10, NF12, and Eagle's medium, and found to be unsatisfactory in promoting reproducible growth of porcine granulosa cells either when used alone or supplemented with 0.4% BSA and 0.2% lactalbumin hydrolyzate. When defined media are used it has been our experience that better cell attachment and later growth occurs when a larger inoculum is used compared to when serum-containing media are employed. For example, in Leighton tube cultures using granulosa cells harvested from small porcine follicles, an inoculum of 2×10^6 cells will grow well in medium Y, but an inoculum of 1 to 1.5×10^7 cells is required for good growth in medium 199D or SMD. The appropriate supplementation with BSA is 0.4%.

Mixing of Media. Medium 199 containing 25 m*M* HEPES buffer and Earle's salts as well as the serumless medium of Neuman and Tytell[53] are purchased as a 1× solution and stored frozen until use. Prior to a series of experiments lasting 1–2 weeks, the media are thawed out and mixed in batches of 100–500 ml with antibiotics, L-glutamine, bovine serum albumin, and any other ingredients. After the components of the medium (Table V) are mixed, the final medium is stored in the refrigerator until used. If a period of more than about 2 weeks must elapse before the medium is used, it should be stored in the frozen state.

Addition of Hormones and Other Agents to the Media. Glycoprotein hormones, such as LH, FSH, and TSH are weighed out in 1-mg quantities in small vials and diluted in 1 ml of deionized distilled water. These stock

[53] R. E. Neuman and A. A. Tytell, *Proc. Soc. Exp. Biol. Med.* **104,** 252 (1960).

a homogeneous cell population in terms of mitoses. Furthermore it is desirable for these cells in suspension not to divide. We have done preliminary investigations using suspension cultures of porcine granulosa cells and found that they can be satisfactorily carried out in plastic Packard scintillation vials. Cells stick to nylon and glass scintillation vials, thus making them unsatisfactory. Alternatively, glass vials may be treated with silicone to prevent adherence of the cells to the glass. Plastic vials are sterilized by exposure to UV light for 30 minutes. Cultures are inoculated with 5×10^6 to 1×10^7 cells in 1 ml of culture medium, usually medium 199D (Table V). After inoculation the plastic vials are gassed with 5% CO_2 in air and stoppered with a No. 1 rubber stopper, and incubated in a warm room, with or without gentle shaking. Porcine granulosa cells can be stored up to 18 hours after harvest in the refrigerator before initiation of suspension cultures. When it is desired to end the suspension culture (any time from 5 minutes to 48 hours later) the suspension is transferred with a Pasteur pipette to a 12×75 mm test tube and centrifuged for 5 minutes at 600 *g*. The medium is decanted and frozen for later steroid analysis. The cells are subsequently assayed for responsiveness to LH in terms of stimulation of cyclic 3′,5′-AMP generation (see below) which is a measure of the ability of LH to cause luteinization.[20] The cells may also be assayed for their hCG receptors (see below). We have used these stationary suspension cultures of porcine granulosa cells for examination of the inhibitory effects of porcine follicular fluid on LH stimulation of cyclic 3′,5′-AMP levels after a 48-hour incubation as well as for studies on the induction of hCG receptors. Cells do not divide under these conditions when incubated for 48 hours in plastic vials in media 199D, SMD, or Y.

If a suspension culture is to be carried out for more than 48 hours, the cells are centrifuged and resuspended in fresh culture medium and reintroduced into the warm room for an additional time period.

Methods to Assess Luteinization in Granulosa Cell Cultures and Suspensions

Morphology

Granulosa cell cultures which luteinize spontaneously, i.e., cultures of cells harvested from preovulatory follicles, or those induced to luteinize *in vitro* by the addition of pituitary hormones or other agents (such as cells obtained from medium-sized follicles), can be observed in the living state at 2-day or shorter intervals using phase contrast microscopy. An inverted microscope is used for this purpose. Cultures observed in this manner can be photographed with a Polaroid camera so that a permanent record of the progress of the culture may be made. Figure 7 illustrates the appearance of luteinized and unluteinized monkey granulosa cell cultures after 6 days of culture. Figure 8 illustrates the effects of LH and of dibutyryl cyclic 3′,5′-AMP upon granulosa cell morphology after 11 days in culture. Alternatively cinephotomicrography of granulosa cells provides perhaps the best record of the luteinization process. At the end of the culture the coverslip can be fixed in Bouin's solution and stained with hematoxylin and eosin (see section on methods for tissue culture above). An example of the appearance of fixed and stained preparations of monkey and pig granulosa cell cultures can be found elsewhere.[6,7,10]

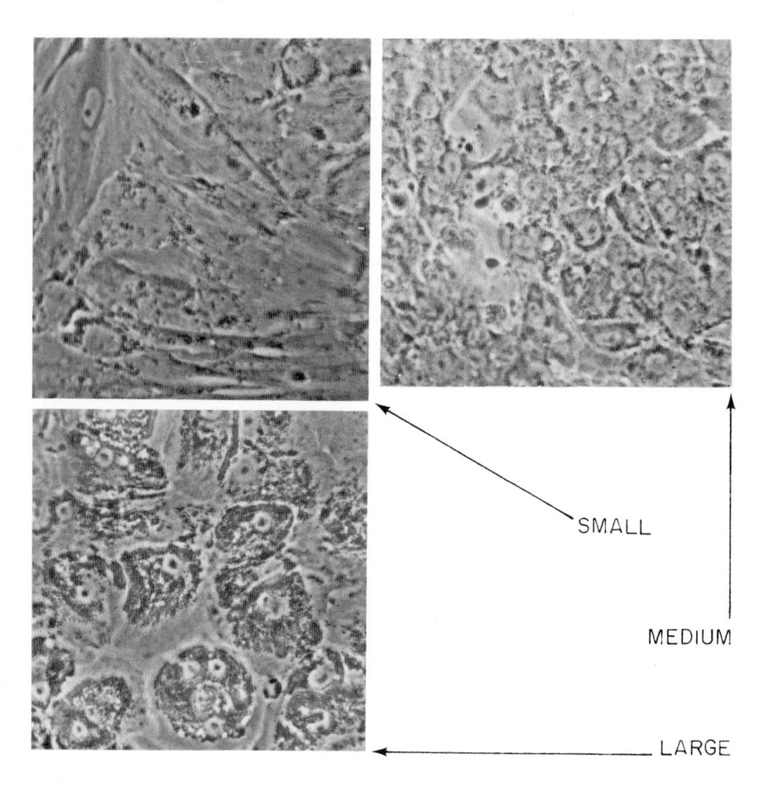

FIG. 7. Phase contrast photomicrographs of a 6-day culture of granulosa cells. *Upper left:* Harvested from small 1-mm monkey follicles. *Upper right:* Harvested from a medium-sized 3.5-mm follicle of a monkey at day 8 of the menstrual cycle. *Lower:* Harvested from a 6-mm preovulatory follicle of a monkey at day 13 of the cycle. The dark regions in the cytoplasm represent discrete granules and the circular light areas are lipid droplets, both morphological characteristics of luteal cells. All ×275.

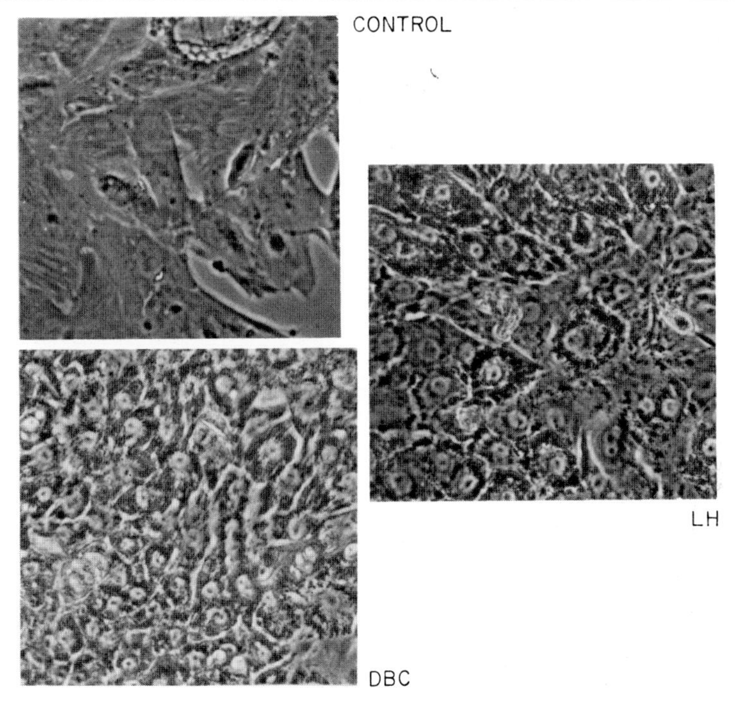

FIG. 8. Effect of LH and dibutyryl cyclic 3,5′-AMP upon the morphology of cultures of granulosa cells harvested from a 5-mm follicle of a monkey at day 9 of the menstrual cycle after culture for 11 days. *Upper left:* Control culture. *Upper right:* Culture incubated in the presence of 0.14 μg/ml human luteinizing (LH) (3.8 × NIH-LH-Sl). *Lower:* Culture incubated in the presence of 0.2 m*M* dibutyryl cyclic 3′,5′-AMP (DBC). All ×275.

Electron microscopy is a useful method for assessment of luteinization of granulosa cells in cultures as demonstrated previously.[1] Granulosa cells are grown in Leighton tubes on carbon-coated coverslips and fixed and embedded as outlined under tissue culture methods above. Coverslips are carbon coated by exposing them to a carbon vapor for 3 minutes followed by hot air oven sterilization for 12 hours at 180°. The hot air oven sterilization serves to stabilize the carbon film. Electron micrographs of a luteinized and an unluteinized monkey granulosa cell culture can be seen in Fig. 9A and B after 6 days in culture. The cultures were grown, fixed, and processed as outlined previously.[1] We are grateful to Dr. Thomas Crisp for providing the electron micrographs. Proper fixation is indicated by the good preservation of mitochondrial structures.

An alternative method is to grow the cells in plastic flasks and embed

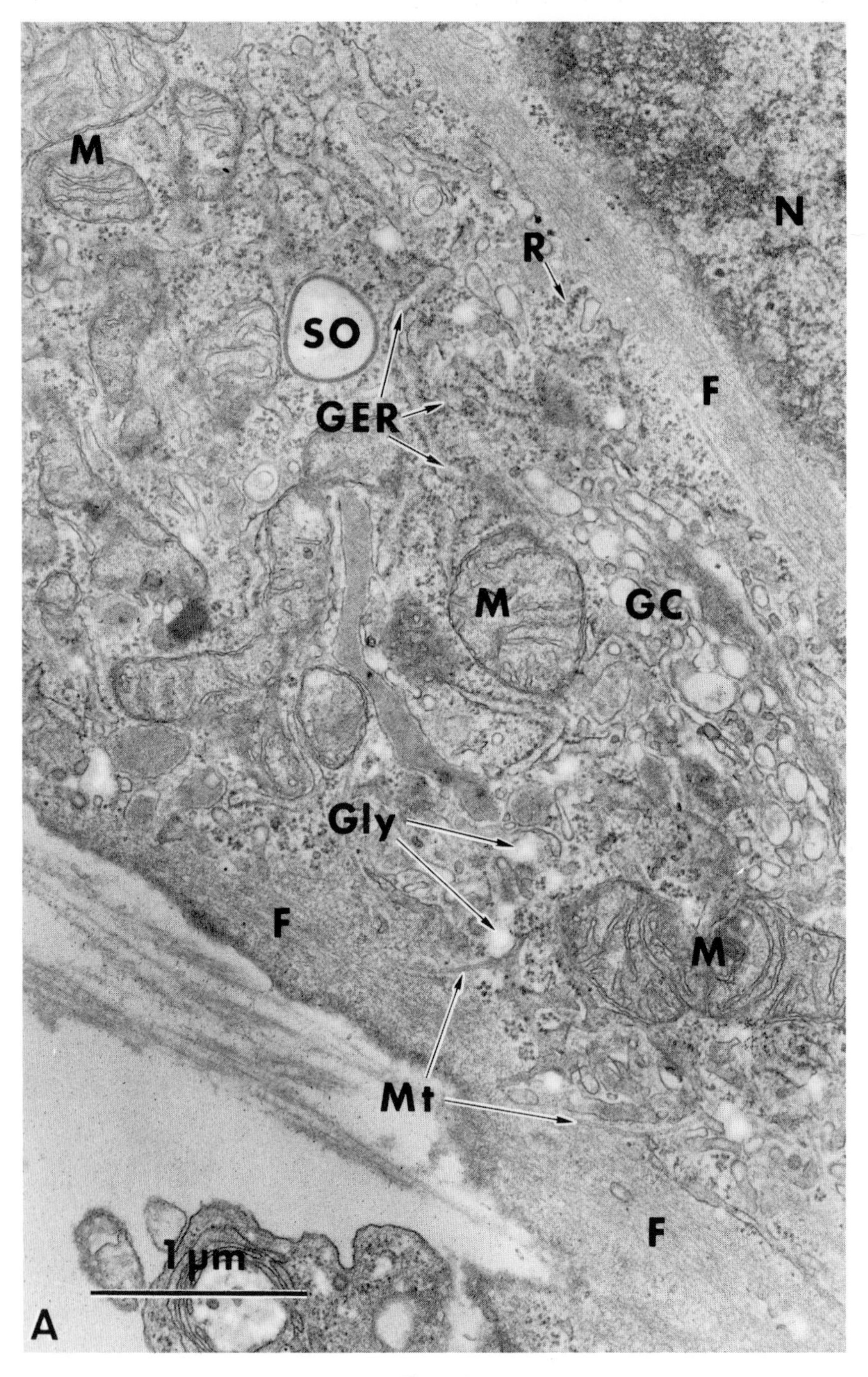

Fig. 9A

them directly on the plastic as outlined by Douglas and Elser.[54] A third method is to grow the cells on a layer of thiolated gelatin, which can readily be detached from the coverslip for subsequent processing for electron microscopy.[55] We found that the thiogel detached from the coverslip during the culture period before we wanted it to, thus hampering its use. Details on general methods for electron microscopy may be found elsewhere.[56]

Histochemistry is also useful for assessing the degree of luteinization provided it is supported by biochemical data. A suitable enzyme to look for is 3β-hydroxysteroid dehydrogenase, which converts pregnenolone to progesterone. This enzyme has been demonstrated histochemically in the cytoplasm of equine granulosa cells[2] and in porcine granulosa cells.[57,58] In both instances the cells have been demonstrated to convert pregnenolone to progesterone biochemically.[4,59] Porcine granulosa cells have been shown to have an α-steroid dehydrogenase histochemically,[45] but biochemical evidence for such an enzyme is lacking. A useful tool for the future is histochemistry at the ultrastructural level. Such a tool is needed since some of the prominent significant ultrastructural changes in spontaneous granulosa cell luteinization are in the cristae in the mitochondria. Once specific steroidogenic enzymes can be localized there, a better idea of cellular compartmentalization of steroidogenic enzymes involved in the luteinization process can be achieved.

[54] W. H. J. Douglas and J. E. Elser, *In Vitro* **8,** 26 (1972).

[55] R. S. Speirs and M. X. Turner, *Exp. Cell. Res.* **44,** 661 (1966).

[56] D. H. Kay (ed.) "Techniques for Electron Microscopy," 2nd ed. Blackwell, Oxford, 1965.

[57] L. Bjersing, Ovarian histochemistry, *in* "The Ovary" (S. Zuckerman, ed.), 2nd ed.

[58] S. Bergman, L. Bjersing, and P. Nilsson, *Acta. Pathol. Microbiol. Scand.* **68,** 461 (1966).

[59] L. Bjersing and H. Carstensen, *J. Reprod. Fertil.* **14,** 101 (1967).

FIG. 9. Electron micrographs of an unluteinized and a luteinized monkey granulosa cell after culture for 6 days. These electron micrographs were generously donated by Dr. Thomas Crisp of the Department of Anatomy, Georgetown University Schools of Medicine and Dentistry, Washington, D.C.

(A) Electron micrograph of part of an unluteinized granulosa cell from a 6-day culture to demonstrate some characteristic features of these cells. Part of the nucleus is in the upper right-hand corner. The cytoplasm contains a well-developed Golgi complex, granular endoplasmic reticulum, numerous ribosomes and helical polyribosomes, microtubules and mitochondria with predominantly lamelliform cristae and electron-lucent matrices. Bundles of fine filaments, particularly numerous in the ectoplasm, "sphaerae occlusae," bordered by annular nexuses and abundant glycogen granules, seen here as negative images (electron-lucent, punctate granules) are typical of such cell cultures which secrete low levels of progestins into the culture medium. ×30,000.

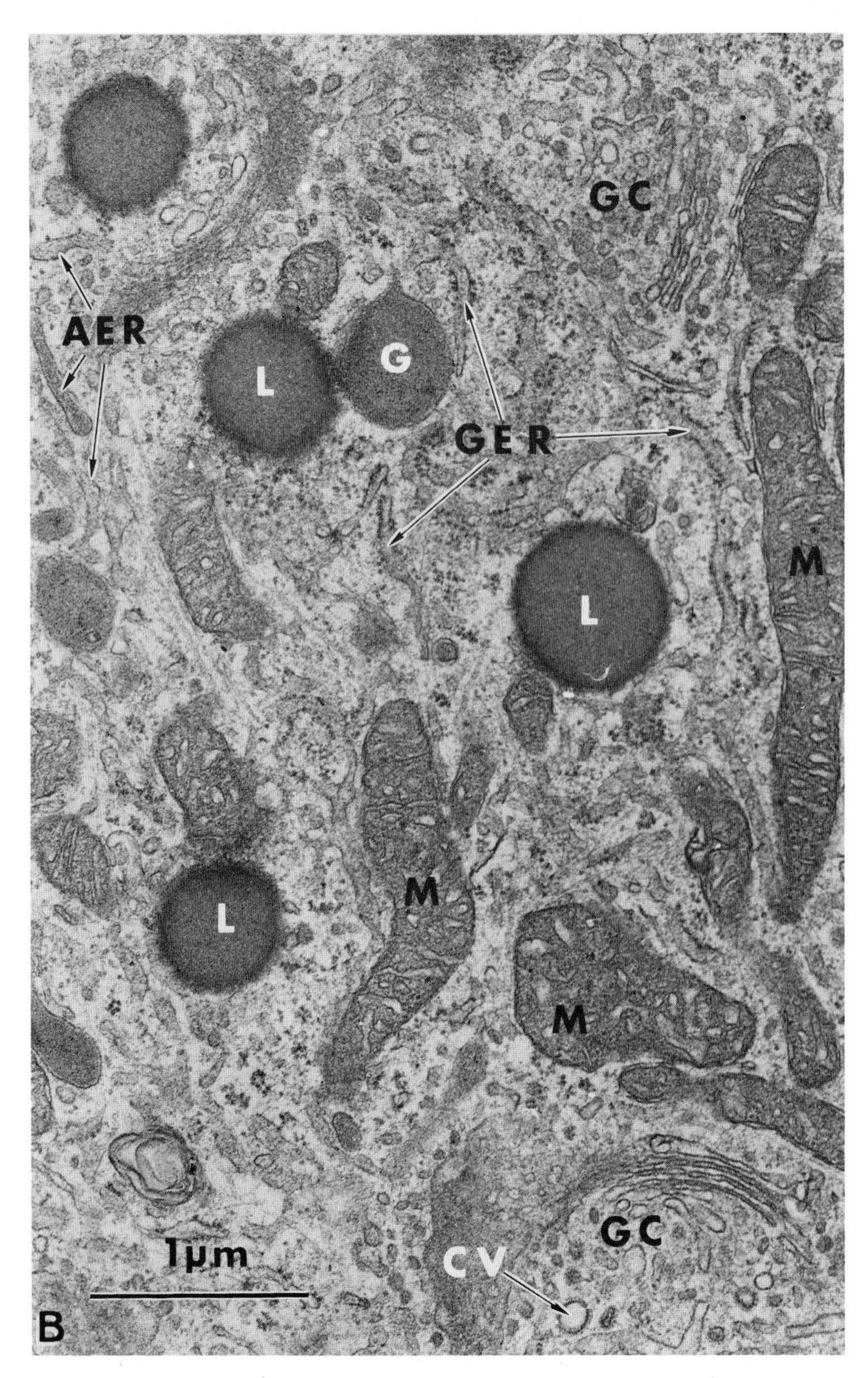

FIG. 9B

Steroidogenesis

In our estimation, the best method of demonstration of luteinization is the observation of characteristic morphological changes in the presence of an elevation in progesterone secretion. The most rapid and accurate way to measure progesterone secretion by granulosa cell cultures or by suspensions is by radioimmunoassay or other methods for measuring progesterone by mass, such as competitive protein binding assay or gas–liquid chromatography. The progesterone in the culture medium is measured after 2, 4, 6, . . . , 30 days in the case of monolayer cultures and the progesterone content of the supernatant medium of cell suspensions can be measured after varying intervals from 5 minutes to 48 hours of incubation. The advantage of measuring the actual mass of progesterone secreted from endogenous precursors, compared to measuring conversion of some labeled substrate to progesterone, is apparent. Differences in pool sizes can affect the conversion of a substrate to progesterone. Furthermore, the exogenous substrate may not properly equilibrate with the endogenous pool of substrate, a situation which would complicate interpretation of data. Two types of rapid progesterone assays can be used for routine analysis of culture media provided that the pattern of progestin secretion has been adequately characterized initially. We have chosen the corticosterone-binding globulin assay for progestins, originally designed by Johansson *et al.*,[31] and more recently the progesterone radioimmunoassay (RIA) designed by Thorneycroft.[60] Both methods are discussed in detail elsewhere.[61,62] Since in the case of porcine,[9] human,[8] equine,[3] monkey[6] granulosa cell cultures it has been demonstrated (using gas chromatographic and thin-layer chromatographic technique) that

[60] I. H. Thorneycroft and S. C. Stone, *Contraception* **5**, 129 (1972).
[61] C. A. Strott, this series, Vol. 36, Chapter 3.
[62] G. D. Niswender, A. M. Akbar, and T. M. Nett, this series, Vol. 36 [2].

FIG. 9B. Part of a luteinized granulosa cell from a 6-day culture to demonstrate abundant pleomorphic mitochondria with electron-dense matrices and tubular or villiform cristae. Such mitochondria are typical of luteinized granulosa cells known to be secreting progestins into the culture medium. The Golgi complex is extensive and consists of numerous isolated regions of smooth-surfaced, fenestrated lamellae and small vesicles, many of which are coated. Note also the presence of numerous lipid droplets, both granular and agranular endoplasmic reticulum, and membrane-bound granules. Such morphological features are characteristic of steroidogenic cells. ×30,000.

Abbreviations: AER, agranular endoplasmic reticulum; CV, coated-vesicle; F, filaments; G, granule; GC, Golgi complex; GER, granular endoplasmic reticulum; Gly, glycogen granules; L, lipid inclusion; M, mitochondria; Mt, microtubule; N, nucleus; R, ribosomes and polyribosomes; SO, sphaerae occlusae.

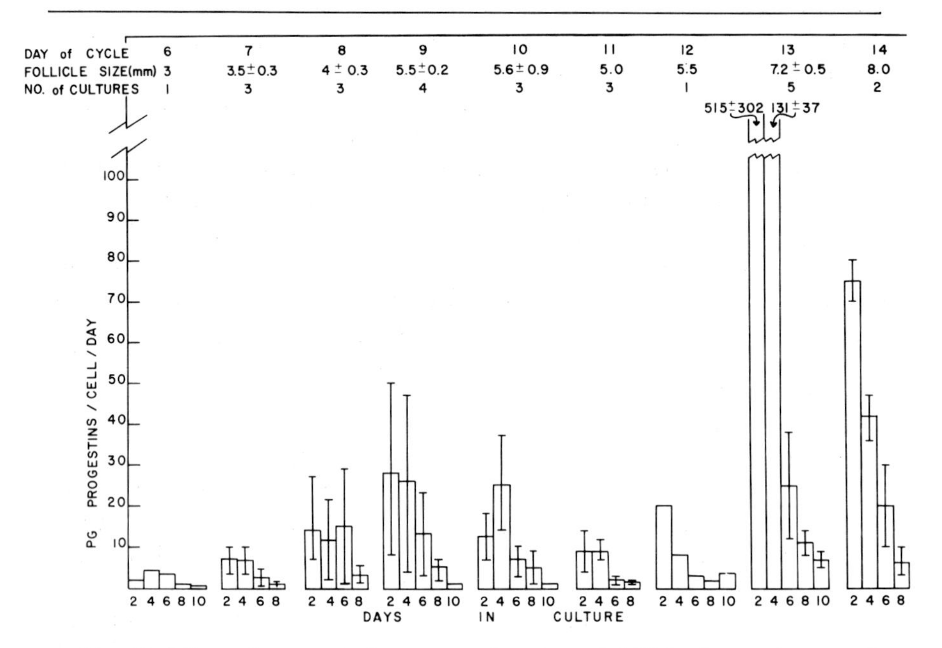

FIG. 10. Effect of stage of the menstrual cycle and follicle size upon progestin secretion by rhesus monkey granulosa cells in culture. Taken from C. P. Channing, *Recent Progr. Horm. Res.* **26,** 589, (1970) with permission.

progesterone represents at least 85% of the total progestins secreted, a more rapid method estimating total progestins may be used to approximate progesterone secretion.

To perform the *corticosterone binding globulin* (CBG) *assay,* a small amount of culture medium (20 μl plus 200 μl of water) is extracted with 2.5 ml of petroleum ether, dried under air and subjected to the assay.[31] Poor success has been achieved in assaying unextracted and undiluted culture medium containing serum since the serum in the medium interferes with the binding reaction. If, however, the sample has sufficient progestins to warrant dilution of the sample (20 μl in 1 ml of water, use of 20 μl in the assay; or 50 μl in 1 ml of water, use of 50 μl in the assay), the sample can be assayed without prior extraction with petroleum ether. A similarly prepared blank culture medium is added to the standard curve in such an assay. Progestin secretion as measured by the CBG assay by cultures of monkey granulosa cells is illustrated in Fig. 10.

A suitable source of CBG for us has been a male dog. A number of male dogs were tested for suitable CBG initially; when one was found, that dog was kept and periodically sampled for blood. We have had the same dog for four years. Alternatively, human serum obtained from a woman on the contraceptive pill may be used as a source of CBG. A limitation of the CBG assay is that not more than about 30 samples can be handled in one assay because of close timing of the assay.

Progesterone analysis by radioimmunoassay has been used more recently and is the method of choice, not only because of its rapidity and ease but also because of its high degree of specificity and high sensitivity. Up to at least 200 samples can be assayed at once, and the timing of addition of reagents is not as critical as in the CBG assay since the reaction between antiserum and tracer takes place overnight. The procedure of Thorneycroft *et al.*[60] is followed using his antiserum at an initial dilution of 1:1000. Culture media samples may be extracted with ether or assayed unextracted. Since the assay is sensitive the samples must be diluted 20 μl in 1 ml, (use 20 μl of diluted sample). With such a dilution the contribution of the serum in the medium to interference with binding is negligible. Such a procedure permits great rapidity in assay of a large number of samples. A similar amount of culture medium blank is added in preparing the standard curve. Using such a protocol the assay in our hands is sensitive to at least 10 pg of progesterone. Periodically, culture medium samples should be chromatographed on an LH 20 column[63] to confirm that progesterone is the major secretory product. The antiserum itself is specific for progesterone, reacting less than 5% with other progestins, such as 17-hydroxyprogesterone.

Granulosa cells of primate origin can be examined for *estrogen* secretion using the radioimmunoassay technique of Hotchkiss *et al.*[30] If influences upon estrogen secretion are to be examined it is essential that the granulosa cells be thoroughly washed to remove traces of follicular fluid which contain high levels of estrogen (up to 10 μg/ml).

Cyclic 3′,5′-AMP Levels

Levels of cyclic 3′,5′-AMP in granulosa cells after a 20-minute incubation are a good measure of how many receptors are occupied with endogenous LH. It has been found that cells such as those harvested from large preovulatory pig follicles and which are able to luteinize spontaneously in culture have higher cyclic 3,5′-AMP levels compared to cells harvested from smaller follicles which fail to luteinize spontaneously in culture.[20] An extension of these studies is that the degree of LH stimulation of cyclic 3′,5′-AMP levels is proportional to the amount of unoccupied receptors. Cells obtained from large (5–11 mm) follicles have 1000-fold more hCG-LH receptors than smaller follicles.[19] Furthermore such

[63] J. Hotchkiss, personal communication.

TABLE VI
RELATIVE RESPONSIVENESS OF GRANULOSA CELLS FROM SMALL, MEDIUM, AND LARGE PORCINE FOLLICLES TO LH IN TERMS OF CYCLIC 3',5'-AMP PRODUCTION[a]

Follicle size	Cyclic 3',5'-AMP (pmoles/5 $\times$ 10^7 cells)					
	Control			Plus 1.0 μg/ml ovine LH		
	1	2	Av	1	2	Av
Small	0.13	0.13	0.13	25.3	26.9	26.0
Medium	8.2	0.4	4.3	68.3	65.0	66.7
Large	67.5	42.1	54.8	853.3	674.7	764.0

[a] The granulosa cells were incubated for 20 minutes in Eagle's medium plus 1% BSA at 37°.

cells are many time more responsive to exogenous LH in terms of stimulation of cyclic 3',5'-AMP levels than cells obtained from smaller follicles[20] (Table VI). Therefore, variations in LH responsiveness in terms of cyclic 3',5'-AMP levels are a useful physiological indicator of LH receptors and adenyl cyclase since cyclic 3',5'-AMP is an essential intermediate in the luteinization process.[17,20] With the advent of a rapid protein binding assay for cyclic 3',5'-AMP designed by Gilman,[64] it is possible to easily measure cyclic 3',5'-AMP accumulation in granulosa cells after short-term (20 minute) incubations with LH[17,20] after a longer-term incubation in the presence of stimulators or inhibitors of luteinization.

Preparation of Porcine Cells and Incubation. Granulosa cells are either incubated right after harvest, after incubation for 48 hours as stationary cultures, or after 2, 4, or 6 days of growing as monolayer cultures. In order to get consistently detectable levels of cyclic 3',5'-AMP in both control and LH-stimulated cells, the following numbers of cells are recommended to be used per incubation flask. The entire batch of cells used in an incubation flask is used for the cyclic 3',5'-AMP determination. Small follicles: 1.5 to 2 $\times$ 10^7 cells; medium follicles: 1 $\times$ 10^7 cells; and large follicles: 4 to 8 $\times$ 10^6 cells.

In the case of cells incubated as a suspension in the presence of various stimulators or inhibitors of luteinization for a 48-hour or shorter period, the cells are transferred from the plastic vial to 12 $\times$ 75 mm tubes and centrifuged for 5 minutes at 600 *g*. The supernatant fluid is saved or discarded, and the cells are resuspended in 1 ml of Eagle's medium con-

[64] A. G. Gilman, *Proc. Nat. Acad. Sci. U.S.* **67**, 305, (1970). See also this series, Vol. 38, [7].

taining Earle's salts, 25 mM HEPES buffer, 1% bovine serum albumin, 50 U of penicillin per milliliter and 50 μg of streptomycin per milliliter (designated hereinafter as Eagle's plus 1% BSA). The cells are transferred with mixing with a Pasteur pipette to glass scintillation vials. Control cultures are incubated with no additions, whereas 0.1 to 1 μg/ml of ovine LH is added to test the responsiveness of the cells to LH in terms of cyclic 3′,5′-AMP levels. Each vial is gassed with 5% CO_2 in air stoppered with a No. 1 rubber stopper, and incubated at 37° with shaking in a water bath. The incubation is usually carried out for 20 minutes. Since LH can stimulate cyclic 3′,5′-AMP levels in granulosa cells within 5 minutes with the stimulation lasting up until 3 hours,[17] any incubation time between 5 minutes and 3 hours is suitable.

In the case of freshly harvested cells, the cells are washed at least twice in medium Y or Eagle's plus 1% BSA. The cells then are diluted in Eagle's plus 1% BSA and pipetted in 1-ml aliquots to glass scintillation vials to which either nothing or ovine LH has been added. The subsequent treatment of the incubation is the same as for the suspension cultured cells outlined above. Cells incubated under these conditions show reproducible responsiveness to LH in terms of cyclic 3′,5′-AMP levels as shown in Table VI.[65] The large follicle cells are more sensitive to the LH compared to the small or medium follicle cells which is a reflection of greater number of LH receptors. Because of their greater responsiveness; smaller numbers of large follicle cells are required per incubation (4 to 8×10^6 cells) compared to small or medium follicle cells (1 to 2×10^7 cells).

In the case of cells incubated as monolayers, preferably in large 250-ml flasks, the following procedure is used. We have found that about 6 Leighton tubes worth of cells are required to get enough cells for one incubation. If a plastic 250-ml flask is covered about 70% with cells, the entire monolayer can be scraped off and divided into four incubations to yield readily detectable cyclic 3′,5′-AMP levels during the subsequent incubation with LH. Monolayer cultures are grown for any length of time, with medium changes on alternate days. When the LH effect upon cyclic 3′,5′-AMP levels is to be examined, the medium is poured off and the cells are washed with Eagle's plus 1% BSA. A fresh aliquot of 4.5

[65] These observations confirm and extend earlier observations made in this laboratory[20] except that the cells of the previous publication were incubated in Eagle's medium containing no bovine serum albumin. In recent experiments we found that the bovine serum albumin slightly increased the effectiveness of the LH (10–20%), probably because it inhibited binding of the LH to the glass and had a generalized protective effect upon the cells in general; 10% pig serum has also been used.

ml of Eagle's plus 1% BSA is added, and the cells are scraped loose with a rubber policeman and mixed thoroughly with a Pasteur pipette. One milliliter of the suspension is added to each of four incubation vials. The remaining 0.5-ml is transferred to a 12×75 mm test tube, centrifuged for 5 minutes at 600 g, washed three times with Hanks' solution, and assayed for protein by the Lowry method.[26] The use of four aliquots of cells per culture conveniently enables two control incubations and two incubations with LH. Subsequent incubation of the cells in the glass scintillation vials is carried out as outlined above, Rather than expressing final results per 5×10^7 cells, as in the case of cell suspensions, the monolayer culture data are expressed in terms of milligrams protein.

Termination of Incubation and Assay of Cyclic 3′,5′-AMP. At the end of the incubation the vials are placed immediately on ice and transferred to 12×75 mm test tubes and centrifuged for 5 minutes at 1200 g. The supernatant medium is decanted with a Pasteur pipette and frozen immediately for later cyclic 3′,5′-AMP assay or progestin assay. In recent preliminary trials we have found that cyclic 3′,5′-AMP can be found in the incubation medium after 20–30, 60, or 120 minutes of incubation of large follicle cells with LH. This unexpected finding has prompted us to save all media for cyclic 3′,5′-AMP analysis. Seelig *et al.* has reported that cyclic 3′,5′-AMP can be secreted into the medium by adrenal cells[66] and Dufau and Catt has shown that testicular cells can secrete cyclic 3′,5′-AMP into the medium,[67] so our observations are not extraordinary.

After decantation of the medium, 0.5 ml of freshly diluted 6% trichloroacetic acid (TCA) is added to the cell pellet, to blank tubes, and to tubes containing 1 pmole of cyclic 3,5′-AMP standard. TCA is made up as a 30% stock solution and diluted to 6% with deionized water immediately before use. The cell pellet is disintegrated by a 3-second sonication at a setting which empirically breaks up all visible particulate material. The TCA extract is centrifuged for 10 minutes at 1200 g and the supernatant fluid is decanted by pouring it into a 12-ml centrifuge tube. The TCA extract is then frozen until analysis of cyclic 3′,5′-AMP by the Gilman protein binding assay[64] as modified by Field.[68] In brief, the assay involves extraction of the TCA extract with ether to remove the TCA, followed by incubation of a 50-μl aliquot of the extract with a mixture of ^{3}H-labeled cyclic 3′,5′-AMP, crude cyclic 3′,5′-AMP-dependent protein kinase, and crude protein kinase inhibitor. After a 1- or 2-hour incubation at 0°, the free and bound ^{3}H-labeled cyclic 3′,5′-AMP

[66] S. Seelig, B. D. Lindley, and G. Sayers, this volume, [29].

[67] M. L. Dufau and K. J. Catt, this volume, [22].

[68] K. Mashiter, G. D. Mashiter, R. L. Hauger, and J. B. Field, *Endocrinology* **92**, 541 (1973).

are separated with charcoal and the bound ^{3}H-labeled cyclic 3′,5′-AMP is counted. The crude cyclic 3′,5′-AMP-dependent protein kinase and the inhibitor are obtained from pig or beef heart. The original description by Gilman for obtaining cyclic 3′,5′-AMP-dependent protein kinase employs beef heart, but we have found that pig heart is just as good a source, if not better. Each unknown is assayed in duplicate. In our laboratory the assay is sensitive from 0.1 to 1 pmole of cyclic 3′,5′-AMP.

Binding of [^{125}I]hCG or [^{131}I]hCG to Granulosa Cells

The most convenient way to study LH-hCG receptors in granulosa cells in various stages of differentiation is to measure binding of the cells to iodinated hCG. Granulosa cells harvested from large preovulatory porcine follicles have many times more receptors, or a higher binding affinity, for iodinated hCG than do cells obtained from small follicles.[19] Interaction of LH with its receptor (LH and hCG share a similar receptor[18]) is essential for initiation of luteinization, and any knowledge of what controls maturation of LH-hCG receptors is therefore of importance. Granulosa cells may be cultured as suspensions or as monolayers in the presence of various agents suspected of controlling maturation of hCG-LH receptors followed by determination of binding to [^{125}I]hCG or [^{131}I]hCG.

Iodination of hCG. The iodination method as outlined by Kammerman and Canfield[69] is followed since it does not alter the biological activity of the hormone. It can be used for either ^{125}I or ^{131}I. It is a modification of the method originally outlined by Greenwood *et al.*[70] The iodination method of Midgley,[49] another modification of the original Greenwood method, can also be used with an iodination time of less than 30 seconds. Midgley's method, as well as those of others, was originally designated for iodination of hCG for immunoassay of hCG and LH and employed a 2-minute iodination time. Since immunologically active iodinated hormone may not necessarily be fully biologically active, a shorter iodination time is used in order to assure biological activity. The method of Kammerman and Canfield employs a 20-second iodination time which iodinates the hormone to yield 70 μCi/μg, or approximately 1 atom of ^{125}I per hCG molecule, assuming a molecular weight of 46,000. The extent of utilization of the ^{125}I is consistently in the order of 70%, demonstrating that a 20-second iodination period is adequate for iodination. A similar degree of utilization of ^{131}I occurs if we perform the iodination according to the Midgley method, using a 30 second iodination period.

[69] S. Kammerman and R. E. Canfield, *Endocrinology* **90**, 384 (1972).
[70] F. C. Greenwood, W. M. Hunter, and J. S. Glover, *Biochem. J.* **89**, 114 (1963).

In the method of Kammerman and Canfield, 5 mCi of carrier-free sodium [^{125}I]iodide in NaOH (Iso-Serve) (20 μl) are added to a mixture of 50 μg of hCG in 50 μl of water and 50 μl of 0.3 *M* sodium phosphate buffer pH 7.4 in a 1-dram rubber-stoppered vial. The hCG used is highly purified (14,000–16,000 IU/mg) and is obtained from Dr. Robert Canfield.[71] After gentle mixing, 175 μg of chloramine T in 25 μl of phosphate buffer are added, followed in 20 seconds by 600 μg of sodium metabisulfite in 75 μl of phosphate buffer in order to stop the iodination reaction. Fifty microliters of 10% solution of bovine serum albumin (BSA) are added to the reaction vessel, and the mixture is applied to a 15 $\times$ 1 cm disposable column of Sephadex G-75 that has been equilibrated with 0.01 *M* sodium phosphate buffer pH 7.4–0.1% BSA. The iodinated hormone is eluted with 0.01 *M* phosphate buffer, pH 7.4. Fractions, 1.0 ml, are collected, counted for radioactivity, and frozen. The iodinated hormone is eluted immediately after the void volume, usually with a peak of radioactivity occurring in fractions 2 or 3, and is clearly separated from unreacted iodide which is eluted later in about tube 8. The peak fractions containing the iodinated hCG are used for the binding experiments within 1 month of preparation in the case of ^{125}I and within 1 week of preparation in the case of ^{131}I. The [^{125}I]hCG has the advantage of a longer half-life and is therefore preferred.

Incubation of Porcine Granulosa Cells with [^{125}I] or [^{131}I]hCG

Preparation of Cells for Incubation. Freshly harvested cells or cells cultured as suspensions in plastic vials or as monolayers in 250-ml plasic flasks may be used for binding studies. The cells are grown and prepared as outlined above (under morphology) for cyclic 3′,5′-AMP studies. In order to get reproducible binding of significant numbers of counts at least 10-fold above background, the following numbers of cells are recommended per incubation tube with [^{125}I]hCG or [^{131}I]hCG: large follicle cells, 1 $\times$ 10^6 to 5 $\times$ 10^6 cells; medium follicle cells, 1 to 1.5 $\times$ 10^7 cells; small follicle cells, 1 to 2 $\times$ 10^7 cells. Each incubation of cells with [^{125}I]hCG or [^{131}I]hCG should be done at least in triplicate.

Incubation Media. Various media may be used for the incubation of the cells with iodinated hCG. A summary of blank values and binding of granulosa cells to [^{125}I]hCG after a 30-minute incubation in various types of media is presented in Table VII. Binding per se is roughly similar in the various media, Eagle's plus 1% BSA, medium Y, and medium Y plus Eagle's; however, the blank values are higher in the Eagle's plus 1% BSA compared to the serum-containing media. Serum is probably better at inhibiting nonspecific binding of the hCG to glass than to BSA.

[71] F. J. Morgan and R. E. Canfield, *Endocrinology* **88**, 1045 (1971).

TABLE VII
BINDING OF [^{125}I]hCG TO LARGE FOLLICLE PORCINE GRANULOSA CELLS UNDER VARIOUS INCUBATION CONDITIONS ($n = 3$)[a]

Medium	Experiment	CPM [^{125}I]hCG bound			
		Blank		Plus 5×10^6 granulosa cells	
		Av	SE	Av[d]	SE
Y[b]	I	291	± 22	36,275	± 2586
Y Eagle's[c]	I	484	± 38	30,600	± 1457
Y plus 1 μg/ml Unlabeled hCG	I	342		466	± 195 ±
Y	II	440	± 93	27,979	± 462
Eagle's plus 1% BSA	II	2796	± 362	26,979	± 1048

[a] Each incubation was carried out at 37° for 30 minutes with 0.1 μg of [^{125}I]hCG per milliliter as outlined before [C. P. Channing and S. Kammerman, *Endocrinology,* **92,** 531, 1973].

[b] Y: 15% pig serum in medium 199 plus Earle's salts, 25 m*M* HEPES buffer, and 50 μg of streptomycin per milliliter and 50 U of penicillin per milliliter.

[c] Y Eagle's: 15% pig serum in Eagle's medium plus Earle's salts, 25 m*M* HEPES buffer, 50 μg of streptomycin per milliliter and 50 U of penicillin per milliliter.

[d] Corrected for blank counts.

Pig or mare serum may be used. The fact that the binding to cells is unaltered by the presence of serum indicates that the serum exerts no unusual influences on the binding compared to the BSA. Routinely we now use Eagle's medium containing 15% pig serum or mare serum plus 25 m*M* HEPES buffer and either Earle's or Hanks' salts and 50 μg of streptomycin per milliliter and 50 U of penicillin per milliliter (Y Eagle's).

Incubation of Cells with Iodinated hCG. The incubation, which usually lasts for 30 minutes, takes place with shaking at 37° in a water bath with a total volume of 1 ml in 12 × 75 mm test tubes using a gas phase of 5% CO_2 in air. Initially the cells are suspended in a large tube in Y Eagle's to yield an appropriate cell number per 0.5 ml of medium and kept on ice until used. Cells may be stored up to 48 hours prior to use. Small incubation tubes are prepared in groups of 20–30 per incubation. The iodinated hormone is diluted to a concentration of 1 μg/ml in Y Eagle's medium and kept on ice. To each small incubation tube is added the appropriate volume of Y Eagle's medium and unlabeled hCG (usually 1 μg in 10 μl). Five tubes are used for blanks containing no cells, and each test batch of cells is run in triplicate or quadruplicate with at least three incubation tubes carried out in the absence and three in the presence of unlabeled hCG to test for binding specificity. Subsequently the cells in 0.5 ml medium are added and the tubes gassed with 5% CO_2 in air stoppered and preincubated at 37° for 10 minutes. The total volume of cells and other reagents equals 0.9 ml. Subsequently 0.1 μg of iodinated hCG should be added at timed intervals in 0.1 ml (the use of a Schwartz Biopette is recommended) to each incubation tube followed by incubation for 30 minutes. A 30-minute incubation period is chosen since that permits close to maximal binding to occur.[19] An example of a typical experimental protocol is presented in Table VIII. If cold, unlabeled hCG or LH or any other agent which is to be examined for possible competition for the binding sites is to be added, it is essential

TABLE VIII
TYPICAL EXPERIMENTAL PROTOCOL FOR INCUBATION OF PORCINE GRANULOSA CELLS WITH IODINATED hCG[a]

Tube No.	Cells (ml)	[^{125}I]hCG[b]	Medium[c]	Unlabeled hCG[d] (ml)	cpm	cpm − blank	Av
1	0	0.1	0.9	0	914	—	—
2	0	0.1	0.9	0	777	—	—
3	0	0.1	0.9	0	242	—	—
4	0	0.1	0.9	0	600	—	—
5	0	0.1	0.9	0	688	—	644
6	0.5 Sample A	0.1	0.4	0	32,787	32,143	—
7	0.5	0.1	0.4	0	31,746	31,102	—
8	0.5	0.1	0.4	0	34,482	33,828	32,357
9	0.5	0.1	0.4	0.01	851	207	—
10	0.5	0.1	0.4	0.01	648	4	—
11	0.5	0.1	0.4	0.01	815	171	127
12	0.5 Sample B	0.1	0.4	0	7,552	6,908	—
13	0.5	0.1	0.4	0	7,250	6,606	—
14	0.5	0.1	0.4	0	7,256	6,612	6,708
15	0.5	0.1	0.4	0.01	1,664	1,020	—
16	0.5	0.1	0.4	0.01	1,396	752	—
17	0.5	0.1	0.4	0.01	1,546	902	891

[a] Sample A consists of 4.8 × 10^6 cells obtained from large porcine follicles. Sample B consists of 8.4 × 10^6 cells obtained from small porcine follicles. The cells were incubated with the iodinated hCG for 30 minutes.

[b] 0.1 μg/ml ^{125}I-hCG (2.3 × 10^6 cpm).

[c] Y Eagle's.

[d] Unlabeled hCG was used at a stock solution concentration of 100 μg/ml and was added prior to the addition of the iodinated hCG to give a final concentration of 1.0 μg/ml.

that it be added *before* the iodinated hormone.[19] If a study on the effects of culturing the cells in the presence of various agents suspected of stimulating formation of gonadotropin receptors is to be done, it is essential to determine the degree of binding of the cells to the iodinated hCG *before* the culture period as well as after the culture period.

Stopping the hCG Binding Reaction. To stop the binding reaction, the incubation tubes are placed on ice and centrifuged immediately at 800 *g* for 5 minutes. The supernatant medium is discarded, and the cells are resuspended in 1 ml of 0.9% NaCl containing 1% BSA.

Washing the Cells. The cells are washed 3 times in 0.9% NaCl plus 1% BSA, pH 7.4. The NaCl-BSA solution is added 1 ml at a time followed by gentle mixing of the cells and centrifugation at 800 *g* for 5 minutes; the supernatant fluid is discarded.

Counting the Iodinated hCG Bound to the Cells. After the third wash, the supernatant fluid is discarded and the contents of the entire tube containing the cell pellet are counted in a well-type autogamma counter.

Calculations. The counts per minute (cpm) for each sample are estimated, and the average blank cpm is subtracted from each value (Table VIII) and the replicate values averaged and normalized per 5×10^6 cells. Further statistical analyses such as calculations of standard errors may be done as required. The binding of iodinated hCG obtained in the presence of unlabeled hCG gives a measure of the nonspecific binding of that sample which is usually very low, ranging from 1 to 10% of the total binding capacity.

Hormonal Stimulation of Differentiation (Luteinization)

In order to examine hormonal control of differentiation of granulosa cells from stage 1 to 2, 2 to 3, or 1 to 3, it is best to grow the cells in the presence of the suspected hormones for 2, 4, 6, or 8 days followed by examination of the cells for their degree of luteinization using methods outlined above under methods to assess luteinization. Such methods include examination of the degree of LH stimulation of morphological luteinization and cyclic 3′,5′-AMP levels as well as the number of LH-hCG receptors present. It is recommended that the hormones suspected to have a role in induction of luteinization be added at each medium change until more is known about their effectiveness in a temporal sense. Granulosa cells harvested from medium-sized monkey follicles of normal, HMG- or PMSG-treated animals gradually lose their responsiveness to LH as illustrated in Fig. 11. This emphasizes the need to add the hormone daily from the start of the culture to get a maximal response. The cells

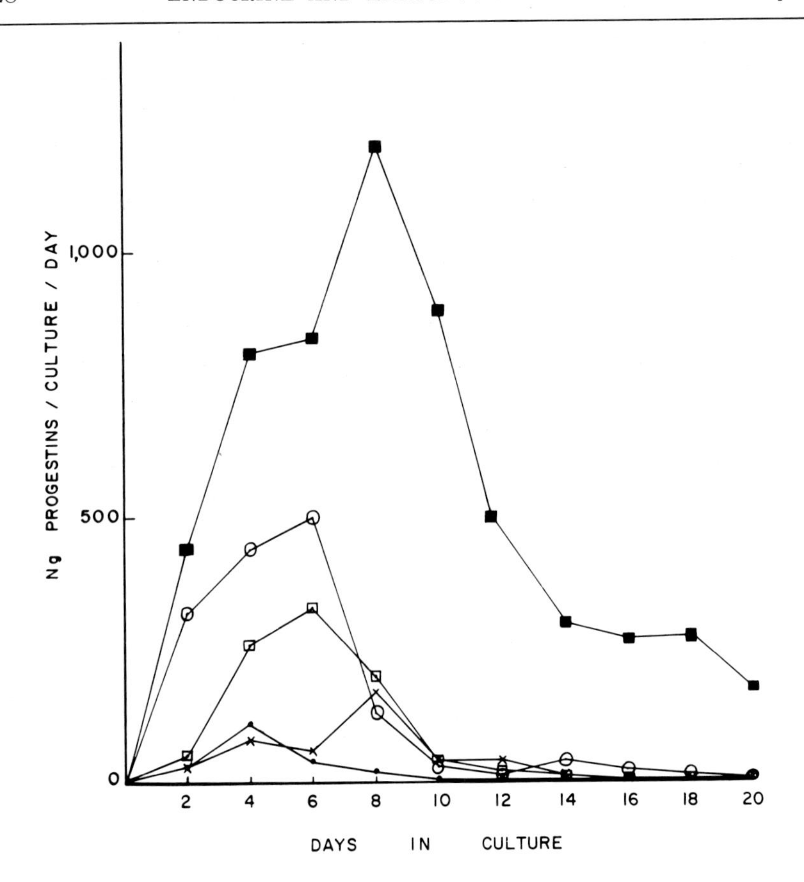

FIG. 11. Effects of various times of addition of human LH (0.1 μg/ml 3.8 × NIH-LH-Sl) (●——●, control; ○——○, days 0–2; □——□, days 2–4; ×——×, days 6–8; ■——■, days 0–20) to cultures of granulosa cells obtained from a human menopausal gonadotropin-treated monkey upon progestin secretion in culture. The cells were grown as outlined previously [C. P. Channing, *Endocrinology* **94**, 1215 (1974)] and medium progestins were measured by protein binding assay.

should be grown in a serumless medium such as SMD or 199D rather than a medium containing serum because such serumless media are endocrinologically defined.

Inhibition of Luteinization

Granulosa cell suspensions and monolayer cultures or a combination of both techniques offer a useful means to examine sites of inhibitory action upon a number of steps in the control of luteinization. In ongoing studies on the inhibitory influence of porcine follicular fluid upon granulosa cell luteinization, cells harvested from large (6–12 mm) pig follicles are incubated in aliquots of 5 to 10 $\times$ 10^6 cells as suspensions in plastic vials for 24–48 hours with 20 or 50% follicular fluid in medium 199 plus antibiotics. At the end of the suspension culture period the cells are centrifuged for 5 minutes at 600 *g* and resuspended in Eagle's medium containing 1% BSA, 50 μg of streptomycin per milliliter and 50 U of penicillin per milliliter. The effect of 0.1 to 1 μg of ovine LH per milliliter upon cyclic 3',5'-AMP levels in the cells during a subsequent 20-minute incubation is then determined. Thus far we find a reproducible inhibitory effect of small follicular fluid, but not of large follicular fluid or of serum upon LH stimulation of cyclic 3',5'-AMP levels after a 48-hour preincubation period. Follicular fluid should be filtered, first through a 0.45 then through a 0.22 μm filter to sterilize it prior to use in such studies.

The *in vitro* inhibitory action of agents suspected as luteolytic agents must be interpreted with caution until their action has been examined for specificity in other cell cultures and suspensions.

Other Ovarian Cell Preparations

Ovarian cells other than granulosa cells can be examined for hormonally mediated differentiation. Difficulties lie in the fact that it is difficult to isolate them as a pure cell type. Furthermore the cells must be treated with trypsin and other hydrolytic enzymes to separate them, a procedure that may alter them in an unknown fashion. Such cell dispersal is feasible, however, since it has been successfully employed for adrenal cells,[72] thyroid cells,[73] and corpus luteum cells.[74] A summary of physical methods available for cell separation following cell dispersal is available[75] and should be referred to. Physiological methods for separation of ovarian and other endocrine cell types on the basis of their dependence on hormones for growth have been presented by Sato *et al.*[76,77]

Summary

The process of induction of luteinization has been divided into three stages depending upon the size of the follicle from which the cells are

[72] R. J. Beall and G. Sayers, *Arch. Biochem. Biophys.* **148**, 70 (1972).

[73] W. Tong, this series, Vol. 37, Chapter 20.

[74] D. Gospodarowicz and F. Gospodarowicz, *Endocrinology* **90**, 1427 (1972).

[75] K. Shortman, *Annu. Rev. Biophys. Bioeng.* **1**, 93 (1972).

[76] G. H. Sato, J. L. Clark, M. Posner, H. Leffert, D. Paul, M. Morgan, and C. Colby, Karolinska Symposium on *in Vitro* Methods in Reproductive Cell Biology, *Acta Endocrinol.*, Suppl. 153, 126 (1971).

[77] J. L. Clark, K. L. Jones, D. Gospodarowicz, and G. H. Sato, *Nature (New Biology)* **236**, 180 (1972).

obtained with the most immature type coming from the smallest follicle. The endocrine control of the luteinization process can be studied at any stage in either short-term cell suspension cultures or in monolayer cultures. Methods for harvesting, long-term culturing, and short-term incubations of granulosa cells have been summarized and updated for the pig and rhesus monkey as well as other species. Luteinization of these cells can be assessed by morphology, steroidogenesis, cyclic 3′,5′-AMP levels, and binding of the cells to iodinated hCG.

Acknowledgments

We are grateful for the contributions of Drs. Jaroslav Kolena, Sandra Kammerman, Thomas Crisp, and Vernon Gay and Miss Margaret Stetson leading toward the development of these methods. The expert technical assistance of Mrs. Viki Tsai, Mrs. Maxine Montgomery, Mrs. Patti Ting, Miss Haina Tu, and Mr. James Cover is appreciated. We thank Drs. Charity Waymouth, Thomas Crisp, and Sandra Kammerman for their helpful criticisms and editorial assistance in the preparation of this manuscript.

[19] A Technique for Perfusion of Rabbit Ovaries *in Vitro*

By KURT AHRÉN, PER OLOF JANSON, and GUNNAR SELSTAM

Several techniques for perfusion of ovaries for metabolic studies have been described in the literature, most of them dealing with steroid metabolism. Ovaries from different species have been used, perfused either *in vivo* or *in vitro*. (For review, see Ahrén *et al.*[1]) Recently, Stähler and Huch,[2] Sturm and Stähler,[3] and Stähler *et al.*[4] presented a technique for perfusion of isolated bovine and human ovaries *in vitro*, in which studies of steroidogenesis were carried out along with investigations of ovarian energy metabolism. In this laboratory studies of ovarian carbohydrate and protein metabolism have been carried out using prepubertal rat ovaries in *in vitro* incubation experiments (for review, see Ahrén *et al.*[5]). To obtain additional information on similar aspects of ovarian metabo-

[1] K. Ahrén, P. O. Janson, and G. Selstam, *Acta Endocrinol.* (*Copenhagen*), Suppl. 158, 285 (1971).
[2] E. Stähler and A. Huch, *Arch. Gynäkol.* **211**, 527 (1971).
[3] G. Sturm and E. Stähler, *Arch. Gynäkol.* **211**, 545 (1971).
[4] E. Stähler, R. Buchholz, and A. Huch, *Arch. Gynäkol.* **211**, 576 (1971).
[5] K. Ahrén, L. Hamberger, and L. Rubinstein, *in* "The Gonads" (K. W. McKerns, ed.), p. 211. Appleton, New York, 1969.

lism, it has been our aim to develop an *in vitro* preparation of the ovary in which nutrition, oxygenation, and regulatory substances are supplied to the organ via an intact vascular bed.

It has been our intention to develop a perfusion technique for ovaries from a common, and from an endocrinological point of view, well defined laboratory animal so as to achieve a simple, standardized dissection procedure and a perfusion apparatus that is easy to run and modify.

Method

Dissection Procedure

Virgin albino Swedish land rabbits, 5–6 months old, were kept on a standardized diet and were not allowed to grow fat. On the day of the experiment their weight ranged from 2.5 to 3.5 kg. They were anesthetized via a marginal ear vein with sodium pentobarbital until surgical anesthesia was reached. The total dosage necessary ranged from 30 to 40 mg/kg. The rabbit was placed on a heating pad and breathed spontaneously. A supplementary local anesthetic was given, and the abdomen was opened from the symphysis pubis to the xiphoid process. The small intestine was drawn out gently and wrapped in a plastic bag, and a thin avascular connection to the sigmoid was divided. The main stem of the inferior mesenteric artery was cut between double ligatures near the aorta. In this way the sigmoid mesentery was divided along the abdominal aorta and the inferior caval vein. To expose the origins of the ovarian arteries from the aorta the loose connective tissue surrounding the great vessels was carefully dissected. This was made in a blunt manner using an iridectomy forceps. The approximated tips of the forceps were placed in the connective tissue sheath between the aorta and the inferior caval vein and were allowed to separate. The exposed lumbar arteries, usually 2–3; were divided between ligatures.

The aorta was cannulated in a retrograde direction just above the bifurcation. A plastic cannula (Venflon, Viggo, Helsingborg) with a diameter of 1.7 mm was used. The tip of the catheter was placed a few millimeters distal to the origin of the ovarian arteries. During the cannulation procedure and when the catheter was tied in place, all organs proximal to it, including both ovaries, had an intact circulation.

The aortic cannula was connected to a perfusion system with a reservoir suspended above the operating table to give a hydrostatic pressure of 75 mm Hg. The perfusion fluid was kept at +4° using cooling jackets and was continuously gassed with carbogen (95% O_2 + 5% CO_2). Perfusion flow was monitored in a drop chamber. The rabbit was heparinized intravenously and the infusion of perfusion fluid into the aorta was started at the same moment that the aorta was tied proximal to the ovarian arteries. The tissues supplied by the ovarian arteries were now perfused *in situ*. The blood of these tissues was rapidly washed out, as indicated by blanching of the structures. The blanching took place in the following succession: fat and connective tissue along the vascular pedicles, fat around the ovary and the oviduct and finally the oviduct itself. One or two minutes after the onset of perfusion all structures of the ovary were completely blanched and cooled.

During the first minute of perfusion there was usually a moderate spasm of the ovarian artery. The spasm could easily be overcome by a temporary increase in perfusion pressure. During continuous perfusion the vascular connections between the ovary and the oviduct, the uterus as well as the perirenal fat were divided without ligatures on both sides. This dissection was continued along the ovarian arteries to the aorta and the ovarian veins were cut close to the inferior caval vein and were left open. During continuous perfusion the aortic segment proximal and distal to the origins of the ovarian arteries was dissected free and the whole preparation, i.e., the aorta, the ovaries and their vascular pedicles, were lifted out from the abdomen.

The aortic segment was cannulated from the opposite direction. The aorta was divided between ligatures between the origins of the ovarian arteries, thus giving two isolated ovarian preparations (see Fig. 1). The cannula of each preparation was slipped into a center hole of a rubber stopper fitting the organ chamber of the perfusion apparatus. The ovary was suspended by a ligature in the ovarian ligament in such a way that the vessels formed an arch. Twisting of the vessels was carefully avoided.

Perfusion Procedure

Nonrecirculating System. Each ovary was rapidly transferred to an organ chamber of the perfusion apparatus (Fig. 2). The perfusion was started in a nonrecirculating system at $+37°$, and this was arranged in the following way: perfusion fluid was introduced at (8) of Fig. 2 and the venous effluent was retrieved at (7) and could then be analyzed. Perfusion pressure to the ovary was adjusted by changing pump speed and regulating screw clamps. This so called preperfusion with nonrecirculation of perfusion fluid is aimed at stabilizing the preparation before starting metabolic studies.

Recirculating System. In many metabolic investigations, however, a recirculating system is desirable. This was achieved in the following way: the desired volume of perfusion fluid was recirculated, warmed, and

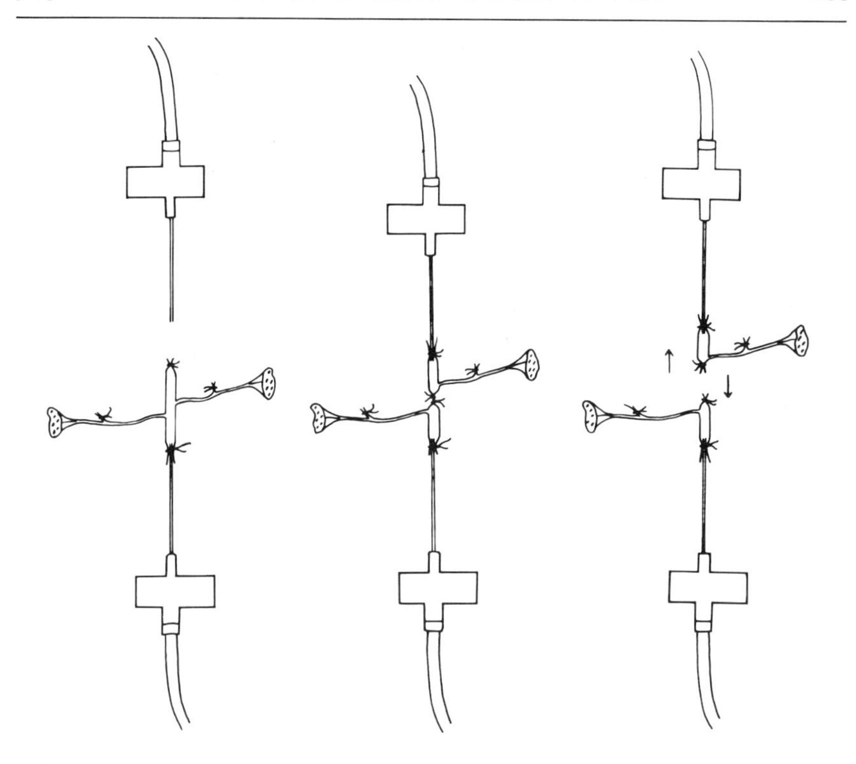

FIG. 1. Sequences of final part of the operation. *Left:* Aorta cannulated in an anterograde direction. *Center:* The aortic segment cannulated from both directions and then ligated with double ligatures between the origins of the ovarian arteries. *Right:* Aorta divided between the two ligatures. During the whole dissection procedure continuous perfusion is carried out. Perfusion fluid is administered from a reservoir above the operating table. The two ovaries are then placed in two separate organ chambers during preperfusion and perfusion.

oxygenated in the apparatus before the ovary was introduced into the organ chamber. The ovary has been perfused for several hours with this system. Arterial and venous samples could be taken during perfusion. The ovary was either frozen or directly homogenized for analysis at the end of the perfusion. The perfusion fluid was collected after the perfusion for analysis.

Perfusion Medium. A medium devoid of blood cells was used. The medium consisted of Krebs bicarbonate buffer with the addition of 1 mg of glucose per milliliter and either dextran with mean molecular weight of 40,000 or Ficoll (prepared by Pharmacia Ltd., Uppsala) with mean molecular weight of 49,300.

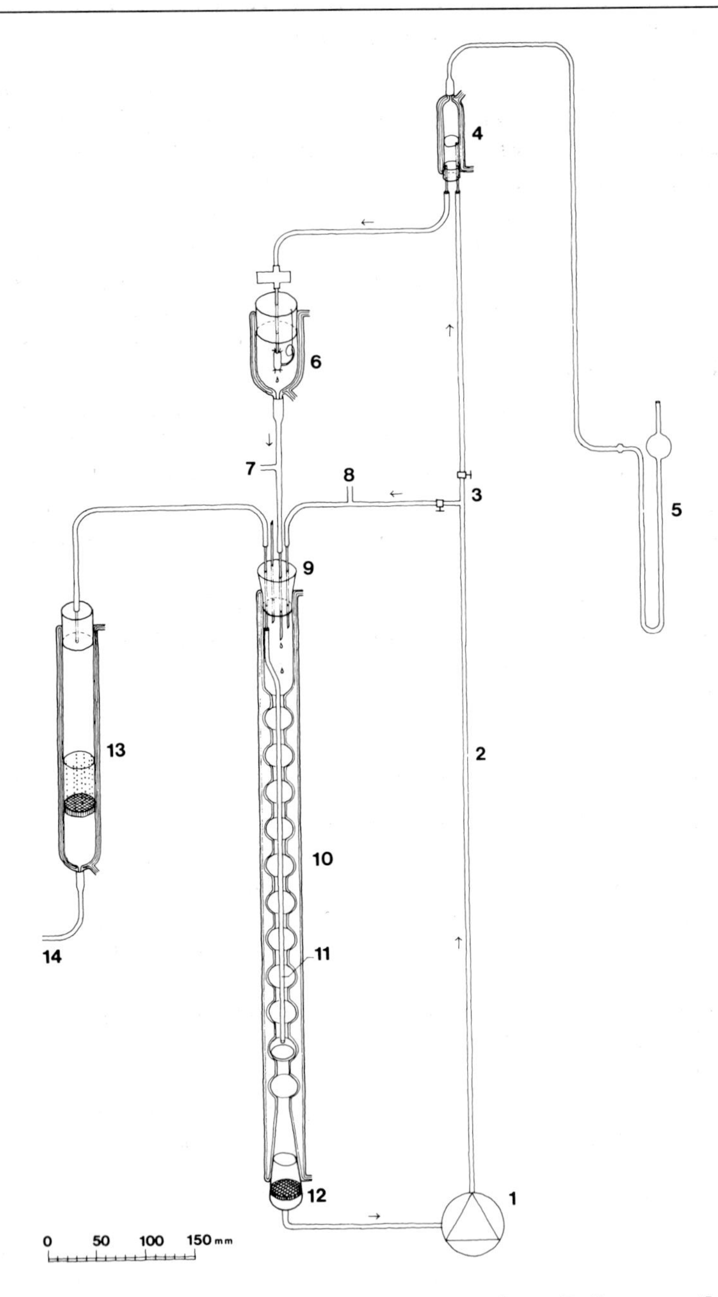

FIG. 2. Perfusion apparatus. All details drawn to scale. 1: Roller pump (Multifix, $\frac{1}{10}$ hp). 2: Silicone rubber tubings (2×4 mm). 3: Overflow system, regulated by two screw clamps. 4: Bubble trap connected to (5) via an air-filled tube: 5: Hg

Technical Details of the Perfusion Apparatus

The perfusion fluid was warmed to +38° by the heating jackets around the oxygenator, bubble trap, and organ chamber. The temperature was then 37° in the ovary due to 1° heat loss of the perfusion fluid in the tubing from the bubble trap to the organ chamber. The volume of perfusion fluid was 25–40 ml. In order to mix, warm, and oxygenate the fluid adequately, the flow rate of the overflow system was set to around 50 ml per minute. The perfusion fluid was gassed with a constant humidified gas mixture, led from a gas tank through a humidifier and via very short tubes to the top rubber stopper of the oxygenator and from there via tubings to the bottom of the oxygenator.

The perfusion fluid, not containing blood elements, spread along the walls of the oxygenator, gassed with 95% O_2 + 5% CO_2 in countercurrent fashion, has been efficiently oxygenated with the oxygen partial pressure at around 500 mm Hg. The evaporation of perfusion fluid from the system is dependent on the gassing. The change in volume of perfusion fluid was 1 ml per hour.

The roller pump used gave a nearly nonpulsatile flow. In a series of organ perfusions, it was shown that nonpulsatile flow was compatible with organ viability, although a pulsatile flow was preferable in a few instances.[6]

Some Criteria of Viability of the Perfused Follicular Rabbit Ovary

Oxygenation. Since a perfusion medium without erythrocytes was used in these experiments it was of utmost importance to test whether there was a continuous and adequate cellular supply of oxygen to the ovary. An enzyme sensitive to anoxia, glycogen phosphorylase, was therefore studied. Enzyme activity was determined according to Bueding *et al.*[7]

[6] E. F. Bernstein, *Acta Endocrinol.* (*Copenhagen*), Suppl. 158, 44 (1971).

[7] E. Bueding, E. Bulbring, H. Kuriyama, and G. Gercken, *Nature* (*London*) **196**, 944 (1962).

manometer. 6: Air-filled organ chamber with ovary attached to rubber stopper. 7: Branch for sampling of venous perfusate. 8: Branch for arterial sampling and addition of substances to perfusion fluid. 9: Rubber stopper pierced with four cannulas (2.0 mm in diameter) for overflow inlet, venous effluent from the ovary, gas inlet, and gas outlet. 10: Oxygenator and reservoir. 11: Silicone rubber tubing leading gas mixture to lower part of oxygenator, ending above perfusion fluid surface. 12: Coarse-porosity sintered-glass filter. 13: Humidifier of gas mixture. 14: Gas inlet. Heating jackets around (4), (6), (10), and (13) serially coupled and flooded with water at a temperature of 38°.

When four follicular ovaries from anesthetized rabbits were frozen *in situ,* the percentage of the enzyme in *a*-form (i.e., activity in absence of 5′-adenosine monophosphate) was 25.1 ± 3.8% (SEM). When ovaries were frozen *in situ* 1 minute after stopping the arterial blood supply, phosphorylase *a* was 2–3 times higher, showing how rapidly this enzyme reacts to anoxia. Ovarian phosphorylase *a* in four ovaries after 2 hours of perfusion with the recirculating system described above was 18.0 ± 1.9%, indicating that the ovary can be adequately oxygenated under these perfusion conditions without erythrocytes in the medium. The phosphorylase experiments were performed with 4% Ficoll in the medium.

Distribution of Tissue Water. The distribution of tissue water between intra- and extracellular compartments of the perfused ovary is regulated by perfusion pressure and osmotic pressure of the medium. It is of importance to keep physiological intra- and extracellular contents of water and electrolytes during the perfusion period. The dry weight of eight follicular rabbit ovaries *in vivo* was 18.7 ± 0.8%. After 2 hours of perfusion with 10% dextran in the medium and at a perfusion pressure of 40 mm Hg, the dry weight of five ovaries was 18.2 ± 1.4%. Under these conditions the extracellular space in eighteen ovaries, as measured with [^{14}C]inulin, was 43.8 ± 1.1%. The *in vitro* values are of the same magnitude as those found in preliminary *in vivo* experiments (Janson and Selstam, to be published).

Adequacy of Perfusion Flow. During the first 20–30 minutes of perfusion at 37° with constant perfusion pressure there was a gradual increase of flow. It is possible that this change of flow was due to a decrease of vascular tonus. This tonus might be acted upon by mechanical factors, hypoxia, temperature changes, osmolarity of the medium, or denervation during the dissection procedure and the initial part of the perfusion. At a pressure of 50 mm Hg the perfusion flow stabilized at around 2.5 ml per minute. Considering the ovarian weight to be 100–300 mg, this flow is very high. A comparison to ovarian blood flow *in vivo* is, however, difficult to make, as, up to now, the magnitude of the ovarian blood flow is largely unknown.

Several factors contribute to the high flow rate of the *in vitro* preparation described here: We have used a perfusion fluid devoid of blood cells and of low viscosity, perfusing a probably maximally dilated vascular bed with a considerable leakage through small anastomoses and small branches of vessels cut along the ovarian artery. An insufficient or inconsistent perfusion of the ovary may sometimes occur and is usually due to thromboses of vessels in the hilar region of the ovary. If one part of the vascular bed is occluded, this is reflected in a low total flow through the preparation and in an extracellular space, markedly divergent from the normal. By studying these parameters and by observing the wash-out of blood corpuscles from the ovary during the dissection procedure, it is possible in each experiment to exclude preparations with inadequate flow.

Concluding Remarks

There are both advantages and disadvantages in using the rabbit ovary as an *in vitro* perfusion preparation. The advantages are that the rabbit is a common and handy laboratory animal, that its reproductive cycle is well known and well characterized and that *in vivo* perfusions of the rabbit ovary have been described.[1] This makes it possible to compare the results of *in vitro* perfusions with those of *in vivo* perfusions. A disadvantage is that the amount of tissue is quite small which makes it difficult or impossible to determine the ovarian uptake of a substance (e.g., substrates or O_2) by measuring arteriovenous concentration differences. For this purpose ovaries from larger animals are more suitable. The perfusion apparatus described here can, however, with only minor modifications be used for larger ovaries. Investigations of *in vitro* perfusions with bovine and human ovaries including determinations of O_2 uptake and substrate uptake from the medium have recently been published.[2,4]

The *in vitro* perfused rabbit ovary has up to now been found most useful in experiments aimed at analyzing the tissue itself, e.g., after adding a labeled substrate or metabolic precursor to the medium. The nonutilizable amino acid α-aminoisobutyric acid (AIB) has been found to accumulate intracellularly in a linear fashion for at least 2 hours, indicating that the kinetics of amino acid transport in the ovary can be explored with this preparation. Protein, RNA, and glycogen metabolism are other aspects of ovarian metabolism in which the mechanism for gonadotropic control is still not clear and where the perfused rabbit ovary can be useful. Regulation of enzymes by gonadotropins can also be analyzed with this preparation, e.g., adenyl cyclase and phosphorylase activities. It is also apparent that some aspects of the hormonal control of ovarian steroidogenesis can be studied with this *in vitro* perfusion preparation.

[20] The Technique of Slice Incubation to Study Corpus Luteum Function

By John M. Marsh, Adalgisa Rojo, and Jorge E. Cidre

The use of the *in vitro* technique of incubating corpora lutea slices has contributed significantly to our understanding of the steroidogenic process in this tissue and its control by gonadotropins. It was first used by Duncan *et al.*,[1] who reported the synthesis of progesterone by slices of sow corpora lutea. Shortly thereafter Mason *et al.*[2] reported the stimulation of steroidogenesis in incubating slices of bovine corpora lutea by luteinizing hormone (LH). Since then, this technique has provided considerable information on the possible site[3] and mechanism of action of this hormone.[4] It has also been used to assess the steroidogenic capacity of corpora lutea of several other species.[5,6] Because of its simplicity and high degree of repeatability, this method is still used extensively to study various biochemical parameters of this tissue and their response to gonadotropins and other substances.

Gonadotropins and Other Chemicals

Bovine and ovine LH preparations have been used predominantly in our studies; both of these hormones, along with the other pituitary hormones tested, were obtained from the National Institutes of Health. Radioactive chemicals and other chemicals were obtained from commercial sources. All solvents were reagent grade and freshly distilled before use.

Collection of the Corpora Lutea

Fresh corpora lutea are used in this procedure because it was found quite early in our studies that slices of frozen corpora lutea did not respond to gonadotropins with an increase in steroid production. Ovaries are collected at slaughter from cows which are in the first 6 months of

[1] G. W. Duncan, A. M. Bowerman, W. R. Hearn, and R. M. Melampy, *Proc. Soc. Exp. Biol. Med.* **104,** 17 (1960).

[2] N. R. Mason, J. M. Marsh, and K. Savard, *J. Biol. Chem.* **236,** PC 34 (1961).

[3] D. T. Armstrong, T. P. Lee, and L. S. Miller, *Biol. Reprod.* **2,** 29 (1970).

[4] J. M. Marsh, *Advan. Exp. Med. Biol.* **2,** 213 (1968).

[5] K. Savard, J. M. Marsh, and B. F. Rice, *Recent Progr. Horm. Res.* **21,** 285 (1965).

[6] D. T. Armstrong, *Recent Progr. Horm. Res.* **24,** 255 (1968).

pregnancy. In the early experiments, corpora lutea from nonpregnant as well as pregnant cows were used, but there was considerable variation in the synthetic capacity and the ability to respond to LH in the two different types of tissues. It is known, for the cow, that the corpus luteum is required for maintenance of pregnancy up to day 200,[7] indicating that at this stage the corpus luteum produces progesterone and is capable of responding to gonadotropin. When pregnant corpora lutea are used, the variation of steroid synthesis is diminished and nearly all the tissues respond to LH *in vitro*. The time between the death of the animal and the collection of the tissue varies between about 15 and 30 minutes. As soon as the ovary is separated from the pregnant uterus it is placed in ice-cold 0.154 *M* NaCl. The stage of the pregnancy is determined by measuring the crown–rump length of the embryo as described by Kristofferson.[8] Some investigators who have had the availability of a veterinary school have obtained their corpora lutea at surgery from nonpregnant and pregnant cows. This procedure is probably preferable when the facilities are available, but the values of steroidogenesis we obtain from control tissues and tissues stimulated with LH, using corpora lutea obtained at slaughter from pregnant cows, are not obviously different from those reported using the surgical approach.[9]

Preparation of Slices

In order to further reduce the variability, a single corpus luteum is used in each experiment. The ovary containing the corpus luteum is transported to the laboratory in the ice-cold 0.154 *M* NaCl and immediately dissected free from adjacent ovarian tissue in a cold room (4°). The connective tissue layer (tunica albuginea) of the ovary is cut over the top of the corpus luteum using a fine pair of scissors. The cut is extended into the ovarian tissue, keeping the blade of the scissors just above the corpus luteum, cutting the rest of the ovary in half. Each half of the ovary is then pulled apart, stripping the adjacent ovarian tissue from the corpus luteum. This procedure was developed to keep the handling of the corpus luteum itself to a minimum. Care should be taken, however, when stripping the ovarian tissue away, not to put excessive strain or pressure upon the corpus luteum. After this procedure a thin capsule of connective tissue still covers the corpus luteum. The gland is rinsed once very briefly in ice-cold 0.154 *M* NaCl to remove blood, and then cut into

[7] L. E. McDonald, S. H. McNutt, and R. E. Nichols, *Amer. J. Vet. Res.* **14**, 539 (1953).

[8] J. Kristofferson, *Acta Endocrinol.* (*Copenhagen*) **33**, 417 (1960).

[9] K. H. Seifart and W. Hansel, *Endocrinology* **82**, 232 (1968).

quarters with a new Stadie-Riggs slicing blade. The blades are routinely cleaned with toluene and with ethanol to remove an oily film, and then dried and rinsed with ice-cold 0.154 *M* NaCl before use. The quarters of the corpus luteum are placed in a petri dish containing a filter paper (Whatman No. 1, 12.5 cm) moistened with 0.154 *M* NaCl. This petri dish rests in a larger tray of crushed ice. The quarters of the corpus luteum are cut into 0.5 mm slices, using a modified Stadie-Riggs hand microtome, starting from the inner tissue and continuing outward until the connective tissue layer is reached. The Stadie-Riggs hand microtome is modified by having the well in the upper plastic block cut to a depth of 0.5 mm. The usual depth in the commercial microtomes is 0.2–0.3 mm, but this produces slices that are very thin and do not respond well to LH. The slicing should be done with only the pressure of the microtome on the tissue and with the blade passing through the tissue in a sawing motion. The slices are picked up with forceps and arranged in a row on the cold moistened filter paper with each slice corresponding to an incubation vessel. When the total number of incubation vessels is reached (this is usually 5 to 9), the next slices are placed on top of the preceding ones, but starting in the reverse order. In this way there is a random distribution of the slices into equal piles, which correspond to the incubation vessels. The tissue is then taken from the cold room in the ice tray and weighed rapidly using a Mettler balance. The piles of tissue are placed on a 2-inch square piece of glassine paper on the balance pan, the tissue is weighed, and then the slices are transferred to dry 20-ml beakers sitting in a tray of crushed ice. The glassine paper is weighed again, and this value is subtracted from the first to obtain the net weight of the slices (usually between 0.3 and 0.5 g per pile).

Incubation Procedure

The slices are transferred on time to another beaker containing 5 ml of the incubation medium and appropriate test substances, and the beakers are placed in a Dubnoff shaking incubator. The slices are usually incubated at 37° for 2 hours under an atmosphere of 95% O_2 and 5% CO_2. Sometimes the incubation time has been extended beyond 2 hours to 4 or 6 hours, but the rate of progesterone synthesis usually falls off significantly after 2 hours. One set of slices is placed in a beaker containing the incubation medium and frozen (−20°) immediately without incubation. This is called the zero time sample, and it is prepared in this way to determine the amount of progesterone and 20β-hydroxypregn-4-en-3-one in the tissue before incubation. Another set of slices is incubated in incubation medium alone to determine the amount of steroids synthesized in the control incubations. The other groups of slices are incubated with LH or other test substances. If there is sufficient tissue, duplicate incubations are carried out. All incubations are terminated by freezing the contents of the beakers (—20°).

The incubation medium is Krebs-Ringer bicarbonate buffer pH 7.4 equilibrated with 95% O_2 and 5% CO_2. In earlier studies the incubation medium also contained 2 mg of glucose per milliliter, but the inclusion of this carbohydrate did not appear to affect the control rate of steroidogenesis in incubating slices or their response to gonadotropin, and therefore was omitted in later studies.

Many different substances have been tested for their effect in this *in vitro* system: the hormones of the anterior pituitary; plasma proteins; cyclic AMP; NADPH; other nucleotides; inhibitors of protein and nucleic acid synthesis; and various labeled precursors of steroids, cyclic AMP, and other compounds. These substances are added to the incubation medium before the slices are incubated. Usually the amounts of the test substances are small enough so that no changes in the Krebs-Ringer bicarbonate buffer have to be made. When relatively large amounts of substances, such as the nucleotides, are added to the medium, an equivalent amount of NaCl is omitted to maintain isotonicity, and the pH is readjusted to pH 7.4 if necessary. Nearly all the substances used in our *in vitro* model system are readily soluble in the Krebs-Ringer bicarbonate buffer. Cholesterol, however, presents a particular problem because it is relatively insoluble in the incubation medium. Sufficient solubilization of radioactively labeled cholesterol has been achieved by adding the sterol as a solution in propylene glycol (50–100 μl) to the incubation medium, which is modified by the addition of bovine serum albumin (5 mg/ml).[10] Other investigators have used detergent Tween 80 for solubilization of this sterol.[11]

Isolation and Measurement of Progesterone

Steroidogenesis is assessed in this *in vitro* model system by two methods: the mass of steroid formed, and the amount of [1-^{14}C]acetate incorporated into progesterone. At the outset of the isolation procedure, a trace amount of [7-^{3}H]progesterone (10 pmoles, 0.1 μCi) is added to each vessel. The contents of the vessel (tissue and media) are homogenized in 5 ml of 1% NaOH using a 55-ml Teflon–glass mortar and pestle tissue homogenizer, and the homogenate is transferred to a 250-ml separatory funnel. The vessel and the homogenizer are rinsed with 10 ml of 0.5% NaOH, and the rinsing is added to the homogenate. The homoge-

[10] N. R. Mason and K. Savard, *Endocrinology* **75**, 215 (1964).

[11] P. F. Hall and S. B. Koritz, *Biochemistry* 3, 129 (1964).

nate is extracted 4 times with 100 ml of diethyl ether, and the ether layer is collected in a 500-ml separatory funnel. The pooled ether extract is backwashed twice with 50 ml of H_2O, and then the ether fraction is collected in a 1-liter round-bottom flash and the solvent is evaporated in a rotary evaporator attached to a H_2O aspirator vacuum line. The temperature of the evaporator is kept at approximately 40°.

The residue of this extract is transferred to a centrifuge tube with 4 rinses of 2 ml of diethyl ether, and the ether is evaporated under N_2 in a water bath at 40°. The residue is rinsed to the bottom of the tube by successive rinsings with a few drops of ether followed by evaporation of the solvent. The residue is then submitted to a two-dimensional thin-layer chromatography system adapted from Armstrong *et al.*[12] Square glass plates 20 × 20 cm are covered with a 0.5-mm layer of silica gel GF-254, air dried, heated for 1 hour at 110° and then cooled. The residue of the ether extract is transferred with very small quantities of diethyl ether to a circular area of about 5 mm in diameter which is centered in one corner of the plate 3 cm from each side. The ether is evaporated carefully with an electric air blower. The chromatogram is developed in the first direction by placing the plate in a thin-layer chromatogram tank containing approximately 150 ml of ligroin–ethyl acetate (5:2) and allowing the solvent to migrate to a line 18 cm from the bottom of the plate. The plate is then dried with the air blower and developed in the second direction for 18 cm using a solvent system of methylene dichloride–diethyl ether (5:1.5). After the chromatogram is dried, the progesterone zone (approximately 6 cm in the first direction and 11 cm in the second) and the 20β-hydroxypregn-4-en-3-one zone (approximately 4 cm in the first direction and 7 cm in the second) are located by viewing the plate with a short wavelength, mineral light, ultraviolet hand lamp. These areas of silica gel are scraped into centrifuge tubes and 4 ml of CH_3OH added. The centrifuge tube is shaken 10 times, allowed to stand 15 minutes and then centrifuged at 1000 *g* for 10 minutes to elute the steroids. The amount of steroid in the methanol eluate is determined by comparing its ultraviolet absorption in a spectrophotometer to that of a methanol eluate from an equal area of silica gel from a blank plate. The absorption of the sample above that of the blank is measured at 240 nm, 230 nm, and 250 nm. A corrected absorption is calculated by means of the following Allen[13] formula to eliminated nonspecific absorption:

$$\text{Corrected OD} = (\text{OD } 240 \text{ nm} - \text{OD } 230 \text{ nm} + \text{OD } 250 \text{ nm})/2$$

[12] D. T. Armstrong, J. O'Brien, and R. O. Greep, *Endocrinology* **75**, 488 (1964).
[13] W. M. Allen, *J. Clin. Endocrinol.* **10**, 71 (1950).

Duplicate 100-μl aliquots are taken and transferred to scintillation counting vials to determine the radioactivity in the methanol eluate. The methanol is evaporated and 15 ml of toluene scintillation fluid added. The ^{3}H and ^{14}C radioactivity are measured simultaneously in a dual-channel scintillation counter. The recovery of [7-^{3}H]progesterone (usually 80–90%) is calculated from the ^{3}H radioactivity, and this value is used to correct the mass amount of steroids for the losses incurred during the isolation procedure. The ^{14}C radioactivity in the progesterone samples is used to calculate the incorporation of [1-^{14}C]acetate into this steroid. The radiochemical purity of the ^{14}C in the progesterone zone has been assessed by successive crystalizations using carrier progesterone to constant specific activity, and it has been found to be 90–100% pure [^{14}C]progesterone in both the control and LH-treated incubations. The purity of the ^{14}C in the 20β-hydroxypregn-4-en-3-one zone has not been determined and therefore is not used as a measure of steroidogenesis.

Criteria for a Successful Incubation

The criteria for a successful incubation is the continued steroidogenesis by the incubating slices and their response to LH. In order to document this criteria, the results of the first 20 incubations, carried out with corpora lutea obtained from pregnant cows, have been summarized and are shown in the table. The stages of pregnancy ranged from 0.8 month

RESPONSE OF BOVINE CORPORA LUTEA SLICES TO LUTEINIZING HORMONE[a]

	Mass measurements (μg/g tissue)		[1-^{14}C]-Acetate incorporation[b] (dpm × 10^{-3}/g tissue)
	Progesterone	20β-Hydroxy-pregn-4-en-3-one	
Content before incubation	44.0 ± 14.3	8.15 ± 10.1	—
Control incubation	105 ± 29.3	31.5 ± 14.6	222 ± 127
LH incubation[c]	228 ± 62.2[d]	50.2 ± 26.1[d]	786 ± 436[d]

[a] These data represent the summarized (mean ± standard deviation) results of the first 20 experiments using corpora lutea of pregnant cows and the incubation conditions described in the text.

[b] 2 μCi/ml of [1-^{14}C]acetate was added to the incubated samples.

[c] LH (NIH-LH-S-1), 2 μg/ml.

[d] Using paired comparison statistics the probability that the difference (between the LH-treated samples and the controls) was due to chance alone was less than 0.001.

to 4 months, and the cows were of several breeds. LH was added at 2 μg/ml and [1-^{14}C]acetate at 2 μCi (0.1 μmole/ml). The incubations were carried out for 2 hours in an atmosphere of 95% O_2 and 5% CO_2. Only one corpus luteum of these first 20 failed to respond to LH with a greater than 100% increase in both the mass of steroid formed and the incorporation of [1-^{14}C]acetate into progesterone. No reason for this failure was established, and after carrying out more than 200 incubations we find that the expected failure rate is about 3%. The mean value ± the standard deviation indicate what range of values can be expected using this procedure, and although there is considerable variability from one corpus luteum to the next, paired comparison statistics of the values from the control and LH-treated slices of the same corpora lutea show that the stimulatory effect of LH is highly significant ($P < 0.001$). There is more variability in the extent of [1-^{14}C]acetate incorporation than in the mass amount of steroidogenesis, which probably reflects differences in the pool size of the intermediates in the steriodogenic pathway.

It has been reported in previous publications that this effect is quite specific for LH containing preparations[5] and that these incubating slices will respond to as little as 2 ng/ml of LH.[14] The response of the slices increases as the dose of LH increases from this value up to a saturating dose of about 0.2 to 2 μg/ml. Other measurements such as the assessment of the endogenous levels of NADPH[4] and cyclic AMP[15] have been carried out on these incubating slices, and the details of these analytical procedures can be obtained from the original papers.

Finally, as an indication of the simplicity and reproducibility of the technique, we can cite the success which has been attained by other laboratories[9,16] and the fact that this procedure has been carried out successfully for several years by graduate students as a laboratory exercise.

[14] J. M. Marsh and K. Savard, *J. Reprod. Fert.*, Suppl. 1, **113** (1966).
[15] J. M. Marsh, R. W. Butcher, K. Savard, and E. W. Sutherland, *J. Biol. Chem.* **241**, 5436 (1966).
[16] D. T. Armstrong and D. L. Black, *Endocrinology* **78**, 937 (1966).

[21] Perfusion of the Placenta *in Vitro*

By L. Cedard and E. Alsat

The ready availability of human placentas explains why numerous attempts have been made to maintain extracorporeally the functions of the full-term human placentas obtained after vaginal delivery or cesarean section as well as those of mid-gestation placentas or of the complete fetoplacental unit.

The placenta is unique as an organ in that two distinct active blood circulations are involved.

The fetal blood enters the placenta through the paired umbilical arteries. Primary arterial branches with their cruciated anastomoses pass through the chorionic plate, giving off smaller branches. Each fetal cotyledon seems to be an independent circulatory unit. Hence it is possible to perfuse the vasculature of only one cotyledon or of a small group of cotyledons. The fetal blood stream is brought nearest to the maternal blood circulating in the intervillous space supporting fetomaternal exchanges through the hemochorial placental membrane.

Maternal placental blood is derived from the anastomosis of uterine and ovarian arteries. The arcuate branches of this anastomosis run parallel beneath the surface of the uterus, sending radial perforating branches through the extensive myometrial venous network. A combination of histomorphology, injection techniques, radioangiography, and comparative studies between the site of the intervillous space and the structure of the fetal cotyledon has identified the maternal spiral artery and the fetal cotyledon as a functional unit. The maternal blood-leaving the central space enters the narrow capillary-like system of the intervillous spaces and returns into the maternal veins distributed beneath the placenta. A gradient exists, with regard to hydrostatic pressure as well as O_2 tension, from the spiral artery through the central space, peripheral intervillous capillary-like system, and maternal veins.

Perfusion of the Fetal Circulation

Propulsion of perfusion fluid through fetal circulation is easy to accomplish with inexpensive equipment, and many people have been interested by this approach.

It is possible to perfuse isolated placental cotyledons[1,2] or the whole organ.[3]

Some authors used a single-cycle perfusion, the fluid passing through the placenta only once. Others used recycling; e.g., the blood obtained from the umbilical vein was recirculated through the placenta for the duration of the experiment.

The perfusion fluid usually employed consists of balanced buffered salt solutions (Tyrode, Krebs-Ringer, or Hanks' solution); sometimes

[1] L. C. Chesley and I. E. McFaul, *Amer. J. Obstet. Gynecol.* **58,** 159 (1949).

[2] M. Levitz, G. P. Condon, and J. Dancis, *Endocrinology* **58,** 376 (1956).

[3] J. Varangot and J. A. Thomas, *C. R. Acad. Sci.* **228,** 132 (1949).

human or bovine blood is added to these solutions; sometimes blood is used alone. More recently, some large molecular material, such as Rheomacrodex, has been employed.[4] Pulsatile perfusion through the umbilical arteries is ensured by artificial heart pumps which mimic the fetal heart rhythm.

The techniques used for the perfusion of the fetoplacental circulation in rodents (rat, guinea pig) or in the bovine species are similar to those used for the perfusion of human material; the slight differences are due to different vascularization.[5] Since the reported general conditions of perfusion are very similar, it seems most convenient to describe in detail the technique used in our laboratory.[6,7]

The placenta obtained at cesarean section or after a natural confinement is collected aseptically. It is immediately free from the membranes. The umbilical blood vessels are dissected free over a length of 5 cm. The two umbilical arteries are catheterized by means of perfusion needles with olive-shaped points fixed by a ligature and a Murphy artery clip. The umbilical vein is cut across and sometimes is also cannulated by a catheter of 3–5 mm. Ten milliliters of heparin are injected via the arteries to prevent clotting.

The placenta is attached to the perfusion apparatus (which is illustrated diagrammatically in Fig. 1) within 10–30 minutes after its removal.

The perfusion equipment can be described as consisting of three different circuits.

The perfusion circuit is set in motion by a finger pump (Sigma motor) which gives a pulsatile movement with adjustable pressure and speed.

The perfusion fluid consists of a liter of Krebs-Ringer solution buffered to pH 7.4 with 0.15 M Na_2HPO_4 to which 2 g of glucose, 40 mg of papaverine, and antibiotics (1 g of streptomycin and 10^8 units of penicillin) are added. The fetal circulation is washed out during 5–10 minutes, after which 200 ml of preserved human blood are added.

The perfusion fluid leaving the placenta through the umbilical vein as well as through the basal plate is collected in a reservoir and recycled.

Oxygenation is carried out in this reservoir by means of a rotating mixer, which maintains the perfusion fluid in a thin layer.

Initally, 5% carbon dioxide in oxygen was used for oxygenation.

[4] E. Klinge, M. J. Mattila, O. Pentillä, and E. Jukarainen, *Ann. Med. Exp. Fenn.* **44**, 369 (1966).

[5] M. Panigel, *Amer. J. Obstet. Gynecol.* **89**, 1664 (1962).

[6] J. Varangot, L. Cedard, and S. Yannotti, *Amer. J. Obstet. Gynecol.* **92**, 534 (1965).

[7] L. Cedard, IV *Acta Endoc.* **158**, 331 (1972).

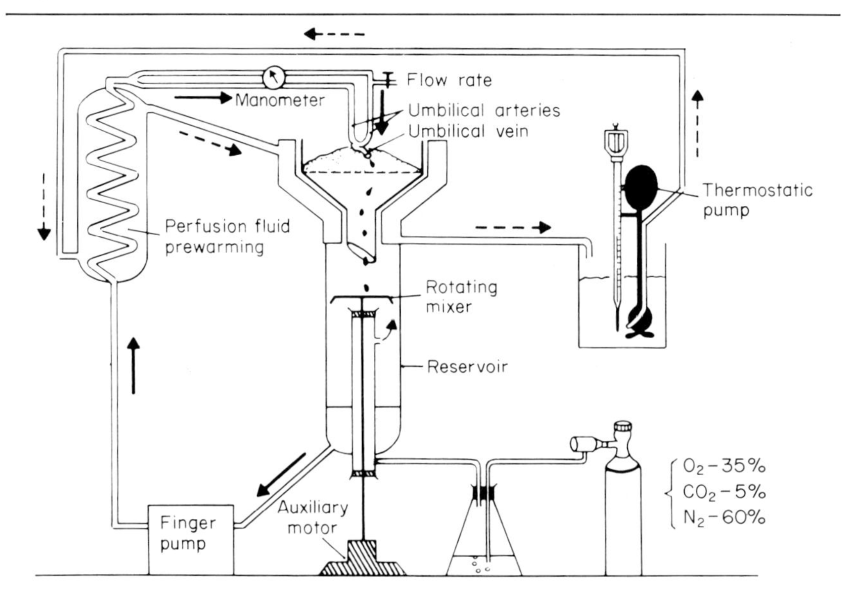

FIG. 1. Schematic diagram of fetal circulation apparatus. From J. Varangot, L. Cedard, and S. Yannotti, *Amer. J. Obstet. Gynecol.* **92**, 534 (1965).

Since two years ago, this has been replaced by a less toxic gas mixture consisting of 60% N_2, 5% CO_2. and 35% O_2. The presence of CO_2 is necessary for preventing vasoconstriction. The temperature of the perfusion fluid as well as that of the placenta is kept at 37° by means of a thermostatic pump (Roucaire) connected to a heating circuit. Prewarming of the fluid ensures a constant solubility of the gases throughout the whole equipment.

The pump frequency is 140–160 strokes per minute, similar to the fetal heart rhythm.

The flow rate ranges from 30 to 100 ml per minute, and the infusion pressure (recorded by a mercury manometer located between the pump and the arterial cannula) is 35–80 mm Hg for the minima and 75–175 mm Hg for the maxima, depending of the finger pump regulation. This pressure is higher than the umbilical pressure measured by Nyberg and Westin[8] at the fetal end of the umbilical cord immediately after birth (87.8/54.5 mm Hg); the umbilical venous pressure is reported to vary between 22 and 34 mm Hg.[9]

[8] R. Nyberg and B. Westin, *Acta Paediat.* (*Uppsala*) **47,** 357 (1958).

[9] P. Wilkin, *in* "Le placenta humain" (J. Snoeck, ed.), p. 158. Masson, Paris, 1958.

The duration of the perfusion is between 2 and 5 hours according to the type of experiment. Usually the venous effluent and the fluid leaving the placenta through the basal plate are mixed and constantly recycled. In kinetic studies the fluid is sampled at the entry of one umbilical artery at regular time intervals.

For certain studies the fetal and maternal fluids are collected separately; in such experiments the fluid passes only once through the placenta.

It is also possible to inject a definite quantity of a stimulating or inhibiting substance directly into one umbilical artery by means of an automatic syringe (Perfusor Braun Melsungen) in addition to the material (labeled or nonlabeled steroids or precursors) which was introduced into the system at the beginning of the experiment.

The efficiency of the fetal perfusion is easy to demonstrate. From the mechanical point of view, at the very onset of perfusion the liquid escapes via the umbilical vein, sometimes in the form of a jet, whereas the arteries that are visible under the amnion become turgescent and pulsatile.

The rapidity of the venous return is not due solely to the presence of superficial arteriovenous anastomoses, which would reflect an incomplete perfusion. Indeed, by introducing dyes into the perfusion liquid (e.g., Prussian blue, methyl green), at a subsequent histological examination, the colored liquid is found in the blood vessels of the villi. Nevertheless, the conditions of the perfusion are not entirely physiological since too high an arterial pressure is required to ensure the venous return.

Sometimes a slight edema of the placenta may be observed at the beginning of the perfusion. Furthermore, there is nearly always intercommunication between the fetal and maternal circuit during the perfusion experiment, and the effluent through the maternal placental surface is usually higher than that through the umbilical vein. However, it is possible that this fetomaternal transfer is—at least in part—physiological.

Development of an Artificial Uterine Chamber and Perfusion of the Intervillous Space

Krantz *et al.*[10] described an original apparatus for the extracorporeal perfusion of the whole organ.

The placenta is placed in a chamber so that the maternal surface is perforated by a series of semirigid polyethylene tubes passing through the basal plate. The blood passing through this system represents the

[10] K. E. Krantz, T. C. Panos, and J. Evans, *Amer. J. Obstet. Gynecol.* **83**, 1214 (1962).

maternal circulation. The fetal surface is exposed to an artificial amniotic cavity filled with artificial or natural amniotic fluid which exerts a hydrostatic pressure similar to the normal amniotic pressure.

The fetal and maternal sides of the placenta are then perfused with fresh, filtered, heparinized blood diluted with perfusion fluid in the maternal circuit which passes through an oxygenator in which a mixture of 95% O_2 + 5% CO_2 circulates.

Initially the authors experienced some difficulties in maintaining the fetal circulation; however, the addition of 40% Rheomacrodex to the fetal circuit increased the osmotic pressure and virtually eliminated fluid transfer from the fetal circulation[11] which was previously inevitable.

The fetal circulation is driven at a flow rate of approximately 120 ml per minute and the maternal circulation is propelled by an artificial heart pump at a flow rate of 600 ml per minute. Recordings and corrections of the pH, pCO_2, pO_2, temperature, gas flow, arterial pressure, and flow rate are made. The maternal pH is maintained at 7.25 to 7.40, and the fetal pH range is 7.1 to 7.3. The pO_2 is kept at 35 to 100 mm Hg for the maternal and 15 to 40 mm Hg for the fetal circuit. The apparatus has been used without any significant modifications by other investigators.[12,13,14a,b] Very recently Krantz *et al.*[11] have published data on glucose utilization and ATP maintenance which demonstrate active metabolism during perfusion *in vitro* and confirm the viability of the placental tissue under these conditions.

However, the introduction of several plastic perfusion tubes through the basal plate into the intervillous space gave rise to some criticism. Thus, using a cineradiographic technique to follow the perfusion flow to which radioopaque dyes are added, Panigel[15] observed that under the above conditions the placental circulation of the perfusion fluid is far from satisfactory and does not at all imitate the normal flow of the maternal blood. Moreover, the fetal vascular tree is usually damaged, as shown by the rapid leakage of radioopaque fluid from the fetal to the maternal side.

[11] K. E. Krantz, F. J. Blakey, K. Yoshida, and J. A. Romito, *Obstet. Gynecol.* **37**, 183 (1971).

[12] F. Charreau, W. Jung, J. Loring, and C. Villee, *Steroids* **12**, 29 (1968).

[13] E. Van Bogaert, *Gynecol. Obstet.* **69**, 387 (1970).

[14a] A. L. Southren, Y. Kobayashi, W. Jung, N. C. Carmody, and A. Weingold, *J. Clin. Endocr.* **26**, 1005 (1966).

[14b] A. L. Southren, Y. Kobayashi, W. Jung, and A. Weingold, *J. Clin. Endocr.* **28**, (1968).

[15] M. Panigel, *The Foeto-placental Unit, Excerpta Med. Found. Int. Congr. Ser.* **183**, 279 (1969).

Panigel has consequently tried to find a way to perfuse the intervillous space of a maternal interseptal compartment corresponding to one or more fetal cotyledonary units on the fetal side. This very delicate technique requires the cannulation of the placental end of a spiral artery on the uterine surface of the placenta with a fine catheter of 200 μm or more.

Recently Nesbitt *et al.*[16] described a newly designed apparatus for perfusion of the whole organ, achieving dual closed circulation. The bottom of the artificial uterus is concave, and 151 holes are bored for the arterial inlets and veinous outlets. Uterine spiral arterioles are simulated by sections of 18-gauge needles which protrude above the bottom surface about 6 mm.

The top of the artificial uterus contains a cavity for amniotic fluid with manometer tap and plug for initially filling this cavity. The tests of functional capability of the apparatus are satisfactory, but the transfer of fluid from the fetal to the maternal circulation was a continuing problem, and the authors think that some cytolytic changes are created by an increase in the level of metabolic toxins generated in the placental tissue.

Interest of in Vitro Perfusion for Studying Hormone Production. Placental perfusion *in vitro* has been successful in extensive studies of the steroidogenic processes.

The results obtained following the perfusion of the isolated placenta confirm and extend the results obtained by simple analysis of placental tissues, by incubation experiments with slices or subcellular preparations, and by *in vivo* techniques reported by many investigators.

The perfusion technique allows us to duplicate the tissue preparation work when we analyze the total perfusate collected over an extended period of time. On the other hand, sampling the umbilical venous outflow at frequent time intervals allows the detection of short-term dynamic changes and permits the performance of longer experiments and more exact kinetic studies than does *in vivo* perfusion.

Prolongation of the survival of the enzymatic activity of the entire placenta in perfusion experiments permits the study of the production of steroids from endogenous material as well as of the metabolism of cold or radioactive precursors; for instance, it has been possible to obtain interesting results concerning estrogen production. The perfusion technique has also made it possible to confirm the formation of estrone and estradiol from C_{19} neutral steroids, to demonstrate the formation of estriol from C_{16} hydroxylated compounds only and to indicate that 16-hydroxy-

[16] R. E. L. Nesbitt, Jr., P. A. Rice, J. E. Rourke, V. F. Torrest, and A. M. Souchay, *Gynec. Invest.* **1**, 185 (1970).

lation is almost nil in human placentas.[6] Furthermore, the formation of equilin and equilenin from Δ_7-dehydroepiandrosterone,[17] the biosynthesis of nonphenolic, aromatic C_{18} steroids by direct aromatization of Δ^{5-7} C_{19} steroids[18] and 6α-hydroxylation of neutral as well as phenolic compounds[19,20] have also been demonstrated. It has also been possible to confirm by *in vitro* perfusion[12] the metabolic pathway from dehydroepiandrosterone sulfate to estrone and estradiol which was established following *in situ* perfusion of mid-term placentas[21] and to explain a paradoxical estrogen insufficiency by a placental sulfatase defect.[22]

Moreover, the possibility of studying the aromatizing ability of the whole organ enabled us to demonstrate that the placental aromatase increases progressively during pregnancy.[23] There is a sudden rise in aromatization between the 12th and 20th week due to a progressive enzymatic maturation rather than to a simple ponderal growth.

The great importance of the placental perfusion method is the possibility of maintaining intact the placental structure and preserving the cellular receptors to some stimuli, such as gonadotropins which have no action in placental preparation.

A stimulating action of human chorionic gonadotropin (hCG) on estradiol hydroxylation was first reported by Troen[24] and Varangot *et al.*[25] Since then we have been able to demonstrate that hCG and luteinizing hormone have a stimulating effect on the aromatization[26] as well as on glycogenolysis.

Cyclic AMP has the same stimulatory action on both parameters, and prostaglandins $F_{2\alpha}$, E_2, and especially E_1 stimulated a significant increase in the estrogens produced from testosterone by the whole human placenta perfused *in vitro*.[27]

Finally, comparative studies of placentas obtained through spontaneous vaginal delivery, by elective cesarean section before the onset of

[17] L. Starka, H. Breuer, and L. Cedard, *J. Endocrinol.* **34,** 447 (1966).

[18] L. Starka and H. Breuer, *Biochim. Biophys. Acta* **115,** 306 (1966).

[19] L. Cedard and R. Knuppen, *Steroids* **6,** 307 (1965).

[20] C. Alonso and P. Troen, *Biochemistry* **5,** 337 (1966).

[21] E. Lamb, S. Mancuso, S. Dell'Acqua, N. Wiqvist, and E. Diczfalusy, *Acta Endocrinol.* (*Copenhagen*) **55,** 263 (1967).

[22] L. Cedard, C. Tchobrousky, R. Guglielmina, and M. Mailhac, *Bull. Fed. Soc. Gynecol. Obstet.* **23,** 16 (1971).

[23] L. Cedard, J. Varangot, and S. Yannotti, *Eur. J. Steroids* **1,** 287 (1966).

[24] P. Troen, *in* "Recent Progress in the Endocrinology of Reproduction" (C. W. Lloyd, ed.), p. 299. Academic Press, New York, 1959.

[25] J. Varangot, L. Cedard, and S. Yannotti, *Excerpta Med. Found. Int. Congr. Ser.* **51,** Abstract 396 (1962).

[26] L. Cedard, E. Alsat, M.-J. Urtasun, and J. Varangot, *Steriods* **16,** 361 (1970).

[27] E. Alsat, and L. Cedard, *Prostaglandins* **3,** 145 (1973).

labor, or by cesarean section during labor for dystocia or fetal distress have shown discrepancies according to the mode of delivery.[28,29]

Perfusion of the entire placenta has enabled us to demonstrate a precocious fall in the synthesis of progesterone occurring before or during the first hours of labor, while the aromatization is maintained at a high level until dislodging of the placenta takes place. This modification of the estrogen:progesterone ratio might be one of the factors responsible for the mechanism of parturition.

Conclusion

Despite the fact that physiological conditions cannot be achieved, it is possible to accomplish easily placental perfusion of the fetal circulation. The advantages of perfusing the whole organ are that the method give a notion of the total enzymatic activity, allows kinetic studies in a closed or an open system, and preserves the cellular receptors.

To obtain dual closed circulation one needs a more sophisticated apparatus; it is more difficult to accomplish, and many placental specimens are too damaged to be used.

[28] L. Cedard, T. Pajszczyk-Kieszkiewicz, and E. Alsat, *C. R. Acad. Sci.* **269,** 223 (1969).

[29] E. Alsat, A. Guichard, T. Pajszczyk-Kieszkiewicz, E. Richard, and L. Cedard, *C. R. Acad. Sci.* **271,** 1783 (1970).

[22] Gonadotropic Stimulation of Interstitial Cell Functions of the Rat Testis *in Vitro*

By M. L. Dufau and K. J. Catt

The responses of the rat testis to gonadotropic hormones *in vitro* have been extensively evaluated in terms of steroidogenesis, activation of adenylate cyclase, and tropic hormone binding. Most studies have dealt with interstitial cell functions, and relatively few have been concerned with the responses of seminiferous tubules to gonadotropins. Because tubular function is believed to be affected predominantly by follicle-stimulating hormone (FSH) and local androgen concentration, and Leydig cell function depends primarily upon luteinizing hormone (LH), studies on the hormonal responses of tubules *in vitro* have usually dealt with metabolic changes during FSH stimulation, and those on Leydig cells have examined steroid biosynthesis and androgen production in response to

LH or human chorionic gonadotropin (hCG). In this article, the relevant literature references to FSH effects *in vitro* will be summarized, and methods for the study of responses of isolated testicular tissue and interstitial cells to LH and hCG will be described.

FSH and Testicular Metabolism

FSH has been shown to produce rapid stimulation of RNA, protein and phospholipid synthesis in the testes of immature rats,[1] and also to enhance adenylate cyclase activity of the seminiferous tubules.[2-4] Demonstration of such effects has been usually performed in testes of immature or hypophysectomized rats, which show maximal responses to FSH. In particular, FSH effects on adenylate cyclase have been difficult to demonstrate in the adult animal, in contrast to the ready stimulation of cyclic AMP production by LH or hCG in both mature and immature rat testes.[3,5-7] However, activation of testicular adenylate cyclase by FSH in adult animals has been reported in the dog and rat in the presence of theophylline,[2] and has been shown to be consistently demonstrable in adult rats when phosphodiesterase activity is inhibited by more potent agents, such as 1-methyl-3-isobutyl xanthine.[8] The binding of tritiated FSH to specific membrane receptors of the germinal epithelium,[9] and the subsequent stimulation of adenylate cyclase and activation of protein kinase, has been recently reviewed by Means.[10]

Gonadotropic Stimulation of Leydig Cell Function

Studies on the stimulation of testicular androgen synthesis by gonadotropins *in vitro* have been largely performed in sliced or perfused rabbit

[1] A. R. Means, *in* "The Human Testis" (E. Rosenberg and A. C. Paulsen, eds.), p. 301. Plenum, New York, 1970.

[2] F. Murad, B. Strauch, and M. Vaughan, *Biochim. Biophys. Acta* **177,** 591 (1969).

[3] F. A. Kuehl, D. J. Panatelli, J. Tarnoff, and J. L. Humes, *Biol. Reprod.* **2,** 154 (1970).

[4] J. H. Dorrington, R. G. Vernon, and I. B. Fritz, *Biochem. Biophys. Res. Commun.* **46,** 1523 (1972).

[5] K. J. Catt, K. Watanabe, and M. L. Dufau, *Nature* (*London*) **239,** 280 (1972).

[6] B. A. Cooke, W. M. O. van Beurden, F. F. G. Rommerts, and H. J. van der Molen, *FEBS Lett.* **25,** 83 (1972).

[7] M. L. Dufau, K. Watanabe, and K. J. Catt, *Endocrinology* **92,** 6 (1973).

[8] T. Braun and S. Sepsenwol, *Endocrinology* **92,** Suppl., A-98 (1973).

[9] A. R. Means and J. Vaitukaitis, *Endocrinology* **90,** 39 (1972).

[10] A. R. Means, *in* "Receptors for Reproductive Hormones" (B. O'Malley and A. R. Means, eds.), p. 400. Plenum, New York, 1973.

testes[11-13] and in the perfused canine testis.[14] The rat testis has not been widely utilized for studies on gonadotropin actions *in vitro,* and the marked sensitivity of the interstitial cells of the rat testis to LH and hCG stimulation was not recognized until quite recently.[15-17] This is partly attributable to the relatively marked loss of responsiveness to gonadotropins which results from simple physical teasing apart of the testis, an effect which has been confirmed by other observers.[18]

The ability of LH and chorionic gonadotropin from various species to stimulate steroidogenesis in the Leydig cells of the isolated rat testis provides a useful *in vitro* bioassay procedure of high sensitivity for such gonadotropins,[15,16] as well as a valuable system for analysis of the mechanisms of action of gonadotropins on testicular steroidogenesis.[5,7,17,19,20] As a bioassay, the steroidogenic response of isolated rat testes or collagenase-dispersed Leydig cells gives high sensitivity, which is unattainable by conventional bioassays,[17] and also provides considerably improved precision and reproducibility (Fig. 1). Existing bioassays for LH and hCG depend upon responses of secondary target tissues to steroids produced by the gonads *in vivo* (e.g., the ventral prostate of the hypophysectomized male rat[21] and the uterus of the immature female rat[22] or mouse[23]) or upon the depletion of ascorbic acid in the ovary of the immature pseudopregnant rat.[24] Such assays are technically difficult and time-consuming, and are sometimes subject to unpredictable fluctuations in precision which can be at least partially improved by computerized covariance analysis of the assay data.[25] A valuable feature of such *in vivo* bioassays

[11] R. O. Brady, *J. Biol. Chem.* **193,** 145 (1951).

[12] P. F. Hall and K. B. Eik-Nes, *Biochim. Biophys. Acta* **63,** 411 (1962).

[13] L. L. Ewing and K. B. Eik-Nes, *Can. J. Biochem.* **44,** 1327 (1966).

[14] K. B. Eik-Nes, *Recent Progr. Horm. Res.* **27,** 517 (1971).

[15] M. L. Dufau, K. J. Catt, and T. Tsuruhara, *Biochim. Biophys. Acta* **252,** 574 (1971).

[16] M. L. Dufau, K. J. Catt, and T. Tsuruhara, *Endocrinology* **90,** 1032 (1972).

[17] K. J. Catt and M. L. Dufau, *in* "Receptors for Reproductive Hormones" (B. O'Malley and A. R. Means, eds.), p. 379. Plenum, New York, 1973.

[18] F. F. G. Rommerts, B. A. Cooke, J. W. C. van der Kemp, and H. J. van der Molen, *FEBS Lett.* In press.

[19] M. L. Dufau, K. J. Catt, and T. Tsuruhara, *Biochem. Biophys. Commun.* **44,** 1022 (1971).

[20] M. L. Dufau, K. J. Catt, and T. Tsuruhara, *Proc. Nat. Acad. Sci. U.S.* **69,** 2414 (1972).

[21] J. W. McArthur, *Endocrinology* **50,** 304 (1952).

[22] A. Albert, *Recent Progr. Horm. Res.* **12,** 227 (1956).

[23] J. A. Loraine and J. B. Brown, *Acta Endocrinol.* (*Copenhagen*) **17,** 250 (1954).

[24] A. F. Parlow, *in* "Human Pituitary Gonadotropins" (A. Albert, ed.), p. 300. Thomas, Springfield, Illinois, 1961.

[25] F. Sakiz and R. Guillemin, *Endocrinology* **72,** 304 (1963).

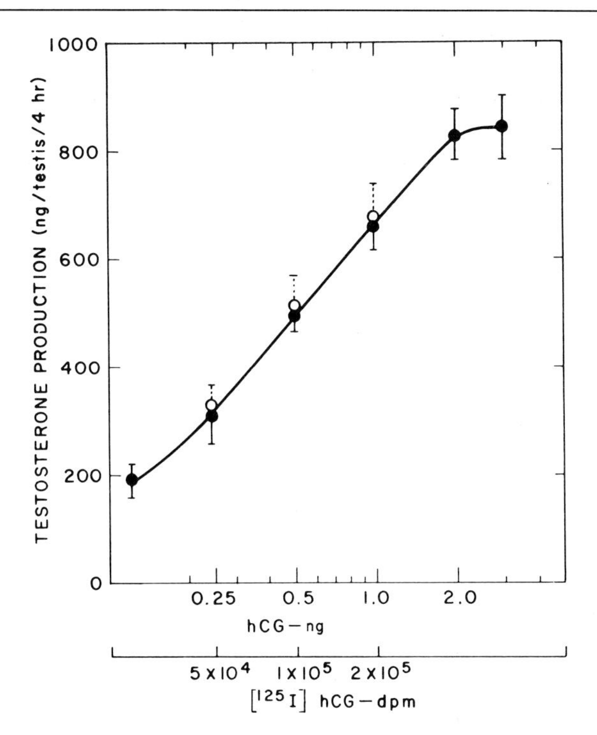

FIG. 1. Bioassay of an ^{125}I-labeled human chorionic gonadotropin (hCG) preparation (○——○) in the isolated rat testis system *in vitro,* by comparison with standards of the unlabeled gonadotropin (●——●). Note the dose-response range of hCG to 0.1–2 ng, equivalent to 1–20 mIU per testis. Each point shows the mean ±SD of quadruplicate estimations.

is their ability to provide an overall estimate of hormonal efficacy in the whole animal, by combining such factors as metabolic clearance and target cell activation in the final response metameter used to determine the specific hormonal activity. Such a quality is useful for certain purposes, such as those concerned with evaluation of *in vivo* responses, and inconvenient for others, particularly when the biological activity of gonadotropins at the target cell level is to be evaluated.

For precise bioassay of the intrinsic activity of small quantities of LH or hCG, the steroidogenic response of isolated rat testes or collagenase-dispersed interstitial cells provides a highly specific and sensitive method which is responsive *in vitro* to physiological concentrations of gonadotropins. The testis assay is sensitive to subnanogram quantities of LH or hCG, giving a dose-response curve between 1 and 20 mIU of hCG, and 5 and 100 mIU of hMG (2nd IRP Standard). The precision of the method is relatively high, the index of precision (λ) being less than 0.1 and the mean regression coefficient (Finney's g) being 0.025. The use of collagenase-dispersed interstitial cells provides an assay of even higher sensitivity and precision. In the presence of phosphodiesterase inhibitors [e.g., 0.1 mM 1-methyl-3-isobutyl xanthine, (MIX)], assay sensitivity is increased to 2 pg (20 μU) of hCG, and precision is significantly improved with a λ of less than 0.04, and a g factor of 0.025. Such bioassays are thus of comparable sensitivity and precision to radioimmunoassay and permit bioassay of endogenous LH levels in plasma.

Testis Preparations

Sliced and Teased Testis Tissue

Although testis slices have been utilized for studies of androgen biosynthesis in the rabbit,[26] the loose anatomical structure of the rat testis makes it unsuitable for such preparations. The seminiferous tubules of the rat testis are surrounded by an unusually abundant plexus of lymphatics, and there are no connective tissue septa as in most other mammals. The use of teased rat testis preparations was introduced by Christensen and Mason,[27] who confirmed that the interstitial cells are the major source of testicular androgens and that the tubule fraction transforms progesterone to 17α-hydroxyprogesterone, Δ^4-androstenedione, and testosterone. Later studies in similar preparations by Hall *et al.*[28] showed that cholesterol is converted to testosterone by interstitial tissue but is not metabolized by seminiferous tubules.

More recent studies have employed the separated compartments of the teased rat testis for experiments on the binding of gonadotropins,[29-32] activation of adenylate cyclase,[2-7] and stimulation of steroidogenesis *in vitro*.[15,18]

Method of Teasing. Decapsulated testes from mature 200–300 g rats are gently teased apart with fine forceps in petri dishes containing cold phosphate-buffered saline pH 7.4. After dispersion of the testes to sepa-

[26] G. M. Connell and K. B. Eik-Nes, *in* "The Gonads" (K. W. McKerns, ed.), p. 491. Appleton, New York, 1969.

[27] A. K. Christensen and N. R. Mason, *Endocrinology* **76,** 646 (1965).

[28] P. F. Hall, D. C. Irby, and D. M. de Kretser, *Endocrinology* **84,** 488 (1969).

[29] K. J. Catt, M. L. Dufau, and T. Tsuruhara, *J. Clin. Endocrinol.* **32,** 860 (1971).

[30] K. J. Catt, M. L. Dufau, and T. Tsuruhara, *J. Clin. Endocrinol.* **34,** 123 (1972).

[31] M. L. Dufau, T. Tsuruhara, and K. J. Catt, *Biochim. Biophys. Acta* **278,** 281 (1972).

[32] N. R. Moudgal, W. R. Moyle, and R. O. Greep, *J. Biol. Chem.* **246,** 4983 (1971).

rate the majority of the tubules, the suspension is filtered through cotton wool or fine nylon mesh to separate the bulk of the tubules from the interstitial cell fraction. The spermatozoa present in the Leydig cell suspensions prepared in this manner can be removed by selective filtration through Sephadex G-10 or by density gradient separation. In addition, a large number of Leydig cells are ruptured by shearing stress during teasing of the tubules. Intact cells are separated from the resulting subcellular particles by centrifugation at 120 g for 20 minutes; this step is important for binding studies on the Leydig cell fraction, because substantial binding of labeled gonadotropin by membrane particles can occur in such teased preparations.[29,30]

Incubation of physically dispersed Leydig cells is performed in Krebs-Ringer bicarbonate buffer, or in tissue culture media such as medium 199, under 95% O_2–5% CO_2 at 34° in the presence of 5 mM glucose. Physically dispersed interstitial cells usually respond poorly to hormonal stimulation by LH or hCG,[15,18] with relatively small increases in androgen production by comparison with those observed in intact testes or enzymatically dispersed interstitial cells (see below).

Enzymatic Dispersion of Interstitial Cells

Suspensions of interstitial cells are readily prepared by brief exposure to collagenase or trypsin as described for other tissues including adipose tissue,[33] adrenal,[34] and corpus luteum.[35] Although trypsin dispersion of rat testes provides a relatively homogeneous interstitial cell fraction with little tubule breakage and very few spermatozoa, the Leydig cells show significant reduction of gonadotropin binding sites by comparison with collagenase-digested cells, and appear to be much more fragile during subsequent incubation studies.[36] By contrast, collagenase-dispersed Leydig cells are responsive to stimulation with low concentrations of LH or hCG[17] and are more stable during incubation for several hours *in vitro*. For collagenase dispersion, decapsulated rat testes are incubated in KRB buffer or medium 199 containing collagenase (Worthington, type 1) 0.25–1 mg/ml and BSA (1 mg/ml). After addition of 6 testes to 6 ml of incubation medium in 45-ml plastic centrifuge tubes, the tubes are gassed with 95% O_2–5% CO_2 and capped. Incubation is performed by shaking the tubes at 100 cycles per minute in their long axis in a metabolic shaker

[33] M. Rodbell, *J. Biol. Chem.* **239**, 375 (1964).
[34] N. D. Giordano and G. Sayers, *Proc. Soc. Exp. Biol. Med.* **136**, 623 (1970).
[35] D. Gospodorawicz, *Endocrinology* **89**, 669 (1971).
[36] C. Mendelson, M. L. Dufau, and K. J. Catt, unpublished observations.

at 34°.[37] For treatment of more than 6 testes, shaking is performed in a 240-ml Nalgene centrifuge bottle, gassed with 95% O_2–5% CO_2 and capped. After about 15 minutes, the individual testes begin to merge into a homogeneous mass of dispersed tubules, surrounded by a turbid suspension of interstitial cells. The tubes are then filled with cold medium 199, inverted gently several times, and allowed to stand for 5 minutes at room temperature; the turbid supernatant solutions are aspirated with a 35-ml plastic syringe connected to a 10-cm length of 2 mm i.d. Tygon tubing and pooled in a 250-ml Nalgene centrifuge bottle. Each batch of residual tubules is then washed by repeated gentle inversion with 5 ml of medium per testis; the combined supernatants are then centrifuged at 200 *g* for 20 minutes to sediment the interstitial cell fraction, which is finally resuspended in fresh incubation medium. By this method, contamination with spermatozoa is usually less than 5% and Leydig cell breakage is minimal in comparison with that caused by physical dispersion. Viability of the dispersed cells as assessed by Trypan blue exclusion is usually close to 100%, and counting by hemacytometer after staining for 3β-ol dehydrogenase[38] shows that about half of the cells possess formazan granules. The usual number of Leydig cells derived from each adult rat testis is approximately 40×10^6. Incubation of dispersed Leydig cells should be performed in plastic vials, such as polyethylene liquid scintillation counting vials (Packard), or in siliconized glass vials. Under these conditions, disruption of the dispersed cells is negligible during incubation at 34° for up to 4 hours.

Leydig cells prepared by collagenase digestion exhibit satisfactory morphological characteristics and respond to gonadotropic hormone with synthesis of cyclic AMP and testosterone.[17-37] However, the binding of labeled gonadotropin is sometimes detectably reduced by collagenase treatment, and variations in hormone responsiveness are apparent after digestion with certain batches of collagenase. The nature of this effect is not yet clear; the presence of proteolytic activity in crude collagenase is a possible contributory factor, but addition of high concentrations of BSA during digestion does not appear to prevent loss of binding sites. Addition of trypsin inhibitor following collagenase digestion of Leydig cells has also been performed on an empirical basis,[37] but no systematic study on optimal conditions for collagenase treatment of the testis has been reported. Such variations are probably due to uncontrolled factors in the preparation of individual batches of collagenase by commercial suppliers.

[37] W. R. Moyle and J. Ramachandran, *Endocrinology* **93**, 127 (1973).

[38] A. H. Baillie, M. M. Ferguson, and D. McK. Hart, "Developments in Steroid Histochemistry," p. 7. Academic Press, New York, 1966.

When collagenase-dispersed interstitial cells are incubated with phosphodiesterase inhibitors during studies on cyclic AMP formation, the concentration of xanthine employed should be lower than that recommended for incubations of intact testes. During hCG stimulation, cyclic AMP recovery is elevated by increasing concentrations of theophylline up to 10 m*M*. However, testosterone synthesis by dispersed Leydig cells is inhibited by 5 m*M* theophylline, and the medium content of theophylline should not exceed 2 m*M* for optimum testosterone production (Fig. 2). When more potent phosphodiesterase inhibitors are employed, accordingly lower concentrations should be used to permit maximum testosterone production; e.g., for MIX, the optimum concentration for

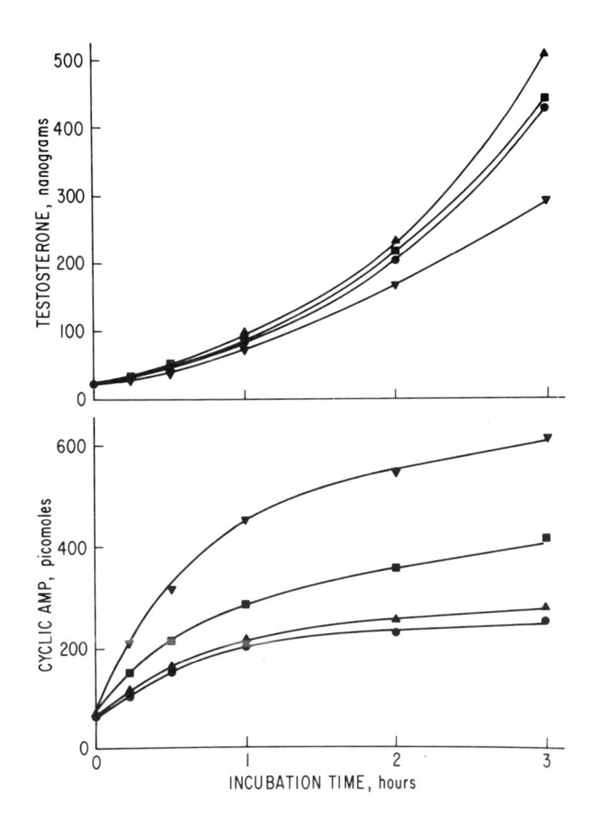

FIG. 2. Effects of theophylline (1–5 m*M*) on testosterone production (above) and cyclic AMP release (below) into the incubation medium of interstitial cells stimulated with 100 μg of human chorionic gonadotropin (hCG). Testosterone production is inhibited by 5 m*M* theophylline, while cyclic AMP release increases with rising theophylline concentrations. ●——●, 100 ng hCG; ▲——▲, ■——■, ▼——▼, 1, 2.5, and 5 m*M* theophylline, respectively.

steroidogenesis is 0.1 mM. For complete inhibition of phosphodiesterase in incubated Leydig cells, up to 10 times higher concentrations of inhibitors are necessary.

Intact Rat Testes

Despite the obvious limitations of intact tissue preparations for incubation *in vitro*, the isolated intact rat testis also provides an extremely sensitive and valuable response system for studies on gonadotropin action *in vitro*. As previously noted, teased testis preparations and physically dispersed interstitial cells are considerably less responsive than the intact testis to LH and hCG *in vitro*. This phenomenon may be due to release of an inhibitory factor during physical separation of the rat testis, and can be partially reversed by extensive washing of the teased preparation.

For incubation of intact testes, the gonads from 250–300 g Sprague-Dawley rats are decapsulated with care to avoid tearing of the soft testis tissue. The decapsulated testes, suspended from the vascular supply, are washed twice in cold phosphate-buffered saline pH 7.4, blotted gently on filter paper, and transferred to individual incubation vials containing 1.9-ml aliquots of KRB-glucose buffer containing bovine γ-globulin, 1 mg/ml. When studies on cyclic AMP formation from [^{14}C]adenine are performed,[3,5] testes are preincubated with 1 μCi [^{14}C]adenine in 1.4 ml of KRB for 1 hour, followed by addition of 0.5 ml of 20 mM theophylline in KRB buffer, and 0.1 ml of hormone solution. When cyclic AMP production is to be determined by radioimmunoassay[5,7,20] incubations are performed in 2 ml of KRB-glucose buffer containing 5 mM theophylline.

The optimum temperature for incubation of the rat testis *in vitro* is 34°,[15] though little difference in steroidogenic response to gonadotropins is apparent at 37°. Stimulation of testosterone production in response to gonadotropins commences after 10–15 minutes, increases in a linear fashion for about 2 hours and reaches a plateau after 4–6 hours.[16] Cyclic AMP release by the testis commences within 1 minute of exposure to high concentrations of gonadotropin, increases to a maximum at 2–3 hours, and then declines to lower values.[7] For correlative studies on cyclic AMP and testosterone responses to various concentrations of gonadotropin *in vitro*, incubations are performed for 2 hours to approach maximum cyclic AMP levels and for 4 hours for maximum testosterone production. The tissue concentrations of cyclic AMP and testosterone have been shown to correlate with the levels released into the incubation medium.[5,7] For many purposes, measurement of cyclic AMP and testosterone accumulation in incubation medium is as useful as tissue estimations and can be more readily performed by direct assay of diluted incubation medium. This method is particularly useful because release of the nucleotide is negligible in the absence of gonadotropin, so that relatively small changes are clearly apparent against the extremely low levels (5–10 pmoles/ml) that accumulate in the incubation media of unstimulated testes. By contrast, such changes are not readily detectable against the relatively high tissue concentration of cyclic AMP (500–600 pmoles/g) which is present in the rat testis.

Preparation of Samples for Assay of Testosterone and Cyclic AMP

After Incubation of Intact Testes

Media are transferred to 12 × 75 mm glass tubes and centrifuged at 1500 *g* for 20 minutes at 4°. Aliquots of the supernatant solutions are taken directly for radioimmunoassay of cyclic AMP, or are diluted 1:50 prior to radioimmunoassay of testosterone. Tissues are washed in cold PBS and blotted, then homogenized in 10 ml of acetic acid:ethanol 1:1; aliquots are dried under an air stream prior to radioimmunoassay of testosterone and cyclic AMP.

After Incubation of Dispersed Leydig Cells

For measurement of testosterone and cyclic AMP release, cells plus media are centrifuged in 12 × 75 mm tubes at 1500 *g* for 20 minutes, then aliquots of the supernatant solution are taken for direct radioimmunoassay of cyclic AMP, and for testosterone assay after 1:25 dilution in PBS.

When total production of testosterone and cyclic AMP is to be determined, an equal volume (usually 1 ml) of 20% acetic acid in distilled water is added to each incubation vial, followed by sonication of the contents to disrupt the interstitial cells. The sonicate is diluted with an appropriate volume of 50% methanol, and suitable aliquots, e.g., 10–100 μl, are taken into glass assay tubes and dried down under an air stream.

Isolation Procedure for Tissue Testosterone[16]

Extraction Procedures. For experiments in which extraction and thin-layer chromatography (TLC) were performed prior to assay, 100-μl aliquots of incubation medium are mixed with internal indicator (1000 dpm of [^{3}H]testosterone in 50 μl of ethanol) and 1 ml of 0.05 *M* NaOH. After extraction with 30 ml of ethyl acetate for 5 minutes in a mechanical shaker and aspiration of the water layer, the solvent extract was washed twice with a 1-ml volume of water and dried under an air stream, then washed down with 0.25 ml of methanol and dried again. Extraction of testes for testosterone assay only is performed by homogenization of each gland in 3 ml of 0.15 *M* NaCl containing 5% ethanol, followed by extraction with 30 ml of ethyl acetate as described above.

Chromatography. Extracts are transferred in methanol–methylene chloride (1:1) to thin-layer chromatography sheets. A marker of 20 μg 17α-hydroxyprogesterone is applied 2 cm from one side of each sheet, and the samples are applied at 4-cm intervals. The sheets were then developed twice in the system, chloroform:ether 80:20, in glass chromatography tanks lined with Whatman 3 MM paper and previously equilibrated for 2 hours. In this TLC system, Δ^4-androstenediol, Δ^5-androstenediol, androstenedione, dihydrotestosterone, and estradiol-17β are adequately separated from testosterone. The use of 17α-hydroxyprogesterone as marker is based on similar mobility to testosterone in this system and negligible cross-reaction in the binding assay and radioimmunoassay. After development, the corresponding sample areas are cut out and eluted with 6 ml of methanol, which is dried under an air stream. Residues are then dissolved in 0.15 *M* NaCl, and appropriate aliquots are taken for assay. Usually, the residue is dissolved in 1.0 ml of saline, and aliquots of 50 μl and 100 μl are made up of 250 μl for assay. A further aliquot of 200 μl is shaken with 10 ml of scintillation solution and counted to determine recovery of testosterone in the individual samples.

Measurement of Testosterone Production by Rat Testes *in Vitro*

Specific measurement of testosterone produced by the rat testis and tissue content of testosterone is performed by competitive protein binding assay or by radioimmunoassay.[16] The higher affinity of antitestosterone sera makes radioimmunoassay a more suitable method of serial determination of testosterone concentration in small samples of biological material,[39] but if antibody is not available, equally reliable results can be obtained by the competitive protein binding technique employing plasma testosterone binding globulin.

Radioimmunoassay

Antiserum. The 3-*O*-carboxymethyl oxime of testosterone prepared by the method of Erlanger *et al.*[40] is coupled to albumin or bovine γ-glob-

[39] M. L. Dufau, K. J. Catt, T. Tsuruhara, and D. Ryan, *Clin. Chim. Acta* **37**, 109 (1972).

[40] B. F. Erlanger, F. Borek, S. M. Beiser, and S. J. Lieberman, *J. Biol. Chem.* **228**, 713 (1957).

ulin by condensation with water-soluble carbodiimide.[39] To prepare conjugates for immunization, 50 mg of bovine α-globulin (Mann Laboratories) are dissolved in 2.0 ml of distilled water by dropwise addition of 0.5 *M* NaOH; the solution is adjusted to pH 5.0 with 1 *M* HCl and mixed with 2.0 ml of dimethylformamide. After addition of 50 mg of testosterone-3-*O*-carboxymethyl oxime in 2.0 ml of dimethylformamide, 50 mg of *N*-ethyl-*N'*-(3 dimethylaminopropyl) carbodiimide HCl (K and K Laboratories, Inc.) are added, and the solution is stirred for 16 hours at 4°. By this method, the degree of coupling of [^{14}C]testosterone added as tracer prior to synthesis of the *O*-carboxymethyloxime derivative is equivalent to 70 residues per mole of γ-globulin, estimated by counting the labeled protein after dialysis and extraction with dioxane. For immunization, the conjugate was emulsified in complete Freund's adjuvant and administered s.c. in doses of 2 mg to sheep at intervals of 2–3 weeks. Antisera suitable for use in radioimmunoassay can usually be obtained after the first few weeks of immunization (i.e., with association constant, K_a, of 0.5×10^9 and titers higher than 1:5000). The specificity of antisera to testosterone coupled via the 3-position is somewhat higher than that of testosterone-binding globulin, which reacts with most steroids bearing a free 17β-hydroxyl group. All antisera to testosterone-3-protein conjugates show moderate to extensive cross-reaction with dihydrotestosterone and with Δ^4- and Δ^5-androstenediol.

Tritiated Testosterone. [1,2-^{3}H]Testosterone (40 Ci/mmole) is diluted with ethanol to a concentration of 50 μCi/ml and stored at 4°. For assay, the stock solution is freshly diluted with phosphate-buffered saline (pH 7.4) containing bovine serum albumin, 1 mg/ml, to a concentration of 80,000 dpm/ml.

Charcoal-Dextran Suspension. Charcoal (Darco G-60, Matheson Coleman and Bell), 250 mg, and 25 mg of Dextran T-70 (Pharmacia) are mixed with 100 ml of distilled water and stored at 4°. For separation of bound and free testosterone, 0.2-ml aliquots (containing 500 μg of charcoal) of the stirred suspension are added to each of the assay tubes.

Radioimmunoassay Procedure. Samples and testosterone standards (15–2000 pg) in 250 μl of saline are added to assay tubes followed by 0.5 ml of testosterone antiserum in phosphate-buffered saline (pH 7.4) containing bovine serum albumin, 1 mg/ml, and 0.5 ml of [^{3}H]testosterone tracer (35,000 dpm, 80 pg) in the same diluent. The final dilution of antitestosterone serum is chosen to give 40–50% binding of tritiated testosterone tracer at the zero point of the assay. After overnight incubation at 4°, bound and free tracer are separated by addition of 0.2 ml of charcoal suspension to each tube, followed by brief agitation on a vortex mixer and standing for 20 minutes at 4°. Tubes are then centrifuged for

15 minutes at 1200 *g* in a refrigerated centrifuge and the supernatants containing the bound tracer are transferred to counting vials. After addition of 1 ml of dioxane to ensure complete extraction of the antibody-bound steroid, [^{3}H]testosterone is partitioned into the counting medium by shaking for 1 minute with 10 ml of scintillation solution. Standards and samples are counted for adequate time to reach a counting error of ±1.5%, and the recovery vials employed during assay of chromatographically purified testosterone are counted for 20 minutes with a mean counting error of ±5%. In all assays, duplicate tubes containing no antibody are set up for estimation of the "nonspecific" counts remaining in the supernatant solutions after charcoal adsorption. This value averages 300–400 cpm and is subtracted from all counts employed in subsequent calculations.

The radioimmunoassay procedure can be employed for direct assay of testosterone in incubation media from isolated testes and interstitial cells stimulated with various gonadotropic hormones *in vitro*. The sensitivity of the method is 10 pg per assay tube, and the mean within-assay precision is ±8% (coefficient of variation). The recovery of testosterone added to incubation media is 95–98%, the blank of the assay is 8 pg/tube, and the mean between-assay precision is less than ±10%.

Competitive Protein-Binding Assay

Binding Protein. Third trimester pregnancy plasma is stored frozen in 0.25-ml aliquots. At the time of use, each aliquot is diluted with 0.15 *M* NaCl containing 0.01 *M* phosphate buffer, pH 7.4.

Binding Assay Procedure. Sample extracts and testosterone standards (0.125–4 ng in 250 μl of saline) are added to assay tubes followed by 1 ml of diluted third trimester pregnancy plasma (1:200) containing approximately 25,000 dpm (65 pg) of [^{3}H]testosterone per milliliter. The assay tubes are vortexed and stood for 75 minutes in an ice bath; separation of bound and free tracer is then performed by addition of 0.2 ml of cold·dextran-charcoal suspension prepared by mixing 250 mg of Darco G-60 charcoal and 25 mg of Dextran T 70 with 100 ml of distilled water. After mixing, tubes are incubated for 45 minutes at 0°, and centrifuged in a model PR6 refrigerated International centrifuge for 15 minutes at 1200 *g*. The bound count is then determined in a liquid scintillation spectrometer after decantation of the supernatant solutions into counting vials and partition of the steroid into 10 ml of liquid scintillation counting solution. The binding assay is sensitive to 125 pg of testosterone per assay tube, with undetectable blank for charcoal-stripped plasma and within-assay precision ±6% (coefficient of variation) over the working range of the assay (125–4000 pg). During assay of chromatographically purified testosterone, the recovery of radioactive testosterone is 57 ± 8% and the corrected recovery of added unlabeled testosterone (2 ng and 4 ng) is 95–105%, while the between-assay precision is ±10%.

Validity of Direct Assay of Testosterone in Incubation Media

Direct measurement of immunoreactive testosterone in dilutions of incubation media can be performed without further purification prior to assay. The cross-reaction of related steroids with antisera to testosterone-3-conjugates indicates that the major potentially interfering steroids are dihydrotestosterone, which has equal potency and slope with testosterone, and Δ^4 and Δ^5-androstenediols, which are approximately 25% as potent as testosterone in the assay. The cross-reactions of dehydroepiandrosterone, epitestosterone, androstenedione, and 5α-androstane-$3\beta,17\beta$-diol are not significant. The slope of the displacement curve obtained with increasing aliquots of incubation medium is identical with that of the testosterone standards (Fig. 3). When individual testis incubation media are

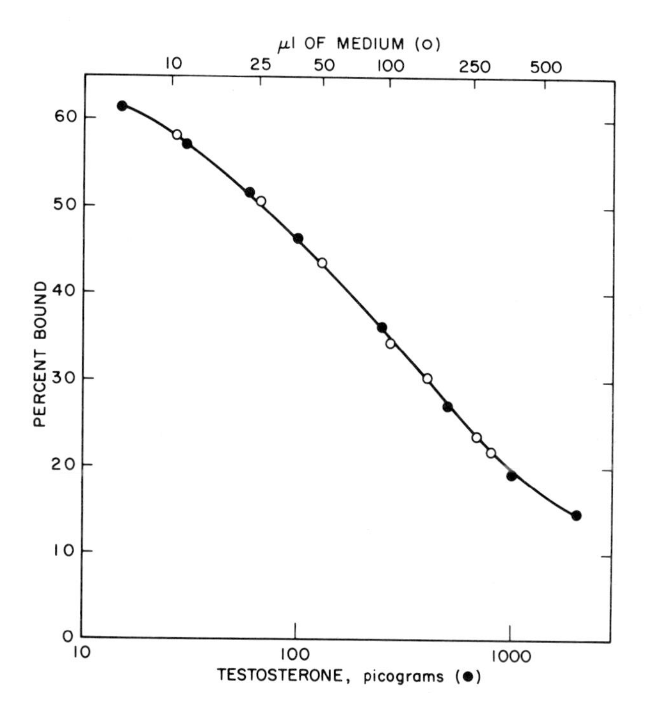

FIG. 3. Standard curve of the radioimmunoassay for testosterone, over the range 15–2000 pg. The slope of the displacement caused by increasing aliquots of a 1:100 dilution of testis incubation medium (○——○) is identical with that of the testosterone standards (●——●).

TABLE I
COMPARISON OF RADIOIMMUNOASSAY AND BINDING ASSAY

Samples	Binding assay ($n = 3$) (mean ± SD)	Radioimmunoassay (direct) ($n = 3$) (mean ± SD)	Radioimmunoassay (thin-layer chromatography) ($n = 3$) (mean ± SD)
Control	89 ± 16	109 ± 11	108 ± 26
HCG, 1 mU	173 ± 4	196 ± 19	219 ± 8
HCG, 10 mU	744 ± 62	744 ± 50	887 ± 32
HCG, 100 mU	742 ± 47	758 ± 92	828 ± 95
Dibutyryl cAMP, 1 mg	1068 ± 200	1071 ± 159	1103 ± 189

subjected to extraction and TLC purification prior to radioimmunoassay or binding assay, the observed testosterone levels are similar to those obtained by direct assay of diluted incubation media. Comparable values are obtained on measurement of testosterone production *in vitro* by specific binding assays and by direct radioimmunoassay of incubation media, and the highly sensitive direct assay is more convenient to follow the response of the rat testis to hormonal stimuli *in vitro* (Table I). The results derived from direct radioimmunoassay are slightly higher than those measured by competitive protein binding assay and slightly lower than those given by radioimmunoassay following TLC (mean difference, 6.8%; $P < 0.05$). In addition, radioimmunoassay of eluates from serial sections of thin-layer chromatograms following analysis of extracts of incubation media confirms that the predominant area of immunoreactive androgen is confined to the region of testosterone.

Measurement of Cyclic AMP Formation *in Vitro*

[^{14}C]Adenine Incorporation

Testes are incubated for 1 hour in 1.4 ml of KRB containing 1 μCi of [^{14}C]adenine of SA 225 mCi/mmole to label the endogenous ATP pool.[3,5,7] After addition of 0.5 ml of 20 mM theophylline in KRB and 0.1 ml phosphate-buffered saline containing bovine γ-globulin, 1 mg/ml, and various concentrations of hCG or LH, incubation is continued for further periods from 1 minute to 4 hours. Aliquots, 1 ml, of incubation medium are then assayed for [^{14}C]cyclic AMP content after addition of unlabeled carrier nucleotides and [^{3}H]cyclic AMP (25 Ci/mmole; 20,000 cpm), with purification by elution from Dowex 50 resin and barium sulfate adsorption as described by Krishna *et al.*[41] for the measurement of adenylate cyclase activity. After thin-layer chromatography of such purified extracts in the system, ethyl acetate:ethanol:0.5 *M* ammonium acetate, 5:5:2, on alumina plates,[42] scanning for ^{14}C showed a single peak of radioactivity coincident with unlabeled cyclic AMP.

Radioimmunoassay of Cyclic AMP

Aliquots of incubation medium (25, 50, and 100 μl) are taken for radioimmunoassay of cyclic AMP by the method of Steiner *et al.*,[43] modified by the use of 70% dioxane instead of second antibody for the separation of bound and free tracer. Aliquots of medium are incubated for 3–16 hours at 4° with 100 μl of antiserum to succinyl cyclic AMP–bovine serum albumin diluted 1:4000 in 50 m*M* phosphate-buffered saline, pH 7.4, and 100 μl of acetate buffer containing 6–10,000 cpm (0.015–0.03 pmole) of [^{125}I]SCAMP-TME tracer. After incubation, 0.2 ml of bovine γ-globulin 5 mg/ml in PBS is added to each tube, followed by 2 ml of 70% dioxane. After centrifugation at 1500 *g* and aspiration of the supernatant solutions, the precipitated bound radioactivity is counted in a automatic gamma spectrometer.

Tissue levels of cyclic AMP are similarly measured by radioimmunoassay after homogenization of washed testes in 5% trichloroacetic acid (TCA) or acetic acid:ethanol (1:1), followed by ether washing of TCA extracts, evaporation under an air stream, and reconstitution in 50 m*M* acetate buffer, pH 6.2. For Leydig cell dispersions, sonication of cells plus medium after addition of 20% acetic acid is performed as described above, followed by dilution with 50% methanol and drying of appropriate aliquots under an air stream. Similar results were obtained by direct radioimmunoassay of cyclic AMP in tissue extracts, and by radioimmunoassay following Dowex 50-barium sulfate purification of tissue extracts with correction for recovery of tracer [^{3}H]cyclic AMP (0.2 picomole) added during homogenization.

Cyclic AMP Synthesis and Release

Comparison of cyclic AMP formation by the [^{14}C]adenine incorporation and radioimmunoassay methods shows parallel release of cyclic [^{14}C]-AMP and unlabeled immunoreactive cyclic AMP during incubation

[41] G. Krishna, B. Weiss, and B. B. Brodie, *J. Pharmacol. Exp. Ther.* **163,** 379 (1968).
[42] G. R. Berg and D. C. Klein, *Endocrinology* **89,** 453 (1971).
[43] A. L. Steiner, D. M. Kipnis, R. Utiger, and C. Parker, *Proc. Nat. Acad. Sci. U.S.* **64,** 367 (1969). See also this series, Vol. 38 [13].

TABLE II
RELEASE OF ^{14}C-LABELED CYCLIC AMP (dpm) AND IMMUNOREACTIVE CYCLIC AMP (pmoles) INTO TESTIS INCUBATION MEDIUM DURING INCUBATION WITH 20 ng hCG (n = 3, mean ± SEM)

Incubation time (min)	[^{14}C]cAMP (dpm)	Immunoreactive cAMP (pmoles)	dpm/pmole
30	2223 ± 628	134 ± 16	16.0 ± 2.0
60	6109 ± 867	384 ± 41	15.8 ± 0.6
120	7909 ± 1021	460 ± 41	14.4 ± 0.8

with hCG (Table II). The concordance between the pattern of release shown by the two methods indicates that newly synthesized cyclic AMP comprises a constant proportion of the total cyclic AMP released *in vitro*. For most purposes, the more sensitive radioimmunoassay procedure provides the most convenient procedure to determine cyclic AMP levels in tissue and medium (Fig. 4).

Gonadotropin Binding by Interstitial Cells and Testis Preparations

Methods for the study of gonadotropin receptors in particular and solubilized Leydig cell fractions have been described in detail elsewhere.[44] In addition to such studies of the binding properties and physical characteristics of such receptors, it is also of considerable importance to evaluate the functional relationships between gonadotropin binding and the subsequent events that express the biological actions of the tropic hormones upon the target cell. For this purpose the binding of radioiodinated LH or hCG can be readily correlated with responses of Leydig cells *in vitro*, in particular with activation of adenylate cyclase and the initiation of steroid synthesis and testosterone production. Valid interpretation of such studies requires that the labeled gonadotropin should be of known biological potency and known specific radioactivity, and that adequate controls are performed in each experiment to determine the degree of any nonspecific binding of the labeled hormone to the tissue under study. With appropriately labeled and undamaged hormone, nonspecific binding should represent at most only a small fraction of the total bound hormone. In all cases, an excess of unlabeled hormone of at least 1000-fold should be employed to ascertain the displaceability of the labeled tracer from

[44] K. J. Catt and M. L. Dufau, this series Vol. 37, [11].

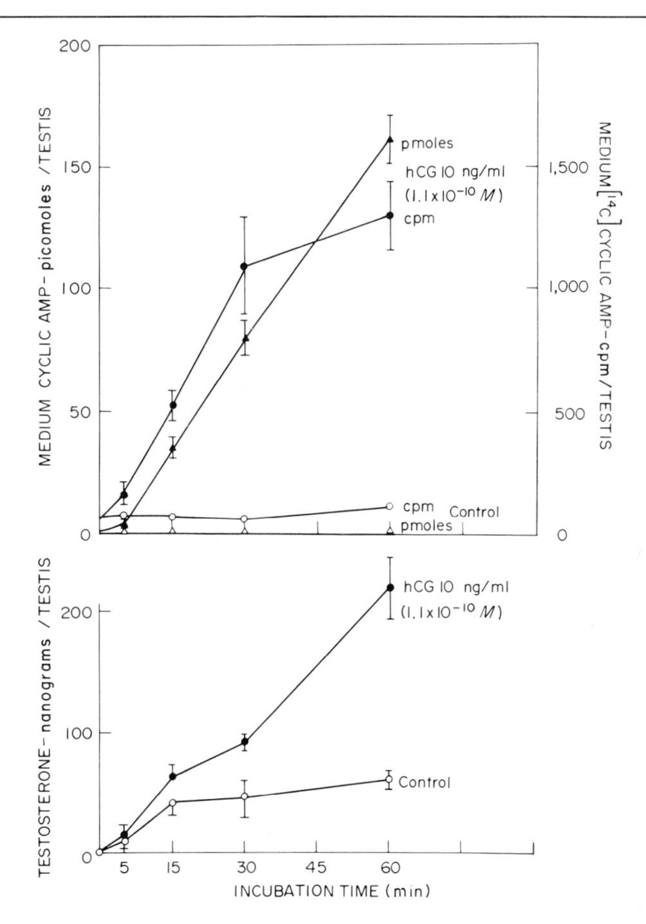

FIG. 4. Release of cyclic AMP. Total immunoreactive cyclic AMP (▲——▲) and ^{14}C-labeled cyclic AMP (●——●), and testosterone (below) into the incubation medium during stimulation of rat testes *in vitro* with 10 ng human chorionic gonadotropin (hCG). Cyclic AMP formation from endogenous [^{14}C]ATP and the total release of immunoreactive nucleotide gave similar patterns of release during incubation. Each point represents the mean ±SD of triplicate determinations.

specific tissue binding sites. Failure of displacement by unrelated hormones and lack of binding of labeled unrelated hormones to the same binding sites are additional essential criteria for the validity of hormone uptake by the specific receptor sites.

In the rat testis, binding of labeled hCG can be employed to quantitate the specific LH receptor sites of the intact testis or of enzyme-dispersed Leydig cells. Specific uptake of ^{125}I-labeled hCG by the intact

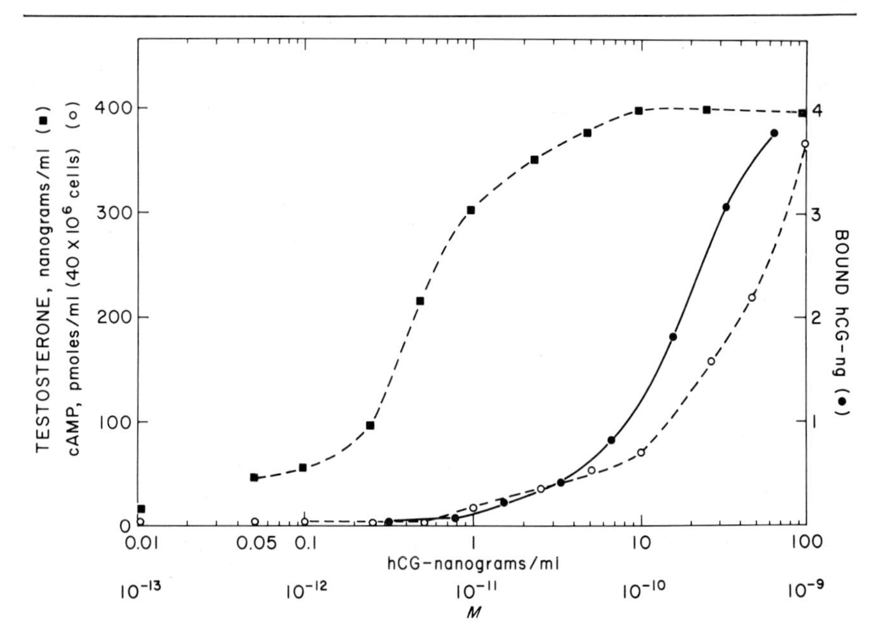

FIG. 5. Correlation of gonadotropin binding (●——●) and cyclic AMP production (○----○) during incubation of collagenase-dispersed interstitial cells with increasing concentrations of human chorionic gonadotropin (hCG). Testosterone production (■----■) is stimulated to near-maximum levels by hCG concentrations which are associated with only a small proportion of the total possible receptor occupancy and cyclic AMP production.

rat testis has been shown to be reversed by exposure to pH 2.5 for short periods, and the eluted hormone was found to display enhanced biological properties in terms of binding, activation of adenylate cyclase, and stimulation of testosterone production.[20] That is, no loss of biological activity occurs during association with the receptor site, and indeed the most active molecules of the labeled preparation are those which become specifically bound to the hormone receptor sites. By contrast, labeled hormone present in the incubation medium undergoes relatively rapid inactivation by an enzyme system which is independent of the receptor site. This appears to be a general phenomenon which applies to a wide variety of hormone-specific binding systems in several endocrine-responsive tissues.

Measurement of tropic hormone binding in relation to cyclic AMP production by isolated rat testes and collagenase-dispersed Leydig cells (Fig. 5) has shown that the mature rat testis contains a large excess of specific gonadotropin receptors which bind much higher quantities of hCG than required for maximum stimulation of steroidogenesis. The capacity of the testis for gonadotropin binding is more than 100 times greater than

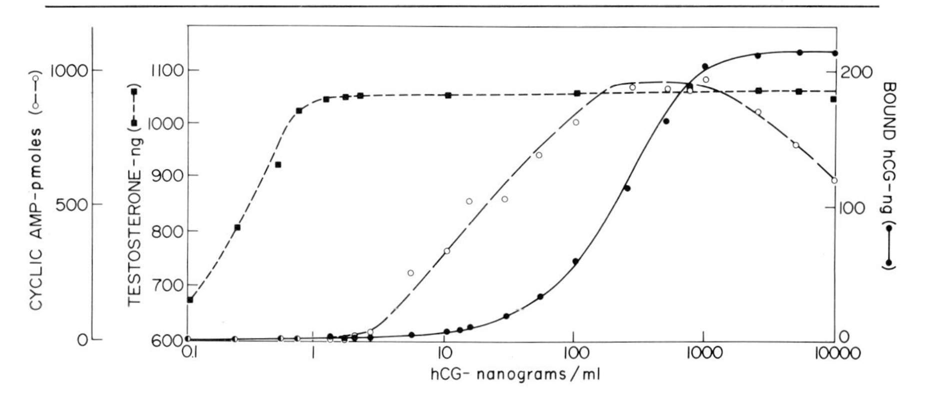

FIG. 6. Simultaneous determinations of human chorionic gonadotropin (hCG) binding (●——●), cyclic AMP release (○----○), and testosterone production (■----■) during incubation of rat testes with increasing amounts of hCG *in vitro*. Testosterone production rose serially with increasing hCG binding, reaching a maximum level at the same hCG level at which cyclic AMP release became detectable. At higher hCG concentrations, cyclic AMP production continued to rise while testosterone production remained constant.

the level which evokes maximum testosterone production. Cyclic AMP formation in isolated testes and Leydig cells is not detectably increased by low concentrations of LH or hCG which stimulate steroidogenesis, but rises pari passu with increasing gonadotropin binding in excess of that required for a full steroid response. Therefore, the role of cyclic AMP as mediator in the actions of LH and hCG at physiological concentrations may not be as unequivocal as formerly believed, or may operate via translocation of minute quantities of the nucleotide in a specific intracellular compartment. There is no doubt that coupling of gonadotropin binding to adenylate cyclase occurs at higher concentrations of the tropic hormone, and that most of the excess or "spare" binding sites of the testis are potentially active receptors with the capacity for increasing cyclic AMP production and release to extremely high levels (Fig. 6). The presence of excess gonadotropin receptors in the testis probably represents a general phenomenon in endocrine tissues which is comparable to the presence of "spare" receptors in drug-responsive tissues as described by Stephenson,[45] and in the adrenal as described by Beall and Sayers.[46] Such spare receptors may increase the sensitivity of the target cell to circulating tropic hormones by enhancing the probability that a given gonadotropin level will result in an adequate degree of receptor occupancy to initiate steroidogenesis.

[45] R. P. Stephenson, *Brit. J. Pharmacol.* **11**, 379 (1956).

[46] R. J. Beall and G. Sayers, *Arch. Biochim. Biophys.* **148**, 70 (1972).

[23] The Infused Testis[1]

By KRISTEN B. EIK-NES

Twenty years have elapsed since West *et al.*[2] demonstrated that blood passing through the testis became enriched in testosterone. Since this observation, steroids secreted into the effluent blood of the male gonad have been measured in numerous species, and this topic has been reviewed.[3] Anesthetized animals have been used in most of these experiments, but spermatic venous blood can be obtained from the conscious human being.[4] Moreover, an animal preparation has been described in the ram[5] where cannulas are placed in the spermatic artery and vein and injections via the spermatic artery can be performed in the conscious, postoperative animal concomitant with sampling of spermatic venous blood from the indwelling cannula in the spermatic vein. This preparation deserves serious consideration in experimental work on testicular secretion of steroids. Secretion rates of testicular steroids can be measured by the principle of isotope dilution[6] (for review see Eik-Nes and Hall[7]). Since discrepancies appear to exist between data on testicular secretion of estradiol-17β in man as measured by this technique[8] and as measured by levels of estradiol in human spermatic vein blood,[9] and since the rate of testicular secretion of testosterone depends to a marked extent on the rate of blood flow in the spermatic artery,[10] testicular preparations where this flow is controlled and where spermatic venous blood samples can be obtained with ease may be useful. Two such preparations are available, far from per-

[1] The work reviewed from the author's laboratory was in part supported by research grant HD-04195 and by research contract NIH-69-2097 from National Institutes of Health, Bethesda, Maryland.

[2] C. D. West, V. P. Hollander, T. H. Kritchevsky, and K. Dobriner, *J. Clin. Endocrinol. Metab.* **12,** 915 (1952).

[3] K. B. Eik-Nes, *in* "The Androgens of the Testis" (K. B. Eik-Nes, ed.), p. 6. Dekker, New York, 1970.

[4] T. Laatikainen, E. A. Laitinen, and R. Vihko, *J. Clin. Endocrinol. Metab.* **32,** 59 (1971).

[5] M. E. Lisano and J. L. Fleeger, *Southwest. Vet.* **23,** 295 (1970).

[6] J. F. Tait, *J. Clin. Endocrinol. Metab.* **23,** 1285 (1963).

[7] K. B. Eik-Nes and P. F. Hall, *Vitam. Horm.* (*New York*) **23,** 153 (1965).

[8] P. C. MacDonald, J. M. Grodin, and P. K. Siiteri, *in* "Control of Gonadal Steroid Secretion" (D. T. Baird and J. A. Strong, eds.), p. 158. University Press, Edinburgh, 1971.

[9] M. R. Jenner, R. P. Kelch, and M. M. Grumbach, *Program 53rd Annu. Meeting Endocrine Soc.,* p. A-46, San Francisco, California 1971.

[10] K. B. Eik-Nes, *Can. J. Physiol. Pharmacol.* **42,** 671 (1964).

fect from a physiological standpoint,[11] but still worth incorporation in the armamentarium of investigators interested in steroid functions of the male gonad.

The Infused Dog Testis

Sexually mature, mongrel dogs (body weight 18–28 kg) are given 30 mg of sodium pentobarbital per kilogram intravenously. Half of this dose is administered instantaneously the remainder is slowly injected until the animal does not react on sharp squeezing of the front paw. The animal is then placed on an operating table, and two preparations[11] are now possible: a preparation of the left testis *in situ* (animal preparation I) or a preparation involving both testes contained in a metabolic chamber (animal preparation II). In preparations I and II, the dog is heparinized and extreme care must therefore be exercised in severing even the smallest arteries and veins between double ligatures.

Animal Preparation I

The left femoral artery is dissected free of connective tissue and neighboring veins and cannulated. This cannula is connected to a Harvard automatic withdrawal-infusion pump. The left spermatic artery and vein are reached through a longitudinal incision from the scrotum to the inguinal region. These vessels are then prepared for cannulation by freeing carefully each vessel from connective tissue and surrounding vessels for a distance of 1.5 cm. In most animals a single spermatic artery and a major spermatic vein are found. If several spermatic veins (or tributaries) are present, the small ones are tied off and the major one is prepared for cannulation. If several spermatic arteries appear to be present, the animal is discarded. One should also look for vascular connections between the two testes. If such are found they should be tied off. So are also vascular connections between the left testis and its epididymis shortly before cannulation of the spermatic artery. At this time a small hole should also be punched in the major lymphatic vessel of the left testis below the area prepared for cannulation of the spermatic artery and vein.

After preparation of the spermatic vessels, the animal is given 10,000 IU heparin intravenously and 10 minutes afterward a cannula is placed in the spermatic vein and tied in permanently. One should employ a can-

[11] K. B. Eik-Nes, *in* "Perfusion Techniques" (E. Diczfalusy, ed.), p. 270. Forum, Copenhagen, 1972.

nula with the largest inside diameter (i.d.) possible for the vein to be cannulated. Working with mongrel dogs it pays to have different-sized cannulas available in the surgical laboratory. The cannula in the spermatic vein is then drained through the chamber of a drop-counter, and the number of blood drops delivered per unit time via the cannula is recorded automatically. The spermatic venous blood is permitted to flow freely into a container following its passage through the drop-counter. A cannula is now tied into the spermatic artery. Again the biggest i.d. cannula which will fit the artery is preferred. In most experiments we used cannulas with an i.d. of 0.023 inch or 0.038 inch. The outside diameters of these cannulas are 0.030 inch or 0.048 inch. The cannula has a blunt tip and is made of rigid material in order to avoid pressure losses due to wall damping.[12] A 3.5 m long, coiled plastic cannula is connected to the outlet channel of the Harvard pump. This cannula is placed in a water bath maintained at 37.5°. The end of this cannula is attached to the cannula in the spermatic artery. Thus, the animal's arterial blood can be infused at a constant rate and at constant temperature via the spermatic artery. A 22-gauge temperature probe is placed under the infused testis. This probe is attached to an automatic temperature recorder and changes of testicular temperature can be controlled by changing the temperature of the small water bath containing the arterial cannula. Experiments are conducted at testicular temperatures of 36.5–37.3°. The infused testis, however appears to secrete about the same amount of testosterone both at higher and at lower temperatures, although at temperatures above 38.8° and below 34° the secretion of testosterone falls.

In some dogs unexpected flow variations of spermatic venous blood are seen. This is caused by arterial or venous contributions outside those controlled via the cannulas. If the flow of venous blood is different from the input of arterial blood via the cannula in the spermatic artery, the experiment should be terminated. It is not advisable to work with this testicular preparation unless the venous return from the infused gland can be determined during the experiments. Whether a pulsatile or a nonpulsatile type blood flow is used in the spermatic artery is of little consequence for testosterone secretion by the infused dog testis. Similar data with regard to type of arterial blood flow and kidney function have been published for the dog.[13]

Following a 30-minute infusion of the left testis (warm-up period), experiments can start. Total elapsed time from anesthesia to collection

[12] W. C. Leith, J. J. Cole, and C. R. Blagg, *J. Biomed. Mater. Res.* **4**, 73 (1970).

[13] G. L. Wolf and R. G. Dluhy, *in* "Microcirculation, Perfusion and Transplantation of Organs" (T. L. Malinin, B. S. Lino, A. B. Callahan, and W. D. Warren, eds.), p. 289. Academic Press, New York, 1970.

of the first spermatic venous blood sample to be used for assay of testicular steroids should not exceed 70 minutes.

Animal Preparation II

After cannulation of the left femoral artery the spermatic arteries on both the left and the right side are prepared as described. The animal is then heparinized and the cannula in the femoral artery is connected to the outlet channel of a Harvard withdrawal-infusion pump. This pump delivers the arterial blood to an inverted 4-liter separatory funnel[14] placed in a constant-temperature chamber maintained at 37.5°; 95% O_2–5% CO_2 is bubbled into the entering blood at the top of the separatory funnel.[14] The oxygenated arterial blood is removed from the bottom of this oxygenation reservoir by two additional Harvard constant-rate withdrawal-infusion pumps placed in the constant-temperature chamber. After priming these pumps (125 ml of arterial blood delivered to the oxygenation reservoir), the spermatic artery is cannulated, the veins of the pampiniform plexus are severed, and the testis is transferred to an organ container placed in a horizonal position in the constant-temperature chamber. This container is made from a 2-liter Erlenmeyer flask. The bottom of the flask is removed and a flange is placed around the ensuing aperture (diameter 9 cm) to hold a rubber glove.[14] The testis is positioned on a specially constructed organ dish in the organ container, and the container has a small opening, 8 cm lateral to the organ dish, for collection of spermatic effluent. The cannula in the spermatic artery is connected via a 1-ml plastic tuberculin syringe (placed through a No. 9 rubber stopper in the neck of the Erlenmeyer flask) to a cannula from the outlet channel of one of the Harvard pumps in the constant-temperature chamber. The organ container is closed by a rubber glove containing a piece of cheesecloth soaked in warm 0.9% saline solution. Presence of this piece of cheesecloth at the end of the organ container is needed to avoid drying of the testis during infusion.

Oxygenated, arterial blood is now infused via the spermatic artery, and spermatic venous blood is permitted to flow freely into a container. After 30 minutes of infusion (warm-up period), the first sample of spermatic venous blood for steroid assay is collected. Total elapsed time from anesthesia to beginning of blood collection should not exceed 100 minutes.

In both animal preparations the testes are infused with blood via the spermatic artery at a rate of 0.18 to 0.20 ml/g per minute. We prefer to work with testes weighing between 17 and 22 g. Small testes (less than

[14] N. L. Vandemark and L. L. Ewing, *J. Reprod. Fert.* **6**, 1 (1963).

10 g) and large testes (larger than 25 g) should not be used. During infusion the animal receives 0.9% sodium chloride solution via a front leg vein at the same rate as that of removal of blood from the femoral artery. The preexperiment spermatic venous blood (warm-up sample) is added to the 0.9% sodium chloride solution and used for the intravenous infusion.

The type of oxygenator[14] used in animal preparation II is rather primitive but will maintain blood pO_2 in an adequate fashion for 120–150 minutes. Testes infused as described in both preparations appear normal when evaluated by low or high resolution microscopy at the end of 150-minute infusion with blood. For blood infusions lasting for many hours, a membrane oxygenator[15,16] should be tried since excellent gas transferral associated with minimal trauma to the constituents of whole blood is possible with this type of oxygenator. With the current oxygenator, however, adequate steroid function is seen in testes,[11] ovaries,[17] prostates[18] and epididymes[19] infused with blood via the arteries for 90–150 minutes.

Testes of animal preparation I are slightly more responsive to gonadotropins via the spermatic artery[20] than testes of animal preparation II. Animal preparation I is thus the preferred one when testing compounds with borderline ability to stimulate testosterone secretion. Moreover, sampling of spermatic vein blood at frequent intervals[21] is more easily done in animal preparation I than in animal preparation II. Finally, spermatic lymph[22] can be collected via a cannula in the main spermatic lymph vessel of the infused left testis in animal preparation I. We have discussed elsewhere why we have found this type of experiment[20] difficult to conduct using animal preparation II. Androgens like Δ^4-androstenedione,[23] testosterone,[22,23] and 5α-dihydrotestosterone[23] can be found in the testicular lymph. Paired testes infused with blood via the spermatic artery (animal preparation II) secrete similar amounts of testosterone into spermatic venous blood and respond in a similar fashion after gonadotropins via the spermatic artery.[24] Thus, in experiments on tropic stimulation of

[15] A. J. Landé, B. Parker, V. Subramanian, R. G. Carlson, and C. W. Lillehei, *Trans. Amer. Soc. Artif. Intern. Organs* **14,** 227 (1968).
[16] E. C. Peirce, *Trans. Amer. Soc. Artif. Intern. Organs* **16,** 358 (1970).
[17] J. A. Engels, R. L. Friedlander, and K. B. Eik-Nes, *Metabolism* **16,** 189 (1968).
[18] G. C. Haltmeyer and K. B. Eik-Nes, *Acta Endocrinol.* **69,** 394 (1972).
[19] J. G. Sowell and K. B. Eik-Nes, *Proc. Soc. Exp. Biol. Med.* **141,** 827 (1972).
[20] K. B. Eik-Nes, *in* "Control of Gonadal Steroid Secretion" (D. T. Baird and J. A. Strong, eds.), p. 76. University Press, Edinburgh, 1971.
[21] K. B. Eik-Nes, *Ciba Found. Colloq. Endocrinol. Proc.* **2,** 87 (1969).
[22] G. C. Haltmeyer and K. B. Eik-Nes, *J. Reprod. Fert.* **36,** 41 (1974).
[23] H. Lindner, *J. Endocrinol.* **25,** 483 (1964).
[24] K. B. Eik-Nes, *Recent Progr. Horm. Res.* **27,** 517 (1971).

testosterone secretion, the one testis can be used as the control, nonstimulated organ. The opportunity to conduct experiments of this nature is often a *must* when working with mongrel dogs where animal age, previous diseases, breed, and environment are unknown to the investigator. Preparations for testicular infusion in rats, where these factors are known, should be encouraged. A good beginning in this respect are the techniques of Suzuki *et al.*[25] and of Free and Jaffe.[26] The cost of raising dogs of known breed to sexual maturity will exceed $200 per animal. With current level of research support few experiments could be done on infused testes in such animals. It is our experience that animals with long legs (German shepherds, greyhounds, golden retrievers) are excellent candidates for the type of testicular infusions discussed in this review.

Dog testes infused with blood via the spermatic artery (animal preparations I and II) secrete the steroids depicted in Fig. 1 in spermatic venous blood.[27,28] Secretion of estradiol-17β is minute, and spermatic venous levels of progesterone are low.[27] Formation of testosterone in the testis can occur via different metabolic pathways (Fig. 1). We do not know what factors direct the use of these pathways or whether species differences exist[27] with regard to testosterone formation from Δ^5-pregnenolone.[24] Of some interest are the observations that testicular experiments *in vivo* strongly favor testosterone formation via pathways not including the steroid hormone progesterone.[3,4,27,29] It has also been observed that the administration of progesterone, but not of testosterone or estradiol-17β, via the spermatic artery of the dog will result in meiotic chromosome alterations of the infused organ.[30]

The Infused Rabbit Testis[31]

The animal is anesthetized with sodium pentobarbital given intravenously. The testis is removed and the spermatic artery is cannulated (with the aid of a binocular dissecting microscope) using a 27-gauge needle attached to a plastic cannula with an i.d. of 0.011 inch and an

[25] Y. Suzuki, M. Takahashi, Y. C. Lin, and T. Asano, *Endocrinol. Jap.* **17,** 431 (1970).

[26] M. J. Free and R. A. Jaffe, *Amer. J. Physiol.* **223,** 241 (1972).

[27] H. J. van der Molen and K. B. Eik-Nes, *Biochim. Biophys. Acta* **248,** 343 (1971).

[28] Y. Folman, G. C. Haltmeyer, and K. B. Eik-Nes, *Amer. J. Physiol.* **222,** 653 (1972).

[29] M. E. Lisano, J. R. Beverly, A. M. Sorensen, and J. L. Fleeger, *Steroids* **19,** 159 (1972).

[30] D. L. Williams, J. W. Runyon, and A. A. Hagen, *Nature* (*London*) **220,** 1145 (1968).

[31] L. L. Ewing and K. B. Eik-Nes, *Can. J. Biochem.* **44,** 1327 (1966).

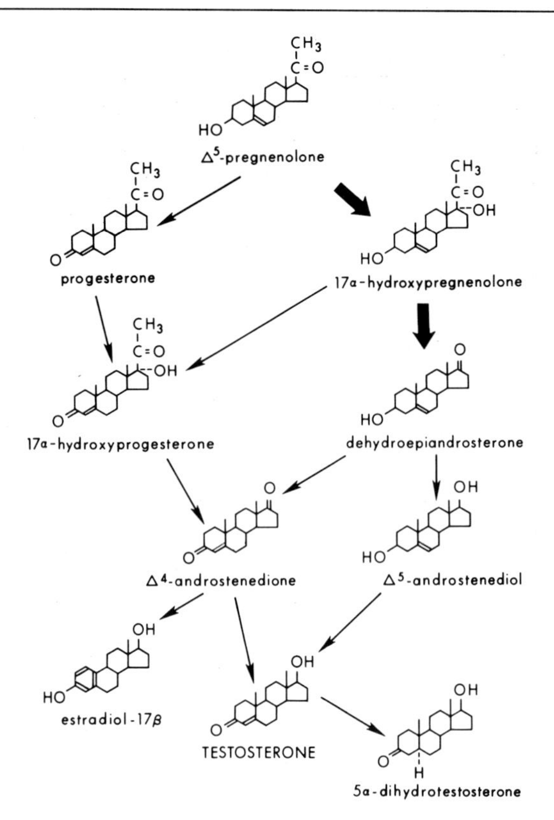

FIG. 1. Steroids secreted by the infused dog testis. Animal preparations I or II were used. Possible metabolic routes from Δ^5-pregnenolone are indicated. These routes were established by infusing radioactive steroids via the spermatic artery and analyzing radioactive steroids in the venous effluent from the infused gland. Only the forward reactions leading to testosterone formation are indicated. Intermediates between Δ^4-androstenedione and estradiol-17β are not given.

outside diameter of 0.024 inch. The organ is flushed with ice cold 0.25 *M* sucrose from an infusion bottle connected to the cannula in the spermatic artery. The testis is now placed in the organ container described under Animal Preparation II. Oxygenation[14] of the infusate is accomplished by the use of a model SP 100 Everett oxygenator (Med. Science Electronics Inc., St. Louis, Missouri). The constant-temperature chamber is kept at 36.8°. The infusate used is either defibrinated rabbit blood[14,31] or a medium[32] containing Krebs-Ringer bicarbonate buffer (pH 7.4), bo-

[32] L. L. Ewing, B. H. Johnsen, C. Desjardins, and R. F. Clegg, *Proc. Soc. Exp. Biol. Med.* **140**, 907 (1972).

vine albumin, fraction V (3%, w/v), glucose (1 mg/ml), crystalline penicillin and dihydrostreptomycin with washed bovine erythrocytes added to a hematocrit of 25%. The infusates are delivered to the testis with a peristaltic pump[31,32] and the pressure at the tip of the infusion cannula is under these conditions 120 mm Hg.[32] A 2.5-g testis will deliver 20 ml of spermatic effluent per hour. After a "warm-up" infusion period lasting from 45 to 60 minutes, the first sample of spermatic effluent fluid to be used for steroid analysis is collected. Infusions lasting 5–7 hours can be done, and at the end of infusions of this duration the organs show a normal histological appearance.

The level of testosterone in effluent fluid of the infused rabbit testis[31] and the infused dog testis (preparations I and II)[24] will fall with the duration of the experiment (Table I). This is also the case when dog testes are infused with oxygenated Krebs-Ringer bicarbonate buffer containing glucose.[24] The reasons for this decline in hormone secretion are unknown, but anesthesia with ether is also associated with decreased testicular production and secretion of testosterone.[33,34] Aono *et al.*[35] have published that blood plasma levels of testosterone will fall in man after anesthesia and a surgical operation. This type of surgical "stress" is associated with increased concentrations of interstitial cell-stimulating hormone (ICSH) in systemic blood.[35] We have observed[36] that with duration of experiment the spermatic venous levels of testosterone will fall while those of Δ^5-pregnenolone will increase in animal preparation II. Thus, the control of secretion of testosterone in these preparations can be a function of the time consumed in establishing conditions for infusion, and the time limits given for animal preparations I and II *must* be followed in order to obtain meaningful control data. Testes secreting less than 0.5 ng of testosterone per minute per gram testis (wet weight) during the first 15 minutes of experiment (animal preparation II) should be viewed with considerable reservation. Moreover, due to the falling rate of testosterone secretion by the dog and the rabbit testis prepared as described, experiments on compounds that per se may inhibit secretion of testosterone are extremely difficult to perform (Table I).

The infused dog or rabbit testis will respond to increased levels of gonadotropins in the spermatic artery with increased secretion and production of testosterone whether the tropins are administered continuously[24,31] or only over 60 seconds (Tables I and II). This effect of

[33] C. W. Bardin and R. E. Peterson, *Endocrinology* **80,** 38 (1967).

[34] B. L. Foriss, T. J. Hurley, S. Hane, and P. Forsham. *Endocrinology* **84,** 940 (1969).

[35] T. Aono, K. Kurachi, S. Mizutani, Y. Hamanaka, T. Uozumi, A. Nakasima, K. Koshiyama, and K. Matsumoto, *J. Clin. Endocrinol. Metab.* **35,** 535 (1972).

[36] R. K. Tcholakian and K. B. Eik-Nes, *Amer. J. Physiol.* **221,** 1824 (1971).

TABLE I

MEAN SECRETION OF TESTOSTERONE IN DOGS BEFORE AND AFTER INJECTION OF 1 ml 0.9% SODIUM CHLORIDE SOLUTION OR 1 ml 0.9% SODIUM CHLORIDE SOLUTION CONTAINING HUMAN CHORIONIC GONADOTROPIN (hCG) VIA THE SPERMATIC ARTERY

No. of testes	Preparation injected	Testosterone (ng per min per gram testis wet weight)					
		Before injection	Minutes after injection				
			15	30	45	60	75
16	0.9% NaCl	1.9 ± 1.2[a]	1.5 ± 0.9	1.3 ± 0.7	1.2 ± 0.7	1.1 ± 0.7	1.0 ± 0.7
15	0.075 μg hCG[b] (± 0.016)[a]	1.9 ± 1.1	2.1 ± 1.2	6.3 ± 4.0	7.4 ± 4.3	5.6 ± 3.4	4.3 ± 2.4
12	1.376 μg hCG[b] (± 0.176)[a]	1.6 ± 1.0	5.4 ± 2.3	26.2 ± 10.1	40.3 ± 12.7	44.5 ± 13.4	42.8 ± 16.6

[a] One standard deviation of mean value.

[b] Mean dose of hCG (μg/g testis wet weight). Animal preparation II was used in this work and hCG was injected via the spermatic artery over 1 minute. The data are from unpublished work by K. B. Eik-Nes and G. D. Meier (1974).

TABLE II
MEAN CONCENTRATION OF TESTOSTERONE IN TESTICULAR TISSUE OF DOGS 78 MINUTES AFTER INJECTION OF 1 ml OF SODIUM CHLORIDE SOLUTION OR 1 ml OF SODIUM CHLORIDE SOLUTION CONTAINING hCG VIA THE SPERMATIC ARTERY

No. of testes	Preparation injected	Tissue testosterone (ng/g testis wet weight)
16	0.9% NaCl	72 ± 25[a]
15	0.075 μg hCG	141 ± 73
12	1.376 μg hCG	1056 ± 365

[a] One standard deviation of mean value. The details of this investigation are described in the legend of Table I. The data are from unpublished work by K. B. Eik-Nes and G. D. Meier (1974).

gonadotropins can be measured within 3–6 minutes.[21] The same response pattern is seen following cyclic 3,5′-AMP via this route.[21] Both animal preparations will secrete radioactive testosterone when radioactive acetate or radioactive cholesterol is infused via the spermatic artery.[24,31] It has proved difficult, however, to saturate the cholesterol pool(s) of the testis by infusing radioactive cholesterol via the spermatic artery.[31] Thus, the infused testis suffers the drawbacks of all infusion techniques when dealing with compounds of low solubility or poor vascular permeability: inability to reach the intracellular effector structures. Cyclic 3′,5′-AMP via the spermatic artery will stimulate testosterone secretion in the dog provided that many times the amount of cyclic 3′,5′-AMP the dog testis can produce under maximal gonadotropic stimulation[37] is infused.

Provided that one infuses compounds of high vascular permeability, the infused testis offers the following advantages over other techniques of studying testicular function *in vivo* (a) in the whole animal or of techniques *in vitro* (b, c) with slices, homogenates, or testicular fractions obtained by high speed centrifugation of homogenates.

a. The test compound is delivered directly to the testicular tissue and is not subjected to metabolic changes before entering this gland. The metabolic fate of the test compound in the testis can be investigated and the distribution of the test compound in the testicular cells can be studied under such conditions.[38]

[37] F. Murad, B. Strauch, and M. Vaughan, *Biochim. Biophys. Acta* **177**, 591 (1969).
[38] H. J. van der Molen and H. W. A. de Bruijn, *in* "Control of Gonadal Steroid Secretion" (D. T. Baird and J. A. Strong, eds.), p. 96. University Press, Edinburgh, 1971.

b. The test compound is delivered to whole testicular cells and the relationship between the test compound and its effect on the vascular situation in the testis can be investigated. Thus, when the catecholamine isoproterenol is infused via the spermatic artery in animal preparations I and II, increased production and secretion of testosterone are seen.[39] No effect of this catecholamine can be observed on testosterone production when added to testicular slices or homogenates. Flow of blood in the testis of the conscious ram will increase after administration of isoproterenol.[40]

c. Cofactors for testosterone biosynthesis are present in the infused testicular tissue in physiological concentrations. The problem of nature of cofactors and their concentration on testicular formation of testosterone *in vitro* can be a serious one. (For review of this problem in the rat testis *in vitro,* see paper by Eik-Nes[11]). Moreover in the infused organ, biosynthetic intermediates are not permitted to accumulate beyond physiological levels and excess intermediates are removed by the spermatic effluent (blood, buffer).

It is clear that certain enzymes in the testis are inhibited by products in the biosynthetic pathways leading to testosterone formation[3] (Fig. 1). Inano *et al.*[41] have demonstrated that $17\alpha,20\alpha$-dihydroxy-4-pregnen-3-one is a competitive inhibitor of the C_{17-20}-lyase in rat testicular tissue *in vitro*. This effect is also seen with 20α reduced 17α-hydroxyprogesterone in rabbit testis *in vitro*,[41] and the rabbit testis secretes radioactive $17\alpha,20\alpha$-dihydroxy-4-pregnen-3-one into spermatic venous blood when infused with radioactive progesterone via the spermatic artery.[42] Still, when $17\alpha,20\alpha$-dihydroxy-4-pregnen-3-one is infused via the spermatic artery of rabbits, no inhibition of the C_{17-20}-lyase in the infused testis can be observed.[42] One may thus seriously question whether we are dealing with a *veritas in vitro* and a *veritas in vivo* in experimental biology. The balanced view on biological properties of an organ must draw equally from data *in vivo* and *in vitro*. The goal for future work in testicular technology *in vivo* is to develop preparations that will permit easy sampling of spermatic venous blood and normal function of testicular cells producing the testicular androgens Δ^4-androstenedione,[43] testosterone,[2,43] and 5α-dihydrotestosterone.[28] In these preparations due respect should also be paid to the other *major* testicular function: *production of spermatozoa.*

[39] K. B. Eik-Nes, *Amer. J. Physiol.* **217,** 1764 (1969).

[40] B. P. Setchell, G. M. H. Waites, and G. D. Thorburn, *Circ. Res.* **18,** 755 (1966).

[41] H. Inano, H. Nakano, M. Shikita, and B. Tamaoki, *Biochim. Biophys. Acta* **137,** 540 (1967).

[42] H. W. A. de Bruijn, and H. J. van der Molen, *J. Endocrinol.* In press.

[43] T. Brinck-Johnsen and K. B. Eik-Nes, *Endocrinology* **61,** 676 (1957).

[24] *In Vitro* Techniques for the Study of Spermatogenesis

By ANNA STEINBERGER

Availability of suitable *in vitro* models for the study of spermatogenesis would eliminate the complex interactions which exist between various organs in a whole animal, allowing evaluation of the direct effects of various factors on spermatogenesis under more rigidly controlled conditions. In studies of human spermatogenesis, the *in vitro* models would also alleviate certain moral and legal considerations that may hinder similar investigations *in vivo*. It should be emphasized that a meaningful study of spermatogenesis *in vitro* presupposes a thorough knowledge of testicular morphology and physiology, particularly of the kinetics of spermatogenesis. Reviews of these subjects have been published recently.[1,2]

Attempts have been made by numerous investigators to define culture conditions that would support the spermatogenic process.[3,4] These attempts have met with a limited degree of success. An organ culture method has been developed which allows maintenance of the tubular structure and viability of Sertoli cells and primitive type A spermatogonia for periods of several months. In addition, several factors have been identified which can stimulate differentiation of gonocytes and spermatogonia to spermatocytes in late pachytene stage of miosis during the initial 3–4 weeks of cultivation. However, no culture methods tested thus far could promote formation of spermatids or maintain differentiation of the diploid spermatogenic cells for longer periods of time.

Differentiation of germinal cells does not seem to take place in cultures of dissociated cells,[5] indicating that certain histotopographical relationships may be essential for the differentiation process. Perhaps this is not surprising in view of the close association of the germinal cells with the Sertoli cells in the intact testes, and the postulated "nourishing," role of the Sertoli cells for the developing germinal cells. Cell culture and other *in vitro* techniques have been very useful in obtaining populations of specific cell types from the testes for studies of their metabolism

[1] E. Steinberger, *Physiol. Rev.* **51,** 1 (1971).

[2] E. Steinberger and A. Steinberger, *in* "Reproductive Biology" (H. Balin and S. Glasser, eds.), p. 144. Excerpta Med. Found., Amsterdam, 1972.

[3] A. Steinberger and E. Steinberger, *in* "The Testis" (A. D. Johnson, W. R. Gomes, and N. L. Van Demark, eds.), Vol. 2, p. 363. Academic Press, New York, 1970.

[4] A. Steinberger and E. Steinberger, *J. Reprod. Fert.,* Suppl 2, 117 (1967).

[5] A. Steinberger and E. Steinberger, *Exp. Cell Res.* **44,** 443 (1966).

and hormonal responses. Since formation of spermatozoa may depend on the integrated function of several or all cell components in the testes, information gained regarding the physiology of individual cell types could contribute to a better understanding of the mechanisms by which spermatogenesis is regulated.

Several *in vitro* techniques which may be useful in the study of spermatogenesis will be described under individual headings.

Organ Culture of Mammalian Testes

Organ culture methods permit maintenance of whole or parts of an organ *in vitro* with minimum disturbance of the normal histotopographical relationships among the tissue components. A method which proved to be most successful for growth and differentiation of the mammalian testes involves cultivation of small (1 mm^3) tissue explants on top of stainless steel wire grids at the surface of liquid culture media.[3,4] This method offers several advantages over the older organ culture methods which utilized plasma clots containing embryo extracts as source of nutrients and for physical support. The tissue enzymes often liquefied the plasma clot, causing submersion of the tissue in the liquid and subsequent necrosis. This could be partly avoided by frequent transfers of the explants to different areas on the same plasma clot or to fresh plasma clots, which in turn induced excessive mechanical trauma. In addition, use of such complex media as plasma and embryo extract precluded the identification of specific nutrients needed for growth and/or differentiation. Growth of tissue explants in liquid media on top of floating pieces of lens paper[6] has many advantages over the plasma clot method but is not very reliable as many lens paper "boats" tend to sink after becoming saturated with the media. Also, lens paper fibers tend to attach to the tissue making histologic sectioning difficult. Use of stainless steel grids for support for organ cultures in liquid media introduced by Trowell,[7] eliminates all these difficulties. A large number of explants can be cultured on a single grid and the culture medium can be replaced, sampled, or altered in its composition at any time without disturbing the tissues. A Millipore filter placed between the grid and the explants further simplifies handling of the explants.

Culture Method

1. One or two stainless steel grids 5 mm high, with a 25 × 25 mm flat platform (Wilks Co., Bethesda, Maryland) are placed inside a small

[6] J. M. Chen, *Exp. Cell. Res.* **7,** 518 (1954).
[7] O. A. Trowell, *Exp. Cell Res.* **16,** 118 (1959).

petri dish and covered with a Millipore filter of 0.47–0.48 μm pore size diameter (Millipore Corp., Bedford, Massachusetts). The Millipore filters can be either purchased in sterile form or should be sterilized strictly according to packaging instructions. A sterile, disposable organ culture dish assembly is also available commercially (Falcon Plastics, Bio Quest).

2. Culture medium is slowly added to each dish until it reaches the level of the filter. Proper level of the culture medium is very crucial, since those parts of tissue which are submerged in the medium become necrotic within 1–2 days of cultivation. Care must be taken to avoid trapping of air bubbles under the grid which may interfere with proper diffusion of products to and from the tissue. A small mirror placed under the petri dish is very helpful during this step.

The culture medium consists of Eagle's minimum essential medium[8] supplemented with 2 m*M* glutamine (in addition to the 2 m*M* glutamine which is already included in the medium formulation), 1 m*M* of sodium pyruvate, and 0.1 m*M* of each of the following nonessential amino acids: alanine, aspartic acid, glutamic acid, glycine, proline, and serine (all available from Baltimore Biological Laboratories, Microbiological Associates, Bethesda, Maryland, and other commercial sources). To reduce the incidence of microbial contamination 100 U of penicillin, 100 μg of streptomycin, and 5 μg of Fungizone (Squibb) per milliliter can be added to the culture medium. The medium is adjusted to pH 7.0 ± 0.1 with 7.5% $NaHCO_3$. Addition of serum to the chemically defined medium, while not absolutely essential, increases its buffering capacity and improves tissue viability during prolonged cultivation. In this respect calf or fetal calf serum in final concentration of 10% appears to be optimal. The prepared culture dishes are kept at 31° in a humidified atmosphere of 5% CO_2 and 95% air for at least 15 minutes prior to use to allow proper pH adjustment.

3. Animals are sacrificed by cervical dislocation or an overdose of ether or chloroform. The testes are removed through an abdominal incision and placed into a petri dish containing several milliliters of the culture medium. After removal of the tunica albuginea, the testes are cut with a pair of sharp blades (Bard-Parker blades No. 11) used in a scissors motion into fragments of 1 mm^3 size. Care should be taken to avoid mechanical damage of the tissue due to excessive manipulation. The germinal cells are easily dislodged from the seminiferous tubules and float out into the medium. Also, the fragments should be cut as uniformly as possible. Undersized fragments become overly distorted by the cutting procedure, and fragments which are significantly larger than 1 mm^3 de-

[8] H. Eagle, *Science* **130,** 432 (1959).

velop "central necrosis" during cultivation due to inadequate oxygen penetration. An inverse relationship exists between the limiting size of the cultured tissue and its oxygen consumption. The limiting size is also directly related to the partial pressure of oxygen in the surrounding atmosphere and the rate of diffusion through the tissue.

4. Up to 25 fragments are arranged on top of each filter-covered grid leaving spaces between them. The covered culture dishes are incubated at 31° in a humidified atmosphere of 5% CO_2 and 95% air. This temperaature and gas composition were found to be optimal for culture of testes from several mammalian species including human.[3,4,9] Incubation temperatures above 31° result in more rapid degeneration of the germinal elements. Increased oxygen concentration is also not beneficial despite the fact that larger explants can be grown without "central necrosis." Higher oxygen concentrations interfere with differentiation and viability of the germinal cells possibly by affecting some crucial oxygen-sensitive enzyme systems.

5. At 3- to 4-day intervals during the incubation, the spent medium is removed with a pipette, and fresh medium is added slowly, taking similar precautions as during the initial preparation of the culture dishes.

Evaluation of Results

Histology. At any time during cultivation one or more explants can be removed for histologic examination. The filter is cut around the desired explants and transfered together with the tissues into freshly prepared Bouin's fluid. The filter usually becomes detached from the tissue in the fixative and is discarded. After 2–4 hours of fixation, the tissues are dehydrated, embedded in paraffin and cut into 3–5 μm sections. The mounted sections can be stained by various methods but the periodic acid–Schiff reaction followed by hematoxylin stain is most useful for microscopic visualization of the acrosomes. Thus, morphological changes related to various experimental conditions can be determined. It must be kept in mind, however, that presence of germinal cells in the cultured tissues may represent cell survival rather than *in vitro* differentiation. On the other hand, appearance of more differentiated germinal cells than present in the tissue prior to cultivation indicates spermatogenic activity in culture. Since considerable variation exists in the microscopic appearance of the cultured explants even when they are grown in the *same* culture dish, several explants should be examined for each experimental point.

[9] A. Steinberger, M. Fischer, and E. Steinberger, *in* "The Human Testes" (E. Rosemberg and C. A. Paulsen, eds.), p. 333. Plenum, New York, 1970.

Presence of "central necrosis" in some cultured explants is a reflection of limited diffusion of oxygen and has no effect on the differentiation of germinal cells or tissue viability in the peripheral portions. The spermatids generally survive for several days, spermatocytes for 3–4 weeks, while some spermatogonia and Sertoli cells remain viable for at least 8 months.[3,4] Cultures, initiated with testes of newborn animals in which the only germinal cells present are gonocytes, show progressive differentiation and formation of late pachytene spermatocytes after approximately 3 weeks of cultivation in chemically defined media supplemented with 2 m*M* glutamine or vitamins A, C, and E, in concentrations of 12 IU (Aquasol, U.S. Vitamin and Pharm. Corp.), 50 μg (ascorbic acid) and 120 μg (tocopherol, Hoffman-LaRoche Inc.) per milliliter, respectively. The pachytene spermatocytes, however, gradually degenerate without giving rise to spermatids. The primitive type A spermatogonia persist and continue to divide in culture for many months.[3,4] These cells retain the capacity to differentiate when placed under suitable conditions, since cultures transplanted into testes of adult, homologous hosts resume spermatogenic activity (see Testicular Transplants).

Radioautography

Differentiation of germinal cells in cultures of sexually mature testes is difficult to assess on the basis of histology alone since the initial tissues contain germinal cells in all stages of spermatogenic development. Radioautography following labeling with [^{3}H]thymidine can be used for identifying the cells which evolve *in vitro*.[10] In this approach, advantage is taken of the fact that germinal cells complete DNA synthesis prior to the leptotene stage of miosis. Only the preleptotene cells incorporate [^{3}H]thymidine during a short period of exposure, while the more advanced germinal cells become labeled only in result of differentiation of the labeled preleptotene cells. Thus, the fate of labeled preleptotene cells in culture can be easily followed by radioautography of the explants following various periods of cultivation.[10]

The preleptotene germinal cells can be labeled either *in vivo* or *in vitro*. For *in vivo* labeling, the animals are given a single subcutaneous or intraperitoneal injection of [^{3}H]thymidine (6.7 Ci/mmole), 1 μCi per gram of body weight. Three hours after the injection the testes are removed and used to initiate cultures.

For *in vitro* labeling (particularly useful for biopsy specimens of human testes) the cultures are grown during the initial 40–60 minutes in medium containing [^{3}H]thymidine (1 μCi/ml). The "hot" medium is

[10] A. Steinberger and E. Steinberger, *J. Reprod. Fert.* **9**, 243 (1965).

then replaced by fresh culture medium and the incubation continued. At time intervals several cultured explants are fixed in Bouin's fluid and 3 μm histologic sections are coated with liquid emulsion (Eastman Kodak NTB-2) by the dipping technique of Kopriwa and Leblond.[11] After a suitable period of dark exposure at 4° the slides are developed, stained with PAS-hematoxylin, and examined microscopically with an oil immersion objective.

Using this technique, differentiation of preleptotene cells to late pachytene spermatocytes was demonstrated in cultures of sexually mature testes from several mammalian species.[3,4,9] The differentiation occurs in chemically defined culture media, providing the media contain 4 m*M* glutamine or vitamins A, C, and E in concentrations of 12 IU, 50μg, and 120 μg per milliliter, respectively. Formulation of labeled spermatids has not been observed in these cultures.

Testicular Transplants

The potential spermatogenic capacity of the cultured tissues can be tested by transplanting the cultures into testes of adult homologous hosts.[3,4] The procedure is as follows:

Testes of an anesthetized animal are exposed through a small abdominal incision. A 2–3 mm slit is made in the tunica albuginea and a cultured tissue fragment is deposited under the tunica using a Barker biopsy needle with a fitted stylet (B-D Co.). The host testes are then returned to the scrotum, and the abdominal incision is sutured. After various time intervals, the testes bearing the transplants are removed, fixed in Bouin's fluid, and sectioned serially at 4 μm. The transplants can be readily identified in the paraffin sections with the naked eye. Selected sections are stained with PAS-hematoxylin and examined microscopically.

After a lag period of about 2 weeks, the transplants increase in size and spermatogenic activity becomes evident in approximately 30% of the seminiferous tubules. Eight to 10 weeks later many tubules contain mature spermatozoa. Using this method, rat testes cultured for 8 weeks were shown to retain spermatogenic capacity.[3,4] It should be noted, that even in a highly inbred strain of animals the transplants evoke an immunologic response which accounts for a certain degree of variability in their spermatogenic activity.

Cell Culture

Cell culture methods provide means for selective isolation of certain cell populations from the testes. Growth in culture, in terms of increased

[11] B. M. Kopriwa and C. P. Leblond, *J. Histochem. Cytochem.* **10**, 269 (1962).

cell number is limited to those cell types which normally divide in the testis and can continue to divide under the *in vitro* conditions, e.g., interstitial and peritubular cells. Nondividing cells like spermatids or mature Sertoli cells can be, at best, maintained in culture for only a limited period of time.

Depending on the objectives being sought the culture techniques are described below.

Culture of "Total" Cell Population

1. Decapsulated testes are placed in a petri dish containing several milliliters of Hanks' balanced salt solution (HBSS) and minced thoroughly with iris scissors. The minced tissue is then incubated with 0.25% trypsin in Ca^{2+}- and Mg^{2+}-free HBSS, pH 7.6, for 15 minutes at 31°. Approximately 30 ml of the trypsin solution are used per 1 g of tissue.

2. After the incubation, the remaining tubules are allowed to settle to the bottom of the flask and the supernatant containing predominantly single cells is transferred to centrifuge tubes containing serum (2% final concentration) or trypsin inhibitor (LBI-Worthington 0.2%). The cells are spun at 500 *g* for 5–10 minutes, resuspended in serum containing culture medium, and kept at room temperature in a tightly stoppered container to prevent rise in pH.

3. The tubules are reincubated for 15 minutes with fresh trypsin, and the liberated cells are again removed, spun and combined with the previous cell fraction. This procedure is repeated several times until most of the tissue becomes dissociated into single cells. Seminiferous tubules of adult testes are more resistant to trypsin digestion than tubules of testes from immature animals most likely due to a larger content of collagen. Therefore, combined action of trypsin and collagenase (used simultaneously or sequentially) may be required for complete dissociation. The collagenase concentration has to be determined experimentally as preparations vary in activity and degree of contamination by proteolytic enzymes.

4. The pooled cells are counted in a hemacytometer and 0.5×10^6 cells per milliliter are placed into culture flasks. The cultures are incubated at 31° in an atmosphere of 5% CO_2 and 95% air. The culture medium is Eagle's minimum essential medium supplemented with 0.1 mM nonessential amino acids, 1.0 mM sodium pyruvate, antibiotics, and 10% calf serum or fetal calf serum. In contrast to organ culture, presence of serum in the culture medium is essential for cell viability and growth in testes cell culture.

In stationary cultures the nongerminal elements become attached to the surface of the culture vessel within 24 hours, begin to divide, and eventually form a confluent monolayer. The germinal cells, on the other hand, remain unattached and can be removed together with the medium. Thus, whenever the culture medium is being replenished, in order to maintain all cells elements in the same culture, the germinal cells have to be recovered from the spent medium by centrifugation and returned to the culture flask.

Cells which form the monolayer are not homogeneous. They represent various proportions of interstitial fibroblasts, Leydig cells, endothelial cells from the blood vessels and lymphatics, myoid cells from the seminiferous tubule basement membrane and Sertoli cells.

The relative proportion of Leydig cells in the monolayer can be increased by *in vivo* pretreatment with hormones.[12,13] The monolayer growth can be maintained for prolonged periods of time by subcultures at 10–14-day intervals. However, cell composition changes during cultivation in favor of the fibroblasts which multiply most vigorously.

The germinal elements survive in culture for varying periods of time: young spermatids for several days, spermatocytes, some spermatogonia and mature spermatids for several weeks.[5] There is no convincing evidence that germinal cells differentiate under these conditions although formation of labeled spermatids from spermatocytes in cell cultures of human testis has been reported. It should be emphasized that presence of mature spermatids in cultures, even after a prolonged period of cultivation, should not be interpreted as evidence for spermatogenesis *in vitro* unless criteria, similar to those previously described for differentiation of germinal cells in organ culture, can be fulfilled.

Owing to their different behavior in culture, the germinal and nongerminal elements can be easily separated. Further separation of some cell types can be accomplished by using velocity sedimentation techniques.

Separation of Germinal Cells by Velocity Sedimentation

Enriched populations of germinal cells at certain stages of spermatogenesis have been obtained by velocity sedimentation at unit gravity.[14–16] The cells are sedimented through a shallow gradient of bovine serum albumin in a "staput" chamber at 4°. While this method has not

[12] O. Vilar, A. Steinberger, and E. Steinberger, *Z. Zellforsch. Mikrosk. Anat.* **74**, 529 (1966).

[13] E. Steinberger, A. Steinberger. O. Vilar, I. I. Salamon, and B. N. Sud, *Ciba Found. Colloq. Endocrinol.* **16**, 56 (1967).

[14] D. M. K. Lam, R. Furrer, and W. R. Bruce, *Proc. Nat. Acad. Sci. U.S.* **65**, 192 (1970).

[15] M. L. Meistrich, *J. Cell Physiol.* **80**, 299 (1972).

[16] R. G. Vernon, V. L. W. Go, and I. B. Fritz, *Can. J. Biochem.* **49**, 761 (1971).

been perfected to yield "pure" populations of all cell types from the testes, it provides a relatively rapid method (several hours) for obtaining enriched populations of several classes of germinal cells. This method has been utilized for cellular localization of various enzymes in the testes and studies of the hormonal control of spermatogenesis. For a more detailed description of the velocity sedimentation method the reader is referred to the original publications.[14-16]

Isolation of Seminiferous Tubules

Seminiferous tubules essentially free of interstitial cells can be isolated by either of two following methods:

Method a. Decapsulated testes are placed into a petri dish containing 10–15 ml of chilled HBSS, pH 7.2, and the seminiferous tubules are teased apart with jeweler's forceps. Bits of tissue which resist teasing are discarded. Well separated tubules are placed on top of a stainless steel mesh grid (70–100 μm mesh opening) and rinsed with a large volume (50–100 ml) of chilled HBSS. The interstitial cells become separated from the tubules by the force of the fluid stream and pass freely through the filter. The seminiferous tubules, on the other hand, are retained by the grid and can be "back-washed" into a clean petri dish. Only occasional residual interstitial cells are revealed by careful histologic or cytochemical examination for 3β-ol-hydroxysteroid dehydrogenase.[13]

Method b. Decapsulated testes are rinsed in several milliliters of Ca^{2+}- and Mg^{2+}-free HBSS and are transferred to an Erlenmeyer flask containing 0.25% trypsin in the above HBSS adjusted to pH 7.6 and prewarmed to 33°. Approximately 15 ml of trypsin solution are used per gram of testis. The flask is incubated at 33° for 20–30 minutes being *vigorously* agitated every 3–4 minutes. After the incubation, the contents are strained through the stainless steel mesh grid. The tubules which are retained by the grid are rinsed with a large volume of cold HBSS containing Ca^{2+} and Mg^{2+} and are then "back-washed" into a clean petri dish.

Method b eliminates interstitial cells more efficiently than method a. Either of these methods is considerably faster and more practical for isolation of seminiferous tubules from large amounts of tissue compared to manual separation of tubules under a dissecting microscope.[17]

Cultures of Peritubular Cells

Growth of homogeneous populations of peritubular cells can be obtained by initiating cultures with seminiferous tubules which have been carefully separated from the interstitial cells, by the procedures described above.

[17] A. K. Christensen and N. R. Mason, *Endocrinology* **76,** 646 (1959).

1. Ten to fifteen tubules are arranged in a petri dish containing 0.5 ml of culture medium plus 10% fetal serum, pH 7.0 ± 0.1. A strip of perforated cellophane is placed over the tubules to prevent them from floating up into the medium. The dishes are incubated at 33° in a humidified atmosphere of 5% CO_2 and 95% air.

After 24 hours of incubation, 3–5 ml of fresh culture medium are added to each dish, and the incubation is continued for 4–5 days.

2. At this time the cultures are examined with an inverted microscope, and, if outgrowth of peritubular cells is apparent along the length of the tubule wall, the cellophane and the tubules are removed with forceps. The culture medium is replenished and the dishes are returned to 33° incubator. The peritubular cells remaining in the dishes continue to divide and eventually form a monolayer. Based on recent ultrastructural studies, the pertibular tissue contains specialized fibroblasts—the myoid cells and endothelial cells from the adjacent lymphatics. Whether one or both of these cell types grow as monolayers in culture has not been determined.

This method provides an effective means for separating peritubular cells from other cellular elements of the seminiferous tubule. It should be pointed out, however, that cultures of peritubular cells initiated with seminiferous tubules from neonatal testes may become "contaminated" by the dividing Sertoli cells which migrate out through the open ends of the tubules. The Sertoli cells do not divide in adult testes[18] and are removed with the tubules. Contamination by germinal cells is of no concern since, not being attached to the culture dish, they are removed together with the medium.

Cultures of Isolated Seminiferous Tubules in Rose Chambers

This method provides an excellent means for direct microscopic observation or cinemicrography of the tubule content without disturbing the cultures.

Several isolated seminiferous tubules are placed with a drop of culture medium on the bottom coverslip of a Rose chamber[19] and covered with a strip of cellophane dialyzing membrane. The culture medium is the same as used for testes organ culture. The chamber is then assembled as shown in Fig. 1.

The chamber is filled with culture medium through one needle while the air is being expelled through the second needle. Both needles are then

[18] A. Steinberger and E. Steinberger, *Biol. Reprod.* **4**, 84 (1971).

[19] G. G. Rose, C. M. Pomerat, T. O. Shindler, and J. B. Trunnell, *J. Biophys. Biochem. Cytol.* **4**, 761 (1958).

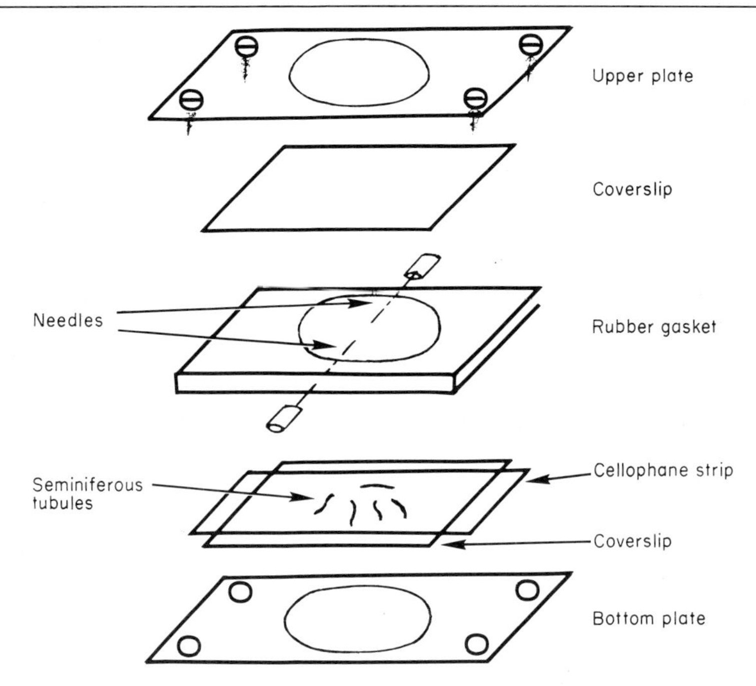

FIG. 1. Assembly of Rose chamber.

sealed with Parafilm and the chamber is incubated at 31°. The culture medium is replenished every 3–4 days through the inserted needles. At any time during cultivation, various test substances can be introduced into the chamber or the chamber examined microscopically.

The cellophane membrane serves several purposes. It exerts a slight pressure on the tubules making the flattened contents optically more suitable for microscope observation or photography. The cellophane membrane also protects the tissue from the disturbing currents which are created when medium is being replenished or substances are introduced into the chamber. The width of the membrane strip can be varied depending on whether all medium components or just the dialyzable part is to come in contact with the tissue.

Under these culture conditions the survival of germinal and Sertoli cells is similar to that described for organ culture of testicular explants. In at least one study, however, completion of the meiotic division has been observed in Rose chamber cultures of seminiferous tubules from a Chinese hamster.[20] Outgrowth of peritubular cells in the Rose chambers is similar to that described for cultures of seminiferous tubules in petri dishes. Out-

[20] D. J. Ellingson and K. T. S. Yao, *J. Cell Sci.* **6**, 195 (1970).

growth of Sertoli cells also occurs from the open ends of the tubules when the cultures are initiated with tubules from neonatal rats.[21] This outgrowth can be utilized to initiate clone cultures of Sertoli cells.

Clone Cultures of Sertoli Cells

Pure cultures of Sertoli cells can be obtained by cloning. However, since Sertoli cells do not divide in adult testes *in vivo* or *in vitro*[18] only embryonal or neonatal testes can be utilized for this purpose. The procedure is as follows:

1. Isolated seminiferous tubules are grown in Rose chambers as described above.
2. After 4–8 days of cultivation, areas near the tubule ends containing large numbers of Sertoli cells are marked with an objective marker (Zieler Instrument Co., Boston, Massachusetts).
3. The chamber is dissassembled, and sterile glass cylinders (3 mm $\times$ 10 mm) are placed over the cells in the marked areas. In order to create a tight seal, the edge of the cylinder is coated with sterile silicone grease (Dow Corning).
4. To each cylinder, 0.1 ml of 0.25% trypsin in Ca^{2+}- and Mg^{2+}-free HBSS, pH 7.5, is added and allowed to remain in contact with the cells for 20 minutes at 37°.
5. After the incubation the contents of each cylinder are triturated with a Pasteur pipette and transferred to a centrifuge tube containing 2% serum or 0.2% of trypsin inhibitor (LBI, Worthington Laboratory).
6. The tubes are spun at 500 *g*, the supernatant is discarded and the cells are resuspended in several milliliters of culture medium containing 20% fetal calf serum, pH 7.0 $\pm$ 0.1.
7. A cell count is performed in triplicate using a hemacytometer. An accurate estimate of the cell concentration and absence of cell clusters are very crucial in the cloning procedure. If the triplicate counts do not agree within 10%, they should be repeated.
8. The cell suspension is diluted with culture medium to 1 cell/100 μl. One hundred microliter aliquots of the diluted cell suspension are dispensed into individual microwells of a tissue culture Microtest II plate (Baltimore Biological Laboratories), and the plates are incubated at 37° in an atmosphere of 5% CO_2 and 95% air.
9. After 16–20 hours of incubation, the microwells are scanned with an inverted microscope at 100 times magnification in order to record the wells containing a single cell. Wells containing multiple cells are considered unsuitable for cloning since they may not be of the same type.

[21] E. Steinberger and A. Steinberger, *Recent Progr. Horm. Res.* **26**, 547 (1970).

10. The wells are rescanned after additional 5–10 days of incubation to ascertain cell proliferation. The Sertoli cells can be identified with phase contrast optics by their nuclear and cytoplasmic characteristics.

11. Clones identified as Sertoli cells can be pooled for biochemical studies or maintained in culture for several weeks for the study of morphological differentiation. The Sertoli cells divide in culture for a period of time comparable to their proliferative period *in vivo*.[18,21] They remain viable for several weeks after division ceases.

There are several major drawbacks to this method. It is time consuming and yields relatively small numbers of Sertoli cells. In addition, the method can only be used for the isolation of Sertoli cells from prenatal or neonatal testes.

Isolation and Culture of Interstitial Cells

The methods used for the isolation of seminiferous tubules are not reliable for obtaining pure populations of interstitial cells. This is because during teasing of the tubules peritubular and germinal cells become mixed with the interstitial cells. While the germinal cells can be subsequently eliminated from the culture by removing the medium, the peritubular cells behave in culture similarly to the interstitial cells and cannot be readily separated from them. Thus, greater purity of the interstitial cells must be achieved by minimizing the degree of contamination by the germinal and peritubular cells. This can be accomplished by either of two following methods.

Isolation Method a

1. The testes are decapsulated carefully so as *not to cut the tubules*, rinsed in Ca^{2+}- and Mg^{2+}-free HBSS and placed in an Erlenmeyer flask containing 0.25% trypsin in Ca^{2+}- and Mg^{2+}-free HBSS. The trypsin solution is adjusted to pH 7.6 with $NaHCO_3$ and warmed to 37°. Approximately 30 ml of the enzyme solution are used per gram of testis.

2. The flask is incubated at 37° for 15 minutes, being gently agitated every 3–4 minutes.

3. After the incubation, the contents are strained through a stainless steel mesh grid (70–100 μm mesh porosity) placed over a beaker and rinsed with 20–30 ml of HBSS. The grid retains the tubules but allows the interstitial cells to pass through. The interstitial cells are recovered from the liquid by centrifugation and resuspended in media containing 0.2% lima bean trypsin inhibitor (LBI, Worthington) or 2% serum.

4. The quality of the preparation is checked microscopically. Good preparations should contain very few germinal elements. Approximately

10–13 million interstitial cells can be obtained from a single testes of an adult (250 g) rat.

Isolation Method b

1. The decapsulated testes are placed directly on the stainless steel mesh grid and the tubules are teased *slightly* with forceps without breaking.

2. The tissue is rinsed with a large volume of chilled HBSS allowed to flow in a forceful stream from a height of approximately 4 inches. The interstitial cells are forced through the grid by the streaming fluid and can be recovered from the fluid by centrifugation.

Method b eliminates cell contact with trypsin but yields considerably fewer interstitial cells. However, it may be the method of choice for certain studies (e.g., hormone binding) where exposure to trypsin may alter some aspects of the cell physiology.

Culture Method. Culture dishes are seeded with 3 to 6×10^5 cells per milliliter of medium containing 10–20% calf serum and are incubated at 33–37° in an atmosphere of 5% CO_2 and 95% air. The culture medium is replenished at 3- to 4-day intervals. More frequent media changes may be necessary with increasing cell densities due to a relatively high rate of carbohydrate metabolism in these cells and the resulting rapid drop in pH.

There is no significant difference in the growth quality of the interstitial cells isolated by either of the two methods. The cells grow well in various commercially available culture media (Eagle's MEM, medium 199, etc.). The growth is more rapid at 37° compared to the more physiological, lower temperatures and can be further accelerated by increasing the serum concentration in the medium from 10% to 20%. Fetal calf serum is generally superior to calf serum or sera of other species in promoting growth of the interstitial cells. However, different serum lots vary considerably in their growth-promoting quality. The interstitial cells remain viable and functional in culture for a period of several weeks as evidenced by their ability to bind LH or HCG and respond to these hormones with increased rate of growth and testosterone secretion. However, the Leydig cells become gradually overgrown by the more rapidly multiplying interstitial fibroblasts.

The isolated interstitial cells provide an experimental system for the study of mechanisms by which hormones and other factors may directly regulate their steroidogenic activity. The system may be also useful for the study of hormone–cell receptor interactions and of the chemical nature of the receptors.

[25] Preparation of Mouse Embryos for the Evaluation of Hormone Effects[1]

By Dianne Moore Smith and John D. Biggers

Mammalian embryos require a maternal environment stimulated by the appropriate steroids in which to undergo early development and implantation. The morphological and biochemical effects of these hormones on maternal tissues are the subject matter of a vast literature. Whether ovarian steroids also affect the metabolism of the embryo is not yet clear. Work is presently in progress to determine whether steroids act directly on the embryo to influence normal preimplantation development, in particular the increase in blastocyst metabolism which occurs prior to implantation.

Techniques for culturing mouse embryos in a chemically defined medium were developed in the early 1960's.[2] They permit preimplantation development of cleavage-stage embryos outside the maternal reproductive tract. The direct action of steroids on various aspects of embryo metabolism can thus be examined in an *in vitro* system where competition from maternal hormones bound by embryos developing *in vivo* is not a complicating variable.

The techniques outlined below are used in the preparation of cultured mouse embryos for evaluation of the effect of estrogen on uptake and incorporation of radioactive amino acids and are similar to those described by Smith and Smith.[3] However, these techniques can easily be modified to study the effects of other steroids on protein or nucleic acid synthesis and to examine the binding of steroid hormones by preimplantation embryos. Since methods have recently been developed for culturing preimplantation rabbit[4-7] and sheep and cattle[8] embryos in chemically

[1] The preparation of this paper has been supported by funds from the Ford Foundation, the Rockefeller Foundation, and the National Institute of Child Health and Human Development.

[2] J. D. Biggers, W. K. Whitten, and D. G. Whittingham, *in* "Methods in Mammalian Embryology" (J. C. Daniel, ed.), pp. 86–116. Freeman, San Francisco, California. 1971.

[3] D. M. Smith and A. E. S. Smith, *Biol. Reprod.* **4**, 66 (1971).

[4] H. Onuma, R. R. Maurer, and R. H. Foote, *J. Reprod. Fert.* **16**, 491 (1968).

[5] M. T. Kane and R. H. Foote, *Biol. Reprod.* **2**, 245 (1970).

[6] M. T. Kane and R. H. Foote, *Proc. Soc. Expl. Biol. Med.* **133**, 921 (1970).

[7] H. Ogawa, K. Satoh, and H. Hashimoto, *Nature (London)* **233**, 422 (1971).

[8] H. R. Tervit, D. G. Whittingham, and L. E. A. Rowson, *J. Reprod. Fert.* **30**, 493 (1972).

defined media, the direct actions of steroids on the metabolism of the embryos of these species may also be evaluated.

Procurement of Embryos

Random-bred Swiss mice, approximately 6–10 weeks of age are injected with gonadotropins to induce ovulation of larger numbers of ova than would be normally released during a natural ovulation. Pregnant mare serum gonadotropin (PMSG) (Equinex, Ayerst; Gestyl, Organon; or PMSG, Sigma), 5–10 IU, are injected intraperitoneally followed 44–48 hours later by 5–10 IU of human chorionic gonadotropin (hCG) (Pregnyl, Organon; Chorionic Gonadotropin, Sigma).[9,10] At the time of hCG injection, females are paired with individually caged males and are considered to have mated if a vaginal plug is observed the next morning. Mice are killed at various times after identification of the vaginal plugs, the length of the interval depending on the degree of embryo development desired. Oviducts are isolated, placed in drops of culture medium in a plastic petri dish, and flushed with medium to obtain the embryos. To flush, a 30-gauge blunt needle attached to a tuberculin syringe containing culture medium is inserted into one end of the oviduct and approximately 0.05 ml of medium is forced through each fallopian tube. The embryos are collected with a finely drawn, orally controlled Pasteur pipette and placed into 1 ml of culture medium under 2 ml of paraffin oil in an embryological watch glass.[11]

Culture of Preimplantation Embryos

Embryos are washed by transferring them through one or two changes of medium and then set up in 50–100-μl drops of culture medium which have been placed under 10 ml of paraffin oil in a 60×15 mm tissue culture dish[12] as described by Brinster.[13] The culture medium is a supplemented Krebs-Ringer bicarbonate solution (see the table) which will support development of certain F_1 hybrid zygotes[14] and 2-cell embryos obtained from most strains to the blastocyst stage.

[9] R. G. Edwards and A. H. Gates, *J. Endocrinol.* **18**, 292 (1959).

[10] A. H. Gates, *in* "Methods in Mammalian Embryology" (J. C. Daniel, ed.) pp. 64–75. Freeman, San Francisco, California, 1971.

[11] Arthur H. Thomas, Inc.

[12] Falcon Plastics.

[13] R. L. Brinster, *Exp. Cell Res.* **32**, 205 (1963).

[14] J. D. Biggers, *in* "The Biology of the Blastocyst" (R. J. Blandau, ed.), pp. 319–327. Univ. of Chicago Press, Chicago, Illinois, 1971.

CULTURE MEDIUM

Component	grams/liter[a]	mM
NaCl	5.52	94.6
KCl	0.356	4.78
KH_2PO_4	0.162	1.19
$MgSO_4 \cdot 7H_2O$	0.294	1.19
Ca lactate $5H_2O$	0.527	1.71
Na lactate (DL)	2.416[b]	21.58
Na pyruvate	0.036	0.33
$NaHCO_3$	2.106	25.07
Glucose	1.0	5.56
Crystalline bovine albumin	4.0	—
K-penicillin G	100 IU/ml	—
Streptomycin SO_4	0.05	—

[a] Gassed in 5% CO_2 in air.
[b] 3.68 ml of 60% syrup.

Embryos are cultured in microdrops of medium under oil in an atmosphere of 5% CO_2 in air, 100% humidity (also according to Brinster[13]) for periods of up to 4 or 5 days. The medium is not changed during this interval. Under these conditions, 80% or more of the 2-cell stages which are collected routinely, develop to blastocysts. The day on which 2-cell embryos are placed in culture is referred to as day 1 of culture and corresponds to the second day of preimplantation development *in vivo*. Embryos at various stages of development may then be harvested from the culture drops: 4-cells on the evening of day 1 or early morning of day 2; 8-cells on the morning or afternoon of day 2; morulae on the evening of day 2 or morning of day 3; early blastocysts on the afternoon of day 3; expanded blastocysts on day 4; and hatched blastocysts on day 5 of culture.

Preparation of Radioactive Precursors

A noncompeting amino acid mixture[15] is prepared by combining 20 μCi of each of 3 amino acids—[^{14}C]valine, [^{14}C]lysine, and [^{14}C]aspartic or [^{14}C]glutamic acid. The mixture is lyophilized to remove the commercial solvent (0.1 M HCl), and the dried material is redissolved in 0.1 ml culture medium (600 μCi/ml).

[15] A. Tyler, J. Piatigorsky, and H. Ozaki, *Biol. Bull.* **130**, 204 (1966).

Preparation of Estrogen-Containing Medium

Crystalline estradiol-17β (Calbiochem, chromatographic standard) is dissolved in absolute ethanol (0.1 *M*). This stock solution is refrigerated in a sealed multiple-dose vial. Small volumes of stock solution are diluted 1:1000 with culture medium and sterilized by passage through a 0.3 μm syringe-adapted Millipore filter. Dilutions of estradiol are then prepared with sterile medium so that the concentration of each dilution is twice that desired for embryo incubation. A small volume of the amino acid mixture (in culture medium) is added to each estrogen dilution so that the final concentration of radioactive material in each dilution is 6.0 μCi/ml. The radioactive amino acids can be added to the estrogen-containing medium before or after the incubation of embryos has begun depending on the desired method for studying hormone stimulation of embryo metabolism.

Labeling and Washing Procedure

A small volume (0.1–0.5 ml) of the amino acid, estrogen-containing medium is added to a small vessel which will support normal development of cleavage-stage mouse embryos for more than 24 hours when embryos are cultured in medium which has *not* been placed under paraffin oil. Several small containers have been tested. The most satisfactory appear to be the tissue culture chamber slides manufactured by Lab-Tek Products.

A group of embryos (usually 20) harvested from culture at a specific stage of development (see above) is added to each chamber in a measured volume of cold, estrogen-free medium equal to the volume of radioactive amino acid, estrogen-containing medium already in the chamber. The medium already present in the chamber is thus diluted 50% by the addition of the embryos. The embryos in the various treatment groups are incubated in a tissue culture incubator (5% CO_2 in air; 100% humidity) for the desired labeling time.

After labeling, embryo metabolism is stopped by adding 2.5% glutaraldehyde to the incubation medium (final concentration 0.25%). The radioactive medium is then withdrawn from the culture vessel, and the embryos are washed with 5 changes of 0.2–0.5 ml of standard culture medium. A final volume of medium is added to the washed eggs. The embryos are removed from the chamber in a measured 50 μl of medium and deposited in one cup of a disposable plastic tray. Each group of 20 embryos is lysed with one drop of 0.5% sodium dodecyl sulfate, and the lysate is placed in a 0.22 μm, 47 mm Millipore filter. The plastic cup is rinsed with 50 μl of culture medium, and the rinse material is added to the corresponding filter disc. For each treatment group, an accompanying 50 μl of final wash medium is processed in an identical manner and used to determine background radioactivity.

All micromeasurements of medium, or medium and embryos, are made with disposable micropipettes[16] fitted with a fingertip suction apparatus,[17] and all manipulations of embryos are carried out under a dissecting microscope.

Treatment of Filter Discs

After embryos or wash medium are added, the Millipore filter discs are air dried. For examination of amino acid (or nucleoside) uptake, the discs are placed into glass scintillation vials, and 20 ml of scintillation fluid (5 g PPO/liter toluene, analytical grade, low background) are added. Vials are counted in a Beckman ambient temperature system. To determine amino acid incorporation into acid-precipitable material, the filter disc already counted for uptake is removed from the scintillation fluid and rinsed in toluene and absolute ethanol to remove the PPO. The filters are then processed in the conventional manner with ice-cold 5% TCA, hot 5% TCA (eliminated in studies on incorporation of RNA precursors), 95% ethanol, absolute ethanol and ether.[18,19] After drying, the filter discs are again placed in scintillation fluid and recounted.

Other Applications

The above techniques can easily be adapted for slightly different types of investigation.

1. To study the effects of other steroids on embryo metabolism. Other hormones of early pregnancy (progesterone, for example) might also have a direct effect on embryo metabolism. Progesterone or other steroid-containing medium can be prepared in a manner similar to that described for estradiol, and the effects of these hormones on the protein and RNA metabolism of cultured embryos evaluated according to the procedures outlined.

2. To study the uptake, binding, and retention of hormones by cultured embryos. Cultured embryos at various stages of development can be incubated in medium containing accurately measured amounts of radio-

[16] Drummond Scientific Company.

[17] Clay-Adams.

[18] R. J. Mans and D. G. Novelli, *Arch. Biochem. Biophys.* **94,** 48 (1961).

[19] A. Tyler, *Biol. Bull.* **130,** 450 (1966).

active hormones of high specific activity (for example, [^{3}H]estradiol-17β, specific activity 110 Ci/mmole), washed as described, and groups of 100–200 embryos in a measured volume of cold medium placed in a scintillation vial to which 1 ml Biosolv BBS-3[20] and 14 ml of scintillation fluid (0.5% PPO in toluene) are added. Background activity is determined by counting an identically treated volume of final wash medium.

[20] Beckman Instruments, Inc.

[26] Superfusion Techniques for Assessment of Steroid Hormone Production in Endocrine Tissue and Isolated Cells (Adrenal)

By SYLVIA A. S. TAIT and DENNIS SCHULSTER

In vitro incubation has been used extensively to examine hormone production and biosynthesis by adrenal glands of a wide variety of vertebrates. Such studies have also included measurement of steroid production following stimulation of donor animals *in vivo* or after addition of the agents *in vitro*. However, although much valuable information has been provided by this approach, the usual procedure which depends on a single analysis after timed incubation has serious limitations when kinetic data are required. In particular, as has been discussed elsewhere,[1-4] quantitative data on the relative importance of various biosynthetic or metabolic pathways cannot be obtained with a single analysis after addition of a tracer amount of precursor at the start of the incubation. If sufficient precursor is added to ensure that the specific activity of precursor and product remains virtually constant throughout the incubation, then, although a complete solution may be obtained with a single analysis, the result could still lead to erroneous conclusions. The biosynthetic pathway may be biased toward the contribution of this precursor or, alternatively, inhibition of various biochemical or metabolic pathways could occur.[2,5,6] Such inhibition can also be encountered under the usual conditions of

[1] D. B. Zilversmit, C. Entenman, and M. C. Fishler, *J. Gen. Physiol.* **26**, 325 (1943).
[2] P. J. Ayres, J. Eichorn, O. Hechter, N. Saba, J. F. Tait, and S. A. S. Tait, *Acta Endocrinol.* (*Copenhagen*) **33**, 27 (1960).
[3] J. F. Tait and S. Burstein, *in* "The Hormones" (G. Pincus, K. V. Thimann and E. B. Astwood, eds.), Vol. 5, p. 441. Academic Press, New York, 1964.
[4] N. M. Kaplan and F. C. Bartter, *J. Clin. Invest.* **41**, 715 (1962).
[5] A. C. Brownie, J. K. Grant, and D. W. Davidson, *Biochem. J.* **58**, 218 (1954).
[6] F. C. Peron, F. Moncloa, and R. I. Dorfman, *Endocrinology* **67**, 379 (1960).

in vitro incubation when the products are allowed to accumulate.[7] Moreover, these products can be further metabolized to compounds not usually encountered under *in vivo* conditions.

It seemed possible that many of these disadvantages could be overcome by the development of an *in vitro* superfusion or continuous flow technique whereby the products of incubation could be removed continuously for analysis. Potentially, such a system would permit the application *in vitro* of some of the principles which had been developed for metabolic studies *in vivo*.[8,9] Such considerations stimulated three groups of workers to develop independently superfusion systems that could be applied to the study of hormone production by endocrine tissue. The first system to be published was that of Orti *et al.*[10] The potentialities of the system are discussed fully, but without including any of the results obtained. In 1967 two groups, Saffran *et al.*[11] and Tait *et al.*[12,13] simultaneously published superfusion systems suitable for analyzing corticoid production by rat adrenal glands including the experimental results obtained when the system was applied.

Suitable Apparatus for Superfusion of Adrenal Tissue

The essential components for a superfusion apparatus are the same for all the systems that have been developed. They comprise a central incubation chamber connected to a pumping system on one side and to an outflow system on the other side. The details will vary according to the requirements of the particular problem under investigation.[10-15] One such system[12] which has proved satisfactory for the incubation of rat adrenal tissue is described in detail.

The apparatus consists of a modified 10-ml Warburg manometer flask, which is used as the central incubation chamber. Details of the flask are shown in Fig. 1. The flask is fitted with a central glass cylinder which contains a platinum wire gauze set 3–4 mm above the base. Perforations

[7] M. K. Birmingham and E. Kurlents, *Endocrinology* **62**, 47 (1958).

[8] J. F. Tait, *J. Clin. Endocrinol.* **23**, 1285 (1963).

[9] R. Horton and J. F. Tait, *J. Clin. Invest.* **43**, 301 (1966).

[10] E. Orti, R. K. Baker, J. T. Lanman, and H. Brasch, *J. Lab. Clin. Med.* **66**, 973 (1965).

[11] M. Saffran, P. Ford, E. K. Mathews, M. Kraml, and L. Garbaczewska, *Can. J. Biochem.* **45**, 1901 (1967).

[12] S. A. S. Tait, J. F. Tait, M. Okamoto, and C. Flood, *Endocrinology* **81**, 1213 (1967).

[13] S. A. S. Tait, M. Okamoto, and J. F. Tait, *Fed. Proc., Fed. Amer. Soc. Exp. Biol.* **26**, 483 (1967) (Abstract).

[14] D. Schulster, S. A. S. Tait, J. F. Tait, and J. Mrotek, *Endocrinology* **86**, 487 (1970).

[15] W. H. Huibregtse and F. Ungar, *Life Sci.* **9**, 349 (1970).

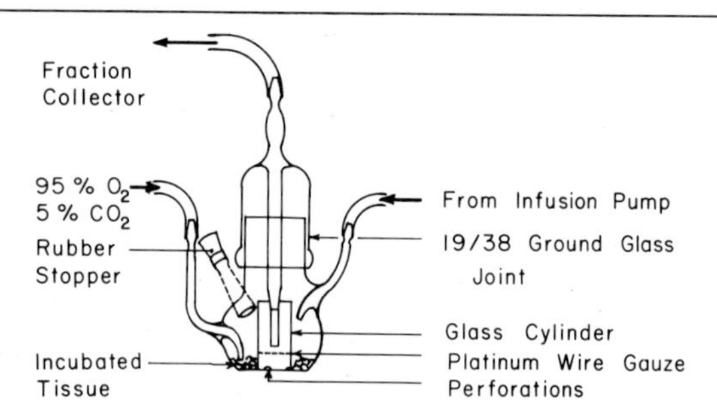

FIG. 1. Constant-flow incubation flask used for superfusion of rat adrenal glands. Adrenal tissue weight 100–600 mg was placed in the bottom of the flask containing 2 ml of Krebs-Ringer bicarbonate solution. Infusion with the same medium was carried out at the rate of 0.8 ml per minute so that the medium was renewed every 8–10 minutes. From S. A. S. Tait, J. F. Tait, M. Okamoto, and C. Flood, *Endocrinology,* **81**, 1213 (1967).

at the base permit free passage of the medium but, together with the wire gauze, prevent removal of the tissue. The flask is sealed by a modified ground-glass joint with a central glass arm, one end of which projects into the interior of the flask. A short piece (1–1.5 cm) of polyethylene tubing is attached to this and adjusted so that the tip is 2–3 mm above the wire gauze. The incubation fluid is infused continuously through one of the side arms of the flask which is connected to a Harvard Apparatus dual infusion pump (model 660/960). A gas supply, usually 95% O_2–5%CO_2, is introduced via the other sidearm, the tip of which opens below the surface of the liquid in the flask. Mixing of the medium and adequate oxygenation of the tissue is assured by placing the flask in a Dubnoff shaker. During infusion the level of the liquid in the flask rises until it reaches above the tip of the polyethylene tubing, thus creating a closed system. Gas pressure then forces the fluid out through the central arm until the level again falls below the tip of the tubing. The effluent is conveyed via plastic tubing attached to the central arm of the stopper to be collected into a siphon of an automatic fraction collector. Test solutions can be introduced through the rubber cap sealing the third arm of the flask. This can be done either as a single pulse or as a continuous infusion without significantly altering the superfusion conditions. A modified superfusion flask[14] which operates on the same principle is shown in Fig. 2. This has been used in some incubations for reasons to be discussed later.

The whole system is operated in a constant temperature room maintained at 37°. Alternatively it can be used at room temperature, but under these circumstances it is essential to precirculate the incubation fluid

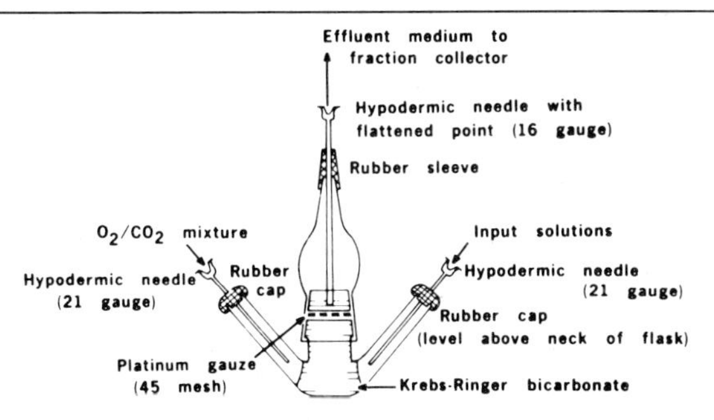

FIG. 2. Superfusion flask of modified design which operates in a similar manner to that previously described [S. A. S. Tait, J. F. Tait, M. Okamoto, and C. Flood, *Endocrinology* **81**, 1213 (1967)]. It consists of 3 glass components constructed from a small ground-glass joint. The main body of the flask, where the glands are incubated in up to 2 ml of medium, is formed from the male joint by sealing on 2 sidearms. The top section of the male joint is cut off to form a retaining collar for a thin disc of platinum gauze which prevents exit of the tissue. The disc and collar are constrained by the female joint, which, at its upper end, is drawn out to a narrow bore just capable of admitting a 16-gauge hypodermic needle. The incubation volume may be adjusted by varying the height of this needle. From D. Schulster, S. A. S. Tait, J. F. Tait, and J. Mrotek, *Endocrinology* **86**, 487 (1970).

through a sufficient length of narrow glass or plastic tubing placed in a water bath at 37° to raise the temperature of the infusion fluid to the required level.

Practical Considerations

The relatively simple system described above has been found to operate reliably in a large number of experiments using rat adrenal tissue.[14,16,17] The use of a separate Harvard Apparatus dual infusion pump for each flask has the advantage that the rate of infusion is constant and reproducible and can be varied over a wide range (32.8 to 0.008 ml per minute) without prior calibration. However, the number of simultaneous incubations that can be carried out will be limited. Alternatively a multichannel peristaltic pump can be used,[10] but this necessitates precalibration of the pump before each run with final adjustments on the

[16] S. Baniukiewicz, A. Brodie, C. Flood, M. Motta, M. Okamoto, J. F. Tait, S. A. S. Tait, J. R. Blair-West, J. P. Coghlan, D. A. Denton, J. R. Goding, B. A. Scoggins, E. M. Wintour, and R. D. Wright, *in* "Functions of the Adrenal Cortex" (K. W. McKerns, ed.), Vol. 1, p. 153. Appleton, New York, 1968.

[17] S. A. S. Tait, D. Schulster, M. Okamoto, C. Flood, and J. F. Tait, *Endocrinology* **86**, 360 (1970).

day of the experiment even if a preparatory run has been made on the previous day. It is also necessary to maintain a constant pressure head of buffer if the infusion rate is to remain constant.[10]

The rate at which the incubation medium is replaced will depend on the effective volume of the superfusion flask and the rate at which the fluid is infused. With an effective volume of 2 ml and an infusion rate of 0.76 ml per minute, used in most of our experiments, 38% of the medium will be replaced per minute and 97% every 9 minutes. However, in order to achieve a uniform rate of sampling it is necessary to maintain a constant gas flow. It is therefore advisable to include not only a bubbler to saturate the aerating gas with water, but also a constant-pressure device and a separate control for each flask to regulate the gas flow. Some slight discontinuity in sampling is inherent in the system, which can lead to variations in volume between fractions. This is not a serious disadvantage except for studies that require collection over short time intervals (1–5 minutes). However, the design of the modified flask (Fig. 2) has largely overcome this problem.

Some of the superfusion systems,[10,15] provide for refrigeration of the infusion fluid and of the fractions collected.[10] In our experience this is not essential for incubations lasting over 5–6 hours but may be desirable for longer time periods of 12–24 hours. Most of the studies reported have used Krebs-Ringer bicarbonate with 200 mg/100 ml glucose added as the superfusion fluid. This can be varied at will provided that proteins, such as albumin, which tend to froth, are not included. Test solutions can be added in the superfusion fluid,[11] although in practice we have found it more satisfactory to introduce these directly into the flask, using a separate pump, which permits more accurate timing of the additions.

The advantages of continuous superfusion for studying steroid production by adrenal tissue have been demonstrated in our own studies[14,16,17] and those of other investigators.[11,15,18] Two of these studies will be described briefly to illustrate the usefulness of this approach.

Steroid Output from Bisected, Capsular, and Decapsulated Rat Adrenals as a Function of Time of Superfusion

Tissue Preparation

Adrenal glands were removed from normal intact female rats (150–180 g) or from rats hypophysectomized 3 or 48 hours previously and transferred immediately to ice-cold saline. The glands were dissected free

[18] M. Saffran and P. Rowell, *Endocrinology* **85**, 652 (1969).

from fat and bisected. Such glands incubated without further dissection are referred to as bisected whole glands. In some experiments the glands were further separated into capsular and decapsulated portions as described by Giroud *et al.*[19] The capsular (capsule plus mainly zona glomerulosa tissue) and decapsulated portions (mainly zona fasciculata-reticularis plus the medulla) were incubated in separate flasks. The tissue was lightly blotted on moist filter paper, transferred to the superfusion flask containing 2 ml of the incubation fluid (Krebs-Ringer bicarbonate buffer containing glucose 2 g/liter, pH 7.4) and weighed. The weight of tissue incubated in each flask was usually 400–600 mg for bisected whole or decapsulated glands and 100–200 mg for capsular glands. The total time (including killing of the rats) taken to the start of the experiment was approximately 30 minutes for bisected whole and decapsulated glands from 12 rats and 45 minutes for capsular glands from 18 rats.

Superfusion

Before starting infusion, the length of the polyethylene tubing attached to the central arm of the flask (Fig. 1) was adjusted so that the tip was level with fluid in the flask. Infusion was at the rate of 0.76 ml per minute unless otherwise stated. All incubations were carried out at 37° under a gaseous phase of 95% O_2–5% CO_2. Glass and plastic surfaces were coated with Desicote (Beckman Corporation) to minimize hormone absorption.

Incubations were continued for 5 hours and fractions were collected in a 5-ml siphon of an automatic fraction collector with a time control. Fractions collected over the following time intervals were bulked for analysis: 0–15, 15–30, 30–60, 60–120, 120–180, 180–240, 240–300 minutes.

Analytical Procedure

The bulked fractions were made up to a given volume and divided into suitable samples for analysis. [1,2-^{3}H]Aldosterone and [4-^{14}C]corticosterone were added to each sample as indicators of recovery through the extraction procedure. Steroids were extracted with methylene dichloride, partitioned between aqueous ethanol and cyclohexane and oxidized with 0.1 *M* periodic acid as described in detail by Tait *et al.*[13] 18-Hydroxycorticosterone (18-hydroxy-B), aldosterone (aldo), 18-hydroxydeoxycorticosterone (18-hydroxy-DOC), and corticosterone (B), four of the major steroids known to be secreted by the rat adrenal, were measured after separation of the etiolactones of the 18-hydroxylated steroids

[19] C. J. P. Giroud, J. Stackenko, and E. H. Venning, *Proc. Soc. Exp. Biol. Med.* **92**, 154 (1956).

from corticosterone etio acid by extraction from aqueous alkali. The lactones and B etio acid were chromatographed on separate paper systems, and the quantity of steroid produced was measured by continuous automatic assay of soda fluorescence[20] directly on paper using a modified Turner fluorometer.[13]

Endogenous Steroid Output from Bisected Whole, Decapsulated, and Capsular Glands

The endogenous steroid outputs as a function of time of superfusion by whole bisected glands are shown in Fig. 3. It can be seen that the outputs follow a distinctive pattern that is very reproducible from experiment to experiment. The shape of the output curves varies with the individual steroid analyzed and also with the physiological status of the donor rats. Incubations of glands from normal intact rats show that although the output of all four steroids decreased with time of superfusion, the decline in output of 18-hydroxy-B and aldosterone (known to be produced mainly by the zona glomerulosa) follow a simple single exponential, whereas those of 18-hydroxy-DOC and corticosterone (produced mainly by the zona fasciculata) decline rapidly during the first 2 hours and then more slowly in the next 3 hours. It is also apparent that there is no essential change in the outputs of 18-hydroxy-B and aldosterone 3 hours after hypophysectomy; even 48 hours after hypophysectomy, although the output decreased 3-fold, the shape of the decay curves is not changed. By contrast, 3 hours after hypophysectomy the outputs of 18-hydroxy-DOC and corticosterone no longer exhibit the biphasic decline observed for glands from intact rats but are virtually constant throughout the 5 hours of incubation. The microgram level was about the same as that from intact rats during the last 3 hours of superfusion. At 48 hours after hypophysectomy, the outputs of 18-hydroxy-DOC and corticosterone are markedly lowered (10- and 5-fold, respectively) but otherwise resemble those from acutely hypophysectomized rats except for a decrease (smaller than that from intact rats) during the first hour of superfusion.

One explanation of these experimental results is that the pattern of steroid output may be determined by the anatomical site (i.e., zona glomerulosa or zona fasciculata-reticularis) at which the steroids are produced. This possibility was examined by incubating the capsular and decapsulated glands in separate flasks.

As can be seen from Fig. 4, the outputs of 18-hydroxy-DOC and corticosterone from decapsulated glands of intact and hypophysectomized rats

[20] I. E. Bush, *Biochem. J.* **50,** 370 (1952).

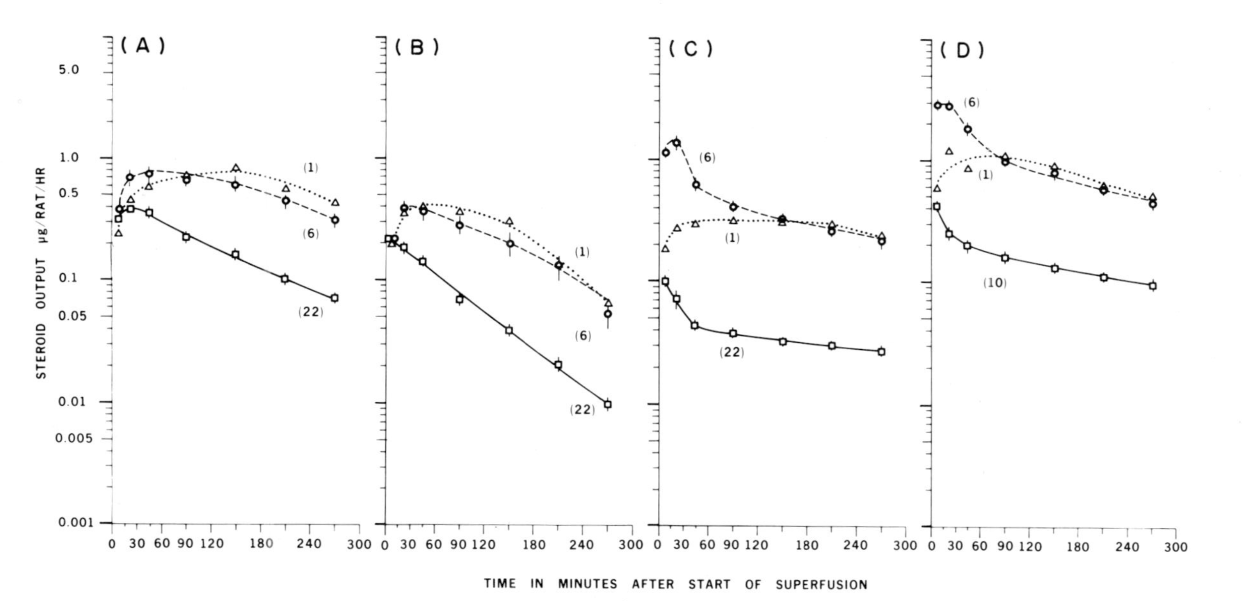

FIG. 3. Steroid outputs as a function of time of superfusion from whole bisected glands from normal intact rats and rats 3 hours and 2 days after hypophysectomy. Steroid analysis in this and subsequent experiments was carried out on superfusate collected over **0–15, 15–30, 30–60, 60–120, 120–180, 180–240,** and **240–300** minutes. (A) 18-Hydroxycorticosterone; (B) aldosterone; (C) 18-hydroxydeoxycorticosterone; (D) corticosterone. ○---○, intact; △···△, hypophysectomized 3 hours; □——□, hypophysectomized 48 hours. Vertical lines indicate mean ±SE; number of incubations is given in parentheses. From S. A. S. Tait, D. Schulster, M. Okamoto, C. Flood, and J. F. Tait, *Endocrinology* **86,** 360 (1970).

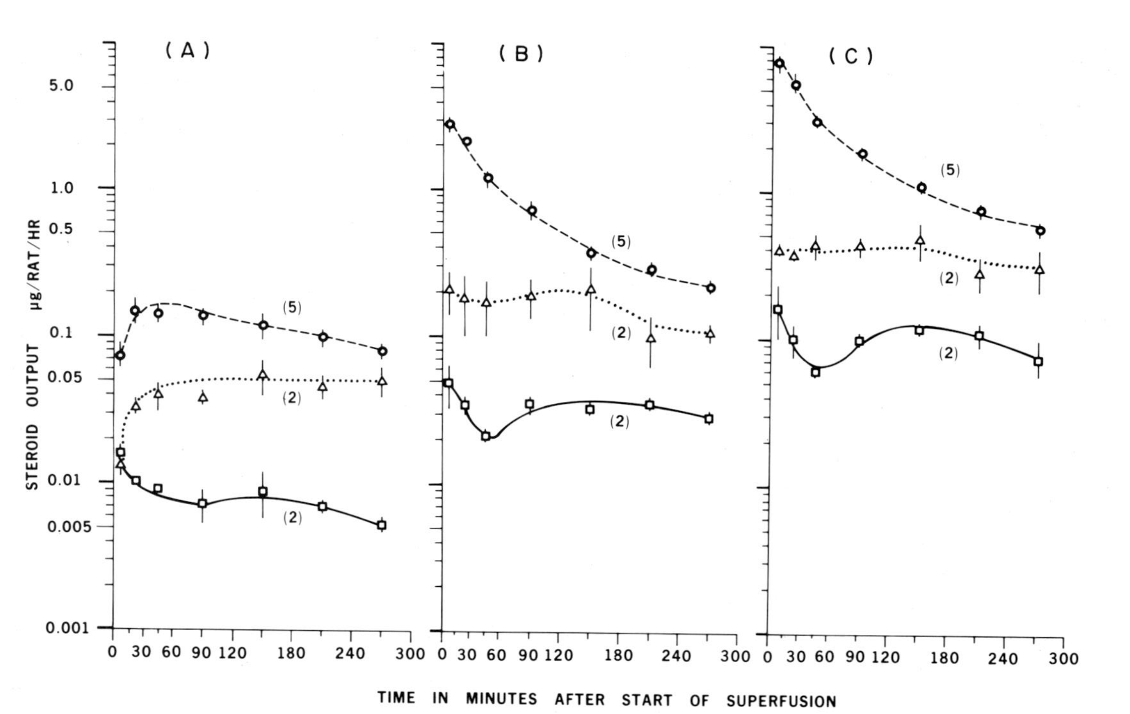

FIG. 4. Comparison of steroid output as a function of time of superfusion of decapsulated adrenals from normal intact (○---○) and hypophysectomized rats 3 (△···△) and 48 (□——□) hours after operation (see legend to Fig. 3). (A) 18-Hydroxycorticosterone; (B) 18-hydroxydeoxycorticosterone; (C) corticosterone. Vertical lines indicate mean ±SE; number of incubations given in parentheses. From S. A. S. Tait, D. Schulster, M. Okamoto, C. Flood, and J. F. Tait, *Endocrinology* **86,** 360 (1970).

are essentially the same as those from bisected whole glands. In particular, the same lowered constant output is observed after hypophysectomy. Aldosterone was not detected in these incubations, but 18-hydroxy-B was present although in somewhat lower amounts than the other two compounds. Although, for reasons which are not yet understood, the output curve of 18-hydroxy-B from glands of intact rats differed somewhat from those of 18-hydroxy-DOC and corticosterone, nevertheless, following hypophysectomy, the output curves for 18-hydroxy-B closely resemble those of the other two steroids, both as regards the marked reduction in microgram output and the constancy of output during the last 4 hours of superfusion.

Incubations of capsular glands, on the other hand, reveal a distinct change in the pattern of steroid output when compared with those of bisected whole or decapsulated glands. As can be seen from Fig. 5, there is a close similarity in the shape of the decay curves for all four steroids (cf. Figs. 3 and 4) which show a rapid decline in steroid output over the first 90–150 minutes of superfusion; after this time the outputs of 18-hydroxy-DOC and corticosterone tend to become constant. The other noteworthy feature is that the shape of the output curves is not affected by hypophysectomy for any of the four steroids examined.

Mechanism of Decay in Steroid Output

Any evaluation of the usefulness of continuous superfusion *in vitro* for assessment of hormone production by endocrine tissue must include a careful examination of the underlying causes for the observed decline with time of steroid output from adrenals in certain situations. If this decline is an artifact arising out of the particular conditions of incubation employed, then the results will be of limited value only. A number of experiments were carried out in an attempt to discover the factors involved in the observed decline.

One obvious explanation was that the rapid initial fall represented the release of preformed steroid from the gland rather than a decrease in de novo synthesis. Assay of 18-hydroxy-B, aldosterone, 18-hydroxy-DOC and corticosterone present in unincubated capsular glands of hypophysectomized rats showed the amount to be only one-fifth of the measured output during the first 15 minutes of superfusion.[17] Similar results were obtained for decapsulated glands of intact and hypophysectomized rats showing that the adrenal stores would be depleted within 10 minutes of the start of superfusion.[14] Hence it can be concluded that the release of stored steroid can contribute but little to the observed decline.

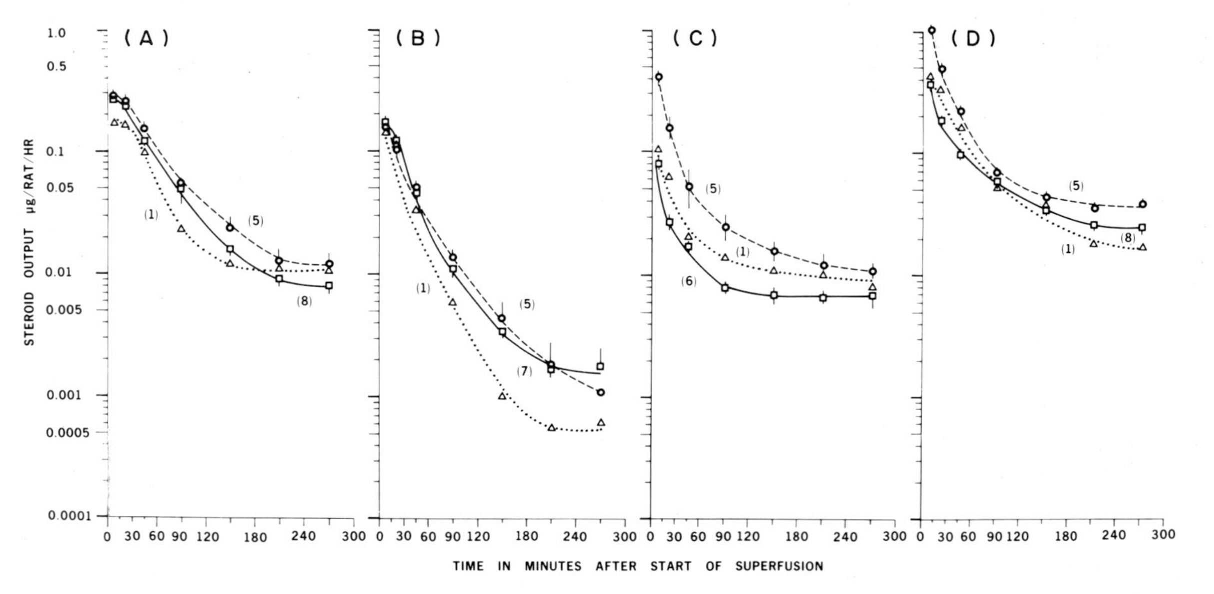

FIG. 5. Comparison of steroid output as a function of time of superfusion of capsular glands from normal intact (○- - -○) and hypophysectomized rats 3 (△···△) and 48 (□——□) hours after operation (see legend to Fig. 3). Vertical lines give mean ±SE; number of incubations is given in parentheses. (A) 18-Hydroxycorticosterone; (B) aldosterone; (C) 18-hydroxydeoxycorticosterone; (D) corticosterone. From S. A. S. Tait, D. Schulster, M. Okamoto, C. Flood, and J. F. Tait, *Endocrinology* **86**, 360 (1970).

It was also demonstrated in a large series of experiments, that the shape of the output curves for aldosterone from whole bisected glands of hypophysectomized rats was independent of the strain of rat used, the concentration of glucose in the medium, and the time taken for preparation of the glands.[16] Another possible explanation, which arises from the very nature of the superfusion technique, is that the continuous renewal of the medium will result not only in the removal of the products of incubation but also in the loss of any substrate, enzymes, or cofactors which diffuse from the tissue into the medium. This possibility was examined by reducing the rate of infusion from 0.76 to 0.38 and 0.08 ml per minute. Figure 6 shows that even a 10-fold reduction in infusion rate did not appreciably alter the shape of the output curves of 18-hydroxy-B and aldosterone from bisected whole glands of hypophysectomized rats although, as might be expected, the time required for the system to reach equilibrium was somewhat longer. Moreover, the same decay in output

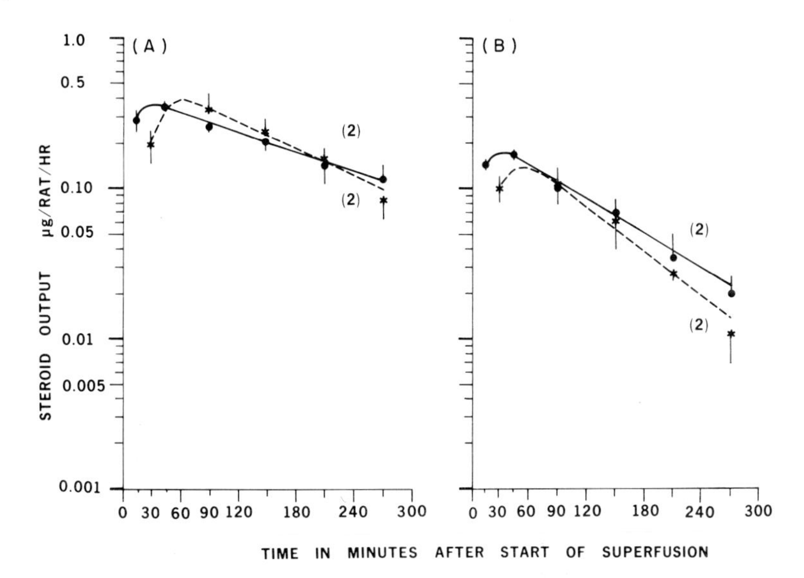

FIG. 6. Effect of altering the rate of superfusion from 0.8 (●——●) to 0.08 (×---×) ml per minute on steroid output, as a function of time of superfusion, by whole bisected glands of rats hypophysectomized 48 hours prior to sacrifice. Analyses during the 0–60 minutes of superfusion were carried out on the pooled effluent from 0 to 30 and 30 to 60 minutes when the infusion rate was 0.8 ml per minute and 0 to 60 minutes with an infusion rate of 0.08 ml per minute. Subsequent analyses were done as described in the legend to Fig. 3. Vertical lines show mean ±SE; number of incubations is given in parentheses. (A) 18-Hydroxycorticosterone; (B) aldosterone. From S. A. S. Tait, D. Schulster, M. Okamoto, C. Flood, and J. F. Tait, *Endocrinology* **86**, 360 (1970).

of these two steroids was observed even when whole uncut glands were used, although it has been shown[21] that substrate and enzyme loss from the tissue to the medium does not occur under these conditions.

Finally, it must be considered whether the decay in output represents a random disintegration of the integrity of the tissue and activity of the enzyme systems with time of superfusion. This explanation appears to be unlikely in view of the reproducibility of the decay curves for individual steroids under different conditions of incubation. In particular it would be difficult to explain, on this basis, the low constant outputs of 18-hydroxy-DOC and corticosterone from decapsulated glands of hypophysectomized rats (see Fig. 4). The possibility cannot be excluded, however, that tissue and enzyme loss may contribute to the more rapid decline in 18-hydroxy-B and aldosterone outputs from capsular than from bisected whole glands (cf. Figs. 3 and 5). An alternative explanation would be that the more gradual decrease in the outputs of these two steroids from bisected whole glands is the result of the availability of an additional supply of corticosterone (a major precursor for 18-hydroxy-B and aldosterone synthesis[2,22]) from the zona fasciculata for the synthesis of 18-hydroxy-B and aldosterone after reentry to the biosynthetic sites in the zona glomerulosa tissue. When capsular glands are used, on the other hand, 18-hydroxy-B and aldosterone production will decline automatically as the availability of this common precursor (i.e., corticosterone) decreases. Nevertheless, whatever the correct explanation, the constant nature of the output curves for steroids, even from capsular glands, suggests that the rate of decay represents some general property of, or a decline in, influence on the tissue rather than to artifactual phenomena due to the conditions of the experiment.

In Vivo Decay Hypothesis

At the present time it seems most likely that a major factor in the decline in steroid output represents a decay *in vitro* in the effect of the stimulatory agent operating *in vivo*. The characteristic output curves of the steroids produced by the zona fasciculata-reticularis, which is known to be under the control of ACTH, could be satisfactorily explained on the basis of the "*in vivo decay hypothesis*."[16] The decline in output of 18-hydroxy-DOC and corticosterone from bisected whole and decapsulated glands of intact rats would then represent the decay *in vitro* of

[21] C. Tsang and A. Carballeira, *Proc. Soc. Exp. Biol. Med.* **122**, 1031, (1966).

[22] J. Stachenko and C. J. P. Giroud, *in* "The Human Adrenal Cortex" (A. R. Currie, T. Symington, and J. K. Grant, eds.), p. 30. Livingstone, Edinburgh and London, 1962.

the effects of ACTH initiated *in vivo*. The lowered constant output of these steroids from the adrenals of hypophysectomized rats could be explained by the decay of the stimulus *in vivo* during the 3 hours which elapsed after operation and before removal of the glands.

The failure of hypophysectomy to alter the output curves of steroids produced by zona glomerulosa tissue is also consistent with the *"in vivo decay"* hypothesis. It has long been recognized from *in vivo* studies,[23-25] that aldosterone secretion, the only hormone produced specifically by the zona glomerulosa, is relatively independent of pituitary control. Also the finding[17] that combined hypophysectomy and nephrectomy did not result in a constant lowered steroid output from capsular glands was not entirely unexpected, as it has not yet been shown that the renin–angiotensin system plays a major role in the control of aldosterone secretion in the rat.[24,26] However, it is generally accepted that alterations in sodium balance have an important effect on aldosterone secretion *in vivo*. We have examined, therefore, the effects of sodium loading on steroid output by superfused adrenal glands.

Effect of Sodium Loading

In these experiments hypophysectomized rats were given either water or physiological saline as drinking fluid for 2 days before the animals were killed. The results obtained for incubations of bisected whole glands are shown in Fig. 7. It can be seen that although sodium loading does not abolish the decay in 18-hydroxy-B and aldosterone outputs with time of superfusion, there is a reduction in output of all four steroids during the first 30 minutes of superfusion, which is greater for 18-hydroxy-B and aldosterone than for 18-hydroxy-DOC and corticosterone. Moreover, whereas the outputs of 18-hydroxy-B and aldosterone from glands of sodium-loaded rats are lowered throughout the whole period of superfusion those of 18-hydroxy-DOC and corticosterone are reversed during the later time intervals so that the outputs of the glands from sodium-loaded rats are higher than those from the control animals.

These results suggest that sodium loading lowers aldosterone secretion *in vivo* by affecting both an early step in the biosynthesis pathway to reduce the corticosterone production and also a late step involving the conversion of corticosterone to aldosterone. During *in vitro* incubation

[23] B. Singer and M. Stack-Dunne, *J. Endocrinol.* **12,** 130 (1955).

[24] E. Eilers and R. E. Peterson, *in* "Aldosterone" (E. E. Baulieu and P. Robel, eds.), p. 251, Blackwell, Oxford, 1964.

[25] N. J. Marieb and P. J. Mulrow, *Endocrinology* **76,** 657 (1965).

[26] B. E. Murphy, *J. Clin. Endocrinol.* **27,** 973 (1967).

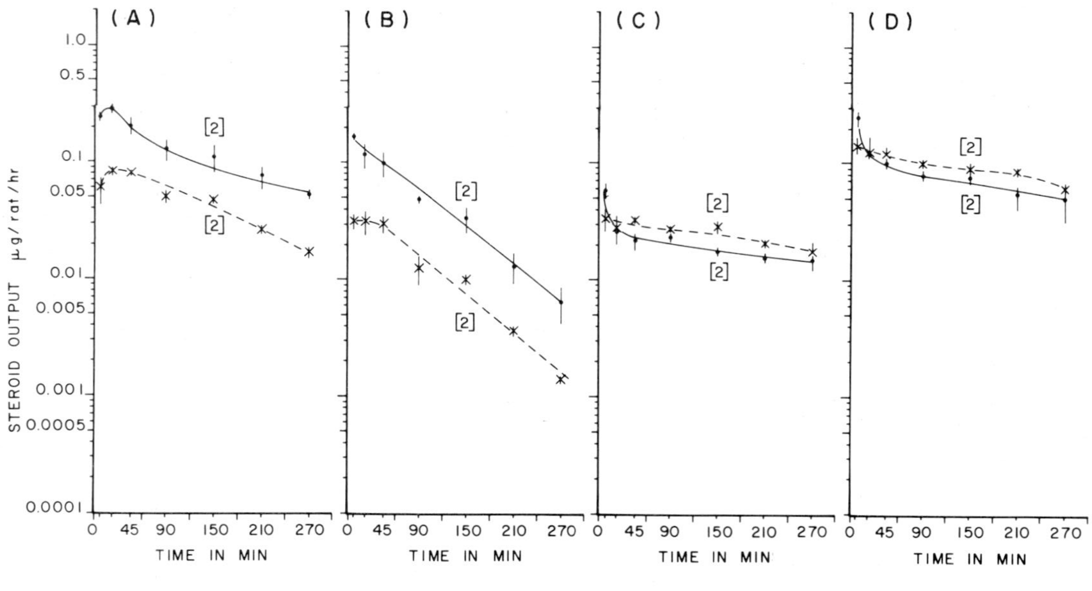

FIG. 7. Effect of sodium loading on steroid output by bisected whole adrenal glands of hypophysectomized rats as a function of time of superfusion. Control hypophysectomized rats (●——●) were maintained on 5% dextrose, and sodium-loaded hypophysectomized rats (×- - -×) on 1% sodium chloride in 5% dextrose for 2 days prior to the experiment. Twelve rats were used per group. Vertical lines indicate mean ±SE; number of incubations is given in brackets. (A) 18-Hydroxycorticosterone; (B) aldosterone; (C) 18-hydroxydeoxycorticosterone; (D) corticosterone. From S. Baniukiewicz, A. Brodie, C. Flood, *et al.*, *in* "Functions of the Adrenal Cortex (K. W. McKerns, ed.), Vol. 1, p. 153. Appleton, New York, 1968.

the effect on the early pathway declines more rapidly than that on the late pathway. The higher output of 18-hydroxy DOC and corticosterone from glands of sodium-loaded rats would then be explained by the decreased conversion of corticosterone to 18-hydroxy-B and aldosterone. Evidence in support of this conclusion was obtained in experiments with infusion of radioactive corticosterone and measurement of conversion to 18-hydroxy-B and aldosterone. Similar experiments using capsular glands gave essentially the same results although the picture presented greater complexity in the form of the decay curves. Although the observed differences cannot be discussed in detail here, the pattern of output curves from capsular glands emphasizes the different rates of decay of the various enzyme systems involved, i.e., in the production of corticosterone, the conversion of corticosterone to 18-hydroxy-B and of corticosterone to aldosterone.[16]

Corticosterone Output from Decapsulated Rat Adrenals Stimulated by Adrenocorticotropin or Cyclic 3′,5′-Adenosine Monophosphate

Tissue Preparation

Preparation of decapsulated adrenal glands from normal intact and hypophysectomized rats has been described earlier. Hypophysectomy was 3–4 hours before sacrifice of the animals and adrenals from 12 rats were used in each flask. Hypophysectomies and decapitations were so arranged that although incubation of the control flask was started 45 minutes after the experimental flask, both flasks contained tissue from rats hypophysectomized for very similar durations.

Superfusion

Superfusion was carried out as previously described except that the modified flask (Fig. 2) was used in some experiments where the medium was collected over short time intervals to reveal the nature and lag time in the acute response after addition of a regulatory factor to the medium.

Infused Solutions and Collection and Assay of Superfusate

The preparation of solutions of regulatory agents such as ACTH, cyclic AMP, and cycloheximide have been detailed.[14] Each of the substances was infused, like the superfusion medium, through individual hypodermic needles inserted through the rubber seal on the right-hand sidearm (Fig. 1). These various solutions were infused continuously at a constant rate (usually 0.04 ml per minute) at chosen time points during the superfusion. Effluent medium was collected over set time periods using an automatic fraction collector and various fractions were combined depending upon the region of the steroid output curve under investigation. Recovery indicator [^{14}C]corticosterone (8000 dpm; 25 ng) in ethanol (0.2 ml) was added to each 40-ml sample in order to evaluate losses during the extraction procedure.[12] Recoveries were usually about 90% and all results were corrected for individual values. Quadruplicate aliquots of the extracted samples were assayed for corticosterone by a competitive protein binding method based on that of Murphy.[26] By virtue of the extensive extraction method and the specific corticoids produced by the rat,[27] these samples were virtually uncontaminated by cholesterol, progesterone, 17α-hydroxyprogesterone, cortisol, and other compounds that can affect the specificity of the assay method.

Decapsulated Adrenals in the Presence and in the Absence of Continuously Infused ACTH

Decapsulated glands from both normal and acutely hypophysectomized rats were superfused for 5 hours, and the corticosterone outputs were compared with those obtained in the presence of continuous infusion of ACTH (64 mU/ml). The output graphs (Fig. 8) represent the adrenal corticosterone secretion rates under the incubation conditions, being plots of micrograms of steroid per hour against superfusion time. Corticosterone and 18-hydroxy-11-deoxycorticosterone are quantitatively the most significant steroids produced by the zona fasciculata–reticularis,[16,28] and the shapes of the kinetic output curves for both these steroids were very similar under a wide variety of superfusion conditions (cf. Figs. 4 and 5). For this reason measurements of corticosterone outputs alone were considered sufficient to evaluate the steroidogenic responsiveness of decapsulated rat adrenals to tropic hormones and other regulatory compounds.

The corticosterone output rate from adrenals of hypophysectomized rats was constant (0.3–0.5 μg per rat per hour) in the absence of infused ACTH, whereas that from adrenals of intact rats, again in the absence of infused ACTH, fell rapidly from high initial values (about 7 μg per rat per hour) and closely approached those found for control adrenals from hypophysectomized rats within 5 hours of superfusion (Fig. 8). Other workers[11,15] have noted a similar decline during superfusion of quartered glands from intact rats.

[27] I. E. Bush, *J. Endocrinol.* **9**, 95 (1953).

[28] H. Sheppard, R. Swenson, and T. F. Mowles, *Endocrinology* **73**, 819 (1963).

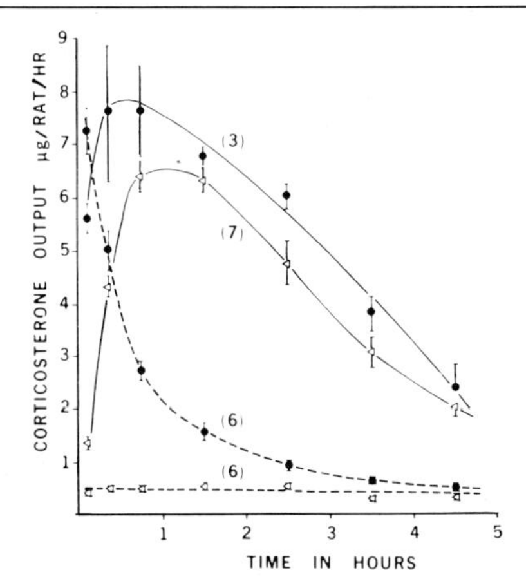

FIG. 8. Corticosterone output rates as a function of time of superfusion for decapsulated adrenals (zona fasciculata-reticularis and medulla) from normal intact rats and rats 3 hours after hypophysectomy. Combined tissue from 12 rats was used for each incubation. Krebs-Ringer bicarbonate buffer was continuously infused at the rate of 0.76 ml per minute. Corticosterone was assayed in superfusate collected over 0–15, 15–30, 30–60, 60–120, 120–180, 180–240, and 240–300 minutes. ●---●, intact; ●——●, intact with ACTH; △---△, hypophysectomized; △——△, hypophysectomized with ACTH. Vertical lines indicate mean ±SE; the number of incubations is given in parentheses. From D. Schulster, S. A. S. Tait, J. F. Tait, and J. Mrotek, *Endocrinology* **86**, 487 (1970).

Glands continuously infused with ACTH, exhibited marked stimulations of corticosterone output rates compared to those superfused in the absence of ACTH. A 13-fold stimulation of corticosterone output by tissue from hypophysectomized rats was observed within 1 hour of ACTH (64 mU/ml) infusion. In further studies it was noted that when ACTH or cyclic AMP is continuously infused 30 minutes after the incubation start, the peak corticosterone output rate (about 7 μg per rat per hour) is consistently achieved within 30–45 minutes of starting infusion of these stimulators. This enhanced responsiveness may be due to allowing the tissue to oxygenate more fully and to equilibrate to 37° during this initial 30 minutes of superfusion. During later time periods of superfusion (2–5 hours), the steroid output rates by glands from intact and hypophysectomized rats, both continuously infused with ACTH, declined similarly with half-lives for the exponential decay of 93 and 102 minutes, respectively.[29] This decline in steroid output during later time periods of continuous ACTH infusion has also been demonstrated by other workers.[15,18]

[29] D. Schulster, S. A. S. Tait, and J. F. Tait, *Int. Congr. Endocrinol. 3rd, Excerpta Med. Found. Int. Congr. Ser.* **77**, 31 (1968) (Abstract).

The studies shown in Fig. 9 demonstrate that this late decline could not be attributed to irreversible changes or cellular damage within the tissue occurring during the superfusion. The peak steroid output rates obtained, when continuous infusions of ACTH (64 mU/ml) to adrenals from intact rats were begun 3, 4, or 5 hours after the start of the superfusion, were all similar to that observed when ACTH was infused continously from the start. Using glands from hypophysectomized rats, a similar result was obtained following continuous infusions of ACTH begun at, or 3 hours after, the incubation start. Other studies using superfused glands from intact rats[11] have similarly shown that the same maximum response was obtainable when low doses of ACTH were infused repeatedly over 30-minute periods to the same tissue. It is clear from these data that glands from intact and hypophysectomized rats remain consistently responsive to ACTH throughout 5-hour superfusions.

Stimulation by ACTH or Cyclic AMP and the Kinetic Response to a Protein Synthesis Inhibitor

The superfusion technique allows the infusion at any time during the continuous incubation, not only of steroidogenic stimulators such as ACTH or cyclic AMP, but also of inhibitors such as cycloheximide—a noted inhibitor of both the steroidogenic response to ACTH and protein synthesis.[30] When constant infusions of cycloheximide (1 m*M*) were introduced 30 minutes after the continuous infusion of either ACTH (64 mU/ml) or cyclic AMP (2 m*M*) and the start of the incubation, an inhibitory effect was apparent within 5 minutes of cycloheximide infusion (Fig. 10). Under these conditions the corticosterone output rate at 30 minutes was stimulated about half maximally and was rising rapidly. A semilogarithmic plot of the data (Fig. 10) demonstrated that the steroid output rate declines exponentially over a 4-hour period with a decay half-life of 49 minutes for continuous infusions of cycloheximide coupled with either ACTH or cyclic AMP.

When cycloheximide was continuously infused 2 hours after the ACTH continuous infusion it was also found to be effective in inhibiting corticosterone synthesis. At this point the steroid output rate had already passed its peak, yet cycloheximide infusion again resulted in an exponential decay in corticosterone output rate with a half-life of 45 minutes. This may be compared with half-lives of 102 and 106 minutes for the late exponential decay in steroid output rate during continuous infusions of ACTH or cyclic AMP, respectively, in the absence of cycloheximide.

[30] J. J. Ferguson *in* "Functions of the Adrenal Cortex" (K. W. McKerns, ed.), Vol. 1, p. 463. Appleton, New York, 1968.

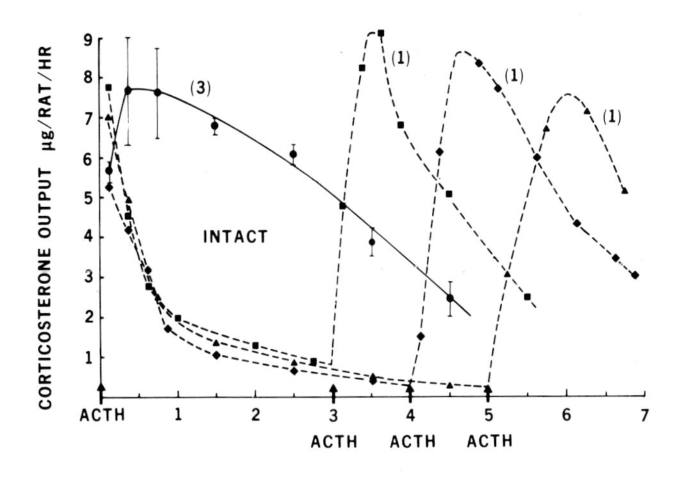

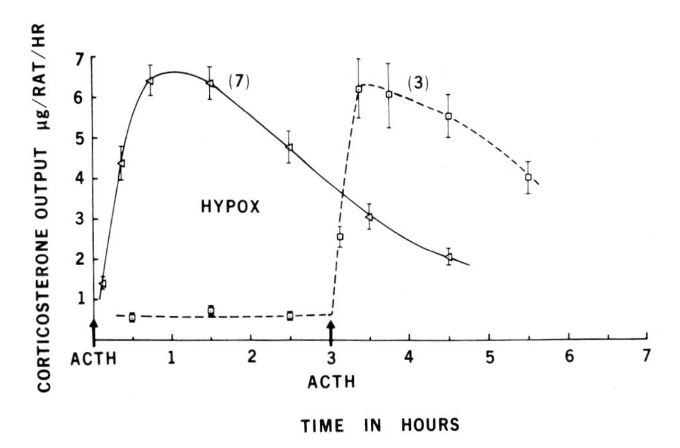

FIG. 9. Effect of continuous infusions of ACTH (64 mU/ml medium), begun at various times, on corticosterone output rate by decapsulated adrenals from normal intact rats (upper diagram) and rats 3 hours after hypophysectomy (lower diagram). ACTH infusions to glands from intact rats were begun at either 0, 3, 4, or 5 hours (arrows) after the start of the incubation. Similar ACTH infusions to glands from hypophysectomized rats were begun at either 0 or 3 hours after the incubation start. Superfusion procedures were as described in legend to Fig. 8, except that immediately after ACTH infusion corticosterone was assayed in superfusion collected over short time periods. *Upper:* ●——●, intact, at start; ■- - -■, intact + ACTH after 3 hours; ◆- - -◆, intact + ACTH after 4 hours; ▲- - -▲, intact + ACTH after 5 hours. *Lower:* ◁——◁, hypophysectomized + ACTH at start; □- - -□, hypophysectomized + ACTH after 3 hours. Vertical lines indicate mean ±SE; number of incubations are given in parentheses. From D. Schulster, S. A. S. Tait, J. F. Tait, and J. Mrotek, *Endocrinology* **86,** 487 (1970).

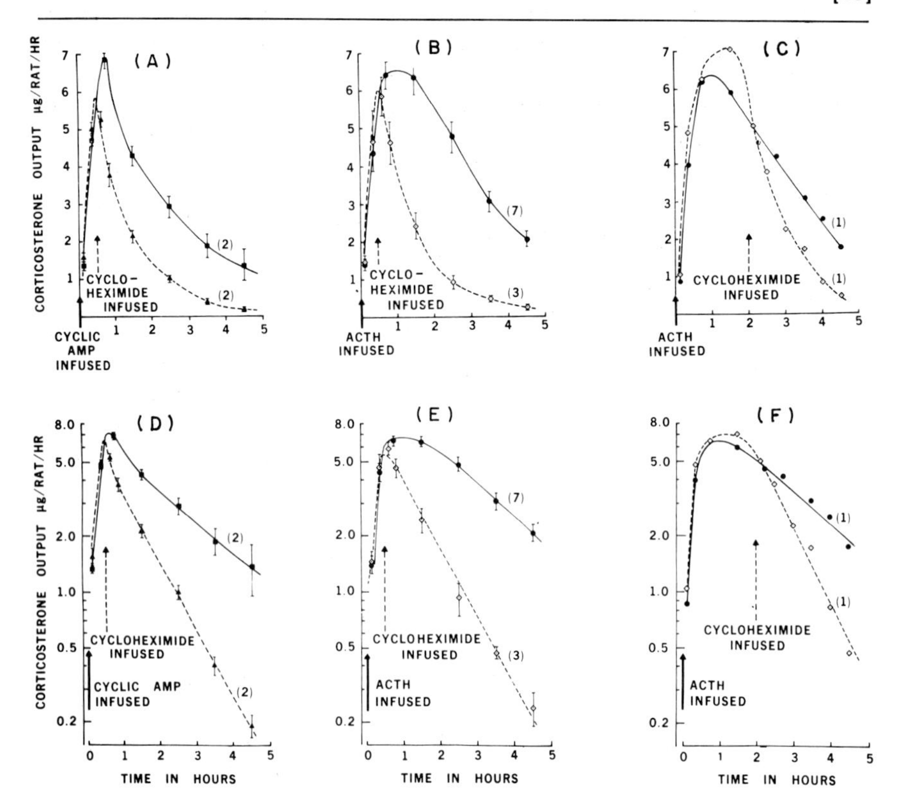

FIG. 10. Effect of continuously infused cycloheximide (1 mM) either 30 minutes after continuous cyclic AMP infusion (2 mM) or 30 minutes or 2 hours after continuous ACTH infusion (64 mU/ml medium). The resulting corticosterone output rates by decapsulated adrenals from rats 3 hours after hypophysectomy are plotted both linearly (upper diagrams) and semilogarithmically (lower diagrams). Cycloheximide, cyclic AMP, and ACTH were all infused at 0.035–0.04 ml per minute. Superfusion procedures were as described in legend to Fig. 8, except that, immediately after cycloheximide infusion, corticosterone was assayed in superfusate collected over short time periods. (A, D) ●——●, cyclic AMP; ▲- - -▲, cyclic AMP + cycloheximide 30 minutes later. (B, E) ●——●, ACTH; ◇- - -◇, ACTH + cycloheximide 30 minutes later. (C, F) ●——●, ACTH; ◇- - -◇, ACTH + cycloheximide 2 hours later. Vertical lines indicate mean ±SE; numbers in parentheses indicate number of incubations. From D. Schulster, S. A. S. Tait, J. F. Tait, and J. Mrotek, *Endocrinology* **86,** 487 (1970).

Conclusions

The present studies of hormone production by rat adrenals have shown that many of the objectives for developing a continuous *in vitro* superfusion technique have been achieved. The dynamic approach has permitted a detailed examination of steroid production by discrete zones of the adrenal cortex with time of superfusion. This has revealed the importance of the physiological status of the donor rat in determining the pattern of the steroid output *in vitro*. It has also been possible to follow the steroidogenic responsiveness of the zona fasciculata-reticularis to ACTH and cyclic AMP under conditions where complicating factors such as alterations in blood flow have been eliminated.

However, the consistent finding that, under most conditions, steroid output declines with time precludes the use of this approach, at least for the adrenal gland, for obtaining quantitative data on the relative importance of biosynthetic pathways if a single sample is analyzed. Nevertheless, the demonstration that there is a differential rate of decay in the various enzyme systems involved in a particular biosynthetic pathway emphasizes the importance of the dynamic approach if valid interpretations are to be made.

Apparatus for the Superfusion of Isolated Adrenal Cells

The superfusion method has previously been applied to the study of quartered, bisected, or whole glands with medium bathing merely the outer surfaces of tissue. Cells located in the interior of such blocks of tissue may have restricted accessibility to medium components, and the comparative insensitivity to ACTH shown by adrenal tissue *in vitro*[31] indicates that ACTH may penetrate to its target sites within the tissue only with difficulty. Such delayed penetration by ACTH may account for the relatively delayed onset of maximum steroidogenic output rate by superfused adrenals (Fig. 9) when compared with the time taken *in vivo*.[32] Moreover a tissue block may have difficulty in clearing itself of metabolites, and products such as corticosteroids may accumulate within superfused tissue fragments with consequent feedback inhibitory effects on the ACTH steroidogenic response.Thus corticosteroid can accumulate within the gland to an extent sufficient to inhibit protein synthesis.[33]

The situation *in vivo*, in which each adrenal cell lies close to the blood supply and is thus well provided with nutrients and an efficient clearance

[31] P. W. C. Kloppenborg, D. P. Island, G. W. Liddle, A. M. Michelakis, and W. E. Nicholson, *Endocrinology* **82,** 1053 (1968).

[32] L. D. Garren, R. L. Ney, and W. W. Davies, *Proc. Nat. Acad. Sci. U.S.* **53,** 1443 (1965).

[33] M. Clayman, D. Tsang, and R. M. Johnstone, *Endocrinology* **86,** 931 (1970).

system, is in this respect poorly mimicked by *in vitro* systems using tissue fragments. The relatively low steroid production rate observed following ACTH stimulation both in conventional static *in vitro* incubations and using superfusion techniques (usually 20% or less of that observed *in vivo*[17]) may be due to the above limitations. However, suspensions of isolated cells may be used to overcome these problems associated with relative inaccessibility of cells within tissue fragments. A variety of workers have reported studies using adrenal cells prepared by disaggregation with either collagenase[31,34–36] or trypsin[37,38] and their subsequent separation[39] into discrete cell types. The development of these techniques has opened the way for dynamic studies into the control systems operating in these cells. To this end, the counterstreaming centrifuge technique, which was originally devised for the separation of cells of different sizes,[40,41] has been adapted for the superfusion of suspensions of isolated adrenal cells.[42]

Adrenal cell suspensions[36] contained in a conical glass tube were spun at low speed (10–40 *g*) in the apparatus shown in Fig. 11. Medium (Krebs-Ringer bicarbonate buffer containing 0.5% w/v albumin and 0.2% w/v glucose; pH 7.4) was equilibrated at 37° under an atmosphere of 95% O_2:5% CO_2 and introduced at a constant flow rate into the conical tube at its apex—the point at which the sedimentation rate due to the centrifugal field is maximal. The direction of liquid flow within the tube is toward the axis of rotation. The cells are held at a distance from the rotation axis when the centrifugal acceleration has that value necessary to oppose the viscous force due to the flow of liquid past the cells. Both the viscous force and the centripetal acceleration decrease with the distance from the axis of rotation. By adjusting rotational speed and the rate of liquid flow through the tube, the forces acting on the suspension of cells may be counterbalanced. Turbulence effects along the walls of the conical tube assist the maintenance of the cells in suspension and oppose pellet formation. When the behavior of the cell suspension is visualized by means of a synchronized stroboscope, a stable sharp boundary may be observed (Fig. 12).

[34] R. Haning, S. A. S. Tait, and J. F. Tait, *Endocrinology* **87**, 1147 (1970).
[35] I. Rivkin and M. Chasin, *Endocrinology* **88**, 664 (1971).
[36] M. C. Richardson and D. Schulster, *J. Endocrinol.* **55**, 127 (1972).
[37] G. Sayers, R. L. Swallow, and N. D. Giordano, *Endocrinology* **88**, 1063 (1971).
[38] A. E. Kitabchi and R. K. Sharma, *Endocrinology* **88**, 1109 (1971).
[39] J. F. Tait, S. A. S. Tait, and R. P. Gould, *J. Endocrinol.* **55**, xxxi (1972).
[40] P. E. Lindahl, *Nature (London)* **161**, 648 (1948).
[41] C. R. McEwen, R. W. Stallard, and E. T. Juhos, *Anal. Biochem.* **23**, 369 (1968).
[42] D. Schulster, *Endocrinology* **93**, 700 (1973).

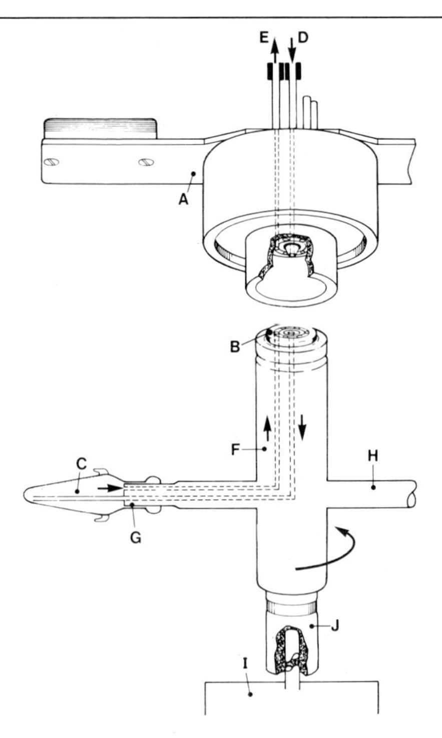

FIG. 11. Apparatus constructed for the incubation of isolated cells in continuously flowing medium. The zonal centrifuge feedhead A (Beckman-Riic Ltd or M. S. E. Ltd) remains stationary and incorporates a rulon liquid seal B which allows medium pumped by an LKB peristaltic pump to enter the rotating conical tube C through the central orifice D and flow out of the apparatus through orifice E into a fraction collector. The suspension of isolated cells is placed in the glass conical tube C, which is attached to one arm of the Teflon rotor F at the ground-glass joint G. A retaining band links C to an identical balance tube (not shown) on arm H and holds the two tubes firmly in place during centrifugation. An electric motor I rotates F via the rubber connector J, and the air in the chamber enclosing the apparatus is maintained at 37°. A thin Teflon tube fixed into the sidearm of the rotor F introduces to the apex of the conical tube C , medium which then flows toward the neck of the tube and out through the upper orifice in the sidearm (→ indicates direction of liquid flow). From D. Schulster, *Endocrinology* **93**, 700 (1973).

Lindahl[43] has shown that the cell radius has a minimum value:

$$r_{\min} = \frac{3.60}{2\pi n}\left(\frac{\eta V}{2\pi(S - S')}\right)^{1/2} \frac{1}{\frac{2}{3}(R/L)Z(\frac{1}{3}Z)^{1/2}}$$

where R is the radius of the cone at the distance L from the apex, Z the distance from the apex to the center of rotation, V the volume of

[43] P. E. Lindahl and E. Nyberg, *IVA* (*Ingenioervetenskapsakad.*) *Tidskr. Tek.-Vetensk. Forsk.* **26**, 309 (1955).

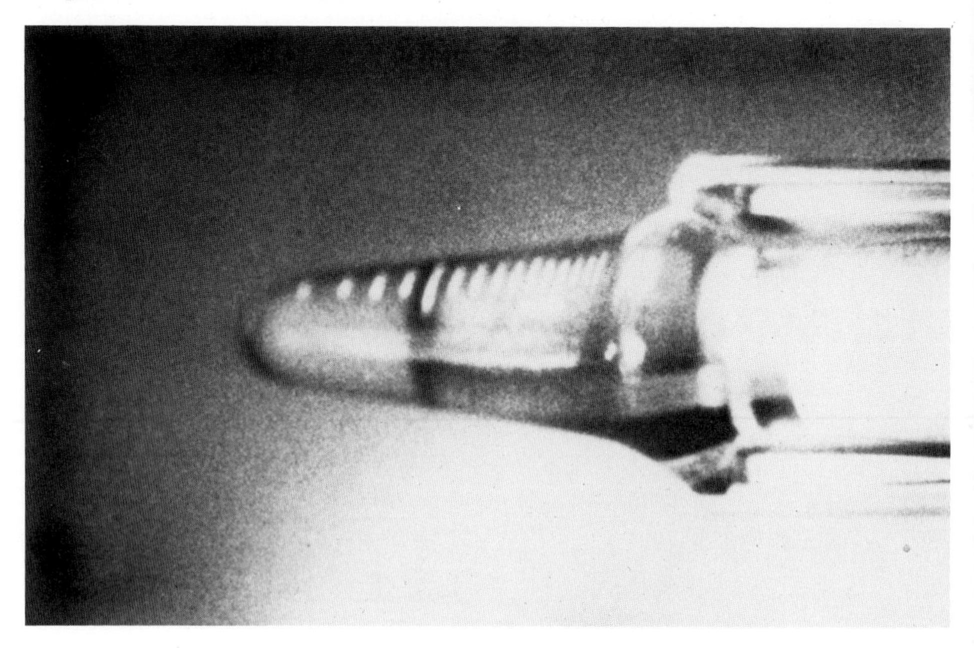

FIG. 12. Superfused cells spinning in the apparatus shown in Fig. 11. Photograph was taken using a synchronized stroboscope. A stable, sharp boundary may be seen, the position of which may be altered by adjusting the rotational speed or liquid flow rate.

medium passing any section of the tube in unit time, S and S' the specific gravity of the cells and the medium, η the viscosity of the medium, and n the revolutions per minute. It follows that only cells of radii less than r_{min} may flow out of the tube with the medium. Furthermore r_{min} is dependent only on the geometry of the tube, being proportional to L/R (i.e., the cone's apical angle), other values remaining constant. Using isolated rat adrenal cells (radius about 5 μm) a tube of 3 cm cone length, 18° apical angle, and 10 cm from apex to center of rotation, has been found convenient for the retention of cells in suspension within the tube, while the cellular products are continuously removed. The tube is rotated in a horizontal plane in order to avoid an effect of the gravitation field on the cellular distribution.

Effect of ACTH on Superfused Suspensions of Isolated Cells

Infusions of adrenocorticotrophic hormone (ACTH) to superfused adrenal cells produced rapid and marked changes in corticosteroid production (Fig. 13). The preparative and analytical methods used have been described.[36,42] The similar maximum response observed at different time periods during the superfusion provides evidence that the isolated cells remain responsive during 3–4 hours of superfusion. A single injection of ACTH (10 mU) through orifice D (Fig. 11) at 35 minutes and 95 minutes after the start of the superfusion produced rapid increases in corticosteroid output rate reaching maxima within 15 minutes and declining rapidly thereafter (half-life = 10 and 12 minutes). A continuous infusion of ACTH (10 mU/ml) was begun 155 minutes after the superfusion start and again produced a rapid initial rise in corticosteroid output rate, peaking within 30 minutes, and followed by a declining steroid output rate (half-life = 42 minutes). This late decline in steroid output rate in the presence of con-

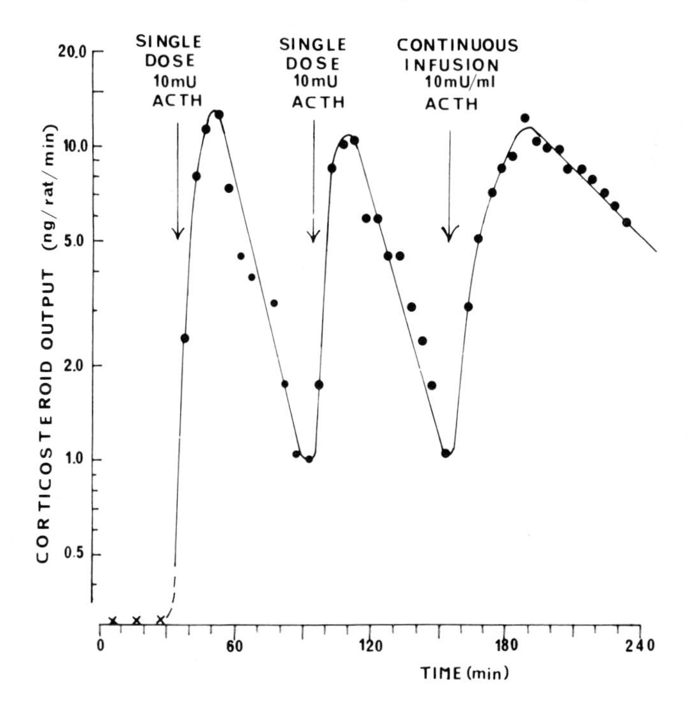

FIG. 13. Effect of ACTH on the superfused adrenal cells prepared from 5 rats. Corticosteroid output rates are plotted semilogarithmically. At 35 and 95 minutes, ACTH (10 mU) was injected as a single pulse dose—care being taken not to disturb the suspension of cells. At 155 minutes the medium was changed to one containing ACTH (10 mU/ml) in the original medium. This was continuously infused until termination of the experiment. Flow rate of medium was constant throughout at 0.5 ml per minute, rotational speed was 350 rpm, and 5-minute fractions were collected for corticosteroid analysis. From D. Schulster, *Endocrinology* **93**, 700 (1973).

tinuously infused ACTH may be compared with the decline of longer half-life observed in previous superfusion studies,[14,15,18] employing quartered, bisected, or whole adrenal glands and which was not attributable to irreversible cellular damage or depletion of steroid precursors.

These experiments have shown that the superfused isolated cell suspension responds rapidly to ACTH stimulation and is a valuable system for examining the dynamic functioning of the adrenal control mechanism.

[27] The Perfused Adrenal Gland

By Siret Desiree Jaanus and Ronald P. Rubin

A wide variety of *in vitro* preparations have been used to study hormone production and release. Although this experimental approach has added greatly to our present knowledge, it suffers from certain major disadvantages. The preparation of slices and homogenates results in damage or disruption of tissue, so that there is loss of cellular integrity and compartmentalization. In addition, the lack of constant flow in most *in vitro* systems makes kinetic analysis of hormone production and release very difficult. During the incubation procedure the accumulation of intermediates and end products may interfere with hormone synthesis (end-product inhibition), and hormone metabolism may be spuriously enhanced by reentry of product into the biosynthetic pathways.

Perfusion of an intact organ *in situ* has the decided advantage that it stimulates the physiologic situation and thus can provide more meaningful data on the actual events which occur *in vivo,* especially in regard to relating other cellular processes to the physiologic response. While *in vivo* experiments present difficulties in controlling the many external factors which can affect a given organ, complicating neural and humoral influences can be minimized by perfusing the intact organ in isolation. As with *in vitro* preparations, saline solution—rather than whole blood or plasma—may be used to perfuse the intact organ when it is necessary to control more precisely the composition of the fluid.

In this section a method is described for perfusing the isolated cat adrenal *in situ.* This technique was originally developed and refined by the authors in collaboration with Dr. William Douglas, while at the Albert Einstein College of Medicine, to study the mechanism of medullary catecholamine release.[1] However, some time later, in our own laboratory, we ascertained that this preparation can also be very productively em-

[1] W. W. Douglas and R. P. Rubin, *J. Physiol. (London)* **159,** 40 (1961).

ployed to study cortical activity.[2] This method permits one to study the kinetics of hormone synthesis and production by measurement of both gland and perfusate. Moreover, both adrenals of the same animal can be perfused simultaneously, so that by comparing paired glands, even small changes in activity can be detected, which would otherwise not be possible owing to the variability in responses from one animal to another.

Equipment

Glassware. Very little specialized glassware is required for adrenal perfusion. All glassware is rinsed with demineralized water before use, and should be provided with a silicone coating if ACTH is to be used; this is to prevent adsorption of the polypeptide onto the glass. The perfusion fluid is contained in aspirator bottles (Kimble-Corning) (500–1000 ml capacity) with an outlet near the bottom of the bottle. Pieces of thick-walled amber Latex tubing are attached to the outlet, and the free end of the tubing is clamped. An 18-gauge needle whose hub has been removed is inserted into the rubber tubing; polyethylene tubing inserted over the cut end of the needle serves as the perfusion cannula.

Perfusion Solutions. The basic perfusion medium is bicarbonate-buffered Locke's solution, which has the following basic composition (mM): NaCl, 154; KCl, 5.6; $CaCl_2$, 2.0; $MgCl_2$, 0.5; $NaHCO_3$, 12; dextrose, 10. The medium is equilibrated with 95% oxygen and 5% carbon dioxide and has a final pH of 7.0. Phosphate buffer (Na_2HPO_4, 2.2; NaH_2PO_4, 0.85) may be substituted for the bicarbonate buffer, but this Locke's solution must be equilibrated with pure oxygen. Although other saline solutions, such as Kreb's Ringer, have not been used, there is no obvious reason why solutions of similar composition would not also be satisfactory. Double-distilled or ion-free water is generally employed to make up all solutions. It does not appear necessary to go to extreme measures to obtain pyrogen-free water, although traces of heavy metals must be removed.

All perfusion fluids are generally kept at room temperature (22–25°), in order to minimize the metabolic demands of the adrenal gland during perfusion and thereby prolong the viability of the preparation. Perfusion with fluid having temperatures of 30–37° does not grossly alter the secretory activity of the medulla.

Perfusion Apparatus. Flow of Locke's solution through the gland is controlled at a constant pressure of about 70 mm Hg by a large reservoir

[2] S. D. Jaanus, M. J. Rosenstein, and R. P. Rubin, *J. Physiol. (London)* **209**, 539 (1970).

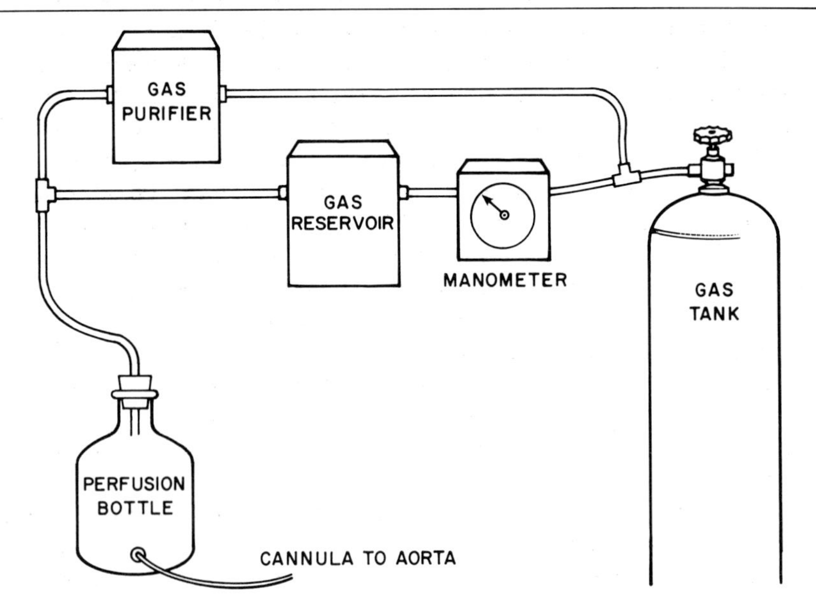

FIG. 1. Schematic representation of adrenal perfusion apparatus.

of compressed air (Fig. 1). The air is fed into the large gas reservoir from the same gas cyclinder which is used to bubble the Locke's solution. By means of a T-tube system, gas can be continuously equilibrating with the fluid in the perfusion bottles while the inflow to the reservoir is clamped. The pressure in the reservoir is transmitted to the perfusion bottles through Latex tubing connected to a one-hole rubber stopper by a small piece of glass tubing.

A constant-flow perfusion pump system has not been employed with this preparation, but may work as well as the flow-control method described. However, a perfusion pump is definitely not necessary, since there are usually only minor changes in the flow rate during the course of a single experiment, unless substances toxic to the preparation are employed. The small decline in flow rate which may occur over prolonged periods of perfusion (2–3 hours) can be readily controlled by a small increase in the perfusion pressure produced by briefly opening the clamp leading to the reservoir from the gas cylinder.

Preparative Procedures for Adrenal Perfusion

All studies are carried out on lean, young, male cats weighing 1.5–2.5 kg. These animals usually possess a minimum of abdominal adipose tissue, so that dissection of the region surrounding the adrenal gland is facilitated. The cats are anesthetized with pentobarbital (30 mk/kg), and a tracheal cannula is inserted and artificial respiration begun with a minute volume just sufficient to depress spontaneous respiratory movements. This procedure serves to avoid asphyxia of the adrenal gland, which may either enhance medullary secretion or impair the viability of the gland.

A midline abdominal incision is then made from the pubis up to the lower end of the sternum, and an evisceration is carried out by ligating and cutting in sequential order: the lower end of the colon, the inferior mesenteric; the superior mesenteric and celiac arteries; the hepatic artery, portal vein, and bile duct; and the lower end of the esophagus. The removal of the gastrointestinal tract provides unimpeded access to the adrenal glands and their associated vessels. The operative procedure is facilitated by retracting laterally the abdominal skin flaps and abdominal muscle.

The arterial supply to the adrenal glands is diffuse and variable (Fig. 2). Branches to the adrenal glands run from (1) the adrenolumbar arteries and the phrenic arteries, (2) the celiac axis and superior mesenteric

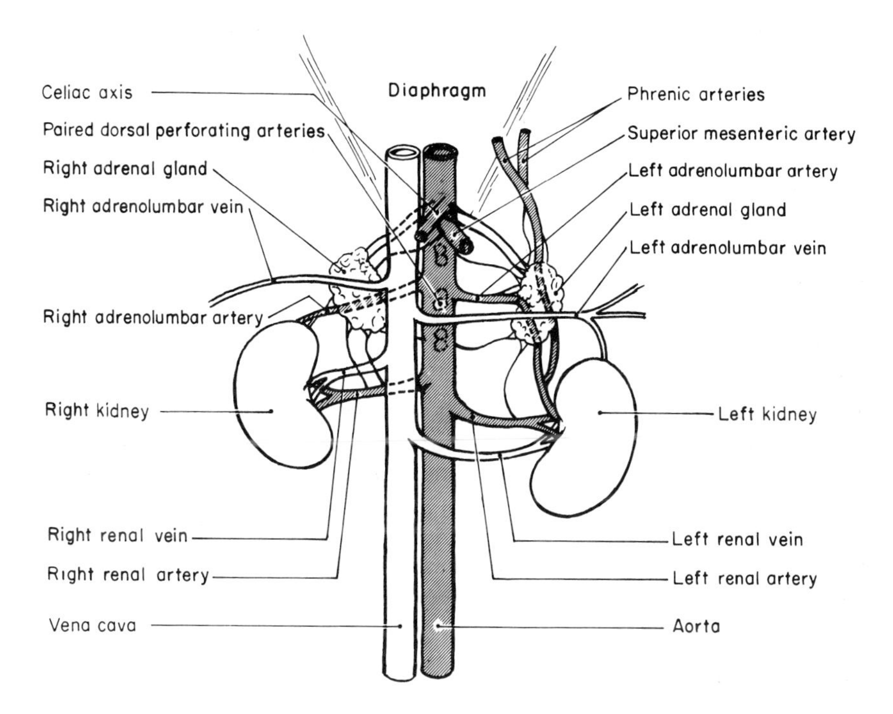

FIG. 2. A view of the abdominal blood vessels of the cat after evisceration, showing their relation to the adrenal glands.

artery near their origins from the aorta, (3) the renal arteries at varying lengths between the aorta and the kidney, and (4) the abdominal aorta at varying distances below the adrenal down to and even below the origin of the renal arteries. Thus, in order to provide perfusion flow approximating the normal, the left gland or both glands are perfused via a cannula placed in the abdominal aorta.

When it is necessary to perfuse only one gland the left adrenal is preferred, since it is much more accessible than the right gland, which is frequently found directly beneath the inferior vena cava. The procedure for preparing the left gland for perfusion is as follows: (1) on the right side major renal vessels and smaller vessels in the perirenal fat are ligated and the right kidney is removed; (2) the right adrenolumbar vein is ligated both lateral and medial to the right adrenal gland; this gland may be removed if desired; (3) loose ligatures are placed around the abdominal aorta, vena cava, left renal artery and vein close to the hilus of the kidney, and the left adrenolumbar vein (both lateral and medial to the gland); (4) by lifting the ligatures around the aorta, two or more pairs of dorsal perforating lumbar arteries arising from the aorta are exposed and ligated; (5) a polyethylene cannula (PE 190) filled with perfusion fluid and connected with the perfusion bottle is tied into the abdominal aorta about 2–3 cm below the left renal artery, pointing cephalically. The aorta is tied immediately above the celiac axis to cut off the circulation of the animal below the diaphragm, and perfusion is immediately begun; the carotid arteries are then severed to allow the animal's blood pressure to fall to zero; (6) a cut is made in the inferior vena cava, and a polyethylene cannula with a short right-angle bend close to its tip is inserted and pushed up to the point of anastomosis with the left adrenolumbar vein. The angled tip of the cannula is then placed into the mouth of the adrenolumbar vein and tied in place. The adrenolumbar vein is next tied just lateral to the gland, so that the adrenal perfusate will not be diluted by perfusion fluid from other sources; (7) a third polyethylene cannula is inserted into the lower inferior vena cava and pushed up to its point of entry into the liver. This cannula collects fluid perfusing the liver and other extra adrenal tissues, and is discarded; development of edema, which may jeopardize the viability of the adrenal gland, is therby curtailed; (8) a fourth polyethylene cannula fitted with a tap is inserted into the stump of the superior mesenteric artery; it provides a means of rapid washout of the dead space as the composition of the medium is altered; (9) neural influences may be terminated either by opening the chest and cutting the sympathetic chain as it runs beneath the thoracic aorta, or by cutting the greater and lesser splanchnic nerves just above the adrenal gland in the vicinity of the celiac ganglion;

(10) perfusion flow is maintained at approximately 1–2 ml per minute by regulation of the perfusion pressure.

Under some conditions it may be necessary to perfuse both adrenal glands simultaneously. The right kidney and adrenal gland are then left *in situ,* and the adrenolumbar artery and vein are tied off just lateral to the right gland. Collection of the effluent may be carried out by means of a cannula inserted into the vena cava after this vein has been tied off below the liver, or by inserting a curved cannula directly into the adrenolumbar vein of one or both glands.

Responsiveness of the Perfused Adrenal Gland

This preparation is very sensitive to a variety of catecholamine-releasing agents, including the physiologic neurotransmitter, acetylcholine, as well as other nicotinic and muscarinic agents. Acetylcholine concentrations of the order of 1–10 μM are capable of augmenting catecholamine secretion; higher concentrations may increase release by more than 100-fold (Fig. 3). The splanchnic nerve also may be isolated, and release be evoked by electrical stimulation. In addition, a variety of other substances, including excess potassium, histamine, serotonin, sympathomimetic amines, polypeptides and ouabain, are medullary secretogogues. (For references, see review by Rubin[3].) By contrast, cyclic nucleotides and inhibitors of phosphodiesterase are either very weak stimulators or are completely devoid of any catecholamine-releasing activity.

Catecholamine secretion rates vary with the concentration of secretogogue or the concentration of calcium in the perfusion medium. The ability of this preparation to yield precise quantitative data is emphasized by the finding that the output of catecholamines can be correlated with the potassium concentration of the perfusion medium up to 84 mM. Moreover, the ability of varying concentrations of potassium to elicit secretion in a graded manner during 1–2 minute exposures indicates that equilibration between the gland and the perfusion medium is quite rapid.

The perfused adrenal also responds to brief exposure to physiologic (microunit) concentrations of ACTH. The time-course of ACTH-induced corticosteroid release is quite different from that of acetylcholine-evoked catecholamine release. Thus, there is a latency of 5–15 minutes before an increase in secretion is observed in response to ACTH, and the maximum effect is not demonstrable until after 30–40 minute (Fig. 3). The secretory response to ACTH remains long after the polypeptide is removed from the perfusion medium. By contrast, the maximum response to acetylcholine is observed during the first 2 minutes of exposure and

[3] R. P. Rubin, *Pharmacol. Rev.* **22,** 389 (1970).

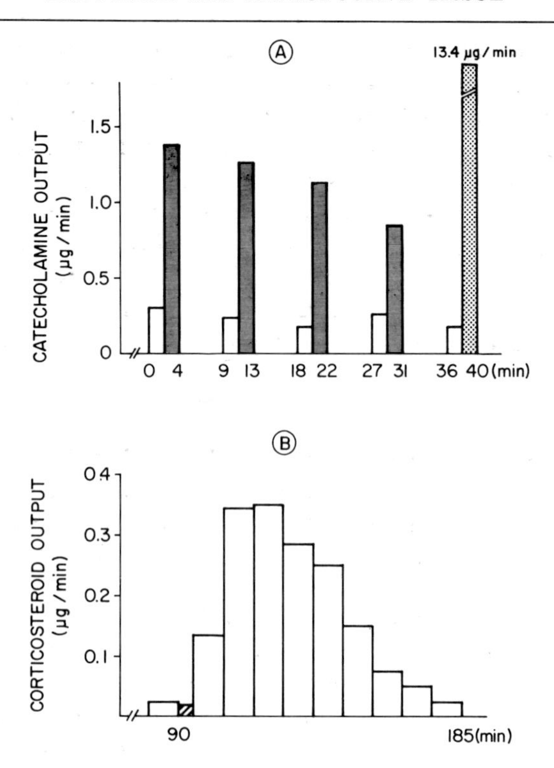

FIG. 3. Adrenal secretory response to (A) acetylcholine (▤, 2×10^{-6} *M*; ▩, 5×10^{-5} *M*) and (B) ACTH (▨, 4 μU/ml). The vertical bars represent outputs of catecholamine and corticosteroid obtained during 2- and 10-minute collection periods, respectively. ACTH was present for 5 minutes.

declines almost immediately after it is removed (Fig. 3). Thus, many responses to acetylcholine may be tested during the same time interval in which a single response to ACTH runs its course. The observed disparities may be explained by the fact that the action of ACTH includes steroid synthesis as well as release, and the action of acetylcholine involves only the release of preformed catecholamine.

The cortex also differs from the medulla in that it is not responsive to substances other than the physiologic secretogogue. Excess potassium and barium can increase corticosteroid release slightly above control levels, but these effects are neither consistent nor significant. Cyclic nucleotides and phosphodiesterase inhibitors are also unable to augment steroid synthesis or release.

Although the emphasis of this discussion has focused on methods for measuring hormone secretion, the *in situ* adrenal preparation can also be productively employed to study more directly activity within the medulla and cortex. Constructive use may be made of the observation that the left and right adrenals within the same animal are of similar size and appear to respond to various challenges in an identical manner. For example, perfused paired glands contain identical catecholamine[4] and steroid concentrations[5]; and when they are exposed to the same stimulus they respond similarly. By employing the left gland as a control and the right gland as the experimental, the problems arising from variability in responses from one animal to another can be minimized; thus, even small changes in such parameters as calcium,[6] steroid,[5] and cyclic nucleotide[7] concentrations can be detected in the adrenal gland.

Comparison of Adrenal Perfusion *in Situ* with *in Vitro* Methods

Other methods for perfusing the intact adrenal gland have been employed in various species, including the cow[8] and dog,[9] and generally involve perfusion through the venous circulation. During our own early studies, a method of retrograde perfusion was also used.[1] The major advantage of this procedure is its simplicity, with consequent ease and rapidity of completion. In these preparations the perfusion cannula is placed in the adrenolumbar vein medial to the gland. The gland with the attached veins and cannula is cut free from the animal and transferred to a glass funnel containing liquid paraffin, and retrograde perfusion begun. Perfusion fluid is collected as it leaks into the funnel from the cut ends of the numerous small arterial twigs.

Although in our hands the retrograde preparation is somewhat easier and quicker to prepare than the arterially perfused preparation, some difficulty is encountered in maintaining the flow of the perfusion fluid; in addition, with the retrograde perfusion technique the adrenal gland appears to deteriorate more rapidly; this may be due to the abnormal direction of perfusion flow, which may not provide as good a circulation as the conventional technique of perfusion through the arteries. Electromicrographs of the arterially perfused glands show excellent preservation of cell structure and organelles, which supports the general impression that this method of perfusion does indeed provide the gland with excellent circulation. However, some of the problems encountered with retrograde

[4] R. P. Rubin, *J. Physiol. (London)* **206,** 181 (1970).

[5] S. D. Jaanus, R. A. Carchman, and R. P. Rubin, *Endocrinology* **91,** 887 (1972).

[6] S. D. Jaanus and R. P. Rubin, *J. Physiol. (London)* **213,** 581 (1971).

[7] R. A. Carchman, S. D. Jaanus, and R. P. Rubin, *Mol. Pharmacol.* **7,** 491 (1971).

[8] O. Hechter, R. P. Jacobsen, V. Schenker, H. Levy, R. W. Jeanloz, C. W. Marshall and G. Pincus, *Endocrinology* **52,** 679 (1953).

[9] R. L. Robinson, *J. Pharmacol. Exp. Ther.* **151,** 55 (1966).

perfusion may be obviated by the use of a Technicon AutoAnalyzer pump.[10]

The bovine adrenal gland, perfused retrograde, has also been commonly used as a test preparation for both cortical[8] and medullary studies.[11-13] The obvious advantage of the bovine adrenal over the cat adrenal involves the relative sizes of the two glands. The average weight of a cat adrenal is of the order of 0.1 g; the bovine gland is some 100 time heavier. Thus, the bovine preparation is generally to be preferred when large quantities of material are to be collected and isolated. On the other hand, the necessity of obtaining bovine glands from an abattoir precludes their use by certain investigators, due to their inaccessibility. Moreover, if indeed an investigator is fortunate enough to be able to obtain cow adrenals, there is still the problem of the time delay between exsanguination, removal of the adrenals, return trip to the laboratory, and the eventual setting up of the preparation. The overall viability of such a preparation must be compromised to a greater or lesser degree, since adrenal parenchymal cells are very sensitive to anoxia. Thus, it appears that a mastering of the *in situ* perfusion technique may be worth the effort if one wishes to simulate more closely conditions *in vivo*.

[10] M. J. Peach, F. M. Bumpus, and P. A. Khairallah, *J. Pharmacol. Exp. Ther.* **176**, 366 (1971).

[11] A. Philippu and H. J. Schümann, *Naunyn-Schmiedebergs Arch. Pharmakol. Exp. Pathol.* **243**, 26 (1962).

[12] P. Banks, *Biochem. J.* **101**, 536 (1966).

[13] F. H. Schneider, A. D. Smith, and H. Winkler, *Brit. J. Pharmacol.* **31**, 94 (1967).

[28] The Rat Adrenal *in Situ*[1]

By WENDELL E. NICHOLSON and ANDRE PEYTREMANN

The rat adrenal *in situ* provides a model for evaluation of the acute effects of pituitary hormones on adrenal function. Two major reasons for its usefulness are the availability of hypophysectomized rats, thus removing endogenous pituitary hormones from the circulation, and the ease with which the adrenal effluent can be sampled. The procedures available for the preparation of hypophysectomized rats are described first, followed by a description of the techniques involved in assessing

[1] Original work included in this report was supported in part by the following grants from the National Institutes of Health: 5-TOI-AM-05092, 5-ROI-AM-05318, and 3-FO5-TW-01650.

the effects of adrenocorticotropic hormone (ACTH) on adenosine 3′,5′-monophosphate (cyclic AMP) and steroidogenesis in the rat adrenal *in situ*.

Transsphenoidal Hypophysectomy

The first procedure is a modification of the transsphenoidal hypophysectomy method first described by Smith.[2] Animals weighing 150–200 g are the most suitable for all techniques described. Ether anesthesia is preferable and is accomplished by placing the animal in a covered beaker or jar containing gauze sponges saturated with ether. Anesthesia is maintained during surgery by placing the animal's nose in a 50-ml beaker containing gauze also saturated with ether. The anesthetized animal is secured on his back to a soft wood board approximately 8×10 inches in size by means of rubber bands looped about the feet and over hooks which have been placed about 1 inch apart around the edge of the board. The head and neck are kept extended by a rubber band looped over the upper incisors. A rubber tube is inserted into the mouth in a manner which depresses the tongue and permits the animal to breathe through the mouth. The neck is clipped, shaved, and then painted with 70% alcohol. The operation is performed with the head of the animal toward the operator. Illumination is provided by a desk lamp. A midline skin incision is made with scissors extending from the mandibular papilla to the lower border of the submaxillary gland. The salivary glands and subcutaneous tissue lying to the right are freed by blunt dissection and retracted to the right with 5-inch Halstead forceps. Longitudinally at the lower border of the larynx and laterally at the border of the trachea the closed tips of a pair of 115 mm curved, fine-tipped forceps are forced through the omohyoid muscle at the point of its insertion under the sternohyoid muscle. The tips of the forceps are directed through the muscle toward the midline and are then spread apart to make a longitudinal division of the muscle extending from the point of its upper attachment to the level of the lower border of the thyroid. Spreading the forceps laterally and retracting toward the head will expose the area to be prepared for drilling. This step of the operation should be carried out without producing hemorrhage and should give a clear view of the area to be prepared for drilling without additional dissection and without damage to the nerves and vessels of this region. The trachea is almost completely closed and therefore this retraction can be maintained for only a few seconds at a time. If respiratory movements cease, they can be readily restored

[2] P. E. Smith, *Amer. J. Anat.* **45**, 205 (1930).

by promptly inserting a rubber tube into the mouth of the rat and blowing air directly into the lungs. The bony ridges representing the occipitosphenoid synchondrosis and the crista occipitalis are exposed by scraping away the overlying muscle and finally wiping the area dry with cotton-tipped applicators.The pharynx must not be entered. The pituitary is exposed by means of a long-shank, round bone-burr attached to a dental drill controlled by a foot-operated rheostat. The size of the burr required is determined by the size of the rat. A No. 9 dental bone-burr is satisfactory for operations on rats weighing 150–200 g. The hole is drilled in the midline on the occipitosphenoid synchondrosis (Fig. 1) in a manner which allows approximately two-thirds of the hole to be cephalad of the suture line. The drill is not allowed to break through the bone, but is carefully controlled so that the last shell of bone is drilled away, preferably without rupture of the membranes covering the gland. A pick suitable for manipulation of the membranes and of the pituitary can be prepared by breaking the shaft of a 22-gauge needle from its base and fixing it in a slender, removable handle. For most purposes the sharp point of the needle should be converted into a fine hook by blunting. The pituitary is exposed by tearing away the dura. The operative field is dried when necessary by the use of cotton-tipped applicators. A transfer pipette cut to fit the size of the drill hole is placed against the pituitary, and the gland is removed by negative pressure (water pump assembly). The strength of the suction must be regulated. The suction should not be stronger than is necessary to draw the gland into the pipette. The pipette is attached by means of a two-hole rubber stopper

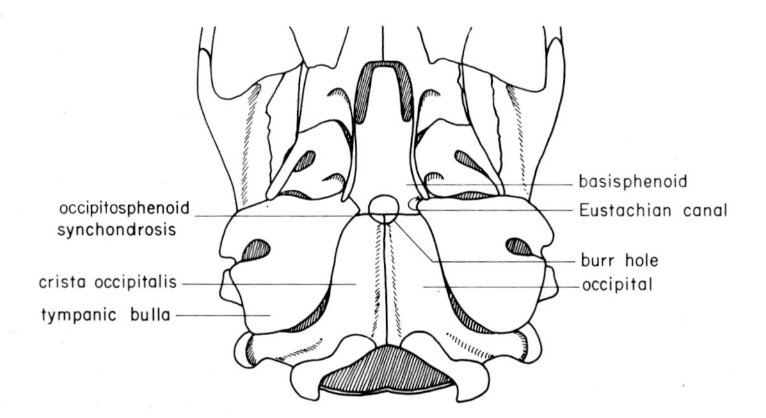

FIG. 1. Ventral aspect of the rat's skull indicating location of burr hole through which the pituitary gland is removed. From H. G. Q. Rowett, *in* "The Rat as a Small Mammal," 2nd ed., p. 7. Murray, London, 1960.

to a glass tube approximately ¾ inch in diameter and 4 inches in length. The tube acts as a suction trap and allows the tissue to be visualized and examined after removal. When properly removed, the anterior lobe is usually broken into two parts, and the posterior lobe is separated from the anterior but is unbroken.

Following the removal of the pituitary tissue the exposed tissues are wiped free of blood and the skin incision is closed by suture or wound clips. The animals should be kept in dry cages in a warm room (28°). The hypophysectomized rat can survive throughout the day of surgery without further treatment. Completeness of the hypophysectomy is confirmed by gross or microscopic examination of the sella turcica after the rats are sacrificed.

Transaural Hypophysectomy

Transaural hypophysectomy of the rat is faster but more difficult to master than the transsphenoidal technique. In the hands of a skilled operator the rate of morbidity, mortality, or completeness of hypophysectomy is similar for the two approaches. The following procedure is a modification of the method first described by Koyama.[3] The method of anesthesia is the same as for the transsphenoidal approach. The animal is placed on his stomach with the head toward the operator. The rat's head must remain in a flat plane throughout the rest of the procedure. The head is grasped between the thumb and middle finger of the left hand with the index finger resting on the back of the neck and the remaining fingers resting against the middle finger underneath the head. An 18 gauge, 1½ inch needle (TLNRS, Becton, Dickinson and Co.), with its bevel reduced from 4 mm to 2 mm and attached to a 5-ml glass syringe containing 0.5 ml of water, is passed via the left external ear canal to the tympanic bulla. As the needle, with the bevel turned down, reaches the outer wall of the tympanic bulla, a slight resistance will be met. By slowly rotating the needle and applying gentle pressure the needle can be passed completely through the tympanic bulla into the sella turcica (Fig. 2). The plane of insertion should be slightly downward, turning toward the rostral edge of the right ear. The total length of the insertion of the needle into the sella should not exceed 2 mm. The pituitary is then aspirated into the syringe by gentle negative pressure (accomplished by withdrawing the plunger 0.5 ml). The contents of the syringe can then be expelled onto a 3 $\times$ 3 gauze sponge for inspection. Since external pressure will result in aspiration of blood via the eustachion canal into the

[3] R. Koyama, *Jap. J. Microsc. Sci. Trans.* **5**, 41 (1931).

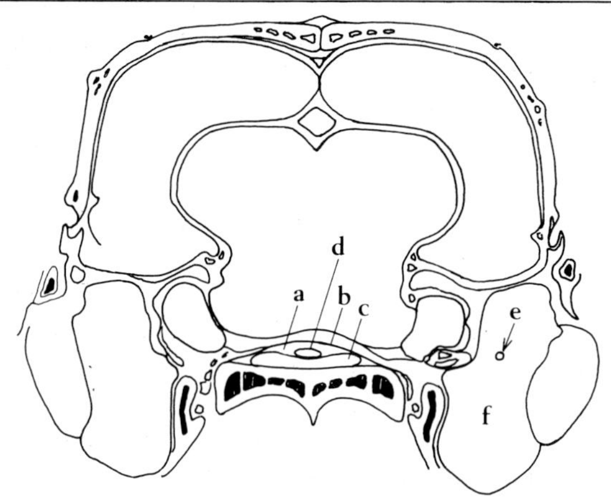

FIG. 2. Transverse section of the rat's skull showing the point (e) at which the needle passes through the tympanic bulla (f) before entering the sella turcica (a) and capsula hypophysis (b) for removal of the anterior lobe (c) and posterior lobe (d) of the pituitary gland. From A. Tanaka, *Shionogi Kenkyusho Nempo* **5**, 678 (1955).

lungs, no attempt should be made to control the bleeding which occurs through the left external ear canal. Postoperative care of the transaurally hypophysectomized rat is the same as that for the animal recovering from the transsphenoidal procedure.

Hypophysectomized Rats from Commercial Sources

For an additional charge, some animal suppliers will make hypophysectomized rats available to investigators within 24 to 48 hours after surgery. The loss in adrenal sensitivity (steroidogenic) to ACTH that occurs following hypophysectomy can be restored by treating these animals with a single subcutaneous injection of 40 mU of ACTH 4 hours before they are to be studied. Plasma levels of ACTH, corticosteroids, and cyclic AMP will be low and steady 4 hours after the ACTH injection. Thus the sensitivity of the 24- or 48-hour hypophysectomized rat to ACTH is improved, while each investigator is relieved of the necessity of developing the technique of hypophysectomy in his own laboratory.

Preparation of the Rat for ACTH Injection

When the rats have been hypophysectomized in the investigator's own laboratory, 2 hours should elapse in order to achieve low and steady levels of ACTH, cyclic AMP, and corticosteroids. The adrenal gland will then maintain a constant level of sensitivity to ACTH up to 12 hours after hypophysectomy. Subsequently, the steroidogenic response begins to decline while the cyclic AMP response remains constant for up to 9 days. Similarly, the 24- or 48-hour hypophysectomized rat should be studied between 4 and 8 hours after preparative treatment with ACTH.

The animal is again anesthetized, this time with an intraperitoneal injection of Nembutal, 3 mg/100 g body weight. Five minutes later the rat is immobilized on his back with the head away from the operator. A 2-cm skin incision is made with scissors on the medial surface of the thigh and parallel to the left inguinal ligament, the operator being careful to avoid opening the peritoneum. By blunt dissection the left femoral vein is exposed and a 1.5 cm segment between the popliteal vein and inguinal ligament is readily available for injection of the test materials. During the course of the dissection the popliteal vein may be torn, but slight pressure for a few seconds will control the bleeding and exposure of the femoral vein can continue. If other injection sites are needed, the right femoral vein can be exposed in a similar manner. After the injection, bleeding is controlled with 5-inch Halstead forceps.

Preparation of ACTH Solutions

ACTH is heat labile and unstable at neutral pH. In addition, the relatively small amounts required in these studies are easily lost on the glassware. Therefore, stock solutions containing at least 200 mU/ml are prepared in polypropylene tubes in acid-albumin saline (0.6 ml of 12 *N* HCl, 1 g of bovine albumin, and 8.5 g of NaCl per liter of water) and kept frozen at −56° in 0.2-ml aliquots for up to 4 months. Each aliquot of stock solution is used once to prepare working solutions on the day they will be tested. The working solutions are prepared in acid-albumin saline so that each animal will receive 0.5 ml containing the appropriate amount of ACTH.

Collection of Adrenal Effluent for Cyclic AMP and Corticosterone Analysis

Sampling of the adrenal effluent in rats was first described by Munson and Toepel.[4] A glass cannula was inserted in the vein, and blood was collected by gravity flow. Later Lipscomb and Nelson[5] collected blood

[4] P. L. Munson and W. Toepel, *Endocrinology* **63**, 785 (1958).

[5] H. Lipscomb and D. H. Nelson, *Fed. Proc., Fed. Amer. Soc. Exp. Biol.* **18**, 95 (1959).

from the rat's adrenal vein through a needle attached to a syringe. The procedure described is a modification of the Lipscomb and Nelson technique.

After intravenous ACTH, cyclic AMP, and corticosterone in the adrenal vein blood rise to a peak within 9 minutes; the cyclic AMP concentration then declines while the corticosterone level remains elevated (Fig. 3). One minute before adrenal vein blood is to be collected, the animal's peritoneum is opened with a 5-cm midline incision. Two transverse incisions beginning in the center of the midline incision and extending 3 cm are made. The abdominal viscera are exposed by retracting and retaining the two upper pieces of abdominal skin outward and upward with 5-inch Halstead forceps. Similarly, the lower pieces are retracted and retained outward and downward. During the course of dissection the slight capillary bleeding that occurs will be controlled spontaneously. The digestive tract and associated structures are displaced and held to the right side of the abdominal cavity with a saline-moistened gauze sponge. The left adrenal and kidney are now clearly visible (Fig. 4), and it can be seen that the left adrenal vein enters the left renal vein at a point about mid-

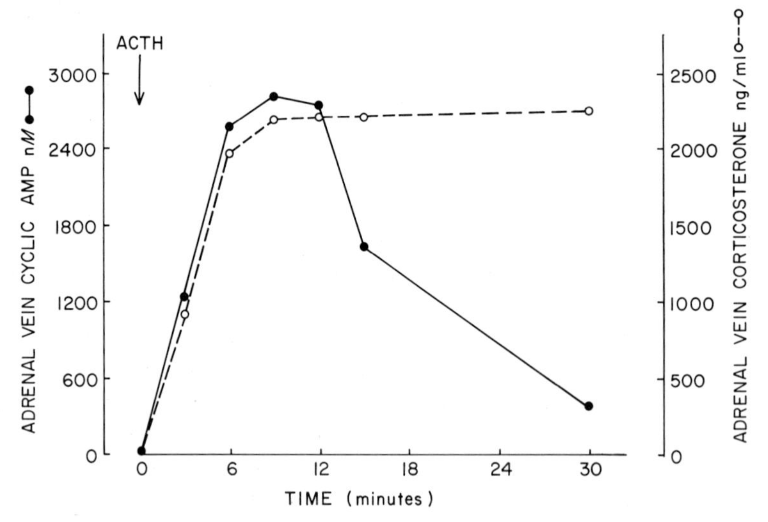

FIG. 3. Cyclic AMP and corticosterone levels in adrenal vein blood of the hypophysectomized rat following intravenous ACTH. Blood was collected continuously from 1 minute before until 1 minute after each time indicated on the abscissa. Each animal received 100 mU of ACTH at time zero. The maximally effective dose of ACTH on adrenal vein cyclic AMP is 100 mU; the minimally effective dose is 0.2 mU. The maximally effective dose of ACTH on corticosterone secretion is 0.2 mU; the minimally effective dose is 0.02 mU.

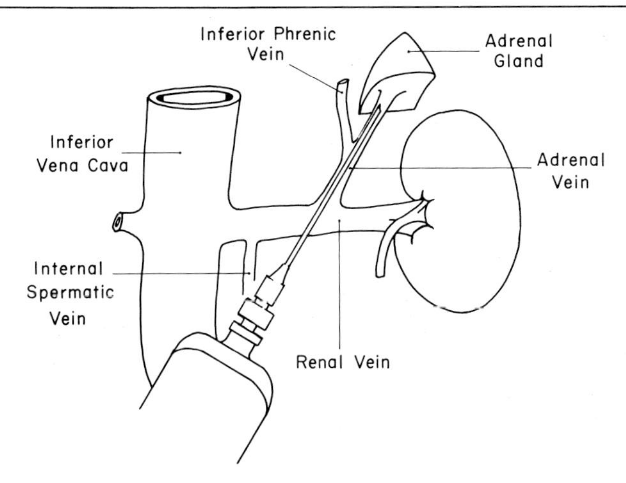

FIG. 4. Cannulation of the left adrenal vein of the hypophysectomized rat with a 27-gauge needle. From H. Lipscomb and D. H. Nelson, *Endocrinology* **71,** 14 (1962).

way between the inferior vena cava and left kidney. The left internal spermatic vein enters the left renal vein at a point between the inferior vena cava and the point where the left adrenal vein enters the left renal vein. Fifteen seconds before sampling of the adrenal effluent is to begin, the left adrenal vein is cannulated by advancing a 27-gauge stainless steel needle through the renal vein into the adrenal vein until it lodges in the vein. The left internal spermatic vein is kept taut by grasping it and surrounding tissue with a pair of smooth-tipped forceps, thus facilitating the entry and advancement of the needle. Adrenal venous blood is drawn into a lubricated (thin film of petrolatum) glass tuberculin syringe attached to the needle. The syringe contains 20 μl of 0.3 *M* EDTA (pH 7.4) as an anticoagulant and to prevent hydrolysis of the cyclic AMP. Heparin should not be used since it usually contains a preservative which interferes with the fluorometric assay of corticosterone and at the same time it affords no protection for the cyclic AMP. The negative pressure used for withdrawal of blood should not be so great as to collapse the wall of the vein against the bevel of the needle, yet it should be strong enough (and held constant) to collect most of the blood flowing through the vein. No attempt is made to ligate the nonadrenal tributaries to the adrenal vein, although some of them are by-passed by advancing the needle as far as possible into the adrenal vein. The blood collected is thus a mixture of adrenal and systemic venous blood. At the end of the collection period (usually 2 minutes) the blood (volume is noted) is immediately expelled into a chilled 12 × 75 mm polypropylene tube containing tritiated cyclic AMP to monitor recovery. An aliquot is removed for microhematocrit determination and subsequent determination of plasma volume. The remainder is centrifuged at 6000 *g* for 10 minutes at 4°. The plasma is removed, an aliquot is taken for cyclic AMP determination, and another aliquot (0.3 ml) is frozen for corticosterone measurement. The plasma for cyclic AMP measurement is immediately deproteinized by the addition of 3 volumes of 0.4 *N* perchloric acid, centrifuged at 6000 *g* for 10 minutes at 4° and the supernatant is stored at 4° until it is used for purification and assay.

Collection of Adrenal Glands for Cyclic AMP Analysis

After the intravenous injection of ACTH, the cyclic AMP content of the rat adrenal promptly increases, reaching a peak within 5 minutes, and gradually declines thereafter (Fig. 5). This rapidly formed cyclic AMP then becomes the intracellular mediator of the steroidogenic action of ACTH. Since the interval between ACTH injection and the subsequent increase in adrenal gland cyclic AMP is shorter than the time necessary

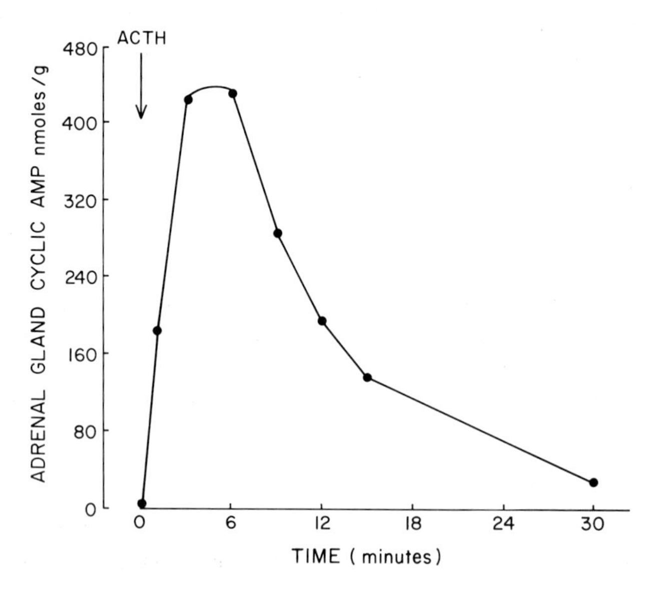

FIG. 5. Cyclic AMP content of the adrenal gland in the hypophysectomized rat after intravenous ACTH. Each animal received a maximally effective dose of 100 mU ACTH, at time zero. The minimally effective dose of ACTH for adrenal gland cyclic AMP is 0.1 mU per animal.

(2 minutes) to collect enough adrenal vein blood for cyclic AMP measurement, additional animals other than those from whom adrenal vein blood is to be collected, will be required for studies of the adrenal content of cyclic AMP during the first 2 minutes after ACTH. With this exception, the cyclic AMP content of the adrenal gland can be determined at the end of each collection of adrenal vein blood.

One minute before the adrenals are to be removed, the peritoneum is opened as previously described and the abdominal viscera are packed to the right. The left adrenal is removed and the abdominal viscera packed to the left for removal of the right adrenal. The adrenals are trimmed free of fat, frozen on dry ice, weighed while frozen on a torsion balance, and homogenized in 20 ml of 0.1 *N* HCl (Waring blender with Eberbach semimicro container), containing tritiated cyclic AMP to monitor recovery. The homogenates are centrifuged at 6000 *g* for 10 minutes at 4°, and the supernatants are stored at 4° until purification and assay for cyclic AMP. The time necessary to remove, trim, and freeze the adrenals should not exceed 20 seconds. The two adrenals from one animal are handled simultaneously by two persons. The adrenals can best be removed by lifting the gland, via the adrenal vein and/or periadrenal fat, with a pair of 125 mm straight, blunt-pointed dissecting forceps, and snipping the gland out with a pair of 125 mm straight, blunt-pointed dissecting scissors. The gland is then placed on a piece of paper where the periadrenal fat is quickly trimmed away with a pair of 105-mm straight, sharp-pointed (narrow blades) dissecting scissors and 115 mm straight, sharp-pointed dissecting forceps. The fat tends to stick to the paper as the adrenal is gently rolled, thus facilitating easy removal.

Cyclic AMP Assays

Several methods for cyclic AMP measurement are described in Volume 38 of this series. The data shown in Figs. 3 and 5 were obtained from samples purified by Dowex-50 column chromatography and assayed by the enzymatic method of Hardman *et al.*[6] as modified by Ishikawa *et al.*[7]

Corticosterone Assay

In rat plasma the predominant adrenal cortical steroid is corticosterone, which can be conveniently measured fluorometrically. The method

[6] J. G. Hardman, J. W. Davis, and E. W. Sutherland, *J. Biol. Chem.* **241**, 4812 (1966).

[7] E. Ishikawa, S. Ishikawa, J. W. Davis, and E. W. Sutherland, *J. Biol. Chem.* **244**, 6371 (1969).

described is basically the Mattingly[8] procedure for the determination of cortisol in human plasma but includes certain modifications that were necessary when the method was adapted to the measurement of corticosterone in rat plasma. In addition, changes in the method which simplify the procedure are also described.

Preparation of Reagents. Dichloromethane (Matheson, reagent) is purified by passing it through a 7 × 125 cm column of silica gel (28–200 mesh). A single resin bed can be used to prepare 10–15 liters of solvent. Sulfuric acid (Dupont, reagent) is used without further purification. Lots of sulfuric acid should be selected that have a native fluorescence equivalent to less than 2 ng of corticosterone. Ethyl alcohol (United States Industrial Chemical Company) can also be used without further purification. The fluorescence reagent is prepared fresh daily (background fluorescence increases even with storage at 4°) by dripping sulfuric acid into ethanol. Three parts of acid are used to 1 part ethanol. The acid is delivered at the rate of 2 ml per minute, via a separatory funnel with pressure equalizing joint, into cold ethanol (methanol–dry ice bath) with continuous stirring of the acid–ethanol mixture. The fluorescence reagent is brought to room temperature prior to use.

Extraction and Development of Fluorescence. Each 0.2-ml sample of plasma is diluted to 0.5 ml with water and then extracted on a horizontal shaker (Eberbach) for 10 minutes at 120 strokes per minute with 7.5 ml of dichloromethane in a 16 × 150 mm glass-stoppered test tube (Ace Glass Co. No. 8645). The aqueous layer is removed by aspiration after the tubes have stood for 5 minutes, and 5 ml of the extract is transferred to a clean tube for fluorescence development. Aliquots, 0.5 ml, of three aqueous corticosterone standards (600, 300, and 150 ng/ml) are similarly processed as standards, and 0.5 ml of water as a blank. The corticosterone standards are prepared by first dissolving the steroid in absolute ethyl alcohol, after which the appropriate dilutions can be made in water. Two milliliters of the fluorescence reagent are added with a 5-ml repipette (Labindustries) to each 5 ml of extract and mixed on a Vortex mixer for 20 seconds. The dichloromethane layer is removed by aspiration, and fluorescence is allowed to develop in the acid layer at room temperature for 30 minutes.

Measurement of Fluorescence. The intensity of the fluorescence is measured in one of several fluorometers which are currently available. The fluorescence developed with corticosterone in ethanolic-sulfuric acid is directly proportional to concentration, up to 200 ng. Samples containing more than 200 ng of corticosterone should be diluted and reassayed. The

[8] D. Mattingly, *J. Clin. Pathol.* **15**, 374 (1962).

actual manipulations necessary to measure the fluorescence will vary depending on the instrument used, but they involve basically the following steps. The fluorometer should be set up to disperse light from the light source into monochromatic radiation (475 nm for corticosterone) incident on the sample. Similarly, fluorescent light from the sample is dispersed into monochromatic radiation (520 nm) incident on the photomultiplier tube. The radiation is then transformed into a weak electrical signal and visualized on a suitable photometer. The most concentrated standard, in this case 600 ng, is arbitrarily assigned a setting on the photometer and the readings from the other standards should provide a rectilinear standard curve after correction for background has been made. The fluorescence intensity (minus background) of each plasma sample is then recorded, and the concentration of corticosterone is computed from the standard curve.

The recovery of corticosterone from aqueous solutions is virtually 100% by this method, while the sensitivity is 5 ng per milliliter of fluorescence reagent or 15 ng per sample. The coefficient of variation is 4%.

[29] A New Approach to the Structure–Activity Relationship for ACTH Analogs Using Isolated Adrenal Cortex Cells[1]

By STEVEN SEELIG,[2] BARRY D. LINDLEY, and GEORGE SAYERS

Suspensions of isolated adrenal cortex cells of the rat produce cyclic AMP and corticosterone upon addition of ACTH and related peptides. The cells respond with a high degree of sensitivity, specificity, and reproducibility. The technique offers an important new tool for the quantitative analysis of the relation between structure and biological activity (SAR). Estimates of the maximum response induced by the analog and of the apparent dissociation constant of hormone receptor interaction are derived from complete log concentration response curves for the natural ACTH and its analogs. We here describe approaches to curve fitting and model building designed to delineate "active centers" and "affinity centers" of the ACTH molecule. We believe the methods and models pre-

[1] This work was supported by National Science Foundation Grant GB-27426 (G.S.) and United States Public Health Service Grants 5 T01 GM-00899 (S.S.) and NS-23016 (B.D.L.).

[2] Predoctoral Trainee.

sented are of general applicability to the study of hormonal peptides and their analogs.

Preparation and Incubation of Suspensions of Isolated Adrenal Cortex Cells of the Rat: Analyses of Cyclic AMP and of Corticosterone

The techniques and methods have been described in detail in another volume of *Methods in Enzymology*.[3] Briefly, 40 rat adrenals are excised and decapsulated. The capsular tissue consists mainly of glomerulosa tissue and may be employed for the preparation of isolated cells which secrete aldosterone, a subject of separate study. The observations described in this chapter are confined exclusively to the core tissue of the adrenal which consists of fasciculata, reticularis, and medulla. Medullary cells of the rat do not survive the dispersion process[4]; the suspensions consist almost exclusively of corticosterone-secreting cells of the two inner zones of the cortex.

The cores are quartered and added to dispersion medium. Thirty microcuries of [8-^{14}C]adenine are added and are incorporated into ATP during dispersion. After dispersion, the cells are centrifuged at 300 *g* for 10 minutes at room temperature. The pellet is resuspended in volumes of 60–80 ml of Krebs-Ringer bicarbonate buffer containing calcium, 7.65 m*M*, glucose, 0.2%, bovine serum albumin, 0.5%, and lima bean trypsin inhibitor, 0.1%. Aliquots of the suspension, 0.9 ml in volume, together with 0.1 ml of vehicle or with 0.1 ml of vehicle to which has been added ACTH, are incubated in an atmosphere of 95% O_2–5% CO_2 at 37°. At the end of 60 minutes of incubation, methylene chloride is added; an aliquot of the methylene chloride phase is analyzed for corticosterone and an aliquot of the aqueous phase is analyzed for cyclic [8-^{14}C]AMP. In all cases, results are based on net cyclic AMP production and net corticosterone production; quantities of cyclic AMP and of corticosterone in aliquots to which no ACTH has been added (blanks) have been subtracted.

Since the maximum response and the concentration of hormone necessary to induce one-half the maximum response varied among suspensions of cells, synthetic $ACTH_{1-24}$ was employed as a reference compound. $ACTH_{1-24}$ induces the same maximum response as natural porcine $ACTH_{1-39}$ and is about 1.4 times as potent as the natural hormone on a molar basis. The synthetic peptides used in these studies were obtained from Professor R. Schwyzer, Swiss Federal Institute of Technology,

[3] G. Sayers, R. J. Beall, S. Seelig, and K. Cummins, this series, Vol. 32 [68].

[4] S. Malamed, G. Sayers, and R. L. Swallow, *Z. Zellforsch. Mikrosk Anat.* **107**, 447 (1970).

Zurich, Switzerland, Dr. W. Rittel, Ciba–Geigy Limited, Basel, Switzerland, and Dr. H. Strade, Organon Ltd., West Orange, New Jersey.

Analysis of Concentration Response Curves

Concentration of ACTH, A, and the biological response of isolated adrenal cortex cells, V, are related by the equation for a rectangular hyperbola

$$V = (V_{\max} A)/(A + A_{50}) \tag{1}$$

where V is either the rate of cyclic AMP production or the rate of corticosterone production, $V_{\max}$ is either the maximum rate of cyclic AMP production ($cAMP_{\max}$) or the maximum rate of corticosterone production ($B_{\max}$), and A_{50} is the concentration of ACTH required to induce $\frac{1}{2}$ $V_{\max}$.

One method to estimate the value of the constants A_{50} and $V_{\max}$ of the rectangular hyperbola is to transform Eq. (1) into a linear form The transformation, however, may introduce unequal weighting of data points.[5,6] Another approach which does not introduce undue weighting is a nonlinear least-squares fit of the data by the Newton-Raphson method based on linearization by a Taylor's series expansion of Eq. (1).[7–9] In brief, the nonlinear method involves initial estimation of A_{50} and $V_{\max}$ by a linear transformation and then refinement of these estimates by the nonlinear method through an iterative procedure designed to minimize the sum of squares of deviations of the observations from the values predicted by Eq. (1).

Initial estimates of A_{50} and $V_{\max}$ are determined by a linear least-squares fit to the equation

$$A_i/V_i = (A_i + A_{50}^{\circ})/V_{\max}^{\circ} \tag{2}$$

where A_i and V_i are the ith values, and A_{50}° and $V_{\max}^{\circ}$ are the initial estimates of the parameters, obtained by well-known methods for linear regression as follows.

$$V_{\max}^{\circ} = \frac{\Sigma A_i^2 - (\Sigma A_i)^2/N}{\Sigma(A_i^2/V_i) - [\Sigma A_i \, \Sigma(A_i/V_i)]/N}$$

$$A_{50}^{\circ} = \frac{V_{\max}^{\circ} \, \Sigma(A_i/V_i) - \Sigma A_i}{N}$$

N is the number of values of V_i, and the summation is over all values of i.

[5] G. N. Wilkinson, *Biochem. J.* **80,** 324 (1961).
[6] J. E. Dowd and D. S. Riggs, *J. Biol. Chem.* **240,** 863 (1965).
[7] G. E. P. Box, *Ann. N.Y. Acad. Sci.* **86,** 792 (1960).
[8] B. D. Lindley, G. E. Bartsch, and B. J. Eberle, *Math. Biosci.* **1,** 515 (1967).
[9] W. W. Cleland, *Advan. Enzymol.* **29,** 1 (1967).

Refinement of the initial estimates involves the linearization of Eq. (1) by means of a Taylor's series expansion

$$V_i \cong f_i + (A_{50} - A_{50}^\circ) f_{A_{50}} + (V_{\max} - V_{\max}^\circ) f_{V_{\max}} \tag{3}$$

where V_i is the observed response at some value A_i, f_i is the value predicted for V_i at A_i from inserting the initial estimates in Eq. (1), and $f_{A_{50}}$ and $f_{V_{\max}}$ are the partial derivatives of the right-hand side of Eq. (1), evaluated for the initial estimates of the parameters and A_i.

$$f_{A_{50}} = -(A_i V_{\max}^\circ)/(A_i + A_{50}^\circ)^2$$
$$f_{V_{\max}} = A_i/(A_i + A_{50})$$

The least-squares fit to Eq. (3) will determine new values for A_{50} and $V_{\max}$.

Minimization of the sum of squares of the deviations of the observed values from those predicted yields two simultaneous equations for the values of the parameters; these equations can be solved by any convenient method to give increments for the parameters.

$$A_{50} - A_{50}^\circ = (1/D)\{\Sigma f_{V_{\max}}^2 [\Sigma(f_{A_{50}}(V_i - f_i))] - [\Sigma(f_{A_{50}} f_{V_{\max}}) \Sigma(f_{V_{\max}}(V_i - f_i))]\}$$
$$V_{\max} - V_{\max}^\circ = (1/D)\{\Sigma f_{A_{50}}^2 [\Sigma(f_{V_{\max}}(V_i - f_i))] - [\Sigma(f_{A_{50}} f_{V_{\max}}) \Sigma(f_{A_{50}}(V_i - f_i))]\}$$
$$D = \Sigma f_{V_{\max}}^2 \Sigma f_{A_{50}}^2 - [\Sigma(f_{A_{50}} f_{V_{\max}})]^2$$

The refined estimates of A_{50} and $V_{\max}$ are obtained by adding the increments to the initial estimates. These refined estimates then replace the initial estimates in Eq. (3), and the process is repeated to give further refined estimates of A_{50} and $V_{\max}$. This process is iterated until the difference between the new estimate and the previous estimate is less than 0.1% of the previous estimate.[10]

The residual variance S^2, is $\Sigma(V_i - f_i)^2/(N - 2)$. The standard errors of A_{50} and $V_{\max}$ are $S[\Sigma(f_{V_{\max}})^2/D]^{1/2}$ and $S[\Sigma(f_{A_{50}})^2/D]^{1/2}$, respectively.

A comparison of the fit of typical experimental observations by three linear transformations and by the nonlinear method is presented in the table. As can be seen, the nonlinear method of fitting provides the "best" fit of the observations, as reflected by the smallest residual variance. An additional advantage of the nonlinear method is that it is applicable to equations which express response of isolated adrenal cortex cells to a com-

[10] It can be noted that the parameter $V_{\max}$ actually occurs linearly, and for a given estimate of A_{50} can be obtained without the iterative method. For more complicated problems this difference may be usefully exploited, but for this simple well-behaved instance we have preserved the symmetry of the procedure, which may be easily generalized to any 2-parameter problem.

COMPARISON OF THE FIT OF OBSERVATIONS BY NONLINEAR METHOD AND THREE LINEAR TRANSFORMATIONS

ACTH concentration[a]	Observed responses		Predicted responses: Nonlinear method		Linear transform[d]		Linear transform[e]		Linear transform[f]	
	cAMP[b]	B[c]	cAMP	B	cAMP	B	cAMP	B	cAMP	B
3.45×10^{-12}	2.94, 0.28[g]	0.039, 0.056	2.2	0.095	2.3	0.077	0.9	0.116	0.5	0.047
1.73×10^{-11}	1.2, 5.0	0.221, 0.307	10.6	0.321	11.3	0.275	4.7	0.363	2.6	0.205
3.45×10^{-11}	13.5, 12.3	0.541, 0.532	20.7	0.458	21.8	0.408	9.3	0.494	5.2	0.355
1.73×10^{-10}	106, 72	0.705, 0.653	88	0.696	88	0.663	43	0.694	28	0.860
3.45×10^{-10}	149, 186	0.792, 0.749	149	0.745	142	0.719	79	0.732	64	1.05
1.73×10^{-9}	303, 320	0.792, 0.800	329	0.788	277	0.772	241	0.765	−11851	1.266
3.45×10^{-9}	381, 402	0.775, 0.749	388	0.794	315	0.779	322	0.769	−485	1.300
1.73×10^{-8}	401, 472	—	453	—	354	—	443	—	−275	—
3.45×10^{-8}	459, 481	—	463	—	359	—	465	—	−261	—
1.73×10^{-7}	471, 497	—	471	—	363	—	484	—	−250	—
	A_{50} (molar)		7.53×10^{-10}	2.57×10^{-11}	5.42×10^{-10}	3.20×10^{-11}	1.78×10^{-9}	1.95×10^{-11}	-1.69×10^{-9}	9.54×10^{-11}
	V_{max} (cpm/60 min or μg/60 min)		473	0.800	365	0.786	489	0.774	−248	1.34
	Residual variance		404.7	0.0028	4890	0.0041	2462	0.0037	1.67×10^{8}	0.1099

[a] Molar concentration of $ACTH_{1-24}$.
[b] Cyclic [8-^{14}C]AMP accumulation in counts per minute for 60 minutes.
[c] Corticosterone production in micrograms per 60 minutes.
[d] This linear transformation is $V = V_{max} - A_{50} \times V/A$ where the slope is $-A_{50}$ and the intercept is V_{max}.
[e] This linear transformation is $A/V = A_{50}/V_{max} + A/V_{max}$ where the slope is $1/V_{max}$ and the intercept is A_{50}/V_{max} (Eadie).
[f] This linear transformation is $1/V = 1/V_{max} + A_{50}/V_{max} \times 1/A$ where the slope is A_{50}/V_{max} and the intercept is $1/V_{max}$ (Lineweaver-Burk).
[g] Observed responses for two aliquots of the isolated cell suspension.

bination of agonists and partial agonists and to a combination of agonists and competitive antagonists.[11,12]

Relation of Structure to Biological Activity

A simple concept has been generally applied to the interpretation of concentration response curves. In essence, a concentration response curve reflects the combined action of amino acids which excite the receptor and of amino acids which modify the affinity of the analog for the receptor. A decrease in V_{max} is interpreted to mean a decrease capacity to excite the receptor; an increase in A_{50} is interpreted to mean a decreased affinity for the receptor.[13]

$ACTH_{1-24}$, $ACTH_{1-16}(NH_2)$, $ACTH_{5-24}$, and $ACTH_{4-10}$ exhibit the same B_{max} (Fig. 1), and in accord with the simple concept contain the full complement of amino acids involved in excitation of the receptor. The "active center" is in the region $Met^4 \cdot Glu^5 \cdot His^6 \cdot Phe^7 \cdot Arg^8 \cdot Trp^9 \cdot Gly^{10}$. The reduced B_{max} of $[Trp(Nps)^9]ACTH_{1-39}$ suggests that tryptophan is a key amino acid in the "active center." The relatively large A_{50}'s for $ACTH_{1-16}(NH_2)$ and $ACTH_{5-24}$ indicate that the sequences $Arg^{17} \cdot Arg^{18} \cdot Pro^{19} \cdot Val^{20} \cdot Lys^{21} \cdot Val^{22} \cdot Tyr^{23} \cdot Pro^{24}$ and $Ser^1 \cdot Tyr^2 \cdot Ser^3 \cdot Met^4$ contain amino acids involved in affinity. The demonstration that $ACTH_{11-24}$ is a competitive antagonist supports the notion that the sequence of amino acids from Lys^{11} to Pro^{24} is exclusively involved in affinity.[14] The exceedingly large A_{50} for $ACTH_{4-10}$ is the result of the loss of a number of affinity sites.

The structural requirements for maximal cyclic AMP production differ from those for maximal steroidogenesis (Fig. 2). $ACTH_{1-24}$ and $ACTH_{1-16}(NH_2)$ exhibit the same $cAMP_{max}$; $ACTH_{5-24}$, a reduced $cAMP_{max}$; and $[Trp(Nps)^9]ACTH_{1-39}$, a markedly reduced $cAMP_{max}$. On the basis of the simple concept of SAR, $ACTH_{1-16}(NH_2)$ contains the full complement of amino acids involved in excitation of the receptor · adenylate cyclase complex. The cyclic AMP curve for $ACTH_{5-24}$, in contrast to the curve for corticosterone production, suggests that amino acids in the region $Ser^1 \cdot Tyr^2 \cdot Ser^3 \cdot Met^4$ are involved in excitation. Furthermore, derivatization of Trp^9 results in a dramatic loss in capacity to excite the receptor, whereas the loss in capacity to excite as revealed by corticosterone production is only slight.

[11] S. Seelig, S. Kumar, and G. Sayers, *Proc. Soc. Exp. Biol. Med.* **139,** 1217 (1972).
[12] S. Seelig and G. Sayers, *Arch. Biochem. Biophys.* **154,** 230 (1973).
[13] E. J. Ariëns and A. M. Simonis, *J. Pharm. Pharmacol.* **16,** 137 (1964).
[14] S. Seelig, G. Sayers, R. Schwyzer, and P. Schiller, *FEBS Lett.* **19,** 232 (1971).

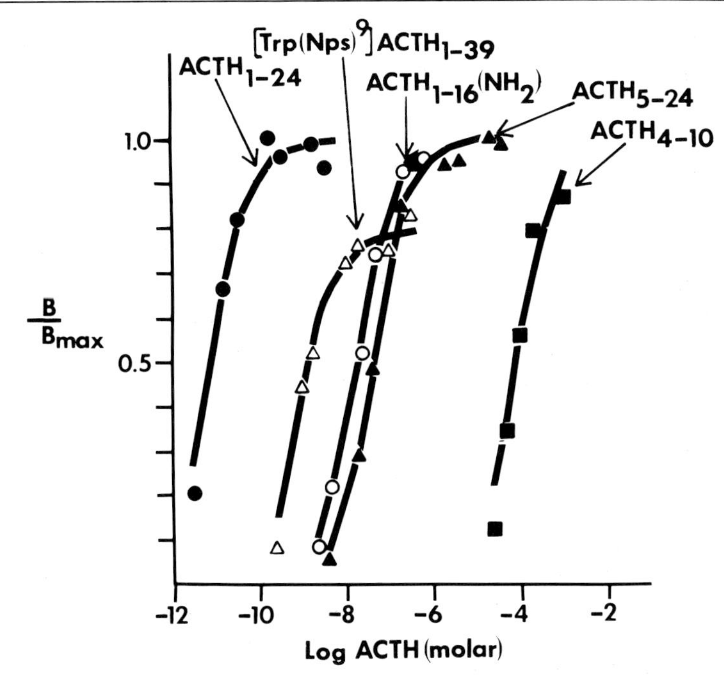

FIG. 1. Log concentration response curves for $ACTH_{1-24}$ and analogs. The abscissa is the log of the molar concentration of analog; the ordinate is corticosterone production (B) divided by the maximum corticosterone production induced by $ACTH_{1-24}$ (B_{max}). The points are the means of analyses on two aliquots of cell suspension; the lines represent nonlinear least-squares fits of corticosterone production.

The differences in structural requirement for cyclic AMP and for corticosterone production have important implications for the approach to the design of analogs with potent inhibitory properties (analogs with a high affinity for the receptor and low capacity to excite the receptor) and to the design of analogs with potent stimulatory properties (analogs with high affinity and/or capacity to excite the receptor). Some of these implications are discussed below.

Models

A *dual receptor* model can accommodate these observations. ACTH acts at a "B" receptor site which when excited stimulates steroidogenesis and at a "C" receptor site which when excited stimulates cyclic AMP production. Both receptor sites respond to $ACTH_{1-24}$, however, the affinity of the "B" receptor for $ACTH_{1-24}$ is about 40 times greater than

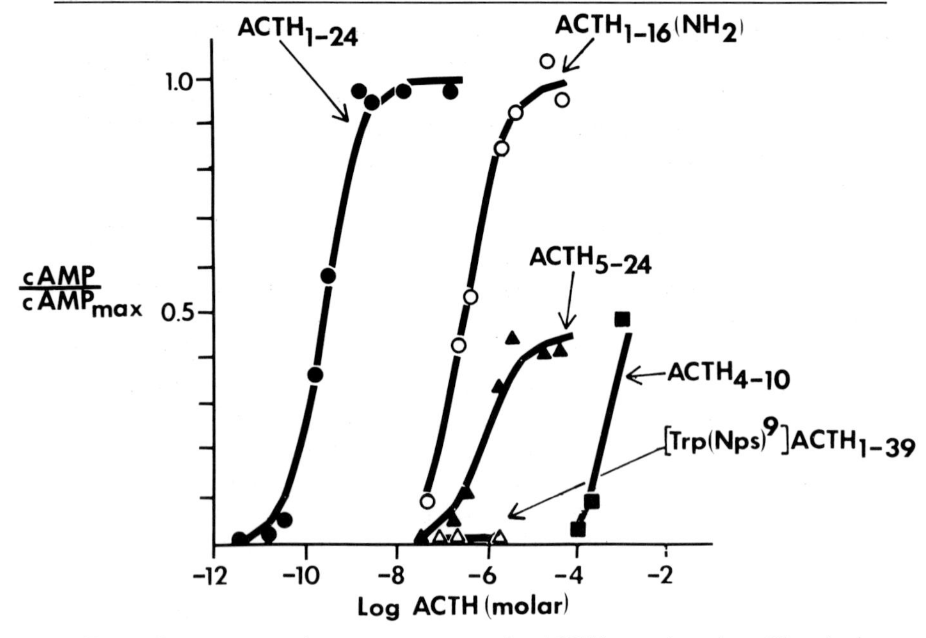

FIG. 2. Log concentration response curves for $ACTH_{1-24}$ and analogs. The abscissa is the log of the molar concentration of analog; the ordinate is cyclic AMP production (cAMP) divided by the maximum cyclic AMP production induced by $ACTH_{1-24}$ ($cAMP_{max}$). The points are the means of analyses on two aliquots of cell suspension; the lines represent nonlinear least-squares fits of cyclic AMP production.

the affinity of the "C" receptor for $ACTH_{1-24}$. This difference between the two sites in affinity for $ACTH_{1-24}$ is sufficiently large such that low concentrations of ACTH capable of inducing a significant increase in corticosterone production are not capable of inducing a measurable increase in cyclic AMP production. Further, the amino acids necessary for full activation of the "B" receptor differ from the amino acids necessary for full activation of the "C" receptor. Full activation of the "B" receptor requires the region $Glu^5 \cdot His^6 \cdot Phe^7 \cdot Arg^8 \cdot Trp^9 \cdot Gly^{10}$; full activation of the "C" receptor requires, in addition to the 5–10 sequence, Met^4 and possibly one or more of the amino acids in the sequence $Ser^1 \cdot Tyr^2 \cdot Ser^3$. Trp^9 is an important amino acid for both receptors. The activation of the "C" receptor is much more dependent on an unencumbered Trp^9 than the activation of the "B" receptor as revealed by characterization of $[Trp(Nps)^9]ACTH_{1-39}$.

According to the *dual receptor* model, the measured increase in cyclic AMP production in response to ACTH is not directly related to steroidogenesis. This implies that the design of analogs with potent inhibitory or stimulatory properties for steroidogenesis should be based on the relation between structure of ACTH and corticosterone production. Further, the dual receptor model implies that certain analogs may stimulate cyclic AMP production but not corticosterone production.

A *single receptor site-receptor reserve* model can also accommodate these observations. ACTH acts at a single receptor site which when excited, stimulates cyclic AMP production. The cyclic AMP produced, acting as an obligatory intermediate, stimulates corticosterone production. The concept of receptor reserve, developed by Stephenson,[15] by Ariëns,[16] and by Furchgott[17,18] accounts for the complex relations among V_{max}'s and among apparent dissociation constants estimated from cyclic AMP and corticosterone production for $ACTH_{1-24}$ and analogs.

Receptor reserve is suggested by the observation that a rate of cyclic AMP production equal to approximately 15% of $cAMP_{max}$ is sufficient for a near maximum rate of corticosterone production. Receptor reserve is reflected by rates of cyclic AMP production greater than 15% of $cAMP_{max}$. Further, a rate of cyclic AMP production equal to approximately 1% of $cAMP_{max}$ is sufficient to induce a one-half maximum rate of corticosterone production.

The receptor reserve model may be expressed by the series combination,

$$cAMP = (e \times cAMP_{max} \times A)/(A + A_{50C}) \quad (4)$$

and

$$B = [(B_{max}) \times cAMP]/(cAMP + k_c) \quad (5)$$

where cAMP is the rate of cyclic AMP production, B is the rate of corticosterone production, $cAMP_{max}$ is the maximum rate of cyclic AMP production induced by $ACTH_{1-24}$, B_{max} is the maximum rate of corticosterone production, A is the concentration of ACTH, A_{50C} is the concentration of ACTH required to induce $\frac{1}{2}\, e \times cAMP_{max}$, and k_c is the rate of cyclic AMP production associated with $\frac{1}{2}\, B_{max}$. To handle analogs which have maximum rates of cyclic AMP production less than $cAMP_{max}$, e is inserted into Eq. (4) and is the intrinsic activity of the analog as determined by the ratio of the maximum cyclic AMP production for the analog to $cAMP_{max}$. The two equations adequately express the relation between concentration of ACTH and cyclic AMP production (Fig. 2) and the relation between cyclic AMP production and corticosterone production (Fig. 3). A key assumption is that Eq. (5) is valid

[15] R. P. Stephenson, *Brit. J. Pharmacol.* **11**, 379 (1956).
[16] E. J. Ariëns, *Advan. Drug Res.* **3**, 235 (1966).
[17] R. F. Furchgott, *Annu. Rev. Pharmacol.* **4**, 21 (1964).
[18] R. F. Furchgott, *Advan. Drug Res.* **3**, 21 (1966).

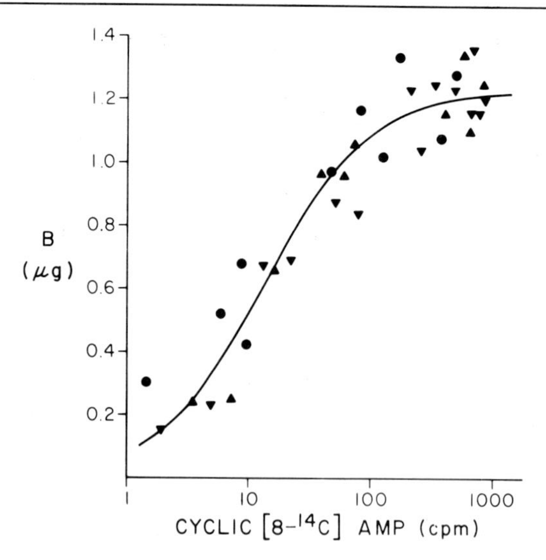

FIG. 3. The relation between observed cyclic AMP production and observed corticosterone production for $ACTH_{1-24}$ (●), $ACTH_{5-24}$ (▲), and carbobenzoxy-$ACTH_{5-24}$ (▼). The points represent cyclic AMP and corticosterone production determined on the same aliquot of cell suspension. The line represents the nonlinear least-squares fit of the points to Eq. (5).

at low concentrations of ACTH where corticosterone production is significant but cyclic AMP accumulation shows only a slight or no measurable change. Whether the model faithfully reflects the actual processes involved must await further developments in methodology designed to test the thesis that cyclic AMP is an obligatory intermediate. Nevertheless, the model serves as the basis for the analysis of the relation between structure and biological activity. Each analog is characterized by an A_{50C} and an e value.

The implications of the *single receptor site-receptor reserve* model are best understood by the combination of Eq. (4) and Eq. (5) which follows:

$$\frac{B}{B_{\max}} = \frac{A}{A\left[1 + \dfrac{k_c}{e \times \mathrm{cAMP}_{\max}}\right] + \dfrac{A_{50C} \cdot k_c}{e \times \mathrm{cAMP}_{\max}}} \tag{6}$$

The effect of e on concentration response curves for cyclic AMP and corticosterone production are shown in Fig. 4. A decrease in e results in an increase in the concentration of analog required to induce one-half the maximum rate of corticosterone production (A_{50B}). The term $A_{50C} \times k_c/e \times \mathrm{cAMP}_{\max}$ of Eq. (6) corresponds to A_{50B} when $k_c/\mathrm{cAMP}_{\max}$ is small and the term $[1 + k_c/e \cdot \mathrm{cAMP}_{\max}]$ is virtually one. The apparent

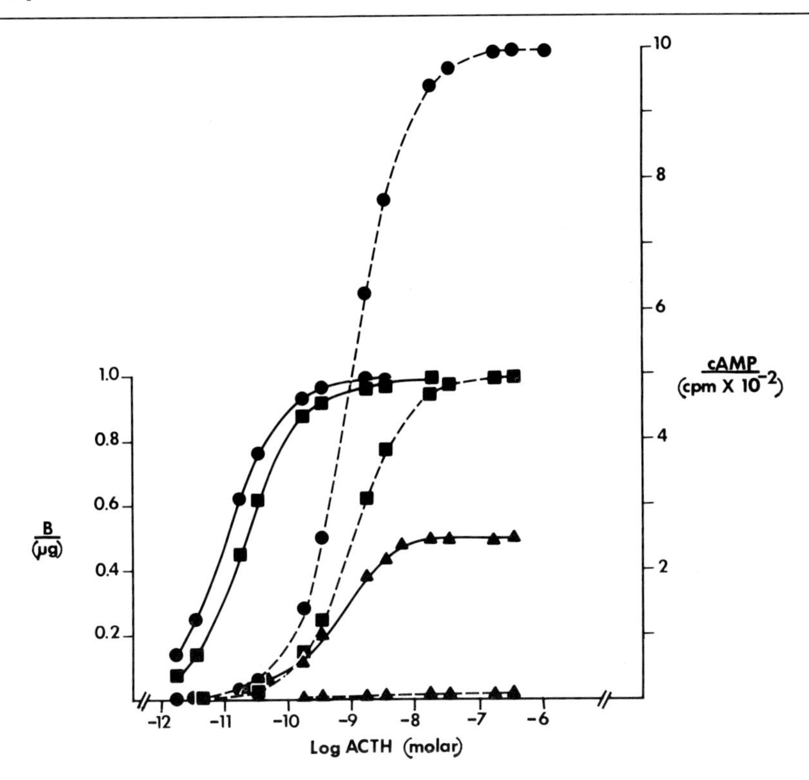

FIG. 4. Log concentration response curves for cyclic AMP production (----) and for corticosterone production (——) generated from Eq. (4) and Eq. (5). The abscissa is log of the molar concentrations of ACTH, the left-hand ordinate is micrograms of corticosterone production per 60 minutes (——) and the right-hand ordinate is counts per minute of cyclic AMP per 60 minutes (----). For the construction of these curves, A_{50C} and $cAMP_{max}$ of Eq. (4) were assigned values of 1.03×10^{-9} M and 1000 cpm per 60 minutes respectively; k_c and B_{max} of Eq. (5), 10 cpm and 1.0 μg per 60 minutes, respectively; e was assigned values of 1.0 (●), 0.5 (■), and 0.01 (▲).

dissociation constant of an analog as revealed by corticosterone production is determined not only by A_{50C}, the apparent dissociation constant of the analog as revealed by cyclic AMP, but also by e, the capacity of the analog to induce $cAMP_{max}$ and by the ratio $k_c/cAMP_{max}$, a ratio that reflects the status of the events post cyclic AMP responsible for corticosterone production. Furthermore, a complex relation exists between a decrease in e and a decrease in the maximum corticosterone production for an analog. The value of e may be decreased to a value of 0.5 without an experimentally detectable decrease in the maximum corticosterone production for an analog. When the value of e is decreased to 0.01, the maximum rate of corticosterone production induced by the analog divided by B_{max} is reduced to 0.5. B/B_{max} of Eq. (6) approaches the intrinsic activity of an analog as the concentration of peptide increases. The limit equals $1/(1 + k_c/e \times cAMP_{max})$. Maximum corticosterone production is determined not only by e but also by $k_c/cAMP_{max}$.

The single receptor site-receptor reserve model indicates that the apparent dissociation constant of an analog as determined by a biological response is in part an expression of the "true" dissociation constant for analog receptor interaction and the capacity of the analog to excite the receptor. The "true" dissociation constant and the apparent dissociation constant would be equal only under highly special experimental conditions. Likewise, changes in capacity of an analog to excite the receptor may not be revealed by changes in the maximum biological response of the analog.

The concentration response curves shown in Fig. 4 simulate remarkably well the observed curves for $ACTH_{1-24}$ ($e = 1$), $ACTH_{5-24}$ ($e = 0.45$) and $[Trp(Nps)^9]ACTH_{1-39}$ ($e = 0.01$) (Figs. 1 and 2). The observed shift of the concentration response for $ACTH_{5-24}$ (Fig. 1) is greater than can be accounted for by a reduction of e from 1.0 to 0.45 (Fig. 4). For this reason, we suggest that the amino acids in the region $Ser^1 \cdot Tyr^2 \cdot Ser^3 \cdot Met^4$ are important for both activation and affinity. Modification of Trp^9 with *o*-nitrophenylsulfenyl markedly reduced the capacity of the hormone to excite the receptor and this reduction can by itself account for the relatively high apparent dissociation constant, A_{50B}, for this derivative. These considerations emphasize the limitations of a simplistic approach to SAR based solely on concentration response curves for steroidogenesis.

Concluding Remarks

Suspensions of isolated adrenal cortex cells respond to the addition of ACTH with increased production of cyclic AMP and of corticosterone. Complete concentration response curves for cyclic AMP and corticosterone production may be constructed for a variety of ACTH analogs. Estimates of a V_{max} and an A_{50} for each analog are determined from these concentration response curves by a nonlinear method with the aid of a computer. With this approach to the study of the relation between structure of ACTH and biological activity we have been able to demonstrate that the structural requirements for maximum cyclic AMP production differ from the structural requirements for maximum corticosterone production. These differences are accommodated by a *dual receptor* model and by a *single receptor site–receptor reserve* model. The single receptor site-receptor reserve model emphasizes the complexity of factors which enter into the determination of V_{max} and of apparent dissociation constants for ACTH and its analogs. Specifically, characterization by steroidogenesis alone may be misleading in the delineation of "active centers" and "affinity centers" of the polypeptide hormone.

[30] Perfusion of the Thyroid Gland[1]

By JUDAH FOLKMAN and MICHAEL A. GIMBRONE, JR.

Many methods for perfusing the isolated thyroid gland have been described since Carrel[2] first perfused chicken thyroid. Several of these methods have been tried in our laboratory. However, the best functional and histologic preservation was obtained with the following technique, which has been used to perfuse more than 100 thyroid glands from rabbit and dog.[3-5]

Perfusion Circuit

Organ Chamber. A sterile circuit is assembled as in Fig. 1. The organ chamber is made from a glass cylinder, 37 mm i.d., 41 mm o.d., and 140 mm long. It has a single outlet of 5 mm o.d., that will fit snugly into gum rubber tubing (1/8 × 3/64 inch). This tubing is used throughout the circuit. The outlet is conical, similar to a glass centrifuge tube. It is etched with volumetric graduations so that flow rate can be measured. A No. 8 silicone rubber stopper is pierced by a short length of glass tube (5 mm o.d.). The arterial cannula is attached to this tube just before the thyroid gland is placed within the chamber. A hypodermic needle is passed through the stopper to serve as an air vent. The arterial cannula is made from glass tubing 5 mm o.d. (Fig. 2). One end is narrowed and beveled so that it will fit easily into the superior thyroid artery. It is

[1] This work was supported by a grant from the National Cancer Institute (No. 1 RO1 CA 14019-01), a grant from the American Cancer Society (No. DT-2A), and gifts from Alza Corporation, Merck Co., and Mr. Morton Bank.

[2] A. Carrel and C. A. Lindberg, "The Culture of Organs." Harper (Hoeber), New York, 1938.

[3] J. Folkman, P. Cole, and S. Zimmerman, *Ann. Surg.* **164**, 491 (1966).

[4] M. A. Gimbrone, R. A. Aster, R. S. Cotran, J. Corkery, J. Jandl, and J. Folkman, *Nature (London)* **222**, 33 (1969).

[5] J. Folkman, and M. A. Gimbrone, Perfusion of the Thyroid, *in* "Fourth Karolinska Symposium on Research Methods in Reproductive Endocrinology" (E. Diczfalusy, ed.). Karolinska Inst., Stockholm, 1972.

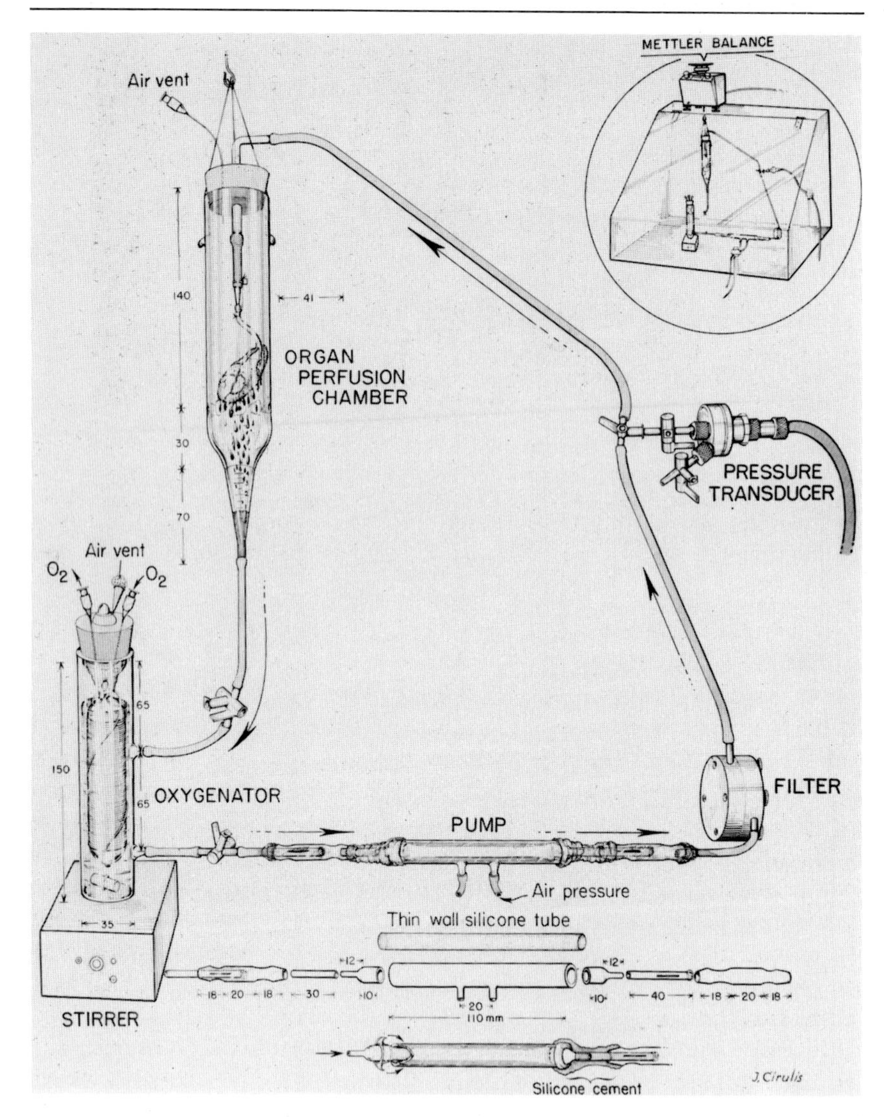

FIG. 1. Diagram of perfusion circuit. Scale 1:4.5. Circuit is housed in a plexiglass cabinet which serves as an incubator and which supports the Mettler balance (see inset).

convenient to have a variety of cannulas with tips of different sizes when the thyroid gland is being dissected. These should range from 0.5 mm up to 2.0 mm. A cannula tip of about 1.5 mm is usually satisfactory. Three small glass hooks are annealed to the outside of the organ chamber so that it can be suspended from the undersurface of a Mettler balance.

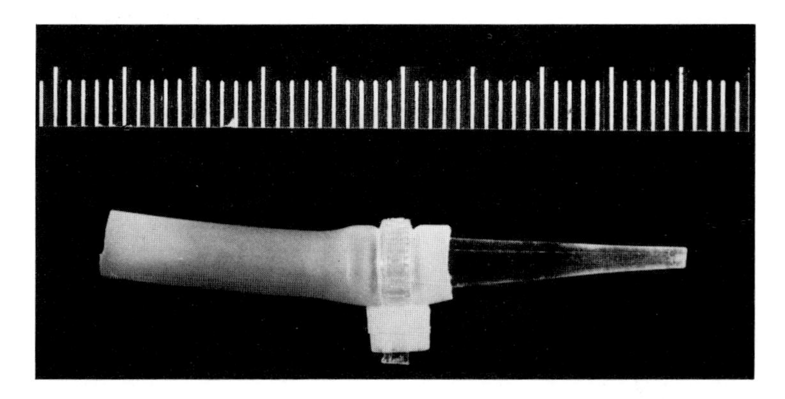

FIG. 2. Glass cannula. Gum rubber tubing is held in place by Silastic adhesive and a crimped nylon band (Nylon Tie, SST-1 from Extracorporeal Medical Specialties, Inc. Pennsylvania). The nylon band serves as an anchor point around which silk ligatures can be tightened after they have been tied around the artery following cannulation. Each division = 1 mm.

Oxygenator. The outer cylinder of the oxygenator is 35 mm o.d., 32 mm i.d. and 150 mm long, with inlet and outlet sidearms of 5 mm o.d. A No. 7 silicone rubber stopper is fitted with a hollow glass core. A thin layer of Silastic[6] cement is applied to this core; the core is then wound with a coil of silicone rubber capillary tubing, 0.025 inch i.d., 0.037 inch o.d., and approximately 30 feet in length. Two No. 23 stainless steel hypodermic needles are driven through the stopper and inserted into the ends of the capillary tubing. The capillary tubes are cemented to the needles and large amounts of Silastic cement are applied to the needle shafts to prevent them from changing position. A mixture of 95% oxygen and 5% carbon dioxide is propelled through one of these needles and escapes from the other. The gas also diffuses through the capillary tubing and oxygenates the perfusate which flows over the coil. The gas is sterilized in this way and air bubbles are eliminated. The space between the coil and the wall of the oxygenator chamber is approximately 2 mm. A third hypodermic needle is passed through the stopper to serve as an air vent. It is capped with sterile cotton or gauze. A magnetic spinner bar is placed in the bottom of the oxygenator. The bar is turned slowly by a magnetic stirrer to increase efficiency of oxygenation. The outlet of the oxygenator is connected through a three-way stopcock, to a pulsatile pump.

Pump. A pulsatile pump is assembled as shown in Fig. 1. Thin wall silicone tubing 10 mm o.d., 9 mm i.d., is stretched through a 110 mm

[6] Silastic Medical Adhesive, Silicone Type A, Dow Corning Corporation, Midland, Michigan.

glass cylinder 14 mm o.d. and 12 mm i.d. The ends of the thin tubing are turned back upon the wall of the glass cylinder and cemented in place. Valves for unidirectional flow are then cemented into each end of the pumping cylinder. These valves are made from silicone tubing 5 mm o.d., 3 mm i.d. and 50 mm long. One end is sealed with Silastic cement and a ½-inch slit is made with a razor in the side wall nearest the sealed end. This end is then inserted into a 60-mm length of glass tubing of 8 mm i.d., the opening of which has been narrowed so that the silicone valve will fit snugly. The valve is cemented in a position that leaves the valve orifice just within the glass tube. The pump chamber is attached to a Harvard small animal respirator[7] with a single length of thick walled Tygon tubing, ¼ inch i.d. The extra sidearm on the pump chamber is occluded, but can be opened slightly with a screw clamp, should a larger respirator be used with a stroke volume that exceeds the volume of the pump chamber. The outflow of the pump is connected by a short length of gum rubber tubing to a filter.

Filter. A filter holder is made of two nylon disks, each 2 inches in diameter and 0.5 inch thick. A shallow well of 0.5 ml is made in each disk. Two layers of unweighted silk mesh are compressed between the two nylon disks. The filter removes cellular debris and platelet aggregates from the circulation.

Pressure Transducer. A transducer head is inserted in the circuit between the filter and the arterial cannula. It is connected to an electronic recorder[8] which measures pressures from 50–150 mm Hg.

All components which touch the perfusate are siliconized.[9] The entire circuit is maintained at 35° in a plexiglass thermostatic cabinet.

Perfusate

Several types of perfusate may be used including tissue culture solutions such as Eagle's medium with 10% calf serum or hemoglobin solutions.[3] However, the best thyroid function and the least edema was obtained by Gimbrone in our laboratory, who employed heparinized, autologous platelet-rich plasma.[4]

This perfusate is obtained by plasmaphoresis of fasting animals 2 hours before thyroidectomy. Large (50 lb) dogs with normal erythrocyte sedimentation rates and platelet counts are used as donors. Under light Surital anesthesia, blood is rapidly collected from a femoral artery can-

[7] Harvard Apparatus Company, Dover, Massachusetts.

[8] Gulton Medical Instruments, Willow Grove, Pennsylvania.

[9] Siliclad, Clay-Adams Co., New York.

nula into a plastic Fenwall bag[10] containing sufficient heparin so that the final concentration will be 2.5 mg/ml. An average volume of 600 ml is collected, representing about one-third of the donor's blood volume. An equivalent amount of Ringers lactate solution is then infused intravenously. Immediately after separation of the platelet-rich plasma by centrifugation, the packed red cells are reinfused into the donor animal. Platelet-rich plasma is prepared by slow centrifugation of blood (1000 rpm = 253 *g*) for 10 minutes at 20° in a bucket-head International Centrifuge. Sterile techniques are used throughout, hemolysis and foaming are avoided, and the plasma is allowed to come into contact with only nonwettable plastic or siliconized surfaces. Platelet counts by phase microscopy routinely vary between 150 and 200 $\times$ $10^3/mm^3$ when plasma is prepared in this way. About 70 ml is needed to prime the perfusion circuit. The remaining plasma is stored at room temperature in plastic bags for later use.

Thyroid Gland

A careful anatomical dissection is carried out in the donor dog. A thyroid lobe attached to its respective superior thyroid artery pedicle is removed. Both thyroid lobes may be removed if two perfusion circuits are available. All veins are allowed to bleed freely, but tiny arterial branches severed during the dissection, are ligated so that normal arterial pressures can be maintained within the gland. Immediately preceding removal, the donor animal is systemically heparinized (2 mg/kg). A glass cannula is then inserted into the superior thyroid artery or into the carotid artery near the branch of the superior thyroid artery, and fixed in position with silk ligatures. The gland is placed in the organ chamber and the cannula attached to the glass inlet tube. The initial venous effluents are contaminated with blood and are discarded before they enter the oxygenator. After the venous effluent has become clear (about 10–15 minutes), it is allowed to enter the oxygenator and recirculate.

Management of Perfusion

The stroke volume is initially adjusted to maintain a systolic pressure of 100–120 mm Hg, but thereafter, both rate and stroke volume are kept constant so that changes in perfusion pressure and flow rate will reflect changes of vascular resistance within the perfused lobe. The flow rate is measured periodically by occluding the outlet of the organ chamber and measuring the time required for 1 ml to accumulate in the conical

[10] Fenwall Transfer Pack, Code No. TA-1.

end of the chamber. During technically satisfactory perfusions, flow rates range between 0.75 and 3.00 ml per gram of tissue per minute. Oxygen tensions of perfusate sampled distal to the filter, average 180 mm Hg (pO_2) and the pH of venous effluents remains within the range 7.30 to 7.55. The oxygenator chamber is drained and refilled with 50 ml fresh plasma every 12 hours. Perfusions can be continued for 5–7 days.

Studies of Function and Morphology during Perfusion

Donor animals were injected with 400 μCi of ^{125}I (as NaI) 48 hours prior to thyroidectomy. This allowed the monitoring of TCA-precipitable and butanol-extractable ^{125}I in the perfusate. In other perfusions, ^{125}I was injected *in vitro* and the percent of iodine incorporated into triiodothyronine was studied.

Mean vascular resistance was estimated as the ratio of effluent flow rate to mean systolic pressure. The increase in mean vascular resistance, expressed as a percentage of initial, averaged 26% after 5 hours of perfusion. After 5 hours thyroid glands perfused with platelet-rich plasma gained an average of 15% in overall weight. All other types of perfusate caused significantly more rapid weight gain. For example when platelet-poor plasma was used, the average weight gain was 41% at 5 hours. Serial histological studies were carried out by both light and electron microscopy. In general, the capillary endothelial cells showed normal preservation up to 5 hours and beyond, the thyroid follicles up to 24 hours in the central portion of the gland and up to 3–5 days in the most peripheral portion of the gland. The best histological preservation was observed in the parathyroid glands which had a normal appearance even after 1 week of perfusion. In other experiments[3,5] tumors implanted into these glands during perfusion, grew well up to diameters of 1–2 mm and then remained alive but dormant and unvascularized.

[31] The *in Vitro* Perfused Pancreas

By GEROLD M. GRODSKY and RUDOLPH E. FANSKA

A technique for removing rat pancreas and maintaining it in a perfusion apparatus was briefly described by Anderson and Long in 1947.[1] Subsequently the technique was modified and used for studying the influ-

[1] E. Anderson and J. A. Long, *Endocrinology* **40,** 92 (1947).

ence of carbohydrates on insulin release.[2] At that time perfusate, containing 20% whole rat blood as the oxygen carrier, was recirculated through the pancreas in a closed system. Closed perfusion systems have the advantage of requiring small amounts of perfusate; however, the continuous buildup of excretion products from the tissue may influence normal tissue activity. This is particularly true in the perfused pancreas, where secretions from the Δ cells can inhibit insulin release.[3] Furthermore, the use of blood cells requires special handling and pumping systems to minimize hemolysis. We showed previously[2] that hemolyzed perfusate rapidly destroys secreted insulin; if not prevented, this destruction precludes accurate evaluation of factors influencing secretion of this hormone.

More recently,[4,5] the pancreatic preparation has been perfused in an open system from which the complete venous effluent was collected after a single passage through the tissue. An open system permits accurate evaluation of transient changes in hormone secretion at fractions of minute intervals. This is particularly important since insulin secretion, even in the presence of a constant level of stimulating agent, is multiphasic.[4,5] The open system also allows the almost instantaneous addition *or deletion* of any particular substance in a variety of patterns (steps, staircases, ramps, etc.).[6] Finally, because of the brief exposure of perfusate to the pancreatic tissue there is less danger of metabolic degradation of stimulatory agents or secreted hormone. With minor modification, we have used the method for perfusing pancreases from mice and Chinese hamsters and for simultaneously measuring insulin and glucagon secretion.[7,8]

In the technique to be described, the adjacent stomach, spleen, and part of the duodenum are retained to minimize difficulties of dissection. Other workers[9,10] recently described systems in which a more isolated

[2] G. M. Grodsky, A. A. Batts, L. L. Bennett, C. Vcella, N. B. McWilliams, and D. Smith, *Amer. J. Physiol.* **205**, 638 (1963).

[3] A. Lernmark, B. Hellman, and H. G. Coore, *J. Endocrinol.* **43**, 371 (1969).

[4] G. M. Grodsky, L. L. Bennett, D. Smith, and K. Nemechek, *in* "Tolbutamide after Ten Years" (W. J. H. Butterfield and W. Westering, eds.), p. 11. Excerpta Med. Found., Amsterdam, 1967.

[5] D. L. Curry, L. L. Bennett, and G. M. Grodsky, *Endocrinology* **83**, 572 (1968).

[6] G. M. Grodsky, *J. Clin. Invest.* **51**, 2047 (1972).

[7] B. J. Frankel, J. E. Gerich, R. Hagura, R. E. Fanska, and G. M. Grodsky, *Clin. Res.* **21**, 273 (1973).

[8] J. E. Gerich, M. A. Charles, S. R. Levin, P. Forsham, and G. M. Grodsky, *J. Clin. Endocrinol. Metab.* **35**, 823 (1972).

[9] A. Loubatieres, *Acta Diabetologica Latina* **6**, Suppl. 1, 216 (1969).

[10] K. E. Sussman, M. Stjernholm, and G. D. Vaughan, *in* "Tolbutamide after Ten Years" (W. J. H. Butterfield and W. Westering, eds.), p. 22. Excerpta Med. Found., Amsterdam, 1967.

pancreas was perfused. Although such preparations appear preferable, they have proved more complicated and, to date, have not given results different from those obtained with the simpler perfusion procedure. A new technique for perifusing isolated islets in an open system has also been described.[11] This technique is useful because it permits measurement of metabolic changes in islet tissue during stimulation, but may be less physiologic in that it no longer possesses the normal vascular system of an organ perfusion.

General Principles and Techniques

Perfusion Media

Two perfusion media can be used which give comparable results. In one, human serum albumin (HSA), fraction V, at a concentration of 4% is the sole source of colloid. In the other, albumin is conserved by using a mixture of 1% HSA with 3% dextran (T-40, Pharmacia Fine Chemicals, Uppsala, Sweden, 40,000 daltons). For either buffer, HSA (about 25%) is dialyzed 4 days at 5° against distilled water; the water is changed twice daily.

One liter of perfusate is freshly prepared for each hour of perfusion. When perfusate containing 4% HSA is prepared, stock salt solutions[12] are added to albumin solution (about 12%) in the following order to achieve the concentrations expressed (mM) KCl, 4.4; $CaCl_2$, 2.1; KH_2PO_4, 1.5; $NaHCO_3$, 29; NaCl, 116; and $MgSO_4 \cdot 7H_2O$, 1.2. The pH is adjusted to 7.4 with 0.1 N HCl, and the mixture is diluted with water to obtain the proper final volume. In the second perfusion medium, sufficient HSA to achieve a final concentration of 1% is added to 30% of the final volume before salt solutions are added; this is useful to prevent precipitation of calcium complexes. An appropriate volume of 15% dextran in water is added, and the pH and volume adjustments are completed. Salt solutions remain the same except that calcium gluconate, 2.1 mM, is substituted for $CaCl_2$.

The perfusate is poured through filter paper and equilibrated for 10 minutes with 95% oxygen and 5% carbon dioxide in the gassing flask of the perfusion system (Fig. 1).

The Perfusion System (Fig. 1)

The level of perfusate in the gassing flask is maintained at a maximum height of 3.5 cm by gravity flow from the reservoir. Humidified

[11] P. E. Lacy, *Diabetes* **21**, 987 (1972).

[12] W. W. Umbreit, R. H. Burris, and J. F. Stauffer, *in* "Manometric Techniques," p. 149, Burgess, Minneapolis, Minnesota, 1957.

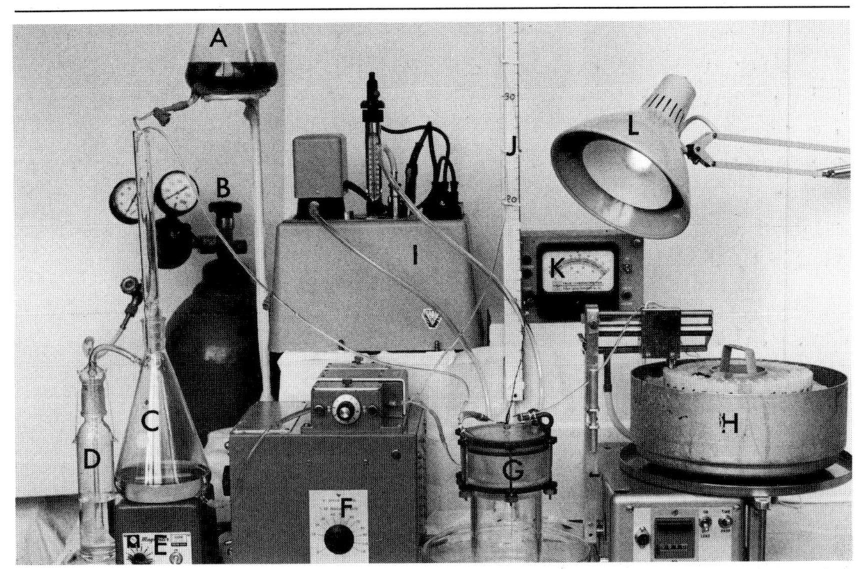

FIG. 1. Perfusion system. (A) Perfusate reservoir; (B) gas source; (C) gassing flask; (D) gas humidifier; (E) magnetic mixer; (F) peristaltic pump; (G) warming chamber; (H) fraction collector; (I) circulating water bath; (J) water manometer; (K) telethermometer; (L) incandescent lamp.

gas mixture is blown into the flask at 3 liters per minute while the perfusing solution is mixed magnetically. The gassed solution is driven toward the tissue in a pulsatile manner with a multispeed peristaltic pump (model 1202, Harvard Apparatus Co., Millis, Massachusetts). Polyvinyl chloride (PVC) tubing, ⅛ inch i.d. (860-06 intravenous set, Cutter Laboratories, Berkeley, California) is installed throughout the system but is interposed in several places. Tygon tubing, ⅛ inch i.d., is used in the peristaltic pump. A "T" joint between the pump and warming chamber allows vertical ascent of the perfusate, forming a water manometer to monitor pressure in the system; although a mercury manometer is adequate, we prefer the more sensitive water manometer particularly when vasoactive agents are perfused. A short section of ⅛ inch i.d. intravenous latex tubing is situated in the line before it enters the warming chamber for easy introduction of small volumes of stimulating substances via hypodermic needle. The warming chamber consists of a transparent plastic container, 11.5 cm in diameter, 6 cm in height, having a removable top sealed with an O ring and metal clamps. It also has inlet and outlet ports for water from a constant-temperature bath to be circulated at 38° (internal temperature of the rat). A hole is bored in the side of the chamber for entry of the PVC perfusion tubing; an additional hole in the top of the chamber allows its exit. Within the chamber, PVC tubing is intersected with a 1/16 inch i.d. glass coil having 4 rings, each 2 inches in diameter. (This immersion coil is sufficient to adjust perfusate temperature to 38° from less than 25° at flow rates even exceeding 10 ml per minute.) A "Y" joint on top of the chamber allows bypass or introduction of perfusate.

An incandescent lamp positioned above the chamber helps maintain external tissue temperature at about 37°. Tissue temperature is continuously monitored by a telethermometer (model 43 TD, Yellow Springs Instrument Co., Yellow Springs, Ohio). An arterial cannula may be fashioned from an 18-gauge hypodermic needle; the tip is slightly beveled and hipped, and it is held in position by a No. 12 copper wire. The venous cannula is a 16-gauge hypodermic needle with blunted point. This cannula is supported by a 1/2 inch diameter wire of pure lead, which is bent with ease to any desired position. Perfusate flow through the venous cannula is collected by an automatic fraction collector (Mini-Escargo Fractionator, Gilson Medical Electronics, Middlesex, Wisconsin) modified to include an ice bath. When large volumes of two different media are required, the system may be duplicated prior to the pump. Alternatively, agents being studied may be dissolved in perfusate and injected into the system by way of syringe and hypodermic needle, driven by an infusion pump (model 903, Harvard Apparatus Co., Millis, Massachusetts). If a ramp-type concentration gradient is desired, the agents can be pumped from a gradient mixer (LKB 8121, LKB-Produkter-Ab, Bromma, Sweden).

Dissection

The object of the dissection is to remove the stomach, spleen, pancreas, and duodenum in one block and to attach cannulas to the celiac axis and portal vein. This is usually completed by a trained operator in 10–12 minutes. Rats weighing 325–425 g, fed or fasted 18 hours, are anesthetized with sodium pentobarbital, 60 mg per kilogram body weight. With the rat supine and the tail pointing toward the operator, the skin is removed from the abdomen using Mayo dissecting scissors. Contents of the peritoneal cavity are exposed by making a midline incision from the pubis to the xiphoid process. Care is taken at this point to avoid producing a pneumothorax. Most of the xiphoid process is amputated, and the abdominal wall may be transected for greater exposure. Using eye dressing forceps and iris scissors, the fusion of duodenal and colonic mesenteries is separated cephalad to the superior mesenteric artery. The mesenteric artery is doubly ligated and severed between the ligatures.

Intestines are moved to the left side of the animal (to operator's right). The colon is freed from the dorsal peritoneal wall, doubly ligated and severed between ligatures. The upper ligature is not shortened but is used to reflect the proximal stump of the colon to the animal's right side. This stump serves as an axis for rotating the intestines in a clockwise direction, exposing the colon transversely across the abdomen. Some of the gastrosplenic portion of the pancreas is now directly below the colon. Pancreas is freed from colon by separating the mesentery which connects both tissues. The colon is held with the fingers directly above the site being cut. A cotton ball saturated with 0.9% saline is used to tease the pancreas down from the colon as the mesentery is severed. Any large interconnecting blood vessels are ligated. Colon, freed of pancreas, is moved to the left, and a single ligature is placed on the intestines. Intestines are severed and discarded. The stomach is freed of its connections to the liver. The esophagus is exposed, ligated above the gastric vein, and severed above the ligature.

Heparin solution (200 units) is injected into the inferior vena cava; a finger is held over the injection site to control bleeding while dissection continues. The aorta is freed of connective tissue from slightly above the origin of the renal arteries to the diaphragm. Subsequently, the aorta is ligated just above the origin of the renal artery and is elevated with the ligature. The aorta is clamped above the origin of the celiac axis with two Kelly-Murphy hemostatic forceps and is severed between these forceps (hypoxia time begins). It is clamped with a third forcep just above the ligature and is then severed between forceps and ligature. The rib cage is depressed and the liver is pulled cephalad over the sternum, exposing the portal vein, which is ligated as close to the liver as possible and severed on the hepatic side of the ligature. The pancreas, proximal duodenum, stomach, spleen, and section of aorta adjoining the patent celiac axis is removed by severing all remaining paravertebral connective tissues while pulling upon two of the hemostatic forceps previously clamped on the aorta.

The *en bloc* preparation is placed on the warming chamber of the perfusion system, and all forceps on the aorta are removed. Patency of the aortal section is restored with a probe and Dumont tweezers; the section is incised lengthwise opposite the celiac axis, which is cannulated to the inflow tubing. The portal vein is partially transected as close as possible to the ligature previously placed on it. Flow of perfusate is started through the tissues (hypoxia time ends) and the portal vein is tied to the outflow tubing using the shortest length of portal vein possible. The superior pancreaticoduodenal vein empties into the portal vein near the liver; if this vein is obstructed the pancreas will become edematous.

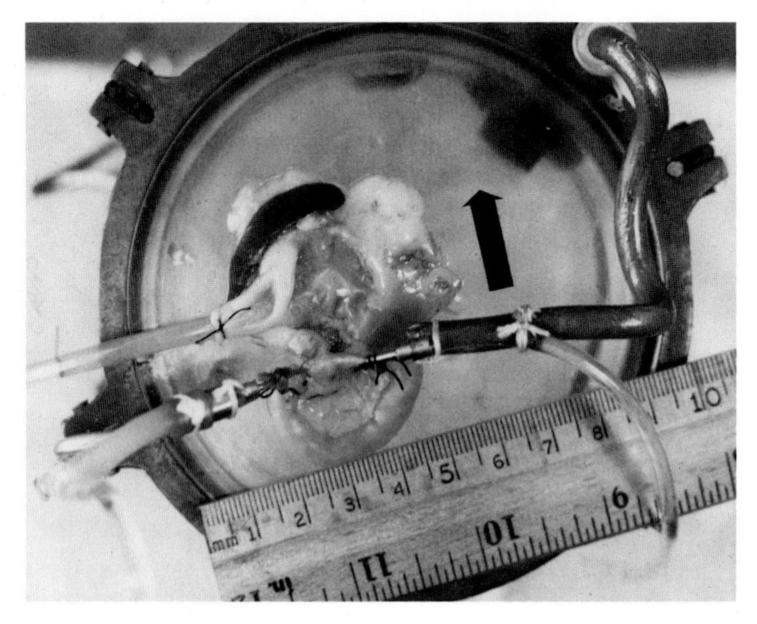

FIG. 2. Proper perfusing position of the *en bloc* preparation. Tissues are gently manipulated as shown by arrow to promote low pressure in the portal vein.

Hypoxia time is usually kept at a minimum of 3 minutes. External temperature of the tissue is maintained close to 37° since impaired insulin secretion has been demonstrated in pancreases at 25°.[13,14] The first 10 minutes of perfusion is considered an equilibration period. During this time a small hole is cut in the cardiac portion of the stomach. The stomach is flushed with warm 0.9% saline solution and a gastric drain is established. Temperature of the pancreas is stabilized, and tissues are positioned to register a low pressure on the manometer between 54 and 95 cm H_2O (Fig. 2). Total flow rate is maintained at 10 ml per minute, and arterial and venous oxygen tension is normally 410 and 175 mm Hg, respectively. Tissues are covered with moist gauze and Parafilm.

Special Considerations

In an open perfusion system, large amounts of albumin are required (up to 0.4 g per minute). The source of albumin is critical; we have found that crystalline preparations of bovine and many human albumins,

[13] D. L. Curry and K. P. Curry, *Endocrinology* **87,** 750 (1970).

[14] D. Baum and D. Porte, *J. Clin. Endocrinol. Metab.* **29,** 991 (1969).

including "Plasmanate" (Cutter Laboratories), cause serious edema detectable within 20–30 minutes of perfusion. The cause of this edema is unknown; albumins with identical viscosities and osmolal characteristics vary widely in their capacity to produce this phenomenon. Although moderate edema may not directly affect islet tissue, we have found impaired release (particularly in the first phase) associated with excessive edema. Albumin sources that have proved satisfactory include normal human serum albumin, 25% (Armour Pharmaceutical Co., Chicago, Illinois) and normal human serum albumin, salt-poor, USP (Cutter Laboratories Inc., Berkeley, California). However, supplies are limited and can be expensive. A satisfactory substitute is 1% albumin + 3% dextran. Pure dextran solutions usually cause edema within 30 minutes of perfusion or impaired insulin release.

Since neither whole blood nor an oxygen carrier is incorporated in either snythetic medium, sufficient oxygen for tissue respiration must be supplied by dissolved oxygen. Hypoxia inhibits insulin release and causes general breakdown of the tissues; thus adequate flow rate must be maintained to provide the tissue oxygen requirement. In this system,[5] oxygen consumption varies proportionally to flow rates up to 10 to 12 ml per minute; oxygen consumption above this flow rate was maximized. Therefore, rates of 10 ml per minute are recommended to ensure greatest oxygen consumption with minimal increase in pressure or unnecessary expenditure of perfusate. With smaller pancreatic preparations (Chinese hamster or mouse) the required flow rates are decreased, more or less, proportionally to the size of the tissue. Flow rates of 1 to 1.5 ml per minute are usually satisfactory with these tissues. At these lower rates one must exercise caution since all forms of synthetic tubing that we have studied permit a significantly rapid diffusion of oxygen. At slow flow rates, it is advisable to substitute glass tubing in the perfusate train wherever feasible.

One must always choose between maintaining constant pressure or flow rate in an organ perfusion system. We prefer to monitor pressure and maintain flow rate, which provides uniform samples for assay. Experiments are discarded if pressure exceeds 160 cm H_2O at any time.

Degradation of protein hormones, such as insulin or glucagon, does not appear to be important in open-perfusion systems, particularly when a synthetic perfusate is used. However, fractions should be collected over ice and promptly frozen to ensure stability of the secreted hormone. If glucagon is to be measured, degradation is reduced by adding 0.07 ml of 15% EDTA to the collecting tubes before collecting and freezing samples. In view of (1) unexplained variation in effectiveness of different albumin preparations (see section Perfusion Media); (2) complicated

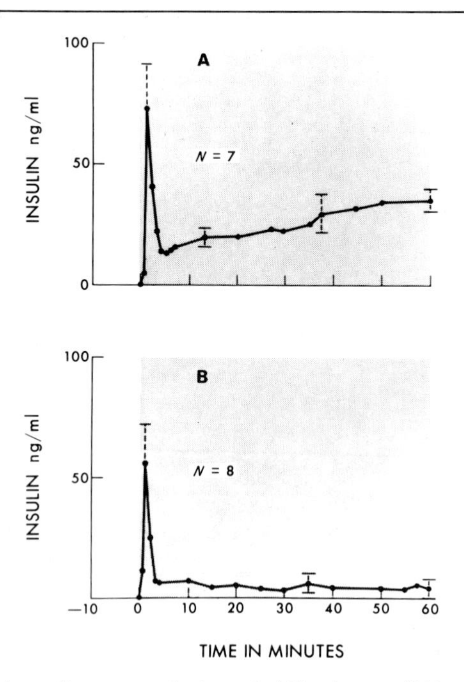

FIG. 3. Effect of continuous perfusion of (A) glucose (300 mg/100 ml) or (B) tolbutamide (300 μg/ml) on insulin secretion.

dissection procedures; (3) associated danger of periods of hypoxia, viability of the preparation should be assessed before novel experiments are performed. Some useful criteria include (1) maintenance of structural integrity based on electron microscopic examination of the tissues after 1–2 hours of perfusion;[15] (2) maintenance of constant oxygen uptake throughout the perfusion period; (3) unimpaired motility of stomach and duodenum; (4) production of the characteristic multiphasic secretion patterns of insulin caused by continuous stimulation with glucose or sulfonylureas (Fig. 3). These patterns were initially established with this preparation and have been confirmed often by others, both *in vitro*[9] and in man.[16] Duplication of patterns indicates the effective functional viability of the perfused pancreatic preparation.

[15] J. C. Lee, G. M. Grodsky, L. L. Bennett, D. F. Smith-Kyle, and L. Craw, *Diabetologia* **6**, 542 (1970).

[16] S. Porte and A. A. Pupo, *J. Clin. Invest.* **48**, 2309 (1969).

[32] Superfusion Techniques in the Study of Insulin Release

By ROBERT A. SHARP and IAN M. BURR

A number of systems have been developed for the study of the release of insulin *in vitro*.[1-6] However, none of these systems have been readily adaptable for the investigation of the dynamic aspects of insulin release. A simple technique for isolation of rat pancreatic islets and perifusion of rat pancreas pieces has been developed.[7-9] This section will describe the perifusion system and the conditions found necessary for optimum utilization.

Methods and Materials

Male Wistar rats weighing between 200 and 300 g are fasted overnight. The rats (3 per experiment) are lightly anesthetized with intraperitoneal Nembutal. The rats are then placed dorsally on a dissecting board (Eberbach Animal Board). The pancreas is distended *in situ* by injection of Hank's balanced salt solution through the common bile duct[10] as follows: The skin covering the abdominal area is removed so as to expose the abdominal wall, which is opened from below upward and laterally to the rib cage on either side. The abdominal wall is then retracted with two hemostats, exposing the abdominal cavity. A suture is placed under the common bile duct (CBD) as it exits from the liver, this suture is retracted superiorly. A second suture is placed from ¼ to ½ inch distally to this and retracted inferiorly. A slit is made in the CBD as superiorly as possible, followed by insertion of a 25-gauge butterfly infusion needle, which is tied in place by the superior suture. Then at the junction of the CBD and duodenum another suture is placed and

[1] P. R. Bowman, *Acta Endocrinol.* (*Copenhagen*) **35,** 560 (1960).
[2] H. G. Coore and P. J. Randle, *Biochem. J.* **93,** 66 (1964).
[3] W. Creutzfeldt, H. Freuchs, and V. Reich, *in* "The Structure and Metabolism of the Pancreatic Islets" S. E. Bolin, B. Hellman, and H. Knutson, eds.), p. 323. Pergamon, Oxford, 1964.
[4] P. E. Lacy and M. Kostianowsky, *Diabetes* **16,** 35 (1967).
[5] W. Malaisse, F. Malaisse-Lagae, and P. H. Wright, *Endocrinology* **80,** 99 (1967).
[6] S. Moskalevski, *Gen. Comp. Endocrinol.* **5,** 342 (1965).
[7] I. M. Burr, W. Stauffacher, L. Balant, A. E. Renold, and G. M. Grodsky, *Lancet* 882 (1969).
[8] I. M. Burr, W. Stauffacher, L. Balant, A. E. Renold, and G. M. Grodsky, *Acta Diabetologica Latina* **6,** 580 (1969).
[9] S. R. Levin, G. M. Grodsky, R. Hagura, and D. Smith, *Diabetes* **21,** 856 (1972).
[10] W. Montigue, and J. R. Cook, *Biochem. J.* **122,** 115–120 (1971).

tied. Cold Hanks' balanced salt solution (Grand Island Biological Co.) is injected slowly into the duct to distend the pancreas which is excised. After removal the pancreas is minced by fine dissecting scissors in a 10-ml beaker, to the particle size of 1–2 mm in diameter. Pancreata are placed into a 50 ml beaker, at which time excess cold Hanks' solution is added and the adipose tissue which floats to the top is removed by syringe suction. This remaining solution is poured into a 50-ml Erlenmeyer flask and 0.005 g (per 3 pancreata) of collagenase (dissolved in Hanks' solution) added to bring the volume up to 20 ml. The flask is placed in a metabolic shaker (Precision Scientific) set at 39° ± 2° at 200 rpm in an artificial atmosphere at 93% O_2–7% CO_2. Duration of digestion is determined according to the mean weight of the rats: 200–250 g, 18 minutes; 250–300 g, 20 minutes; and 300 g or more, 22 minutes plus. After digestion is complete, remove and sediment and tissue 3–5 times in cold Krebs-Ringer-bicarbonate buffer (KRBB) to dilute the collagenase while keeping the flask on ice. The islets are "picked" under a dissecting microscope (Bausch and Lomb) with a Pasteur pipette from a black-bottomed petri dish and collected in another petri dish, which is used for the final preparation. Selected, clean islets free of acinar tissue are transferred to the perifusion chambers. The time taken from removal of the pancreata to transfer is approximately 45 minutes.

An alternative method is to remove the pancreas under light ether anesthesia and immediately place in buffer (KRBB).[8] Pieces of 100 mg size are weighed and returned to buffer for mincing with scissors prior to transfer to the perifusion chambers. The time taken from removal of pancreas to commencement of the perifusion averaged 6 minutes. One of the major disadvantages inherent in this system is the possibly modifying effect of released exocrine enzymes. This has been overcome, almost entirely, by the addition of a kallicrein-trypsin inhibitor, Trasylol, and the rapid freezing of collected samples. A second disadvantage is the commercial cost of Trasylol, which can run extremely high over a period of time.

The perifusion apparatus is represented diagrammatically in Figs. 1 and 2. Buffer (KRBB) is drawn from the desired flasks as directed by valve settings of the twin 4-way valves (Chromatronix Inc.) and passed through the chambers via plastic tubing (Chromatronix or LKB) with diameters of 0.063 inch o.d. × 0.012 inch i.d. at a rate of 2.0 ml per minute by means of twin continuous infusion pumps (Werner Meyer). The chambers are inverted cones of approximately 1 ml with a volume buffer flow being from apex to base over which is clamped a Millipore filter (14 μm). Four exit channels are provided in the plastic clamp, these converge to a single exit tube for each chamber. The temperature of the

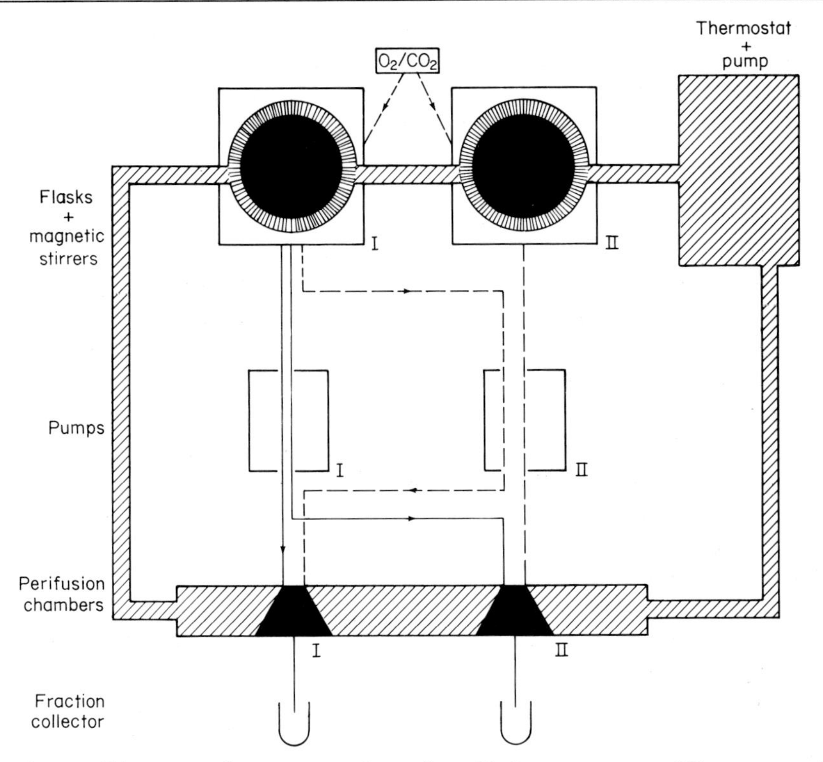

FIG. 1. Diagrammatic representation of perifusion apparatus. When pump I is in operation, perfusate flow is as indicated by continuous lines and arrows. Flow when pump II is in operation is indicated by dashed lines and arrows.

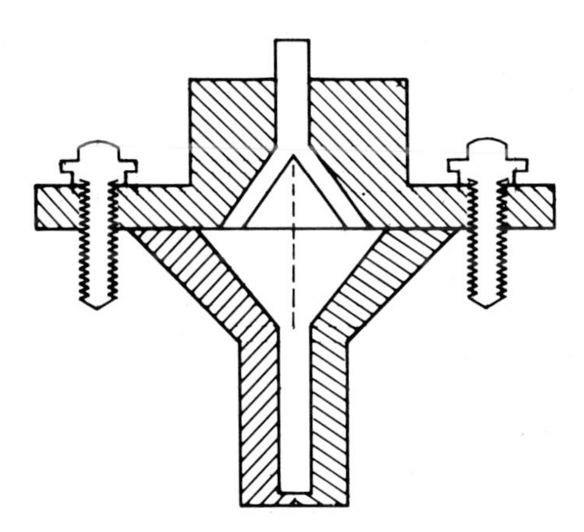

FIG. 2. Diagrammatic representation of perfusion chamber.

system is maintained at 37.5° by a thermostatic pump (Thermomix II), which circulates water flow around the flasks and chambers. The samples from the chambers are collected using an LKB (Colora Cold Box) fraction collector at 4°.

The buffer used throughout is Krebs-Ringer-bicarbonate with 0.5 g of albumin per 100 ml. Oxygenation and pH (7.35) are maintained by constant gassing with 93% O_2–7% CO_2. In all experiments islets are washed with buffer for 35 minutes prior to stimulation for 55 minutes. The samples are immediately sealed with parafilm and stored at —4° for assaying later. The samples are analyzed for insulin content by means of a double antibody radioimmunoassay system.[11]

Comment

This system allows for an *in vitro* assessment of the factors influencing the dynamics of insulin release and for simultaneous testing of controls and test substances in a regulated *in vitro* environment.

[11] C. N. Hales and P. J. Randle, *Biochem. J.* **88,** 137 (1963).

[33] Methods for Assessing the Biological Activity of the Mammalian Pineal Organ

By Daniel P. Cardinali and Richard J. Wurtman

Secretions of the mammalian pineal organ participate in a variety of neural and neuroendocrine mechanisms, including the control of sleep, ovulation, gonadotropin release and various biologic rhythms. The hormone melatonin, first described by Lerner and his associates in 1959, apparently mediates most of these pineal effects.[1] Melatonin was the first methoxyindole to be identified in a mammalian tissue; subsequently it has been shown that the pineal can also synthesize such related compounds as 5-methoxytryptophol which may also exhibit biologic activity. The synthesis of melatonin is accelerated by the release of norepinephrine from the abundant sympathetic neurons that terminate in the pineal organ; the norepinephrine combines with β-receptors on the parenchymal cells, activating adenyl cyclase and thereby increasing the levels of cyclic 3′,5′-AMP present in these cells. Considerable information is now avail-

[1] A. B. Lerner, J. D. Case, and R. V. Heinzelman, *J. Amer. Chem. Soc.* **81,** 6084 (1959).

able concerning the control of melatonin biosynthesis; however, relatively less is known about the precise function of the pineal hormone in the body. This chapter describes various indices of pineal secretory activity. In general, these approaches measure processes related to the synthesis and fate of melatonin. However, no existing method allows the investigator to measure *in vivo* melatonin synthesis or secretion directly, or even to quantify the concentrations of this substance in body fluids. When such advances in methodology are made, they will certainly accelerate the growth of our knowledge of normal and abnormal pineal function.

The mammalian pineal organ is a neuroendocrine transducer[2-4]; it converts neural signals, generated by the responses of retinal photoreceptors to light or darkness, to a hormonal output, the secretion of melatonin. The importance of its sympathetic nervous input is illustrated by the failure of rat pineals to vary melatonin synthesis in response to lighting conditions and time of day, when these neurons are damaged.[5] This contrasts with the responses to denervation of such true endocrine glands as the parathyroid or adrenal cortex; these organs continue to function normally as long as they retain an adequate circulation. The failure of many scientists to recognize the lack of homology between the pineal and true glands contributed to the accumulation over many years of an ambiguous body of data on pineal responses to such experimental procedures as transplantation; these ambiguities undoubtedly delayed the entry of the pineal into contemporary canonical endocrinology.

Melatonin and Other Pineal Indoles

Biosynthesis, Metabolism, and Biological Effects

Melatonin is the prototype of a family of biologically active compounds, the methoxyindoles, produced by the pineals of all vertebrate species thus far examined. Forty years before the structure of melatonin was determined, McCord and Allen showed that the skins of frogs and tadpoles rapidly blanched when the animals were fed extracts of bovine pineals[6]; in 1959, Lerner and his co-workers purified the melanophore-lightening principle from bovine pineals, and identified its chemical struc-

[2] R. J. Wurtman and J. Axelrod, *Sci. Amer.* **213**, 50 (1965).

[3] R. J. Wurtman and F. Anton-Tay, *Recent Progr. Horm. Res.* **25**, 493 (1968).

[4] R. J. Wurtman, J. Axelrod, and D. E. Kelly, "The Pineal." Academic Press, New York, 1968.

[5] R. J. Wurtman, J. Axelrod, and J. E. Fischer, *Science* **143**, 1329 (1964).

[6] C. P. McCord and F. P. Allen, *J. Exp. Zool.* **23**, 207 (1917).

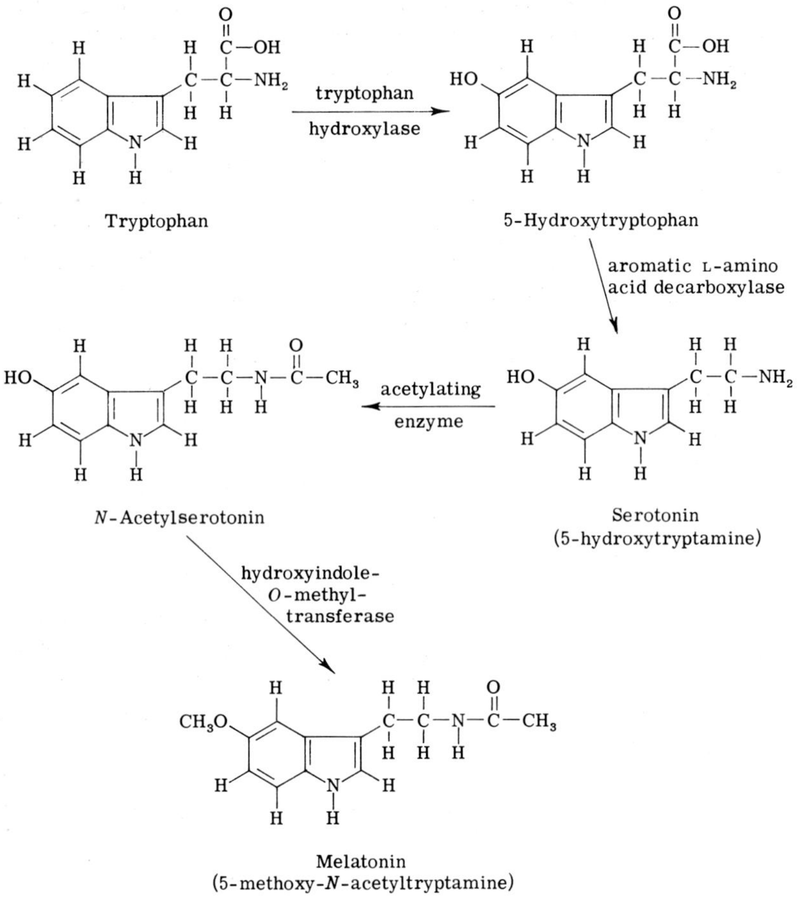

Tryptophan —tryptophan hydroxylase→ 5-Hydroxytryptophan

5-Hydroxytryptophan —aromatic L-amino acid decarboxylase→ Serotonin (5-hydroxytryptamine)

Serotonin (5-hydroxytryptamine) —acetylating enzyme→ N-Acetylserotonin

N-Acetylserotonin —hydroxyindole-O-methyltransferase→ Melatonin (5-methoxy-N-acetyltryptamine)

FIG. 1. Biosynthesis of melatonin (5-methoxy-N-acetyltryptamine) from tryptophan.

ture as 5-methoxy-N-acetyltryptamine.[1] As little as 10^{-13} g/ml of melatonin produced visible changes in amphibian skin[7]; this extreme potency is used to follow the compound through its initial fractionation, and allows melatonin to be measured by bioassay (see below). Melatonin apparently has no overt effect on the cells usually responsible for skin pigmentation in mammals; in place of dermatologic actions, melatonin exerts profound effects on the neuroendocrine apparatus.

The biosynthesis of melatonin in the mammalian pineal organ is initiated by the uptake of the amino acid tryptophan from the circulation

[7] A. B. Lerner and R. M. Wright, *Methods Biochem. Anal.* **8**, 297 (1960).

(Fig. 1); once within the pineal, some of the tryptophan is utilized for protein synthesis, while a larger fraction is converted to serotonin and its derivatives. The initial enzymatic step in melatonin formation involves the oxidation of tryptophan at the 5-position to form the amino acid 5-hydroxytryptophan. This process, which is catalyzed by the enzyme tryptophan hydroxylase, requires the presence of O_2, ferrous iron, and a reduced pteridine cofactor.[8] The substrate K_m for tryptophan hydroxylase is quite high compared with the levels of free tryptophan likely to be present in the pineal organ. Hence, the enzyme in pineal, as in brain,[9,10] may normally function in an unsaturated state, such that physiologic changes in pineal tryptophan concentration could influence the rate at which tryptophan is hydroxylated. The second step in melatonin biosynthesis, the conversion of 5-hydroxytryptophan to serotonin, is catalyzed by aromatic L-amino acid decarboxylase ("dopa" decarboxylase).[11,12] This enzyme is ubiquitous and also participates in the biosynthesis of catecholamines; it requires pyridoxal phosphate as a cofactor. Its K_m for 5-hydroxytryptophan is less than one-hundredth that for tryptophan, and its V_{max} for the hydroxylated amino acid is several times higher. Hence, aromatic L-amino acid decarboxylase probably catalyzes the decarboxylation of little if any tryptamine, and tryptamine is not likely to be a physiological precursor for pineal serotonin. Some of the serotonin produced in the pineal is metabolized by oxidative deamination, catalyzed by monoamine oxidase (MAO).[4] Another portion of the serotonin is converted to melatonin: first, the serotonin is *N*-acetylated by serotonin-*N*-acetyltransferase; a methyl group is then transferred from *S*-adenosylmethionine to the 5-hydroxy position of *N*-acetylserotonin, yielding melatonin (5-methoxy-*N*-acetyltryptamine) (Fig. 1). This latter reaction is catalyzed by the enzyme hydroxyindole-*O*-methyltransferase (HIOMT). Serotonin-*N*-acetyltransferase is not found only in the pineal; HIOMT activity has been detected in mammals only in the pineal,[13] the retina,[14] and the Harderian gland.[15] The pineal and retinal enzymes exhibit similar kinetic properties[16]; those of Harderian gland

[8] W. Lovenberg, E. Jequier, and A. Sjoerdsma, *Advan. Pharmacol.* **6A,** 21 (1968).

[9] R. J. Wurtman and J. D. Fernstrom, *in* "Perspectives in Neuropharmacology" (S. H. Snyder, ed.), p. 143. Oxford Univ. Press, London and New York, 1972.

[10] J. D. Fernstrom and R. J. Wurtman, *Science* **173,** 149 (1971).

[11] W. Lovenberg, H. Weissbach, and S. Udenfriend, *J. Biol. Chem.* **237,** 89 (1962).

[12] S. H. Snyder, J. Axelrod, R. J. Wurtman, and J. E. Fischer, *J. Pharmacol. Exp. Ther.* **147,** 371 (1965).

[13] J. Axelrod and H. Weissbach, *J. Biol. Chem.* **236,** 211 (1961).

[14] D. P. Cardinali and J. M. Rosner, *Endocrinology* **89,** 301 (1971).

[15] G. Vlahakes and R. J. Wurtman, *Biochim. Biophys. Acta* **261,** 194 (1972).

[16] D. P. Cardinali and R. J. Wurtman, *Endocrinology* **91,** 247 (1972).

HIOMT differ considerably, suggesting that the true physiological role of this enzyme involves other substrates.

In the last 10 years, experimental data have accumulated that indicate that the biochemical activity of the pineal organ depends upon the lighting under which the animal is maintained. Exposure of rats to constant illumination depresses the weight of the pineal,[17] obliterates the normal diurnal rhythms in HIOMT[18,19] and serotonin-*N*-acetyltransferase[20] activities, and destroys the diurnal variations in pineal melatonin content.[21] The photic inhibition of pineal HIOMT in the rat displays a characteristic action spectrum similar to the absorption spectrum of retinal rhodopsin.[22,23] The inhibition is mediated via a complicated anatomical pathway involving the retinas, the inferior accessory optic tracts, descending central sympathetic neurons; and the preganglionic fibers to, and postganglionic fibers from, the superior cervical ganglia.[4] Sympathetic nerve terminals within the pineal stimulate melatonin biosynthesis by releasing norepinephrine,[24] which acts on a β-receptor[25] to activate an adenyl cyclase, thereby increasing cyclic 3′,5′-AMP synthesis.[26,27] The electrophysiological activity of the postganglionic sympathetic neurons to the pineal is depressed when light strikes the retinas.[28]

The amounts of melatonin present in the rat pineal are very small (i.e., 0.5–7 ng) and undergo diurnal variations entrained by the photoperiod.[29] Melatonin ultimately enters the general circulation and is excreted into the urine[4]; however, it is not yet certain whether it is secreted directly into the blood or reaches the circulation via the cerebrospinal fluid. Semiquantitative measurements of melatonin concentrations in human and rat plasma suggest a 24-hour rhythm which peaks during darkness.[30] Melatonin is extremely lipid-soluble, and thus, in plasma, it

[17] V. M. Fiske, *Endocrinology* **29,** 187 (1941).

[18] J. Axelrod, R. J. Wurtman, and S. H. Snyder, *J. Biol. Chem.* **240,** 949 (1965).

[19] C. A. Nagle, D. P. Cardinali, and J. M. Rosner, *Endocrinology* **91,** 423 (1972).

[20] D. C. Klein and J. L. Weller, *Science* **169,** 1093 (1970).

[21] C. L. Ralph, D. Mull, H. J. Lynch, and L. Heldlund, *Endocrinology* **89,** 1361 (1971).

[22] D. P. Cardinali, F. Larin, and R. J. Wurtman, *Proc. Nat. Acad. Sci. U.S.* **69,** 2003 (1972).

[23] D. P. Cardinali, F. Larin, and R. J. Wurtman, *Endocrinology* **91,** 877 (1972).

[24] J. Axelrod, H. M. Shein, and R. J. Wurtman, *Proc. Nat. Acad. Sci. U.S.* **62,** 544 (1969).

[25] R. J. Wurtman, H. M. Shein, and F. Larin, *J. Neurochem.* **18,** 1683 (1971).

[26] B. Weiss and E. Costa, *J. Pharmacol. Exp. Ther.* **161,** 310 (1968).

[27] S. J. Strada, D. C. Klein, J. Weller, and B. Weiss, *Endocrinology* **90,** 1470 (1972).

[28] A. N. Taylor and R. W. Wilson, *Experientia* **26,** 267 (1970).

[29] H. J. Lynch, *Life Sci.* (I) **10,** 791 (1971).

[30] H. J. Lynch and R. J. Wurtman, unpublished results.

is largely bound to serum albumin.[31] Circulating [^{3}H]melatonin is taken up and concentrated within several organs, especially the gonads.[32] Isotopically labeled melatonin injected into the cerebrospinal fluid localized unevenly within the brain; highest concentrations are found in the hypothalamus and midbrain.[33,34] Circulating [^{3}H]melatonin disappears rapidly from both the bloodstream and the brain. Although the central nervous system can metabolize melatonin (through yet undefined pathways),[34] the liver appears to be the body's major site of melatonin inactivation; the methoxyindole is first hydroxylated to 6-hydroxymelatonin by a microsomal NADPH-requiring enzyme, and then conjugated with glucuronic or sulfuric acid.[35,36] In the rat, 70–80% of a trace dose of radiolabeled melatonin is excreted in the urine, and 20% in the feces; 10% of the 6-hydroxylated melatonin is excreted unchanged, whereas the remainder is first conjugated. Of these urinary conjugates, 70–80% are ethereal sulfates, and 5% are glucuronides.[36] No data are yet available on the metabolism of endogenous melatonin.

Melatonin injections have been shown to inhibit various aspects of gonadal, thyroid, and adrenal function; these effects are, in general, opposite to those observed after removal of the pineal organ.[4] Melatonin administration decreases ovarian weight, delays puberty, and reduces the frequency of vaginal estrus in both normal and constant estrus rats.[37] Pineal ablation is associated with increases in the concentrations of LH (luteinizing hormone) and FSH (follicle-stimulating hormone) of the pituitary gland[38]; implants of melatonin in the hypothalamus exert the opposite effect. Similarly, in rats, the perfusion of the third ventricle with melatonin lowers plasma LH[39] and FSH[40] concentrations, delays the onset of puberty, and blocks ovulation.[38] The direct perfusion of the hypophysis with melatonin does not modify plasma gonadotropin concentrations; hence, melatonin acts somewhere within the brain to suppress

[31] D. P. Cardinali, H. J. Lynch, and R. J. Wurtman, *Endocrinology* **91,** 1213 (1972).
[32] R. J. Wurtman, J. Axelrod, and L. T. Potter, *J. Pharmacol. Exp. Ther.* **143,** 314 (1964).
[33] F. Anton-Tay and R. J. Wurtman, *Nature (London)* **221,** 474 (1969).
[34] D. P. Cardinali, M. T. Hyyppä, and R. J. Wurtman, *Neuroendocrinology* **12,** 30 (1973).
[35] I. J. Kopin, C. M. B. Pare, J. Axelrod, and H. Weissbach, *Biochim. Biophys. Acta* **40,** 377 (1960).
[36] I. J. Kopin, C. M. B. Pare, J. Axelrod, and H. Weissbach, *J. Biol. Chem.* **236,** 3072 (1961).
[37] R. J. Wurtman, J. Axelrod, and E. W. Chu, *Science* **141,** 277 (1963).
[38] F. Fraschini, R. Collu, and L. Martini, *in* "The Pineal Gland: A Ciba Foundation Symposium" (G. E. W. Wolstenholme and J. Knight, eds.), p. 259. Churchill Livingstone, London, 1971.
[39] I. A. Kamberi, R. S. Mical, and J. C. Porter, *Endocrinology* **87,** 1 (1970).
[40] I. A. Kamberi, R. S. Mical, and J. C. Porter, *Endocrinology* **88,** 1288 (1971).

the secretion of LH- and FSH-releasing factors. Intraperitoneal injections of the hormone causes rapid elevations in the serotonin contents of the hypothalamus and midbrain.[41] This suggests that melatonin may influence the releasing-factor cells in the hypothalamus indirectly, i.e., by acting presynaptically on other neurons that utilize serotonin as their neurotransmitter. The antigonadal effects of melatonin could also result from direct actions on the gonads themselves, or on accessory organs, inasmuch as melatonin inhibits *in vitro* testicular androgen production,[42-44] and suppresses the effects of exogenous testosterone on the prostate and seminal vesicles.[45]

Pineal removal increases thyroid weight,[46] ^{131}I uptake, and the rate at which the thyroid loses ^{131}I (presumably an index of the thyroid hormone secretion)[47]; these effects are counteracted by the administration of melatonin.[48,49] Pinealectomy has also been reported to cause a 3-fold increase in the aldosterone concentration of rat adrenal venous blood, and a corresponding increase in plasma corticosteroid concentrations.[50] Acute injections of melatonin into the lateral ventricles of rats significantly decreases plasma corticosterone.[51] The intravenous administration of melatonin to rats causes a rapid depletion of pituitary MSH (melanocyte-stimulating hormone)[52]; high MSH levels are observed in pinealectomized animals or rats kept in continuous light. This suggests that the effects of injected melatonin mimic those of the endogenous hormone.

Determination of Melatonin and Other Pineal Indoles

Fluorometric Methods

Despite the large amount of melatonin-oriented research, there is no simple analytical method for measuring melatonin in tissues or body

[41] F. Anton-Tay, C. Chou, S. Anton, and R. J. Wurtman, *Science* **162**, 277 (1968).
[42] D. P. Cardinali and J. M. Rosner, *Steroids* **18**, 25 (1971).
[43] F. Peat and G. A. Kinson, *Steroids* **17**, 251 (1971).
[44] L. G. Ellis, *Endocrinology* **90**, 17 (1972).
[45] L. Debeljuk, V. M. Feder, and O. A. Paulucci, *Endocrinology* **87**, 1358 (1970).
[46] A. B. Houssay and J. H. Pazo, *Experientia* **24**, 813 (1968).
[47] G. Csaba, J. Kiss, and M. Bodoky, *Acta Biol. Acad. Sci. Hung.* **19**, 35 (1968).
[48] D. V. Singh and C. W. Turner, *Acta Endocrinol.* **69**, 35 (1972).
[49] N. D. DeProspo, R. J. Safinski, L. D. De Martino, and E. T. McGuiness, *Life Sci.* (I) **8**, 837 (1969).
[50] G. A. Kinson, A. K. Wahid, and B. Singer, *Gen. Comp. Endocrinol.* **8**, 445 (1967).
[51] M. Motta, O. Schiaffini, F. Piva, and L. Martini, *in* "The Pineal Gland: A Ciba Foundation Symposium" (G. E. W. Wolstenholme and J. Knight, eds.), p. 279. Churchill Livingstone, London, 1971.
[52] A. J. Kastin and A. V. Schally, *Nature (London)* **213**, 1238 (1967).

fluids. Melatonin can be measured chemically by extraction into organic solvents such as chloroform[13] and *p*-cymene[53]; or separated from *N*-acetylserotonin, 5-hydroxyindoles, and 5-methoxyindoleacetic acid by extraction into chloroform.[13] Melatonin is then assayed fluorometrically in 3 *N* HCl at wavelengths 300 nm (excitation), 540 nm (emission). However, these methods are not sensitive or specific enough to measure the small amount of melatonin in the pineals of most species.

Maickel and Miller[54] have described the reaction of *o*-phthaldehyde (OPT) with various 3,5-substituted indoles. Under conditions of heat and strong mineral acid, compounds such as serotonin, *N*-acetylserotonin, 5-methoxytryptamine, and melatonin react with OPT to yield highly fluorescent compounds. Fluorescence is measured in a spectrophotofluorometer at wavelengths of 360 nm (excitation) and 470 nm (emission). After a preliminary solvent extraction, this procedure detected melatonin, *N*-acetylserotonin, and 5-methoxytryptamine in pineals of dogs and rats, but not in other brain areas.[55] In confirmation of previous reports,[53,56,57] a high concentration of serotonin was also detected in the pineal.

OTP Fluorescence Assay.[55] Pineal organs are homogenized in 1.0 ml of 0.05 *N* NaOH with a glass homogenizer tube and a motor-driven Teflon pestle. The homogenate is first transferred to a glass-stoppered, 13-ml centrifuge tube containing 6 ml of chloroform, then shaken and centrifuged. At this time, the organic phase contains 5-methoxytryptamine and melatonin; the aqueous phase contains serotonin, *N*-acetylserotonin, 5-hydroxyindoleacetic acid, and 5-methoxyindoleacetic acid.

An aliquot (4.0 ml) of the chloroform phase is transferred to a glass-stoppered centrifuge tube containing 0.3 ml of 0.1 *N* HCl. After being shaken and centrifuged, the organic phase contains melatonin, while the aqueous phase contains 5-methoxytryptamine. An aliquot (0.2 ml) of the aqueous phase is next isolated to react methoxytryptamine with OPT; 0.6 ml of 10 *N* HCl containing 15 mg/100 ml OPT is added and mixed. The mixture is heated for 10 minutes in a boiling water bath, cooled, and its fluorescence measured in a spectrophotofluorometer at wavelengths 360 nm/470 nm.

An aliquot (2.5 ml) of the chloroform phase containing melatonin is shaken with 8.0 ml of *n*-heptane and 0.3 ml of 5 *N* HCl; melatonin passes to the HCl phase. The OPT reaction for melatonin is performed with a 0.2-ml aliquot of the acidic phase as described above.

[53] W. B. Quay, *Anal. Biochem.* **5**, 51 (1963).
[54] R. P. Maickel and F. P. Miller, *Anal. Chem.* **38**, 1937 (1966).
[55] F. P. Miller and R. P. Maickel, *Life Sci.* (I) **9**, 747 (1970).
[56] N. J. Giarman and N. Day, *Biochem. Pharmacol.* **1**, 235 (1959).
[57] S. H. Snyder, J. Axelrod, and M. Zweig, *Biochem. Pharmacol.* **14**, 831 (1965).

An aliquot (0.6 ml) of the original aqueous phase containing serotonin, *N*-acetylserotonin, 5-hydroxyindoleacetic acid, and 5-methoxyindoleacetic acid is shaken with 10 ml of ethyl acetate and 0.2 ml of 1.2 *N* HCl. At this time, the ethyl acetate phase contains *N*-acetylserotonin, 5-hydroxyindoleacetic acid, and 5-methoxyindoleacetic acid, while the aqueous phase contains serotonin. The serotonin-OPT reaction is carried out in a 0.2-ml aliquot of the aqueous phase as described above.

Four milliliters of the ethyl acetate phase are now shaken with 6.5 ml of *n*-heptane and 0.3 ml of 1 *N* HCl. After centrifugation, *N*-acetylserotonin passes to the acidic phase and is measured in a 0.2-ml aliquot (see above). Finally, an aliquot (8.0 ml) of the ethyl acetate–heptane phase is collected in a centrifuge tube containing 0.5 ml of 0.2 *N* NaOH. After shaking and centrifugation, an aliquot (0.4 ml) of the aqueous phase, which contains 5-hydroxyindoleacetic acid, is drawn off for reaction with OPT. 5-Methoxyindoleacetic and 5-hydroxyindoleacetic acids cannot be separated by this procedure. Overall recoveries by this method ranged from 58% for *N*-acetylserotonin to nearly 100% for methoxytryptamine.

Sensitive fluorometric methods for measuring serotonin in the pineal have been reported. In these assays, serotonin is first extracted in *n*-butanol and then returned to an aqueous medium. The native fluorescence of serotonin can be measured directly in concentrated HCl.[58] However, the sensitivity of the fluorometric assay can be increased severalfold by condensing the amine with ninhydrin[57] or with OPT,[55] as described above. The sensitivity of the OPT method can be improved by extraction of the aqueous phase with chloroform prior to reading blank and assay tubes.[59]

Melatonin Bioassay

The bioassay of melatonin depends on the ability of the hormone to blanch isolated skin of *Rana pipiens,* or of intact, live larvae of *Rana pipiens* or *Xenopus laevis.* This assay uses either of two forms: (a) the photometric detection of changes in the transmission or reflectance of incident light by isolated frog skin previously darkened with MSH or caffeine,[60] or (b) in intact amphibian larvae, the microscopic estimation of the degree of melanin dispersion within the dermal melanophores according to the Hogben index.[61,62]

[58] D. F. Bogdanski, A. Pletscher, B. B. Brodie, and S. Udenfriend, *J. Pharmacol. Exp. Ther.* **117,** 82 (1956).
[59] J. H. Thompson, C. A. Spezia, and M. Angulo, *Experientia* **26,** 327 (1970).
[60] W. Mori and A. B. Lerner, *Endocrinology* **67,** 443 (1960).
[61] W. B. Quay and J. T. Bagnara, *Arch. Int. Pharmacodyn. Ther.* **150,** 137 (1964).
[62] C. L. Ralph and H. J. Lynch, *Gen. Comp. Endocrinol.* **15,** 334 (1970).

Bioassay methods for melatonin are extremely sensitive and can detect as little as 10^{-13} g/ml of the methoxyindole. However, until recently, melatonin bioassays were only semiquantitative screening tests that indicated the presence or absence of melatonin. In 1970, Ralph and Lynch reported a quantitative bioassay for melatonin utilizing the dermal melanophore response of live, larval *Rana pipiens*.[62] This method requires a tadpole-culturing program for a constant supply of uniform test animals. Ovulation is induced in frogs, and the eggs are inseminated and cultured; tadpoles are used at the 12th day after fertilization, which corresponds to developmental stage 25. Each assay estimates the melanophore response of 10 tadpoles exposed to 2 ml of melatonin solution or to aqueous homogenates of single pineals. Dermal melanophores in a triangular area below the eye are subjectively assessed by the five-stage Hogben melanophore index. Exposure to melatonin concentrations ranging from 0.1 to 2.5 ng per milliliter of solution obtains the greatest accuracy and precision. A calibration curve using known concentrations of melatonin is run concurrently with each assay; this curve is plotted on semilog paper with melatonin concentrations on the log scale. A linear relationship exists between the log of melatonin concentration and the melanophore index, with concentrations ranging from 0.1 to 2.5 ng of melatonin per milliliter of medium; at higher concentrations, the sensitivity diminishes. The melanophorotropic constituent of pineal homogenates has chromatographic mobility similar to authentic melatonin in gel filtration columns and thin-layer chromatography.[62] It is as yet unestablished if other, nonindolic, pineal constituents also exhibit tropic activity on melanocytes in amphibian skin.

The relative results with *Rana pipiens* and *Xenopus laevis* larvae in the quantitative assay of melatonin have recently been compared.[63] Preliminary studies indicated that with suitable quantitative assessment of the *Xenopus* melanophore response to melatonin, a highly sensitive dose-response curve might be produced that extended the range of the melatonin bioassay at least an order of magnitude (i.e., 0.01 ng/ml).

Gas Chromatography–Mass Spectrometry

Considering their specificity, sensitivity and rapidity, gas chromatographic procedures would appear to be ideally suited to the assay of compounds such as melatonin. A number of investigators have assayed melatonin directly with gas–liquid chromatography,[64-67] but only a few have

[63] H. J. Lynch, R. J. Wurtman, and G. Greenhouse, *Amer. Zool.* **11**, 135 (1971).
[64] C. J. Brooks and E. C. Horning, *Anal. Chem.* **36**, 1540 (1964).
[65] M. Greer and C. M. Williams, *Clin. Chem. Acta* **15**, 165 (1967).
[66] S. N. Pennington, *J. Chromatogr.* **32**, 406 (1968).
[67] E. R. Cole and G. Crank, *J. Chromatogr.* **61**, 225 (1971).

first converted the indole to derivatives suitable for measurement by the more sensitive electron capture detection. Two such methods have recently been described.[68,69] In one, final quantification is done with mass spectrometry by measuring the ion density of the specific molecular fragments; specificity is thus based on both the chromatographic retention time and the quantities of specific fragments formed.[69]

Degen *et al.* have reported the reaction of melatonin with *N*-heptafluorobutyrylimidazole in the presence of triethylamine which proved to be a suitable catalyst for the removal of melatonin's *N*-acetyl group.[68] After chloroform extraction, followed by the derivatization procedure with *N*-heptafluorobutyrylimidazole, melatonin can be detected in pineal tissue down to 100 pg. A standard curve was obtained by making derivatives of amounts ranging from 350 to 1400 pg of melatonin with 20 μl of *N*-heptafluorobutyrylimidazole and 2 μl of triethylamine. The reaction mixtures were heated for 150 minutes at 90°, cooled and extracted with 20 μl of water and 20 μl of *n*-heptane. The extraction with water seemed essential since it removed excess reagent from the derivative (both *N*-heptafluorobutyrylimidazole and free imidazole produced very large solvent peaks). Aliquots of the heptane phase (35–140 pg) were injected into the gas chromatograph. Gas chromatographic conditions were as follows: 6 ft, 2 mm i.d., glass columns, packed with 3% SE-30 on Varapak 30, 100–200 mesh; N_2 flow, 20 ml per minute; ^{63}Ni electron capture detector. Temperature: injector, 230°; column, 170°; detector, 310°. The response of the electron capture detector was linear; 73–77% of melatonin was converted to the corresponding heptafluorobutyryl derivative. To determine melatonin in pineal homogenates, two pineals were combined and homogenized in 0.2 ml of chloroform. The homogenates were then centrifuged and the pellet discarded. The chloroform phase was washed with 0.2 ml of 0.05 *N* NaOH and with 0.2 ml of 0.1 *N* HCl, dried over sodium sulfate and taken to dryness; the residue was converted to the derivative and measured as described above.

A gas chromatography–mass spectrometry assay of four indoleamines (serotonin, *N*-acetylserotonin, 5-methoxytryptamine, and melatonin) in single rats pineals has recently been described by Cattabeni *et al.*[69] To form compounds with the appropriate vapor pressure for gas chromatography, the pineal indoles are reacted with pentafluoropropionic anhydride to obtain acetylation of hydroxyl and primary and secondary amine groups. The reaction is performed with pentafluoropropionic anhydride (100 μl) in ethyl acetate (20 μl) at 60° for 3 hours. The excess of reagent

[68] P. H. Degen, J. R. DoAmaral, and J. Barchas, *Anal. Biochem.* **45**, 634 (1972).
[69] F. Cattabeni, S. H. Koslow, and E. Costa, *Science* **178**, 166 (1972).

is evaporated under a stream of N_2, and the residue dissolved in ethyl acetate (10 μl); reaction yield ranges from 60 to 100%. Gas chromatography conditions were: 9 ft, 2 mm i.d., glass columns packed with OV-17, 3%, on Gas-Chrom Q, 100–120 mesh. Temperature: injector, 290°; column, 210°. Helium flow, 15 ml per minute. All the compounds were completely resolved from each other in the gas chromatographic column. Mass spectrometry conditions were: molecular separator, 250°; ion source, 290°; electron energy, 80 electron volts; trap current, 60 μA; electron multiplier, 3.7 kV. For quantitation of the ion density, the base peak of each indole was recorded as it was eluted from the gas chromatography column. The linearity obtained (over a range of 1 to 600 pmoles) supports the contention that the ion density of the recorded fragments is proportional to the concentration of indole. This assay was applied to measure the indole content in single rat pineals. The pineal was homogenized in 50 ml of 0.1 *M* $ZnSO_4$, and, after neutralization with 50 ml of 0.1 *M* $Ba(OH)_2$ and centrifugation, an aliquot of the supernatant was processed as described above. Pineal melatonin concentrations obtained with this method approximate those determined by bioassay.[62]

Assay of Pineal Enzymes Involved in the Metabolism of Indoles

Enzymes That Catalyze the Biosynthesis or Metabolism of Pineal Serotonin

Tryptophan Hydroxylase (EC 1.14.16.4). Three methods for assaying tryptophan hydroxylase have been reported: (a) Assays using ring-labeled [^{14}C]tryptophan: In these methods, the common requirement is the isolation of the substrate, [^{14}C]tryptophan, from [^{14}C]5-hydroxytryptophan or [^{14}C]serotonin. This has been accomplished by thin-layer chromatography,[70] or through further conversion of [^{14}C]5-hydroxytryptophan to [^{14}C]serotonin, by adding an excess of purified aromatic-L-amino acid decarboxylase.[71] The last assay is based on tryptophan's poor affinity for the decarboxylase, as compared to 5-hydroxytryptophan's (see above); it requires the final measurement of the specific activity of the [^{14}C]serotonin formed by the pineal homogenates. (b) Methods using [^{14}C-carboxy]tryptophan: These methods are based on the trapping of $^{14}CO_2$ by adding an excess of aromatic-L-amino acid decarboxylase, after the enzymatic hydroxylation of [^{14}C-carboxy]tryptophan to [^{14}C-car-

[70] R. Hakanson, M. N. Lombard des Gouttes, and C. Owman, *Life Sci.* **6**, 2577 (1967).

[71] W. Lovenberg, E. Jequier, and A. Sjoerdsma, *Science* **155**, 3759 (1967).

boxy]5-hydroxytryptophan.[72-74] (c) Assays using DL-[5-^{3}H]tryptophan: More recently, a simple and sensitive assay for tryptophan hydroxylase has been described involving the measurement of the tritiated water produced during the enzymatic hydroxylation of [5-^{3}H]tryptophan.[75] This method is based on the intramolecular migration of hydrogen that occurs in all aromatic hydroxylation reactions ("NIH shift"[76]); thus, for tryptophan, the product of the enzymatic hydroxylation of [5-^{3}H]tryptophan is [4-^{3}H]5-hydroxtryptophan. In order to block further conversion of 5-hydroxytryptophan to serotonin and thus to 5-methoxy derivatives (which increase ^{3}H release into water), a decarboxylase inhibitor must be added to the incubation mixture.

Aromatic-L-Amino Acid Decarboxylase (EC 4.1.1.28). Aromatic-L-amino acid decarboxylase is optically specific for L-isomers, but it is not specific for any particular aromatic amino acid, since it catalyzes the decarboxylation of L-dopa, L-5-hydroxytryptophan, L-tryptophan, L-tyrosine, L-phenylalanine, and L-histidine.[11] Hence, aromatic-L-amino acid decarboxylase is probably involved in all the decarboxylation steps in the biosynthesis of biogenic amines. In the pineal glands of rats, cows, pigs and rabbits,[77,78] this enzyme's activity changes with light sympathetic nervous activity and drugs.[12,79] The activity of pineal aromatic-L-amino acid decarboxylase has been determined by measuring the [^{14}C]-serotonin formed after incubation of pineal homogenates with [^{14}C]hydroxytryptophan.[77]

Monoamine Oxidase (MAO) (EC 1.4.3.4). This enzyme, which catalyzes the oxidative deamination of a variety of biogenic amines (e.g., serotonin, norepinephrine, epinephrine, tyramine, and dopamine), deaminates compounds in which the amine group is attached to the terminal carbon group. Depending on the nature of the substrate, the corresponding aldehyde is then either reduced to form an alcohol metabolite, or oxidized to an acidic metabolite. Thus, pineal serotonin undergoing oxidative deamination is first converted to 5-hydroxyindoleacetal aldehyde,

[72] A. Ichiyama, S. Nakamura, Y. Nishizuka, and O. Hayaishi, *Advan. Pharmacol.* **6A,** 5 (1968).
[73] A. Ichiyama, S. Nakamura, Y. Nishizuka, and O. Hayaishi, *J. Biol. Chem.* **245,** 1699 (1970).
[74] T. Deguchi and J. Barchas, *Nature (London)* **235,** 93 (1972).
[75] W. Lovenberg, R. E. Besinger, R. L. Jackson, and J. Daly, *Anal. Biochem.* **43,** 269 (1971).
[76] G. Guroff, J. Daly, D. Jerina, J. Renson, B. Witkop, and S. Udenfriend, *Science*
[77] S. H. Snyder and J. Axelrod, *Biochem. Pharmacol.* **13,** 805 (1964).
158, 1524 (1967).
[78] R. Hakanson and C. Owman, *J. Neurochem.* **13,** 597 (1966).
[79] A. Pellegrino de Iraldi and G. R. Arnaiz, *Life Sci.* **3,** 589 (1964).

an unstable intermediate which is then either oxidized to 5-hydroxyindole acetic acid or reduced to 5-hydroxytryptophol.[4] Several species of MAO have been described, in the pineal and other tissues, which differ in substrate specificity, heat stability, electrophoretic mobility, and response to different inhibitors.[80,81] Pineal MAO activity is relatively high[82]; a significant fraction of this enzyme appears to localize within sympathetic nerve terminals inasmuch as denervation of the pineal causes its MAO activity to decline by about half.[83] In these experiments, MAO activity was determined by measuring deamination of [^{14}C]tryptamine.[84] After the incubation mixtures were acidified with HCl, the deaminated metabolites were extracted into toluene and then counted in a liquid scintillation spectrophotometer.

Serotonin-N-Acetyltransferase (EC 2.3.1.5)

Serotonin-*N*-acetyltransferase, which catalyzes the conversion of serotonin to *N*-acetylserotonin, was partially purified from bovine pineal homogenates.[85] It has been increasingly useful in the study of pineal physiology, since this enzyme exhibits dramatic changes diurnally and in response to environmental lighting or to treatments that modify pineal sympathetic tone. Two methods are now used to assay pineal serotonin-N-acetyltransferase. One[86] utilizes bidimensional thin-layer chromatography to separate [^{14}C]*N*-acetylserotonin from [^{14}C]2-serotonin and its other metabolites produced by rat pineal homogenates *in vitro;* the other[87] utilizes solvent partition to measure the acetylation of tryptamine by [^{14}C]1-acetyl enzyme A.

Thin-Layer Chromatography Method.[86,88] A single pineal is homogenized in 20 μl of 0.1 *M* sodium phosphate buffer (pH 6.8) containing 10 nmoles of [^{14}C]serotonin (specific activity 40–60 μCi/μmole). The homogenate is incubated for 30 minutes at 37° in the homogenizer, and the reaction is stopped by the addition of ethanol–1 *N* HCl (1:1, v/v) containing serotonin, *N*-acetylserotonin, melatonin, 5-hydroxytryptophol,

[80] M. B. H. Youdim, G. G. S. Collins, and M. Sandler, *Nature (London)* **223**, 626 (1969).

[81] C. Goridis and N. H. Neff, *Neuropharmacology* **10**, 557 (1971).

[82] R. J. Wurtman, J. Axelrod, and L. S. Phillips, *Science* **142**, 1071 (1963).

[83] S. H. Snyder, J. E. Fischer, and J. Axelrod, *Biochem. Pharmacol.* **14**, 363 (1965).

[84] R. J. Wurtman and J. Axelrod, *Biochem. Pharmacol.* **12**, 1439 (1963).

[85] H. Weissbach, B. G. Redfield, and J. Axelrod, *Biochim. Biophys. Acta* **54**, 190 (1961).

[86] D. C. Klein, G. R. Berg, and J. Weller, *Science* **168**, 978 (1970).

[87] T. Deguchi and J. Axelrod, *Anal. Biochem.* **50**, 174 (1972).

[88] D. C. Klein and A. Notides, *Anal. Biochem.* **31**, 480 (1969).

5-hydroxyindoleacetic acid, 5-methoxytryptophol, and 5-methoxyindoleacetic acid, all at a concentration of 1 m*M*. A 20-μl sample of this mixture is chromatographed in a bidimensional thin-layer chromatography system (A: chloroform–methanol–glacial acetic acid, 93:7:1, v/v/v; B: ethyl acetate). The plates are developed twice with solvent A, rotated through 90°, and developed with solvent B. The chromatography is performed in darkness and under N_2 atmosphere. The plates are then sprayed with methanol and concentrated HCl (1:1, v/v) for fluorescent visualization of indoles under ultraviolet light. [^{14}C]*N*-Acetylserotonin and [^{14}C]melatonin areas are then scraped into vials, scintillation fluid is added, and the sample is counted in a liquid scintillation spectrophotometer. Radioactivity in both zones is taken as an estimation of the serotonin-*N*-acetyltransferase activity. Thin-layer chromatography has been found suitable for studying serotonin metabolism in rat[88] and duck pineal organ cultures,[89] and in short-term incubations of the rat retina.[90]

Solvent Partition Method.[87] Recently, a considerably simpler method for measuring the activity of serotonin-*N*-acetyltransferase has been described by Deguchi and Axelrod.[87] The principle of the assay involves acetylation of tryptamine with [^{14}C]acetyl coenzyme A, and extraction of enzymatically formed [^{14}C]acetyltryptamine into a nonpolar solvent system.

Single pineals are homogenized in 1-ml glass homogenizers with 70 μl of a reaction mixture containing 2.5 μmoles of potassium phosphate (pH 6.5), 0.1 μmole of tryptamine, and 3.4 nmoles of [^{14}C]1-acetyl coenzyme A (specific activity about 60 mCi/mmole). The mixture is incubated for 10 minutes at 37° in the glass homogenizer and the reaction is stopped by the addition of 0.5 ml of 0.5 *M* borate buffer, pH 10. The reaction mixture is then transferred to a glass-stoppered, 13-ml tube, containing 6 ml of toluene–isoamyl alcohol (97:3, v/v), and mixed thoroughly for 30 seconds. After centrifugation, 2 ml of the organic phase is mixed with 10 ml of Bray's solution in a scintillation vial, and the radioactivity measured in a liquid scintillation spectrophotometer. When enzyme activity is low, as in the daytime, 5 ml of the organic phase is first evaporated to dryness at 80° in a scintillation vial. The residue is then dissolved with 1 ml of ethanol, and the ^{14}C radioactivity measured after adding 10 ml of Bray's solution. Blank values are obtained from incubation mixtures without the enzyme. The radioactive acetylated product had a similar chromatographic mobility to authentic *N*-acetyltryptamine. Radiolabeled acetyltryptamine increased almost linearly

[89] J. M. Rosner, J. H. Denari, C. A. Nagle, D. P. Cardinali, G. D. Perez Bedes, and L. Orsi, *Life Sci.* (II) **11**, 829 (1972).
[90] D. P. Cardinali and J. M. Rosner, *J. Neurochem.* **18**, 1769 (1971).

during the first 10 minutes of incubation. Optimum pH was 6.5, with sharp decreases of enzyme activity on either side of this pH. Tryptamine was a better substrate for *N*-acetyltransferase than serotonin or 5-methoxytryptamine. In addition, when serotonin was used as substrate, it was necessary to use a polar solvent (e.g., isoamyl alcohol) to extract *N*-acetylserotonin; this procedure resulted in very high blank values, thus reducing the sensitivity of the assay.[87]

Hydroxyindole-O-Methyltransferase (HIOMT) (EC 2.1.1.4)

HIOMT, localized in the pineal cytosol fraction, was originally purified 20-fold from bovine pineals.[13] The purified enzyme showed an absolute requirement for *S*-adenosylmethionine and, unlike other *O*-methyltransferases, it did not require metallic ions. Enzyme activity could be destroyed by treatment with *p*-chloromercuric benzoate, thus indicating the presence of essential sulfhydryl groups. Optimal activity occurred in the pH range from 7.5 to 8.3. Multiple forms of HIOMT have recently been observed in the bovine pineal.[91] Although *N*-acetylserotonin is by far the best substrate for HIOMT, pineal preparations can also catalyze the *O*-methylation of a variety of 5-hydroxyindoles.[13] The rate at which serotonin is *O*-methylated is less than one-tenth that of *N*-acetylserotonin; this suggested that *N*-acetylation precedes *O*-methylation in the formation of melatonin *in vivo* (Fig. 1).

HIOMT activity has recently been detected in homogenates of two aditional mammalian tissues, the retina[14] and the Harderian gland of the rat.[15] HIOMT's in rat pineal and in retina appear to be closely related enzymes; both have a similar pH optimum, both are labile when stored at −20°, and both are inhibited by sulfhydryl blocking agents.[16] The enzymes also exhibit similar substrate specificity, and do not require divalent cations; their apparent K_m values for both *N*-acetylserotonin and *S*-adenosylmethionine are similar (about 10 μM) as are the extents of their suppression by specific HIOMT inhibitors. In contrast, two major differences were evident between the Harderian gland enzyme and HIOMT in the pineal and retina, i.e., the absolute requirement for divalent cations exhibited by the Harderian gland HIOMT, and its high K_m value for *N*-acetylserotonin (1.2 m*M*).[16]

HIOMT Assay. HIOMT is assayed, as described by Axelrod and Weissbach,[13] by measuring the transfer of a [^{14}C]methyl group from [^{14}C-methyl]*S*-adenosylmethionine to *N*-acetylserotonin. A single pineal is homogenized in 500 μl of 50 m*M* phosphate buffer, pH 7.9. An aliquot

[91] R. L. Jackson and W. Lovenberg, *J. Biol. Chem.* **246,** 4280 (1971).

of 200 μl of the homogenate is incubated for 30 minutes at 37° in a glass-stoppered, 13-ml tube containing *N*-acetylserotonin (1 m*M*) and [^{14}C-methyl]*S*-adenosylmethionine (6–9 μ*M*) (specific activity 50–60 mCi/mmole) in a final volume of 300 μl. The incubation is stopped by the addition of 1 ml of 0.2 *M* borate buffer, pH 10, and the [^{14}C]melatonin formed is extracted into 8 ml of chloroform. The organic phase is washed twice with 1 ml of borate buffer, and 4 ml is transferred to a scintillation vial. After evaporation, 1 ml of absolute ethanol and 10 ml of toluene phosphor are added to each vial and the ^{14}C radioactivity is determined in a liquid scintillation spectrophotometer. The radioactive methylated product is identified as melatonin by thin-layer or paper chromatography. Blank values are obtained by omitting the enzyme from the incubation medium. Some R_f's of melatonin in various thin-layer chromatography systems[14] are listed in the table.

R_f's OF MELATONIN IN THIN-LAYER CHROMATOGRAPHY SYSTEMS

TLC system	Melatonin R_f
Isopropanol–25% aqueous ammonia (v/v)–water (85:5:15, v/v/v)	0.82
Chloroform–methanol–glacial acetic acid (93:7:1, v/v/v, 2 runs)	0.62
Methanol	0.71
Ethyl acetate	0.28
Diethyl ether	0.15

Techniques for Studying Pineal Indole Metabolism in Organ Culture

Pineal organ cultures provide an excellent model for studying the input-output relationships of the pineal *in vitro*. The use of such cultures made possible the demonstration that norepinephrine is the neurotransmitter released by pineal sympathetic nerve ending terminals which control pineal indole metabolism,[24] that the catecholamine acts by stimulating a β-adrenergic receptor,[25] and that its effects are simulated by dibutyryl cyclic 3′,5′-AMP.[92] The cultured pineal organ of the rat is able to convert [^{14}C]tryptophan both to [^{14}C]protein[93] and to such characteristic [^{14}C]indoles as [^{14}C]serotonin, [^{14}C]melatonin, and [^{14}C]5-hydroxy-

[92] H. M. Shein and R. J. Wurtman, *Science* **166**, 519 (1969).

[93] R. J. Wurtman, H. M. Shein, J. Axelrod, and F. Larin, *Proc. Nat. Acad. Sci. U.S.* **62**, 1015 (1969).

indoleacetic acid. The [^{14}C]indoles are released into the incubation medium; the [^{14}C]proteins are largely retained within the pineal itself. The addition of norepinephrine to the culture medium markedly increased the rate of synthesis of all [^{14}C]tryptophan derivatives[24,93]; it also increases the adenyl cyclase activity and cyclic AMP content of the pineal.[26,27] Dibutyryl cyclic AMP reproduces the stimulation of the [^{14}C]indole synthesis (but not of [^{14}C]protein synthesis) by norepinephrine. The action of norepinephrine in stimulating melatonin biosynthesis is blocked by propranolol but not by phenoxybenzamine.[25]

Environmental lighting directly modifies avian (but not mammalian) pineal biosynthetic activity. Duck pineal explants maintained under low intensities exhibit higher HIOMT activity,[94] and convert more [^{14}C]serotonin to [^{14}C]melatonin, than those under higher intensities.[89] The incorporation of uridine to RNA in explants of duck cerebral cortex is not modified by light.[89] These observations suggest that the avian pineal may function *in vivo* as a photoendocrine transducer.[95]

The following procedure has been used for studying indole metabolism in cultures of rat pineal organs. Pineal explants obtained by sterile dissection are either clotted to the walls of a Wasserman tube,[96] or added to the culture medium.[97] The following chemically defined nutrient media (0.5 ml) have been found suitable for rat pineal organ culture: (a) a mixture of Puck N 16 (75%), Evans NCTC 109 (5%) fetal calf serum (10%), and heat-inactivated horse serum (10%)[96]; (b) modified BGJ medium supplemented with 1 mg/ml of bovine serum albumin.[97] (Duck pineal explants have been cultured in TC 199 medium.[94]) The radioactive precursor (i.e., tryptophan, 5-hydroxytryptophan, or serotonin) is added to the medium, and the cultures are maintained for 24–96 hours at 37°, either under a 5% oxygen–95% carbon dioxide atmosphere, or in sealed roller tubes. The following procedure has been described for assaying the [^{14}C]indoles (i.e., [^{14}C]serotonin, [^{14}C]melatonin and [^{14}C]5-hydroxyindoleacetic acid) produced from [^{14}C]tryptophan or [^{14}C]5-hydroxytryptophan.[98]

An aliquot (0.1 ml) of the nutrient medium is transferred to a glass-stoppered centrifuge tube containing 2 ml of 0.1 *M* borate buffer, pH

[94] J. M. Rosner, G. D. Perez Bedes, and D. P. Cardinali, *Life Sci.* (II) **10,** 1065 (1971).

[95] D. P. Cardinali and R. J. Wurtman, *in* "Pineal Physiology" (M. D. Altschule, ed.). MIT Press, Cambridge, Mass., in press.

[96] H. M. Shein, R. J. Wurtman, and J. Axelrod, *Nature* (*London*) **213,** 730 (1967).

[97] D. C. Klein and J. Weller, *In Vitro* **6,** 197 (1970).

[98] R. J. Wurtman, F. Larin, J. Axelrod, H. M. Shein, and K. Rosasco, *Nature* (*London*) **217,** 953 (1968).

10, and the [^{14}C]melatonin is extracted in 6 ml of chloroform. After centrifugation, a 1-ml sample of the aqueous phase is placed in a second glass-stoppered centrifuge tube and the [^{14}C]serotonin is extracted into 6 ml of a mixture of butanol and chloroform (1:3, v/v). The rest of the aqueous phase in the first tube is removed by suction. The chloroform is washed with 2 ml of borate buffer, pH 10, and 4 ml is evaporated in a counting vial; the radioactivity is measured in a liquid scintillation spectrophotometer after the addition of toluene phosphor. The butanol–chloroform phase in the second centrifuge tube is washed with 2 ml of borate buffer, pH 10, and 4 ml of the washed mixture is evaporated in a separate vial; its radioactivity is determined after the addition of phosphor. A further 0.1 ml of the culture medium is transferred to another glass-stoppered centrifuge tube containing 1 ml of 0.2 *N* HCl, and the [^{14}C]5-hydroxyindoleacetic acid is extracted into 6 ml of diethyl ether. The ether layer is drawn off to another centrifuge tube, and the 5-hydroxyindoleacetic acid is reextracted into 2 ml of 0.1 *M* phosphate buffer, pH 7.9. The ether layer is removed by suction, and the aqueous phase is washed again with 6 ml of diethyl ether. The aqueous phase is then acidified with 2 ml of 2 *N* HCl, and the 5-hydroxyindoleacetic acid is reextracted into 6 ml of diethyl ether. A 4-ml sample of the organic phase is now evaporated to dryness in a scintillation vial and counted after the addition of phosphor. Labeled melatonin, serotonin, and 5-hydroxyindoleacetic acid were chromatographically identified in the corresponding fractions.[98] The recoveries of serotonin, melatonin, and 5-hydroxyindoleacetic acid in this procedure were approximately 80%, 90%, and 50%, respectively. The sum of the three radioactive products was taken as an estimate of the total amount of serotonin produced in that tube.

The conversion of [^{14}C]serotonin to [^{14}C]5-hydroxyindoleacetic acid, [^{14}C]5-methoxyindoleacetic acid, [^{14}C]5-hydroxytryptophol, [^{14}C]5-methoxytryptophol, [^{14}C]*N*-acetylserotonin, and [^{14}C]melatonin by rat pineal organ cultures has been examined with the bidimensional thin-layer chromatography system described above.[88]

A double isotopic procedure for the simultaneous determination of [^{14}C]serotonin metabolism and [^{3}H]RNA synthesis (from [^{3}H]uridine) by duck pineal explants has been described.[89] The pineals were cultured for 24 hours with [^{14}C]2-serotonin; 3 hours prior to termination of the culture, [^{3}H]uridine was added to the medium. The metabolites of [^{14}C]serotonin were examined in the nutrient medium by thin-layer chromatography, and the specific activity of [^{3}H]RNA was determined in the pineal explants by a modification of Munro and Fleck's method.[99]

[99] H. N. Munro and A. Fleck, *Biochim. Biophys. Acta* **55,** 571 (1962).

Synthesis and Metabolism of Radiolabeled Melatonin

Radioactive melatonin has been synthesized with an isotopic label in three positions: [^{14}C]2-melatonin, [^{14}C-methoxy]melatonin, and [^{3}H-acetyl]melatonin.[36] To prepare [^{14}C]2-melatonin, [^{14}C]2-serotonin is acetylated with acetic anhydride in the presence of triethylamine to form [^{14}C]2-*N*,*O*-diacetylserotonin. After separation of this compound by paper chromatography, the *O*-acetyl group is removed by hydrolysis with sodium carbonate. The resulting [^{14}C]2-*N*-acetylserotonin is extracted into ethyl acetate and further purified by paper chromatography; [^{14}C]2-melatonin is then prepared enzymatically by incubating [^{14}C]2-*N*-acetylserotonin with HIOMT from bovine pineals, and *S*-adenosylmethionine. The labeled methoxyindole is extracted into chloroform and isolated by paper chromatography. [^{14}C-Methyl]melatonin is prepared by incubation of unlabeled *N*-acetylserotonin with HIOMT and [^{14}C-methyl]-*S*-adenosylmethionine. [^{3}H-Acetyl]melatonin can be synthesized by acetylating 5-methoxytryptamine with acetic [^{3}H]anhyride in the presence of ethyl acetate and triethylamine. These forms of labeled melatonin are also purified by extraction into chloroform and by paper chromatography; [^{3}H]melatonin has been recrystallized to constant specific activity from ethanol–water (1:1, v/v).[31]

The metabolic fate of radiolabeled melatonin was examined in rats given the indole labeled at the [^{14}C]2-melatonin, [^{3}H-acetyl]melatonin or [^{14}C-methoxy]melatonin positions[36,100]; urine was collected and subjected to paper chromatography. The pattern of metabolites observed was largely independent of the position of the label. Three radioactive peaks could be demonstrated, none of which corresponded to melatonin, 5-methoxytryptamine, or 5-methoxyindoleacetic acid. These results indicated that exogenous circulating melatonin was entirely transformed in the body by pathways other than deacetylation and deamination and that the metabolites of melatonin retain both the 5-methoxy and the *N*-acetyl groups. The 6-hydroxylated derivative of melatonin was subsequently identified as the major metabolite of exogenously injected melatonin.[35]

Metabolism of Catecholamines and Cyclic Nucleotides in the Pineal

Norepinephrine has been found in relatively large amounts (up to 14 μg/g) in the rat pineal.[101,102] Its concentration exhibits a 24-hour

[100] S. Kveder and W. M. McIsaac, *J. Biol. Chem.* **236,** 3214 (1961).
[101] R. J. Wurtman and J. Axelrod, *Life Sci.* **5,** 665 (1966).
[102] A. Pellegrino de Iraldi and L. M. Zieher, *Life Sci.* **5,** 149 (1966).

rhythm with maximal values during the night period.[101] Essentially all the norepinephrine in the pineal is located within the sympathetic nerve terminals. The rat pineal organ also contains large amounts of dopamine which, unlike norepinephrine, is probably located within pineal parenchymal cells.[102] Both catecholamines in the pineal can be fluorometrically measured after column chromatographic separation on alumina, followed by oxidation.[103] All the enzymes involved in the biosynthesis of norepinephrine appear to be present in the pineal organ. Tyrosine hydroxylase (EC 1.14.16.2), the enzyme that catalyzes the conversion of tyrosine to dopa, has been found in the rat pineal,[104] as well as aromatic-L-amino acid decarboxylase (see above).[12,70] Moreover, the administration of L-dopa was associated with a moderate increase in pineal norepinephrine,[105] which indicates that dopamine-β-hydroxylase (EC 1.14.17.1) is also present in the rat pineal. Two enzymes involved in the metabolism of norepinephrine, MAO (see above) and catechol *O*-methyltransferase (EC 2.1.1.6), have been detected in pineal homogenates.[82,106]

Some of the effects of pineal norepinephrine in stimulating melatonin synthesis (see above) are probably mediated by cyclic AMP. The adenyl cyclase activity and the cyclic AMP content of the rat pineal are increased by norepinephrine[26,27]; adenyl cyclase is not stimulated by other biogenic amines or polypeptide hormones that activate this system in other tissues. Sex hormones influence the response of pineal adenyl cyclase to catecholamines. Norepinephrine stimulates pineal adenyl cyclase activity more in male than in female rats.[107] Previous ovariectomy increases the response of the adenyl cyclase to norepinephrine, while estrogen treatment reduces the norepinephrine-induced activation of the adenyl cyclase. A cyclic AMP-dependent kinase that phosphorylates histones has recently been described in the rat pineal, and may mediate some of the norepinephrine effects *in vivo*.[108]

A method has recently been developed for the isolation and concentration of melatonin from human urine. The method involves the use of Amberlite XAD-2, a nonionic polymeric adsorbent employed in columns in a manner similar to ion exchange resins, to adsorb hydrophobic solutes from polar solvents. This procedure, in conjunction with the melatonin

[103] U. S. von Euler and F. Lishajko, *Acta Physiol. Scand.* **51,** 348 (1961).

[104] E. G. McGeer and P. L. McGeer, *Science* **153,** 73 (1966).

[105] L. M. Zieher and A. Pellegrino de Iraldi, *Life Sci.* **5,** 155 (1966).

[106] J. Axelrod, P. D. MacLean, R. W. Albers, and H. Weissbach, *in* "Regional Neurochemistry" (S. S. Kety and J. Elkes, eds.), p. 307. Pergamon Press, Oxford, 1961.

[107] B. Weiss and J. W. Crayton, *in* "Role of Cyclic AMP in Cell Function" (P. Greengard and E. Costa, eds.), p. 207. Raven Press, New York, 1970.

[108] J. A. Fontana and W. Lovenberg, *Proc. Nat. Acad. Sci. U.S.* **68,** 2787 (1971).

bioassay technique described previously, provides a method for the quantitative assessment of urinary melatonin excretion. (With further development, it may provide a means of monitoring mammalian pineal function *in vivo*.)

Fifty- to 300-ml aliquots of urine, adjusted to pH 10 and made 1 *M* in sodium chloride, are passed over columns of hydrated Amberlite XAD-2. Melatonin, adsorbed onto the resin through van der Waal's interactions, is eluted in a small volume (8 ml) of chloroform/isopropanol 3:1. The organic eluate is then washed successively with salt-saturated alkali and salt-saturated acid, and evaporated to dryness. The residue is dissolved in water, and the melatonin content estimated by bioassay. This procedure yields a cleaner extract than simple organic solvent extraction techniques, and many compounds which might interfere with subsequent evaluation are either reduced or eliminated altogether. Because the dermal melanophore response of test animals exposed to extracted melatonin is not identical to that of animals exposed to simple aqueous solutions of melatonin, it is necessary to subject the graded quantities of melatonin used in preparation of the bioassay calibration curve to the same extraction procedure. Using this method, a clear daily rhythm has been demonstrated in the excretion of melatonin into human urine: the amounts of melatonin excreted between 11 PM and 7 AM are many times greater than between 7 AM and 3 PM or 3 PM and 11 PM.

Nonindolic Pineal Compounds

Some evidence has accumulated during the last few years that the pineal contains nonindolic compounds with antigonadal properties in mammals.[109-112] It has been suggested that these compounds include peptides of low molecular weight. The blockade of the compensatory ovarian hypertrophy that follows unilateral ovariectomy in mice has been utilized as an assay for antigonadotropic activity in melatonin-free extracts of bovine pineals.[111]

Acknowledgment

These studies were supported in part by grant USPHS-AM-11709 from the National Institutes of Health.

[109] I. Ebels, A. Moszkowska, A. Scemana, and M. R. Courrier, *C. R. Acad. Sci.* **260**, 5126 (1965).
[110] I. Ebels, A. Moszkowska, and A. Scemana, *J. Neuro-Visceral Relations* **32**, 1 (1971).
[111] B. Benson, M. J. Matthews, and A. E. Rodin, *Life Sci.* (I) **10**, 607 (1971).
[112] B. Benson, M. J. Matthews, and A. E. Rodin, *Acta Endocrinol.* **69**, 257 (1972).

[34] Pineal Gland Organ Culture Techniques

By HARVEY M. SHEIN

Organ culture techniques which maintain intact mammalian or avian pineal glands for 24 hours or longer after explantation permit a wide range of biochemical studies to be carried out on these glands *in vitro*.[1-5]

Organ Culture Methods

Principles. Two different techniques of pineal organ culture have been employed. Method 1: The gland is clotted to the wall of a culture tube and incubated with nutrient medium on a slowly turning roller wheel in an air atmosphere at 37° with the culture tube sealed. Method 2: The gland is incubated at 37° in a stationary position on a platform at the interface between the nutrient medium and a continuous, water-saturated, gassing atmosphere of 95% O_2 and 5% CO_2. The plasma clot-roller wheel technique can be said to be more "physiological" in that it employs an air atmosphere; however, the continuously replenished enriched O_2 atmosphere technique prevents the development of central necrosis in the cultured gland, which occurs reproducibly when the gland is cultured in an air atmosphere. The small volume of central necrosis which develops with the air atmosphere method is sufficiently reproducible among pineals prepared from inbred rats of the same strain, age, sex and weight so that the variation in synthesis of radiolabeled indoles and protein among a group of six to ten cultured pineal glands remains within a standard error of 15% of the means of the totals during 2 days of incubation. Nevertheless, the absence of necrosis with method 2 makes this technique preferable for most biochemical studies.

Method 1

This technique has been used to maintain intact rat pineal glands in organ culture for periods up to 4 days.[6]

[1] J. Axelrod, H. M. Shein, and R. J. Wurtman, *Proc. Nat. Acad. Sci. U.S.* **62**, 544 (1969).

[2] H. M. Shein and R. J. Wurtman, *Science* **166**, 519 (1969).

[3] D. C. Klein, G. R. Berg, and J. Weller, *Science* **168**, 979 (1970).

[4] I. T. Lott, R. H. Quarles, and D. C. Klein, *Biochim. Biophys. Acta* **264**, 144 (1972).

[5] J. M. Rosner, G. Declerq de Perez Bedes, and D. P. Cardinali, *Life Sci.* **10**, 1065 (1971).

[6] H. M. Shein, R. J. Wurtman, and J. Axelrod, *Nature (London)* **213**, 730 (1967).

TABLE I
COMPOSITION OF N16 MEDIUM[a]

Component	mg/liter	Component	mg/liter
NaCl	7400.00	L-Aspartic acid	30.00
KCl	285.00	L-Proline	25.00
$Na_2HPO_4 \cdot 7H_2O$	290.00	Glycine	100.00
$MgSO_4 \cdot 7H_2O$	154.00	L-Glutamine	200.00
$CaCl_2 \cdot 2H_2O$	16.00	L-Tyrosine	40.00
KH_2PO_4	83.00	L-Cystine	7.50
Glucose	1100.00	Hypoxanthine	25.00
L-Arginine HCl	37.50	Thiamine · HCl	5.00
L-Histidine HCl	37.50	Riboflavin	0.50
L-Lysine HCl	80.00	Pyridoxine · HCl	0.50
L-Tryptophan	20.00	Folic acid	0.10
β-Phenyl-L-alanine	25.00	Biotin	0.10
L-Methionine	25.00	Choline	3.00
L-Threonine	37.50	Ca pantothenate	3.00
L-Leucine	25.00	Niacinamide	3.00
DL-Isoleucine	25.00	*i*-Inositol	1.00
DL-Valine	50.00	$NaHCO_3$	1200.00
L-Glutamic acid	75.00		

[a] T. T. Puck, S. J. Cieciura, and A. Robinson, *J. Exp. Med.* **108**, 945 (1958).

Reagents

Hanks' balanced salt solution
Chicken plasma[7]
Chick embryo extract (50% v/v in isotonic saline)[7]
N16 culture medium (composition in Table II)
NCTC-135 culture medium (composition in Table II)
Lactalbumin hydrolyzate (5%) in Earle's balanced salt solution
Fetal calf serum, heat-inactivated at 56° for 30 minutes
Horse serum, heat-inactivated at 56° for 30 minutes
Penicillin G
Streptomycin
Amphotericin B

Procedure. All steps are carried out under aseptic conditions to maintain sterility of the pineal glands, the culture vessels and their contents. Rat pineal glands are removed by sterile dissection following decapitation and placed in a petri dish containing Hanks' balanced salt solution. Each intact pineal gland is then clotted to the glass wall of a

[7] Prepared as described in R. C. Parker, "Methods of Tissue Culture," 3rd ed., Chapter 10. Harper (Hoeber), New York, 1961.

TABLE II
COMPOSITION OF NCTC-135 MEDIUM[a]

Component	mg/liter	Component	mg/liter
NaCl	6800.0000	Niacin	0.0625
KCl	400.0000	Niacinamide	0.0625
$CaCl_2$ (anhyd.)	200.0000	D-Ca pantothenate	0.0250
$MgSO_4$	100.0000	Biotin	0.0250
$NaH_2PO_4 \cdot H_2O$	140.0000	Folic acid	0.0250
D-Glucose	1000.0000	Choline Cl	1.2500
L-Alanine	31.5000	Vitamin B_{12}	10.0000
L-α-Amino-*n*-butyric acid	5.5000	*i*-Inositol	0.1250
L-Arginine HCl	31.2000	*p*-Aminobenzoic acid	0.1250
L-Asparagine · H_2O	9.2000	Vitamin A	0.2500
L-Aspartic acid	9.9000	Calciferol	0.2500
L-Cystine	10.5000	Menadione	0.0250
D-Glucosamine HCl	3.9000	Disodium α-tocopherol phosphate	0.0250
L-Glutamic acid	8.3000		
L-Glutamine	135.7000	Sodium glutathione	10.0000
Glycine	13.5000	Ascorbic acid	50.0000
L-Histidine HCl · H_2O	26.7000	Diphosphopyridine nucleotide	7.0000
Hydroxy-L-proline	4.1000		
L-Isoleucine	18.0000	Triphosphopyridine nucleotide (monosodium)	1.0000
L-Leucine	20.4000		
L-Lycine HCl	38.4000	Coenzyme A	2.5000
L-Methionine	4.4000	Cocarboxylase	1.0000
L-Ornithine HCl	9.4000	Flavin adenine dinucleotide	1.0000
L-Phenylalanine	16.5000	Sodium uridine triphosphate	1.0000
L-Proline	6.1000	Deoxyadenosine	10.0000
L-Serine	10.8000	Deoxyguanosine	10.0000
L-Taurine	4.2000	Thymidine	10.0000
L-Threonine	18.9000	5-Methylcytosine	0.1000
		Tween 80	12.5000
L-Tryptophan	17.5000	D-Glucuronolactone	1.8000
L-Tyrosine	16.4000	Sodium glucuronate · H_2O	1.8000
L-Valine	25.0000	Sodium acetate · $3H_2O$	50.0000
Thiamine · HCl	0.0250	Ethanol · for solubilizing lipid components	40.0000
Riboflavin	0.0250		
Pyridoxine · HCl	0.0625	Phenol red	20.0000
Pyridoxal · HCl	0.0625	$NaHCO_3$	2200.0000

[a] V. J. Evans, J. C. Bryant, H. A. Kerr, and E. L. Schilling, *Exp. Cell Res.* **36**, 439 (1964).

Wasserman tube, which has been coated previously with a thin film of chicken plasma (dispensed from a Pasteur pipette) by the application of a drop or two of chick embryo extract (dispensed from a Pasteur pipette).[6,7] Fifteen minutes later, when the clot is sufficiently firm, 0.5 ml of complete nutrient medium (see below) is added to the culture. The culture tube is then sealed with a rubber stopper and incubated in a standard tissue culture tube roller wheel at 37°C.[6,7] The complete nutrient medium[8] consists of 75.5% N16 medium, 4.0% NCTC-135 medium, 10.0% heat-inactivated fetal calf serum; 10.0% heat-inactivated horse serum, 0.5% of 5% lactalbumin hydrolyzate (Earle's), penicillin G (50 units per milliliter of medium), streptomycin sulfate (50 μg per milliliter of medium), and amphotericin B (1 μg per milliliter of medium). All the constituents of the complete nutrent medium, as well as the chick embryo extract, the chicken plasma, and the Hanks' BSS, can be obtained as standard preparations from Grand Island Biological Co., Grand Island, New York. For studies in which serotonin is to be assayed by spectrophotometric techniques, the media are prepared without phenol red, because this indicator dye interferes with the assay.

Method 2

Stationary Incubation at Interface of the Nutrient Medium and an Enriched O_2 Atmosphere. This technique has been used to maintain intact rat pineal glands in culture without evidence of necrosis for periods up to 6 days.[9,10]

Reagents

Modified BGJb culture medium (composition in Table III)
Bovine serum albumin, Fraction V

Procedure. All steps are carried out under aseptic conditions to maintain sterility of the pineal glands, the culture vessels, and their contents. Pineal glands are obtained by sterile dissection following decapitation. One or two intact pineal glands are incubated on a stainless steel screen (see below) in a specially designed culture vessel (see below) at 37° in the presence of 0.5 ml of chemically defined, modified BGJb culture medium (Table III) supplemented with fraction V bovine serum albumin[9,11,12] (Armour Pharmaceutical Co.) at a final concentration in the complete medium of 1 mg/ml, and are continually gassed with a mixture

[8] G. Yerganian, H. M. Shein, and J. F. Enders, *Cytogenetics* **1**, 314 (1962).
[9] D. C. Klein and J. Weller, *In Vitro* **6**, 197 (1970). This paper provides a sketch of the culture vessel with a pineal gland in place.
[10] R. B. Berg and D. C. Klein, *Endocrinology* **89**, 453 (1971).
[11] L. G. Raisz, *J. Clin. Invest.* **44**, 103 (1965).
[12] L. G. Raisz and I. Niemann, *Endocrinology* **85**, 446 (1969).

TABLE III
COMPOSITION OF MODIFIED BGJb MEDIUM[a]

Component	mg/liter	Component	mg/liter
Calcium lactate	555.00	L-Alanine	250.00
Dihydrogen sodium orthophosphate	90.00	L-Aspartic acid	150.00
		Glycine	800.00
Glucose	10,000.00	L-Proline	400.00
Magnesium sulfate · $7H_2O$	200.00	L-Serine	200.00
Potassium chloride	400.00	Nicotinamide	20.00
Potassium dihydrogen phosphate	160.00	Thiamine hydrochloride	4.00
		Calcium pantothenate	0.20
Sodium bicarbonate	3,500.00	Riboflavin	0.20
Sodium chloride	5,300.00	Pyridoxal phosphate	0.20
L-Lysine	240.00	Folic acid	0.20
L-Histidine	150.00	Biotin	0.20
L-Arginine	175.00	*p*-Aminobenzoic acid	2.00
L-Threonine	75.00	α-Tocopherol phosphate	1.00
DL-Valine	65.00	Choline chloride	50.00
L-Leucine	50.00	Inositol	0.20
L-Isoleucine	30.00	Vitamin B_{12}	0.04
L-Methionine	50.00	Sodium acetate	50.00
L-Phenylalanine	50.00	Ascorbic acid	50.00
L-Tryptophan	40.00	Phenol red	20.00
L-Tyrosine	40.00	Streptomycin	50.00
L-Cysteine HCl	90.00	Penicillin	100,000 units
L-Glutamine	200.00		

[a] L. G. Raisz and I. Niemann, *Endocrinology* **85**, 446 (1969).

of 95% oxygen and 5% carbon dioxide (v/v) in a water-saturated atmosphere.[9,11] The stainless steel screen maintains the culture at the interface of the medium and the gassing atmosphere. The culture vessel is a U.S. Bureau of Plant Industry watch glass (The A. H. Thomas Company), fitted with a 1.5 mm mesh size stainless steel expanded screen (The Expanded Metal Company, Stranton Works, Hartlepool, England) platform fabricated in a machine shop to a 1.7 cm diameter with a curled edge 1 mm deep.[9] To maintain a water-saturated atmosphere during incubation, up to seven of the culture vessels are placed on a 9-cm disc of Whatman No. 3M paper, previously soaked with 3 ml of incubation medium, in a 10-ml petri dish. Each petri dish is then incubated on one of five shelves (spaced 1 inch apart), in a small gas-tight Lucite chamber (with space for a total of 20 petri dishes), fabricated to these specifications in a machine shop, which is perfused continuously (0.3 liter per minute) with the gassing mixture.[9,10]

Intact pineal glands from 10-month-old domestic ducks have also been maintained in organ culture for 24 hours without necrosis by the technique (Method 2) described above, with the modifications that each gland is cultured individually, and a different nutrient medium; i.e., medium 199[13] is employed at a slightly higher temperature, 38°C.[5]

[13] J. F. Morgan, H. J. Morgan, and R. C. Parker, *Proc. Soc. Exp. Biol. Med.* **73**, 1 (1950).

[35] Use of O_2 Microtechniques for Assessment of Hormone Effects on Isolated Endocrine Cells

By LARS HAMBERGER, HANS HERLITZ, and RAGNAR HULTBORN

Methods for microgasometric measurements in biological material have been used and developed during the last four decades. In 1937 Linderstrøm-Lang introduced the Cartesian diver principle for this purpose.[1] The sensitivity of the original device was in the order of 10^{-1} to 10^{-2} μl per hour, which means a considerable increase in sensitivity compared to the Warburg technique.[2] In 1953 another Danish scientist[3] introduced the ampulla or microdiver respirometer with which cellular respiration could be determined in the order of 10^{-4} to 10^{-5} μl per hour.[3] An even more sensitive technique, the magnetic diver microgasometer, was introduced in 1964[4] and recently further developed.[5] This technique gives reliable measurements in the magnitude of 10^{-6} to 10^{-7} μl per hour, specifically interesting when taking into consideration that single tissue cells respire with intensities approximately 10^{-6} μl per hour.

One essential problem in this connection is concerned with procedures for alteration of the medium composition during an experiment. Numerous ingenious arrangements for addition of reactants and inhibitors to the incubation medium have been described for the Warburg technique.[2] However, in microscale this problem is even more intricate. Some of the techniques for addition of reactants and inhibitors in use for standard-divers and ampulla divers will be described in the following.

[1] K. Linderstrøm-Lang, *Nature (London)* **140,** 108 (1937).

[2] W. W. Umbreit, R. H. Burris, and J. F. Stauffer, "Manometric Techniques: A Manual Describing Methods Applicable to the Study of Tissue Metabolism." Burgess, Minneapolis, Minnesota, 1964.

[3] E. Zeuthen, *J. Embryol. Exp. Morphol.* **1,** 239 (1953).

[4] M. Brzin and E. Zeuthen, *C. R. Trav. Lab. Carlsberg* **34,** 427 (1964).

[5] S. Oman and M. Brzin, *Anal. Biochem.* **45,** 112 (1972).

A principally new technique for determination of oxygen consumption in microscale has recently been developed in our laboratory,[6] and this microspectrophotometric technique will be described and compared to the Cartesian diver techniques.

All the above-mentioned techniques have been widely used for determinations of cellular respiration in many different types of cells, but in this presentation we will illustrate the usefulness of the techniques only in some endocrine cell types.

The Standard Diver Technique

The technique is suitable for measurements of gas consumption in the order of 10^{-1} to 10^{-3} μl per hour.[7,8] The standard diver (microliter diver) in use in our laboratory is made from borosilicate glass (Friedrich & Dimmock, Millville, New Jersey, e.g., type KG-33) which has a density of 2.23 g/cm^3. The divers are cylindrical and are coated with silicone (MS 200, 350 cS; Midland Silicones Ltd., London, SW 1) in a 2% (v/v) solution of toluene. As can be seen from Fig. 1, each diver is charged with 1% (w/v) potassium hydroxide (KOH) to absorb the CO_2, which is liberated by the respiring cells. Small drops with reactants or inhibitors are placed on the inside wall of the diver below and above the incubation medium containing the cell sample. Paraffin oil (Uvasol, E. Merck AG, Darmstadt) is placed above the upper drop to prevent the exchange of water between the aqueous solutions in the diver and the strong salt solution (flotation medium) surrounding the diver during the experiment. This medium, composed by Holter,[8] is a solution that contains 46.1 g of $NaNO_3$, 23.2 g of NaCl, and 0.34 g of sodium taurocholate in 100 ml of water (density = 1.326) and has an extremely low solubility for gas. The flotation medium in the top of the diver therefore serves as a barrier against the escape of gas. The source of oxygen for the respiring cells is air.

Divers. Most of the methods for making divers that have been described adhere to the principle used by Boell *et al.*[9] in which the pressure of the air volume enclosed in a sealed piece of capillary tubing is employed in blowing up the bulb of the diver. The method is well suited for making relatively thick-walled divers with a volume larger than 10 μl. Another method of blowing divers has been described by Holter.[8] The

[6] R. Hultborn, *Anal. Biochem.* **47,** 442 (1972).

[7] H. Holter, *in* "General Cytochemical Methods" (J. F. Danielli, ed.), Vol. 2, p. 93. Academic Press, New York, 1961.

[8] H. Holter, *C. R. Trav. Lab. Carlsberg* **24,** 399 (1943).

[9] E. J. Boell, J. Needham, and V. Rogers, *Proc. Roy. Soc. Ser. B.* **127,** 322 (1939).

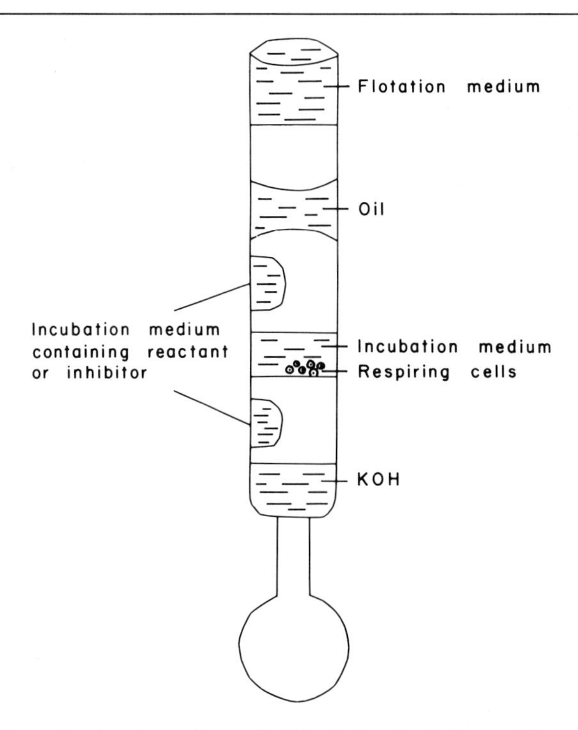

FIG. 1. Schematic drawing of a cylindrical standard diver. Dimensions: length between 15 and 20 mm, diameter 1.0 and 1.3 mm. The diver is filled from the bottom by means of volume-calibrated micropipettes. Small drops of medium containing reactants or inhibitors are placed on the inside wall of the diver above and under the medium containing the cells. After determination of the control respiratory rate the pressure around the diver is either decreased or increased. This procedure leads to mixing of the small drops with the incubation medium. The design makes it possible to introduce two alterations of the composition of the incubation medium during the experiment. Modified after C. Hellerström, *Endocrinology* **81**, 105 (1967).

blowing pressure here is supplied by mouth through a piece of rubber tubing. The bulb of the diver is situated between the neck and the tail, and this is considered the conventional type of standard diver. In our laboratory the divers are cylindrical[10,11] and made in the following manner: One end of a capillary tube with known diameter is collapsed in a flame. This tube becomes the body of the diver. The glass destined for the diver tail is taken from a waste glass capillary with one end sealed. The two sealed ends of the capillaries are then fused together during con-

[10] C. Hellerström, *Biochem. J.* **98**, 7 c (1966).
[11] C. Hellerström, *Endocrinology* **81**, 105 (1967).

tinuous rotation over a microflame. The tail end is heated strongly and shortened to a bulb. The neck is cut at desired length. The diver thus consists of an upper hollow part and a tail of solid glass. The inner diameter of the divers used is between 1.00 and 1.30 mm and varies from 15 to 20 mm in length. One must choose glass tubing so thin that the buoyancy of the diver makes it possible to place the center of gravity so low that the diver floats in an upright position by means of solid glass in the tail. The wall thickness should be about 1/9 of the inner diameter of the tube.

The most frequently used diver sizes are one with an inside diameter of 1.00–1.05 mm and another with a corresponding value of 1.25–1.30 mm. The gas volume of a charged diver at equilibrium pressure for the two diver types is around 7 and 13 μl, respectively.

Filling of the Diver. The divers are filled by means of thin capillaries capable of measuring volumes of 0.2–2 μl with an accuracy of about 99%. They are operated by pressure from the mouth. The pipettes are calibrated spectrophotometrically by using nicotinamide. The different solutions are introduced into the diver by the following procedure. (The volumes which will be mentioned concern the smaller type of diver.) First, 0.75 μl of 1% (w/v) KOH is inserted at the bottom of the diver. Then a side drop consisting of 0.25 μl Tris·HCl buffer containing a test substance is placed on the wall just above the bottom drop. The diver is then submerged under distilled water maintained at 37° in a small thermostatic bath mounted on a stand that can be raised or lowered by means of a rack and pinion.[12] The medium (0.75 μl) is then placed as a neck seal which is held by surface tension. Before the pipette with the buffer solution is introduced into the diver some air is drawn into the tip of the pipette to prevent mixing with the surrounding distilled water. The cell sample, which is dissected free under a stereomicroscope is then transferred to the medium together with some buffer by means of a braking pipette.[7,13] The width of this pipette is chosen according to the size of the tissue sample to be used. The principle of this pipette is based on regulating the rate of intake and discharge, not by the dimensions of the mouth of the pipette, but by slowing down the air current which simultaneously runs through the upper opening of the pipette. The brake consists of a very narrow capillary channel forming the upper end of the pipette. The same amount of buffer that accompanies the tissue in the braking pipette must be sucked back into the pipette when the tissue has left it to keep the medium volume constant. This can be done by means of a

[12] J. Kieler, *J. Nat. Cancer Inst.* **25**, 161 (1960).
[13] C. L. Claff, *Science* **105**, 103 (1947).

horizontal microscope equipped with a micrometer screw. Another side drop is then placed above the medium. The filling of the diver is completed by introducing a neck seal consisting of 0.80 μl of paraffin oil. Finally the mouth is temporarily filled with distilled water.

The diver is then transferred to a flotation vessel in a thermostatically controlled water bath (±0.01°), whereafter the distilled water in the mouth is replaced by flotation medium with a pipette. The length of the mouth seal is then adjusted with a narrow-tipped braking pipette until the diver equilibrium pressure is a few centimeters of Brodie's solution below atmospheric pressure. The flotation vessel is then connected by a manifold to a manometer containing Brodie's solution. By means of compression screws the pressure can be adjusted so that the diver is made to float at a certain level in the flotation medium. After a temperature equilibration time of 10 minutes the manometric registrations of the equilibrium pressure are performed every 10 minutes. When the equilibrium pressure has been followed for a control period a positive pressure is applied so that the lower side drop containing the test substance mixes with the incubation medium containing the cells.[14,15] After a new period of manometric registrations a negative pressure is applied to the system. The upper drop then will mix with the medium. With this technique it is consequently possible to introduce two alterations in the composition of the incubation medium during an experiment. After each experiment the diver is picked out of the flotation vessel and the tissue sample is recovered under a stereomicroscope by careful flushing of the upper part of the diver with distilled water from a fine pipette. The tissue is then placed on a platinum foil, and weighed on an ultramicrobalance (UM7) sensitive to 0.1 μg. The platinum foil is then weighed alone, and the weight of the tissue fragment can be determined. The oxygen consumption is calculated from the slope of the equilibrium pressure-time curve according to Holter.[7,8]

The Microdiver Technique

Divers. The divers are made of thin-walled Pyrex glass capillaries about 1 mm in diameter. The thickness of the wall should permit the charged diver to be floated, and as a thumb-rule they should easily crack between two fingers by moderate pressure. The ampulla of the diver is formed by heating the capillary in both ends, drawing it out and leaving the central part unheated (Figs. 2 and 3). The length of the diver is around 15 mm, with a dry weight of 0.3–1.0 mg. Each diver is weighed,

[14] C. B. Anfinsen and C. L. Claff, *J. Biol. Chem.* **167**, 27 (1947).
[15] J. C. Waterlow and A. Borrow, *C. R. Trav. Lab. Carlsberg* **27**, 93 (1949).

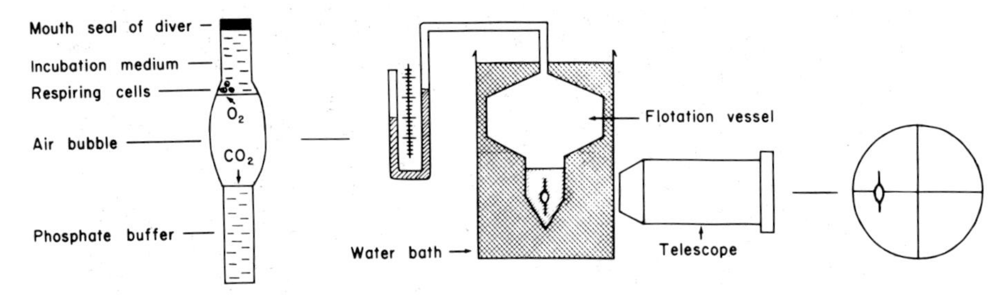

FIG. 2. *Left:* Schematic drawing of a microdiver. *Middle:* Diver in incubation position. *Right:* Telescope for registration of the movements of the diver in the flotation medium.

and the "tail" of the diver is then fastened in a small rubber stopper introduced in a fitting glass tube which also serves as a handle.[16]

Filling of the Diver. The tip of the diver is introduced into a drop of incubation medium containing the cell sample. The cell sample is isolated under a stereomicroscope in a hollow transparent glass cup through which ice-cold water is circulated. Under the stereomicroscope the cell sample is introduced by capillary force into the diver together with approximately 0.3 μl of incubation medium (Fig. 3). The tip of the diver is then sealed with beeswax heated to melting point (about 60°). The diver is then detached from its handle and adjusted to flotation equilibrium in 0.1 *M* phosphate buffer, pH 7.4. This implies that the air bubble in the ampule (the thickest central part of the diver) is bordered by the incubation medium containing the cell sample in the sealed end and by the phosphate buffer in the open end (Fig. 2). The size of the air bubble (0.1–0.4 μl) is such that the diver is just able to remain afloat in the phosphate buffer. The diver is then ready to be transferred to a flotation vessel connected with a manometer (Fig. 2). The flotation vessel is filled with 10 ml of the same phosphate buffer as that used for adjusting the divers since it has been demonstrated[17] that the solubility of carbon dioxide in the phosphate buffer is sufficient for the relevant quantities in this connection.

Water Bath. The flotation vessels are immersed in a thermostatically regulated water bath with good isolation heated by two 250-W lamps dipping down into the water. The temperature variations should not exceed ±0.005°.

After a temperature equilibrium period of 20–30 minutes, manometric readings of the equilibrium pressure are made every 10–15 minutes.

[16] M. Brzin and E. Zeuthen, *C. R. Trav. Lab. Carlsberg* **32,** 139 (1961).

[17] A. Hamberger, *Acta Physiol. Scand.* **58,** Suppl. 203 (1963).

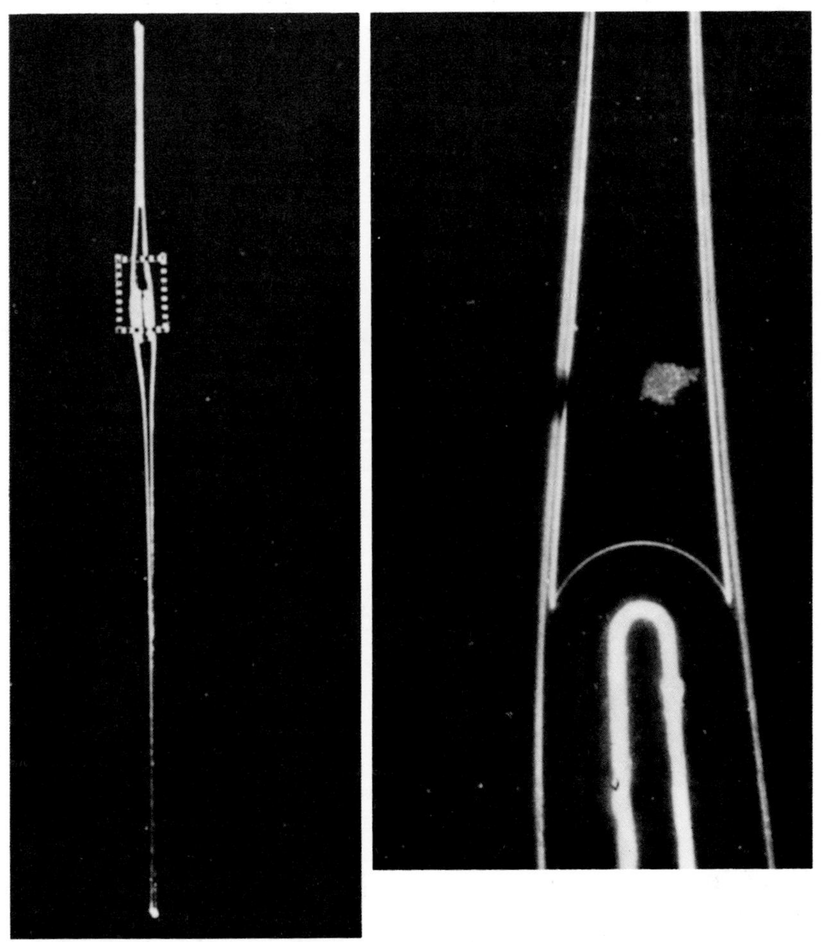

FIG. 3. Photograph of microdiver (left). The dotted rectangle represents the part shown to the right. Position of the cell sample in the diver. At bottom, air bubble.

The equilibrium pressure is plotted against time, and a linear relation is generally found during 3–4 hours and often for a longer period. Blank divers containing incubation medium but no cells must be run in parallel in each experiment. The technique permits registrations in the magnitude of 10^{-3} to 10^{-4} μl per hour. The mathematical calculations for the technique are described in detail elsewhere.[3,18] The error of the microdiver technique has been found to be $\pm 5\%$ both in theoretical calculations

[18] H. Holter and E. Zeuthen, *in* "Physical Techniques in Biological Research," 2nd ed. Vol. III A, (A. W. Pollister, ed.), p. 251. Academic Press, New York, 1966.

and experimentally when gas consumption or production is determined in order of magnitude of 10^{-4} μl per hour.[19]

Methods for Mixing Solutions in the Microdiver. Addition of substrates or metabolic inhibitors during an experiment has been solved in various ways by different laboratories working with the microdiver technique. The microdiver can, after a control registration period, be removed from its flotation vessel in the thermostatically controlled bath and opened; the incubation medium can be replaced by a fortified medium.[16] After this the diver is sealed and replaced in the flotation vessel and the manometric registrations can be continued after a 30-minute thermoequilibration period.

Methods for mixing solutions in the microdiver without removing the diver from its flotation vessel are, however, preferable, and in 1967 Chakravarty[20] described a technique with a small drop containing the reactant placed in the air bubble space of the diver. The drop can be brought to mix with the incubation medium by initiation of an increase of the manometric pressure. Owing to the relatively large size of the air bubble space necessary, the sensitivity of this method is of the order of 10^{-2} μl per hour.

A technique for addition of reactants without removing the diver from the flotation medium and with preserved sensitivity is in use in our laboratory.[21] This so called "two compartment micro-diver" is schematically described in Fig. 4.

Preparation and Filling of the Diver. The diver is made of the same type of glass as described above, but the mouth of the diver should be slightly broader. The cell sample is introduced in the diver together with 0.4–0.6 μl of incubation medium. The diver is then sealed in its upper end by beeswax and adjusted to flotation equilibrium in phosphate buffer (pH 7.4). The beeswax is then removed to enable the introduction of a small capillary (diameter around 70 μm) filled with the reactant to be studied. The agent in the capillary, placed there by dipping the capillary quickly in the fortified medium, is surrounded by two air bubbles of various sizes. The capillary is sealed in its upper end by beeswax and then carefully washed. After introduction of the capillary in appropriate position with its lower end close to the cell sample and its upper end outside of the diver, the diver is sealed again by wax which also serves as a fixative for the capillary. The diver is then transferred to a flotation vessel without further adjustments and after a thermoequilibration period of

[19] J. Zajicek and E. Zeuthen, *in* "General Cytochemical Methods" (J. F. Danielli, ed.), Vol. 2, p. 131. Academic Press, New York, 1961.

[20] N. Chakravarty, *Exp. Cell Res.* **47,** 278 (1967).

[21] L. Hamberger, *Acta Physiol. Scand.* **74,** 91 (1968).

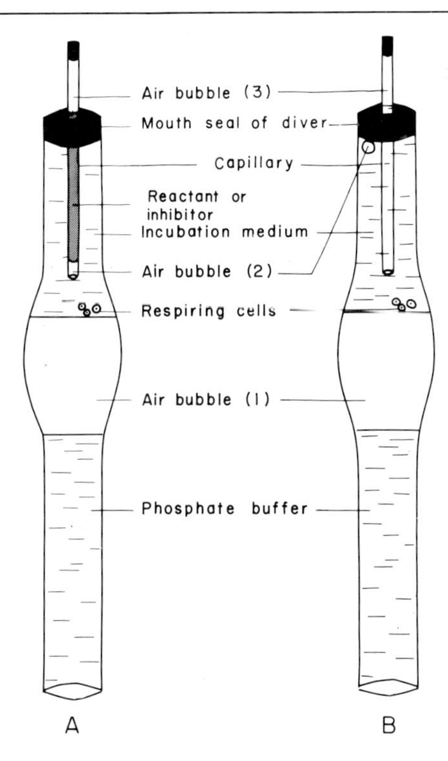

FIG. 4. Schematic drawing of a two-compartment microdiver. Dimensions: length between 20 and 25 mm, diameter 0.3–0.5 mm. The microdiver is filled in its upper part with incubation medium containing cells, and in its lower part with phosphate buffer. A small capillary (diameter approximately 70 μm) containing the reactant or inhibitor, the effect of which is to be studied, is introduced into the upper end of the diver and fixed by beeswax (mouth seal of diver). The small capillary is sealed in its upper end by beeswax, and in both ends there are air bubbles. Air bubble 2 serves as a lock and prevents the content of the capillary to mix with the incubation medium (A). After determination of the control respiratory rate, the pressure around the diver is decreased. This leads to an expansion of air bubble 3, which thus pushes away air bubble 2, allowing the reactant or inhibitor to reach the cells (B). The sizes of air bubble 2 and 3 should be in the ratio 1:4 in order to avoid drastic alterations of the pressure. From L. Hamberger, *Acta Physiol. Scand.* **74**, 91 (1968).

20–30 minutes the manometric readings can start. After measurement of the respiratory rate during a control period, the pressure is decreased in the flotation vessel. This consequently leads to an expansion of all air bubbles, and the small air bubble seal in the lower part of the capillary (Fig. 5) escapes, and thus the content of the capillary can mix with the incubation medium. The procedure does not in itself influence cellular

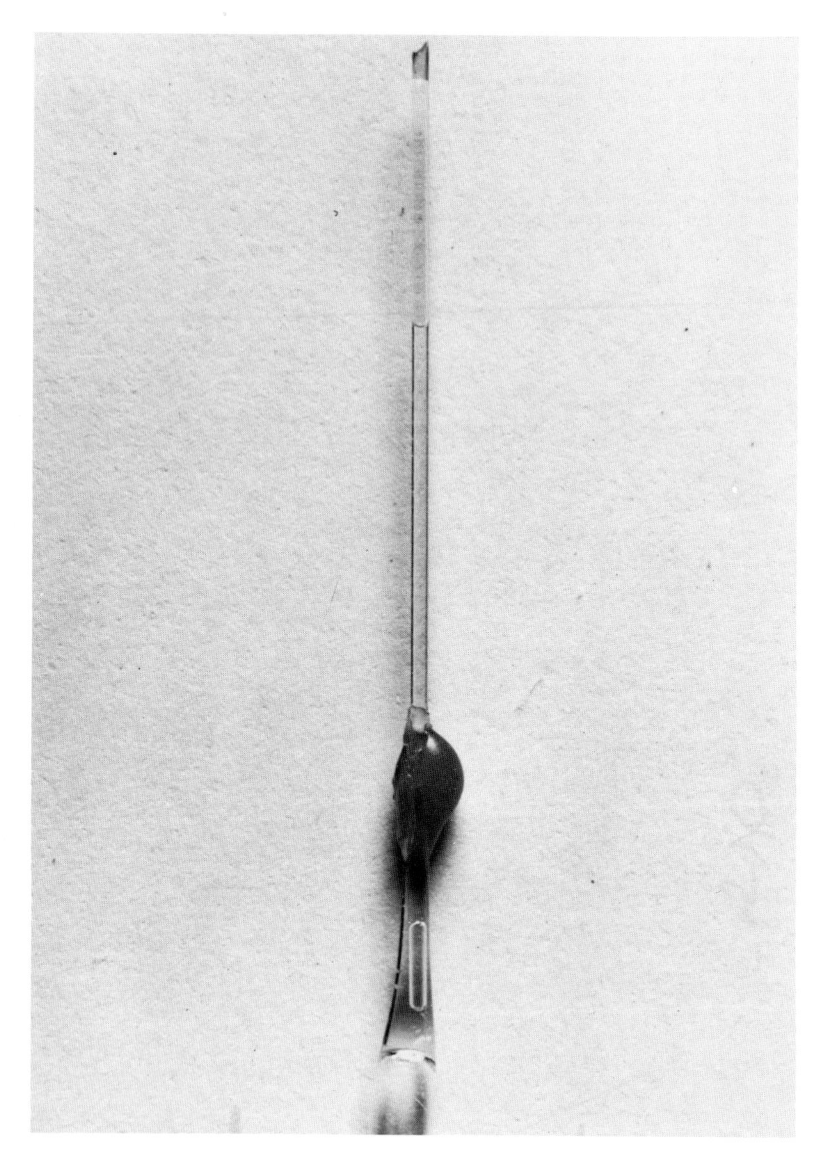

FIG. 5. Photograph of the upper part of a two-compartment microdiver. The capillary is fixed by beeswax in the mouth of the diver. Note the two air bubbles in the capillary. The cell sample is situated close to the big air bubble in the ampule of the diver.

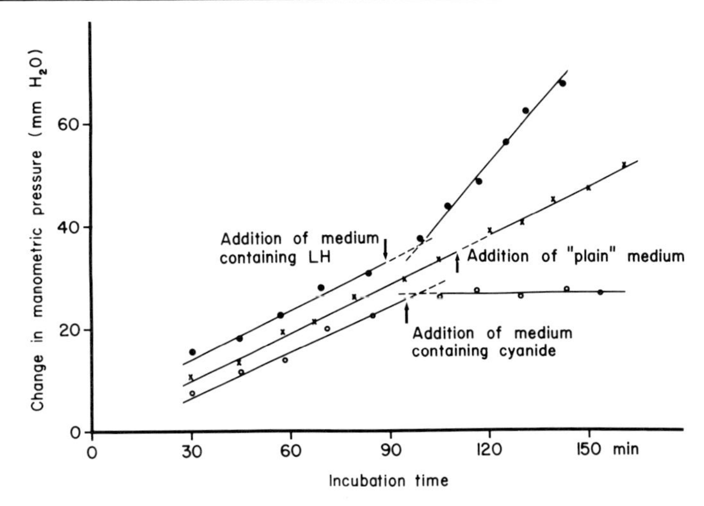

FIG. 6. Experiments performed on isolated granulosa cells obtained from prepubertal rat ovaries applying two-compartment diver technique. The cells were incubated in a medium containing succinate as substrate. The figure shows three different experiments where the oxygen consumption (expressed as change in manometric pressure) is plotted against incubation time. The first 30 minutes of the incubation period is used for the thermoregulation, and readings are then performed every 10 minutes. In a "control" diver, where medium containing no reactant or inhibitor was added from the capillary, there was no change in respiratory rate during the incubation period. Addition of luteinizing hormone (NIH-LH-B3) in a final concentration of approximately 50 μg/ml increased the respiratory rate of the granulosa cells 141% above the control level within 10–15 minutes. Addition of 10 m*M* KCN completely blocked the oxygen uptake within 10 minutes after its addition. From L. Hamberger, *Acta Physiol. Scand.* **74,** 91 (1968).

respiration (Fig. 6 , middle curve), and after a short pressure equilibration period of 2–3 minutes the manometric readings can be reinitiated. The effect of the agent introduced can often be registered after 10–15 minutes (Fig. 6, upper curve). For quantitative purposes the cell sample has to be removed from the diver and weighed after the registration period, but since the influence of various changes in the incubation medium during an experiment often has a more qualitative interest, the two-compartment diver can be used with a minimum of calculations if the effect is expressed as an increase or decrease in percentage of the control respirations (Fig. 6).

Microspectrophotometric Procedure

A review of the micromanometric methods available for studies on minute biological samples has previously been presented. This section

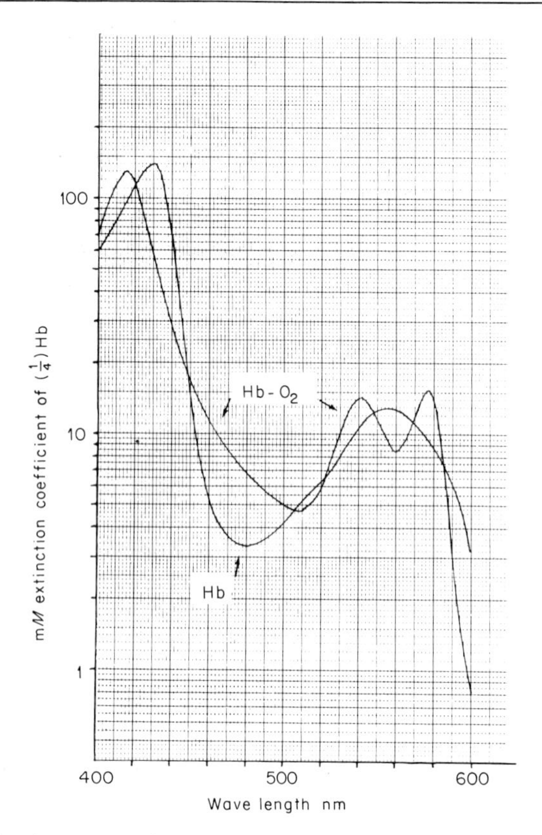

FIG. 7. Extinction curves for oxygenated and deoxygenated hemoglobin. Millimolar extinction coefficients for one heme and globin moiety are given. Any wavelength region where there is a great difference between extinction coefficients of the two types of hemoglobin can be used for obtaining absorbance shift curves.

will deal with a recently introduced technique[6,22] for evaluation of oxygen consumption of tissue fragments of the same dimensions as those studied with the "standard diver" technique, as well as of single isolated cells demanding a sensitivity down to the range of 10^{-6} μl per hour.

The principle of this technique is to incubate the biological material to be studied in a closed chamber containing an oxyhemoglobin solution. Hemoglobin will continuously be reduced by cellular respiration leading to removal of oxygen from the incubation medium. This reduction can be followed and recorded spectrophotometrically since the molecular extinction coefficients of oxygenated and deoxygenated hemoglobin differ at specific wavelengths (Fig. 7). This technique thus utilizes a natural

[22] H. Herlitz and R. Hultborn, *Acta Physiol. Scand.*, **90**, 594 (1974).

biological pigment as indicator of oxygen tension as well as donor of oxygen.[6,22-24] Natural pigments, other than hemoglobin, have been used by other investigators, e.g., cytochromes for studies on fast reaction kinetics in the macroscale.[25,26]

Theoretical Considerations. The oxygen content of a hemoglobin solution in a closed system is divided into two fractions: one physically dissolved and another bound to the hemoglobin. The relative proportions of these fractions depend on the hemoglobin concentration, i.e., at low concentrations the dissolved fraction will dominate and vice versa (see Fig. 12).

If the system contains an oxygen-consuming element, the hemoglobin will serve as both an oxygen donor and an indicator of oxygen tension. However, as follows from the above, at very low hemogloblin concentrations the oxygen contribution from the hemoglobin will be negligible and thus the hemoglobin then serves almost exclusively as an indicator of oxygen tension. Conversely, at high hemoglobin concentrations ($>10^{-3}$ M) the physically dissolved fraction can be disregarded and hemoglobin may be considered to serve as the sole oxygen source and as an indicator of oxygen tension (Fig. 12).

Unlike the physically dissolved oxygen fraction, which is directly proportional to the oxygen tension, the oxygen dissociation curve of hemoglobin is nonlinear to the oxygen tension, most marked at high degrees of saturation. The total oxygen content of the system will thus be nonlinear with respect to the oxygen tension, yet it will be linear for time, provided the rate of oxygen consumption remains constant. Accordingly, the oxygen dissociation curve will be changed when plotted against time and at high hemoglobin concentrations (>1 mM) it will be virtually linear. At very low hemoglobin concentrations the oxygen dissociation curve will retain the shape it has when plotted against oxygen tension. Since a change in hemoglobin saturation is directly proportional to the change in absorbance characteristics of the solution, a grasp of the aforementioned relationships is essential for correct interpretation of absorbance shift curves.

The oxygen consumption can be calculated from the absorbance shift curve which represents the content of the hemoglobin bound oxygen in various ways. One principal approach is to assume that at zero time the solution is equilibrated with air and the partial pressure of oxygen is the volume percent O_2 in air times the barometric pressure corrected for

[23] O. Barzu and V. Borza, *Anal. Biochem.* **21,** 344 (1967).

[24] O. Barzu, L. Muresan, and G. Benga, *Anal. Biochem.* **46,** 374 (1972).

[25] B. Chance, this series, Vol. 4, p. 273.

[26] B. Chance, this series, Vol. 24 p. 322.

water vapor pressure. The partial pressure of oxygen will be 155 mm Hg at a barometric pressure of 760 mm Hg. Since hemoglobin is fully saturated at this pressure, the oxygen bound to hemoglobin can be calculated at zero time taking the oxygen capacity of hemoglobin as 1.34 ml O_2 per gram of hemoglobin. From the solubility coefficient of oxygen the physically dissolved oxygen pool can also be determined. As previously indicated, the absorbance shift curve should be almost linear from an oxygen tension above 50 mm Hg to 0 mm Hg, but since respiration diminishes asymptotically at very low oxygen tensions, the absorbance shift curve must be extrapolated to zero oxygen content using the linear part of the curve. The time value of the abscissa at this point represents the extrapolated time required to consume all the oxygen present in the solution.

Another way of calculation is to determine beforehand the slope of the dissociation curve $[d(\% \text{ sat.})]/[d(pO_2)]$ by means of tonometry in the region of 50% O_2 saturation of the hemoglobin batch used. As earlier indicated, the hemoglobin dissociation curve is linear at 50% saturation and thus for an arbitrary Δ (% sat.) measured on the absorbance shift curve both Δ time and (Δ Hb-bound $O_2 + \Delta$ phys. diss. O_2) can be determined.

It should be pointed out, however, that in the ultramicro-modification of this technique using high hemoglobin concentration (>1 mM) the physically dissolved fraction can be disregarded, and thus calculations will be simpler.

General Instrumentation

A brief survey of the different components of the basic equipment used for microspectrophotometric determinations of oxygen consumption will be presented below. It should be emphasized that most parts of the set-up are commercially available and that the whole equipment can be used for other purposes than oxygen determinations, such as ordinary microscopy, microphotography, cytophotometry, etc. without any alterations of the arrangement.[27] In Figs. 8 and 9 the experimental set-up used is illustrated. In this equipment Zeiss components have been compiled together with other components manufactured at the workshop of this laboratory.

Light Source. Since this technique utilizes the visible range of the spectrum to which photomultiplier tubes are extremely sensitive, and since

[27] A. W. Pollister, H. Swift, and E. Rasch, *in* "Physical Techniques in Biological Research" 2nd ed. Vol. III C, (A. W. Pollister, ed.), p. 201. Academic Press, New York, 1969.

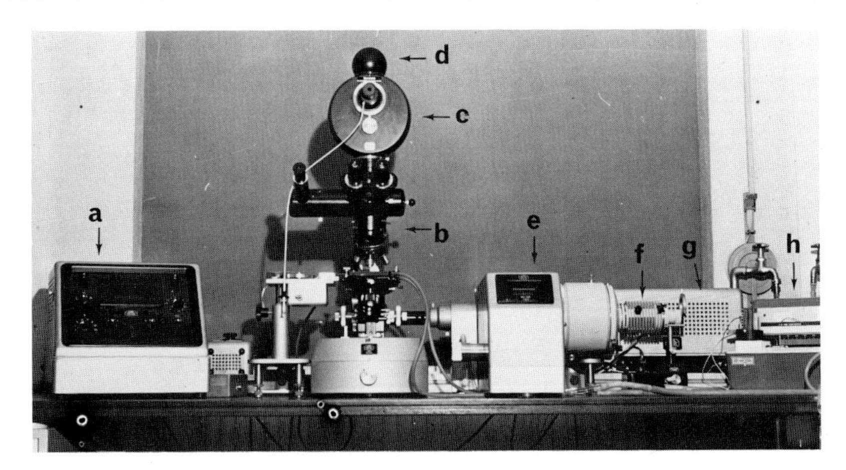

FIG. 8. Complete view of the Zeiss microspectrophotometric equipment used. From left to right: amplifier (a), microscope (b) with photometer head (c) and photomultiplier (d), monochromator (e), light source (f) with voltage stabilizer (g) and recorder (h).

the measured areas are not too small, the light source does not need to be of unusually high intensity. A point of special importance when working with a single-beam instrument in studies of time-dependent reactions is to achieve a light source giving a constant emission. For this reason arc lamps of various types are less suited since the intensity and position of the arc frequently change, and thus the zero and 100% transmission settings cannot be relied upon. Ordinary filament (incandescent) lamps are recommended. When using low voltage sources it is important that connections with the voltage supply are adequate (preferably soldered) so that no resistance changes due to heating or vibrations occur. Voltage stabilizers should be used in all cases.

Monochromator. Various devices such as filters and grating and prism monochromators can be used for spectral isolation. In the present situation, where analyses of absorbance differences of hemoglobin compounds, with rather abrupt absorption maxima and minima (Fig. 7), are to be performed, it is advisable to use instruments providing facilities for changing band widths and proper setting of wavelengths. Thus, interference filters are less suited in this type of work. Most commercially available grating or prism instruments fully meet the above-mentioned demands.

The light source and the monochromator should be adjusted so that no stray light can pass through the monochromator.

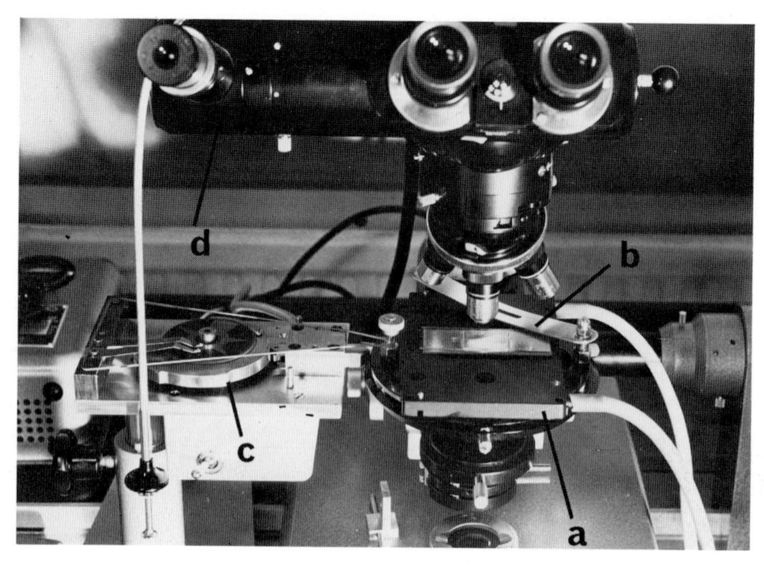

FIG. 9. Close-up view of the equipment. Observe the temperature-regulated object stage (a) with clamp (b), and to the left an automatic cuvette-changing device (c). The extra side-viewing tube (d) to the left can easily be adapted for a camera.

Microscope. Any mechanically stable microscope with a side-viewing tube can be used. There should be convenient possibilities to insert various field iris diaphragms. Centering and focusing facilities of the condensor, preferably a low power objective, must be available. On the object stage, movable in one or two planes there should be ample space to place a temperature-regulated plate (Fig. 9).

The optics do not need to be of quartz since no measurements are made in ultraviolet light. Apochromatic objectives are preferred since measurements are made from 576 nm to 435 nm, but achromatic lenses are sufficient since continuous recordings are made at a single wavelength (435 nm). No high-power objectives are needed since relatively large areas are to be measured.

Photometer Head and Detector with Amplifier. The photometer head should provide facilities to inspect the image of the specimen to be measured. Furthermore there should be different apertures so that the area measured can be adjusted. Electron multiplier tubes are excellent as detectors for measurements in visible light. The amplifier should provide possibilities to vary the voltage of the photomultiplier tube as well as to change the amplification of the photocurrent. The instrument should also be equipped with a galvanometer.

Recorder. Since the *absorbance change* of the preparation studied is proportional to the relative amounts of the two hemoglobin compounds, it is convenient to have a logarithmic transformer of the transmission signal fed from the amplifier. Thus, the graphs obtained from the recoder can be calculated directly. The recorder should have an electromagnetically controlled chart-penlifter easily made by the workshop of the laboratory, if an automatic cuvet changing equipment is used (see below).

Temperature-Regulated Object Stage. Since tissue oxygen consumption is highly temperature-dependent it is essential to be able to control incubation temperature within narrow limits. There are various electrically heated object stages available, but most of them do not permit easy handling of the preparation. Thus it has proved most suitable to use a stage with circulating water (see Fig. 9) from a thermostatically regulated bath. Temperature is calibrated using melting substances within incubating chambers [e.g., Eicosan, $CH_3(CH_2)_{18}CH_3$, melting point 36.8°, must be exactly determined for each batch purchased]. Side tubings to cold tap water should be available.

Hemoglobin Preparation. This protein which serves both as indicator of oxygen tension and as donor of oxygen is preferentially prepared at the laboratory. Commercially available preparations are frequently highly contaminated by methemoglobin and may contain other impurities which may be harmful to cellular metabolism.

Hemoglobin solutions are prepared from human whole blood (initial volume 300 ml). The cells are rinsed with 0.9% NaCl and centrifuged three times in the cold (1200 *g* for 10 minutes), whereupon they are rinsed with 1% NaCl to decrease the cell volume. The cells are lysed by adding half the packed cell volume of redistilled water, lysis being enhanced by shaking and freezing. Cell debris is removed as a pellet after centrifuging at 50,000 *g* for 1 hour. The purified hemolysate is then placed on a Sephadex G-25 column and eluted with the desired buffer solution. Taking the molecular weight of hemoglobin as 68,000, hemoglobin concentrations from 0.5 m*M* to 3.0 m*M* can be obtained directly by fractionated collection of the effluent from the Sephadex column. The solutions are kept frozen at −80° until used. A gradual autooxidation to methemoglobin is seen on storage. Spectrophotometric characterization of the solutions for concentration determinations of oxyhemoglobin are made with the two-wavelength method using extinction coefficients obtained from van Assendelft[28] (Fig. 7).

[28] O. W. van Assendelft, "Spectrophotometry of Haemoglobin Derivatives." Royal VanGorcum, Assen, The Netherlands, 1970.

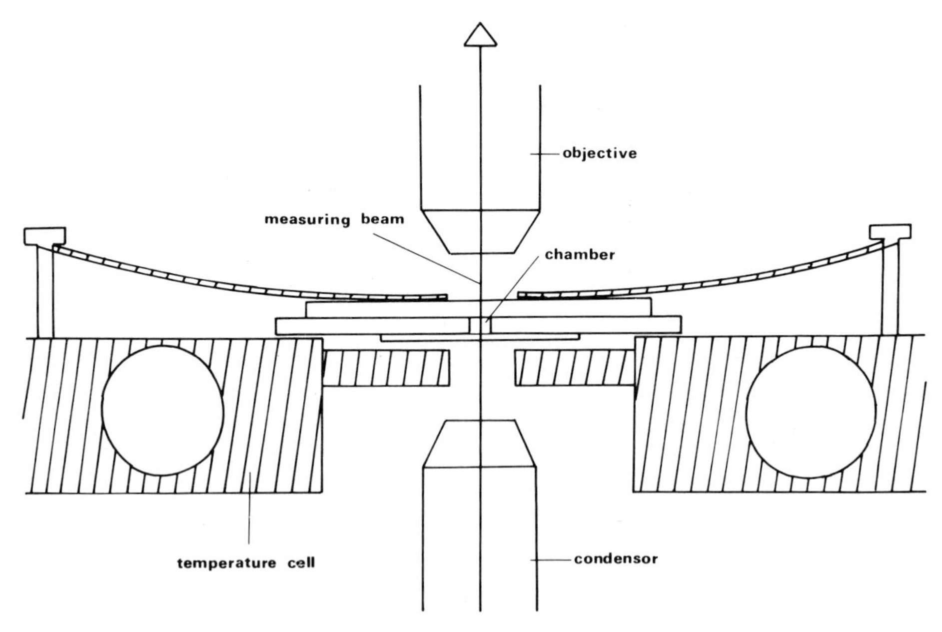

FIG. 10. Section through the temperature-regulated object stage with chamber in position. From H. Herlitz and R. Hultborn, *Acta Physiol. Scand.* (1973).

Semimicromodification

This semimicrotechnique is comparable in sensitivity to the standard-diver technique.[22] Tissue fragments with a dry weight of 3–20 μg can be suitably incubated in chambers having capacities of approximately 1 μl. A suitable experimental time from full to zero oxygenation of the incubate is 30 minutes to 1 hour (hemoglobin concentration 0.1 m*M*). By changing the hemoglobin concentration, the chamber volume or the amount of tissue the incubation time can easily be changed (Fig. 11).

The incubating chambers are produced by drilling holes (approximately 1 mm in diameter) in object slides. The penetration of the glass slides is made by a sintered carbide drill, mounted in an engraving machine. A coverslip is cemented on to one side of the penetrated glass and thus forms the bottom of the chamber. Nobecutan (Bofors, Nobel-Pharma, Mölndal, Sweden), a nontoxic solution of polymeric β-ethoxyethyl methacrylate in ethyl acetate is used as cementing substance. To cover the chamber, an object slide is firmly pressed on top of the penetrated

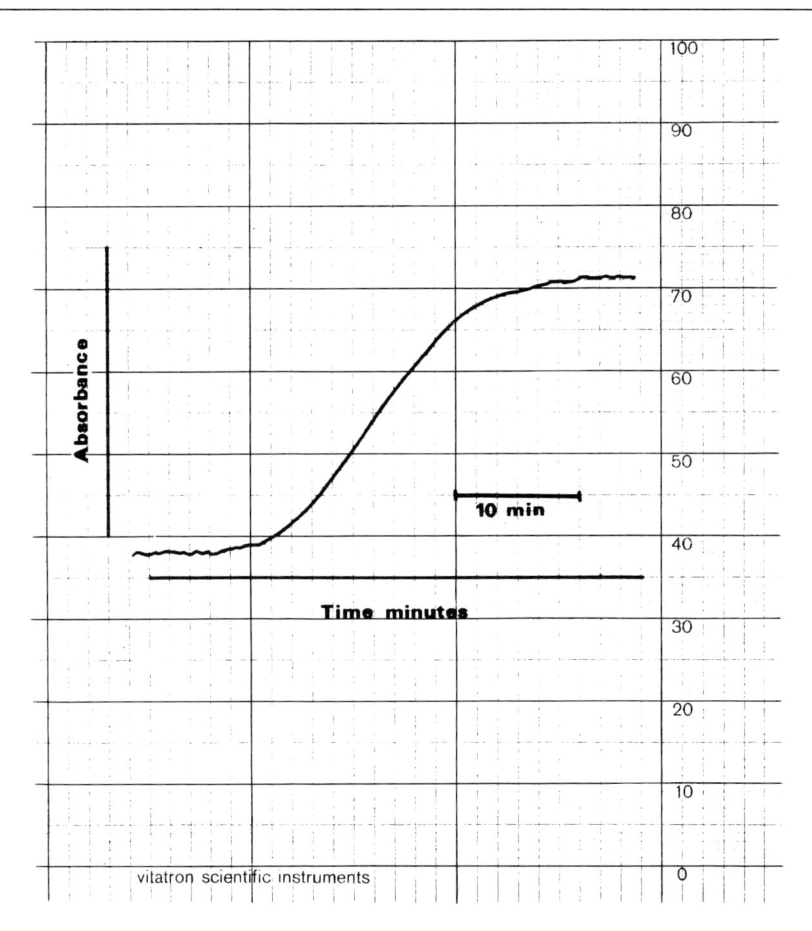

FIG. 11. A recording of an absorbance shift curve from a semimicrotechnique experiment.

glass. The surfaces of the penetrated and the covering slides are siliconized (see Standard Diver Technique).

The chambers are filled by placing a large supply drop (20 μl) of the hemoglobin solution to be used in and above the rim of the chamber. The tissue fragment dissected out is placed in the supply drop and is manipulated under a stereomicroscope into the chamber by use of stainless steel needles. After checking that no air bubbles have been trapped in the chamber, the covering glass is placed above the chamber, whereby the surplus hemoglobin solution is pressed away. The preparation is transferred to the cooled temperature-regulated object stage and is fastened onto it with a firm clamp (Figs. 9 and 10). The preparation is centered and focused and the measuring beam is adjusted so as to occupy only a small fraction of the chamber area. When recording is to be started the object stage is heated to 37°.

To increase the measuring capacity, an automatic cuvet-changing device has been developed.[29,30] The object stage is moved back and forth in six discrete steps every second minute, monitored by a synchronous motor (Fig. 9). Thus oxygen consumption in six chambers can be assayed simultaneously. An electronic device for chart-pen lifting and light-chopping during chamber changes has been developed.

Procedures for introducing reagents, such as hormones, during an experiment are at present under development in our laboratory. The principle is to combine two chambers, one with the sample to be studied and the other with the reactant. After a period of basal "respiration" the chambers are brought in contact with each other either by means of mechanical sliding of the chambers or by splitting a coverslip between the two chambers using a Q-switched laser beam.[29,31]

Ultramicromodification

The ultramicrotechnique can be used for assays of oxygen consumption down to the range of 10^{-6} μl O_2 per hour, i.e., as sensitive or even more sensitive than the microdiver techniques. The method thus makes analyses on single cells possible, which is of special interest when studying functional properties of heterogeneous tissues.

Incubating chambers are manufactured by microetching object slides with hydrofluoric acid. The procedures of manufacturing and volume determination (10^{-4} to 10^{-5} μl) of the bowl-shaped chambers are rather cumbersome and are described in detail by Hultborn.[6] A new type of drilled chambers are therefore at present being tested in our laboratory.[29]

Filling of the chambers must be performed under the microscope, but otherwise the procedure is similar to that described under semimicro modification. It should be pointed out that in this technique the cuvette pathlength is short (below 50 μm), and consequently a high hemoglobin concentration must be used ($5–10 \times 10^{-4}$ M). As indicated above, calculations of the oxygen consumption can be simplified without introducing too much error, by omitting the physically dissolved fraction (Fig. 12).

[29] R. Hultborn, to be published.

[30] Manufactured by ing. Yngve Källström, Department of Physiol., Fack, S-40033, Göteborg, Sweden.

[31] A. Hamberger and B. Tengroth, *Exp. Cell Res.* **37**, 460 (1965).

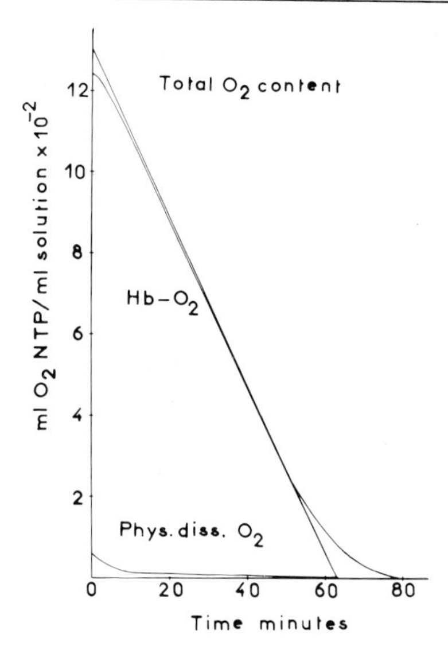

FIG. 12. A graph of an ultramicroscale experiment, showing the relationships between the oxygen pools of the incubate. From R. Hultborn, *Anal. Biochem.* **47**, 442 (1972).

Applications

In our laboratory the three above-described techniques have been mainly used for studies of enzymatic activities of isolated cells from endocrine organs. The first studies[32] concerned succinoxidase activity in granulosa cells from ovarian follicles obtained from rat ovaries. Because of the relatively small size of these cells (diameter 7–8 μm), they were gathered in samples of 50 cells in each microdiver. In the presence of succinate in a high concentration (25 mM), the respiratory activity was approximately 4×10^{-6} μl per hour per cell. The dry cell mass determined by X-ray absorption technique[33] was calculated to 2×10^{-7} mg. The sensitivity of the microdiver technique thus permits multiple analysis within one follicle, and comparisons can also be made between cells situated in various part of the follicle, i.e., cells close to the follicular wall to cells close to the follicular oocyte. In this study a comparison was made between the repiratory rate in granulosa cells incubated in the absence or in the presence of luteinizing hormone. It was found that this gonadotropic hormone could dramatically increase the respiratory rate in most of the experiments by about three times.

[32] K. Ahrén, A. Hamberger, and L. Hamberger, *Endocrinology* **77**, 332 (1965).
[33] B. H. O. Rosengren, *Acta Radiol.* Suppl. 178 (1959).

The microdiver technique has also been applied in studies of other ovarian cell types, such as theca cells, interstitial cells,[34] follicular oocytes,[35] and corpus luteum cells.[36] In these studies the two-compartment microdiver was used in order to study the acute influence of gonadotropic hormones on cellular respiration (see Fig. 6, upper curve). In cellular material with considerable biological variations in enzymatic activities, it is of definite advantage to study the same cell sample before and after addition of a reactant or inhibitor. The two-compartment divers also permit study of the time interval between the addition and the enzymatic response. Also, the duration of an effect can often be studied since linear respiration generally is present for several hours.

Succinate has been the most commonly used substrate in our own studies on ovarian cells, but in experiments on follicular oocytes and corpus luteum cells glucose, pyruvate and oxaloacetate have been tested in various molar concentrations in an attempt to judge which substrate is preferable for the respective cells.

In experiments on isolated testicular cells,[37] the succinoxidase activity of Leydig cells was studied in relation to gonadotropic hormones added *in vitro*. The microdiver technique was especially useful, since the sensitivity of the technique permitted determinations on few cells and it was thus possible to get very pure and well-defined cells. In experiments on isolated parathyroid cells from the rat[38,39] both the microdiver technique and the standard diver technique have been used. The respiratory activity of these cells was measured with succinate as substrate, and the incubation medium was charged with respect to the ionic strength of calcium and magnesium. In the microdiver only 50 cells were needed for each determination, while in the standard diver only two divers could be charged from each parathyroid gland. The standard diver technique was of considerable use in these studies since it was possible to make two alterations in the medium content of the respective ions during one experiment.

Hellerström[40] has used the standard diver technique for evaluating the energy metabolism of the β-cells of isolated islets of Langerhans. The influence of various substrates on the cellular respiration of the β-cells

[34] L. Hamberger, *Acta Physiol. Scand.* **74,** 410 (1968).
[35] T. Hillensjö, L. Hamberger, and K. Ahrén, to be published.
[36] L. Hamberger and H. Herlitz, to be published.
[37] L. Hamberger and V. W. Steward, *Endocrinology* **83,** 855 (1968).
[38] C. G. Hansson and L. Hamberger, *Endocrinology* **86,** 1158 (1970).
[39] C. G. Hansson and L. Hamberger, *Endocrinology,* **92,** 7582 (1973).
[40] C. Hellerström, S. Westman, N. Marsden, and D. Turner, *Struct. Metab. Pancreatic Islets Proc. Int. Wenner-Gren Symp. 16th, 1970* p. 315 (1970).

was tested. Hormonal effects (epinephrine, glucagon) on the oxygen consumption of isolated islets were also investigated.

The semimicro modification of the spectrophotometric method for assay of oxygen consumption was recently tested on small samples (dry weight 5–30 μg) of corpora lutea from 44-day-old rats.[22] Respiration was determined in the absence and in the presence of succinate (25 mM) as substrate. In this series of experiments a comparative evaluation was made of the results obtained using this technique and the results achieved using the standard diver technique. With the spectrophotometric method, the oxygen uptake in presence of 25 mM succinate was 8.61 $\pm$ 0.72 μl O_2 per milligram dry weight per hour (mean $\pm$ SEM, 11 determinations) and the endogenous respiration was 1.47 $\pm$ 0.12 μl O_2 per milligram dry weight per hour (10 determinations). Compared to the results obtained by the standard-diver technique, the respiratory rate in the presence of succinate (11.68 $\pm$ 1.04, 11 determinations) was slightly higher than for the spectrophotometric method, while in the absence of substrate (1.18 $\pm$ 0.17, 19 determinations), there was no difference between the two methods. No differences in precision were found between the two methods.

Since hemoglobin is present in the incubation medium when the spectrophotometric method is used, there is a possibility for interaction between the hemoglobin and various factors (substrates, hormones, metabolic inhibitors, etc.) in the medium. Such interactions may lead to alterations in the slope of the hemoglobin dissociation curve that do not correspond to changes in oxygen uptake. However, by dilution of the hemoglobin and by studying the effects in dose–response experiments, we think that such misinterpretations can be avoided. The simplicity of the spectrophotometric method, combined with a high capacity compared to the Cartesian diver techniques, makes this technique especially exciting for the future.

Section VI

Neural Tissue

[36] Electrophysiological Techniques for the Study of Hormone Action in the Central Nervous System

By B. J. Hoffer and G. R. Siggins

For optimum interpretation of results, analysis of the electrophysiological actions of neurohormones and cyclic nucleotides on central nervous elements requires application of these substances in the immediate environment of a given neuron while simultaneously recording extracellular or transmembrane potentials.

Microiontophoresis of small quantities of ionized drugs from multibarreled microelectrodes has evolved into the most practical method for testing the electrophysiological actions of putative neurohormones. This chapter will deal with the various technical aspects of microiontophoresis of neurohormones and cyclic nucleotides, with the acquisition of data from such studies, and with some limitations of these techniques.

Design of Microiontophoretic Circuitry

If a potential of a given polarity is applied to an ionized solution in a micropipette barrel, ions of the same polarity will be ejected in accordance with Faraday's law, $M = nIT/ZF$, where M = moles of drug ejected;

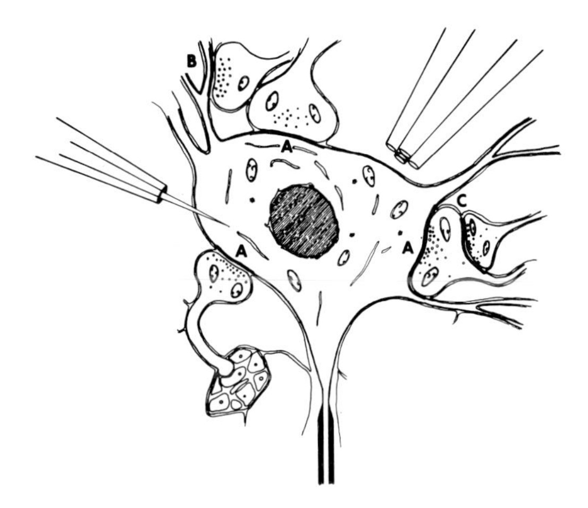

Fig. 1. A diagrammatic representation of a single neuronal unit in which axosomatic (A), axodendritic (B), and interaxonic (C) junctions are exhibited as possible sites of drug action. The tip sizes of concentric dual pipettes and multibarrel pipettes are drawn to approximate scale.

Z = equivalents/mole; F = Faraday's constant; I = amperes of ejection current; T = seconds of ejection time; and n = transport constant. Conversely, potentials of opposite polarity applied to the same drug barrel may be used as "holding currents" to minimize diffusion of drug.

Although ejection of the drug from the pipette appears to vary linearly with charge in any given pipette, the transport constant varies widely from one pipette to the next, making the prediction of absolute amounts of drug release practically impossible.[1,2] These considerations have led to specifications of ejection currents of drugs in microiontophoretic studies rather than moles of drug released.

Since the dc resistance of the drug pipettes may reach 500 megohms, the electronic microiontophoretic devices must be designed to pump constant currents (up to several hundred nanoamperes) through extremely high resistances. Moreover, in order to minimize electrotonic effects of the iontophoretic currents, provision should be made for current neutralization by passage of an equal but opposite current through a barrel filled with sodium chloride (balance barrel; see below). Two such designs are shown in Fig. 2.

The circuit[3] in Fig. 2A makes use of vacuum phototubes (P1, P2) whose current output may be varied by changing the illumination (L). The output will be largely independent of the inserted series resistance (R). The circuit[4] shown in Fig. 2B makes use of an operational amplifier wired in a "Howland current pump" configuration. Both these circuits use large output voltages (>100 V) to permit high currents through large external resistances. It is generally advantageous to use low noise power sources, such as batteries or highly filtered and shielded power supplies, to avoid electrical interference with the recorded bioelectric signal.

Electrodes for Microiontophoresis

Since the placement near neurons of drug and recording electrodes cannot be performed under direct microscopic visualization in the central nervous system, all micropipettes must have their tips fixed in close proximity prior to insertion in the tissue, in order to ensure drug application at the site of neuronal recording. For extracellular recording, the 5-barreled pipette is most widely used. It is constructed by fusing 5 pieces of Pyrex tubing 3 mm in diameter so that 4 of the pieces are radially placed around the fifth. The tubes are then fused and initially drawn

[1] B. J. Hoffer, N. H. Neff, and G. R. Siggins, *J. Neuropharmacol.* **10**, 175 (1971).
[2] K. Krnjevic, R. Laverty, and D. Sharman, *Brit. J. Pharmacol.* **30**, 491 (1964).
[3] G. Salmoiraghi and F. Weight, *Anesthesiology* **28**, 54 (1967).
[4] H. Geller and D. Woodward, *EEG Clin. Neurophysiol.* **33**, 430 (1972).

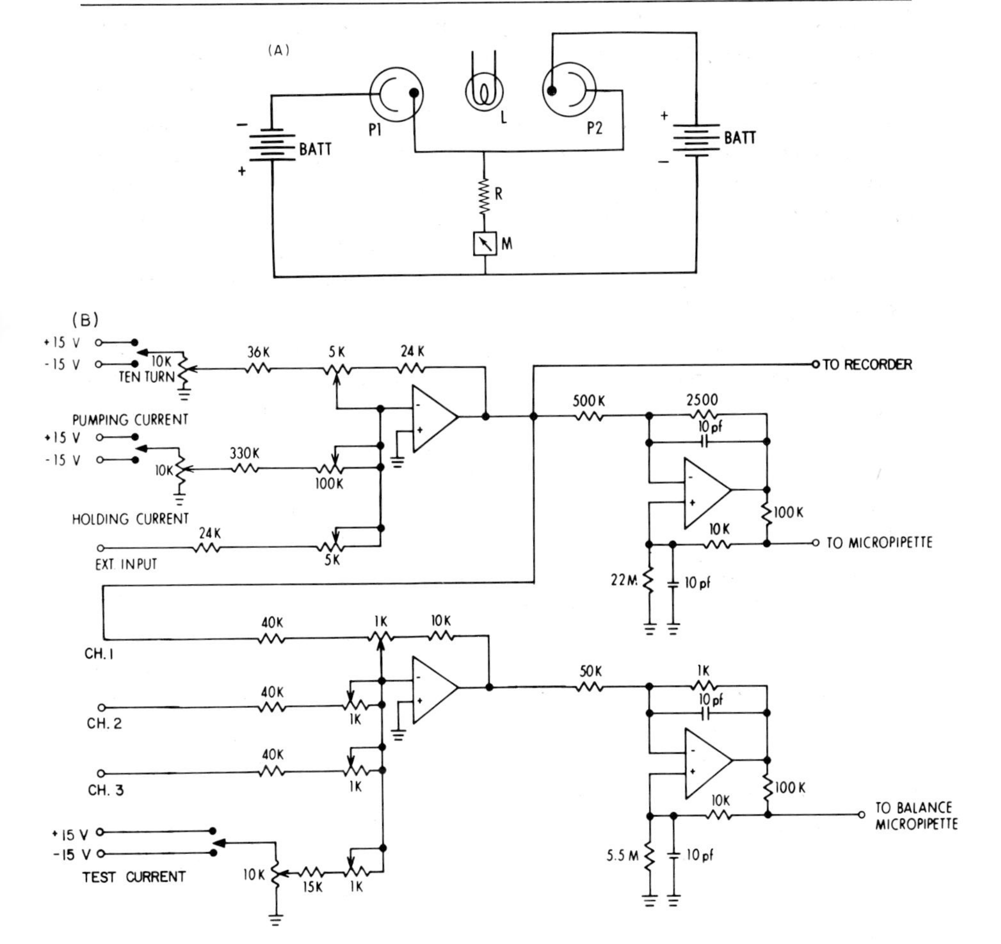

FIG. 2. (A) Schematic showing the principle of the electrophoresis circuit. (B) Schematic circuit for one of three identical drug channels. The input operational amplifier (top) sums three possible control voltage signals, pumping and holding currents, and external inputs; the outputs of this operational amplifier go to the input of the pumping circuit, to a chart recorder and to one input of the balance channel (bottom). Control voltages from the three drug-injection channels are summed, and the resulting voltage controls the balance current source.

to a blunt tip by hand. The tubes are then redrawn in a vertical micropipette puller so that the resulting tip is 1 μm or less (Fig. 3).

After these pipettes are pulled, they can be filled to the fine tip by boiling in distilled water for about 30 minutes and can be stored in this condition for up to 2 weeks. Just prior to use, the distilled water is removed to the "shoulder" of each pipette tip by aspiration with PE 10 tubing, and the appropriate drug and salt solutions are placed in the 5

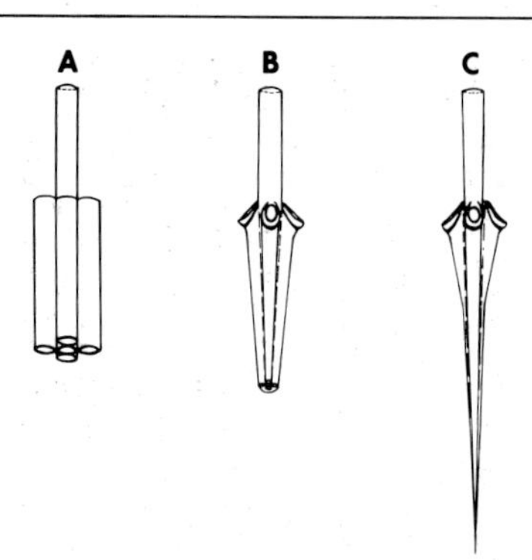

FIG. 3. Steps in the preparation of multibarrel glass micropipette electrodes. The outer glass capillaries are fused to a central barrel (A) and partially drawn out by hand (B). In the final step, the electrode is further drawn out to tip sizes less than 0.5 μm by commercially available electrode pullers (C). When filled and ready for use, the overall tip diameter is increased to 3–6 μm by gently tapping the assembly.

barrels. The central barrel, used for recording neuronal activity, is filled with 5 *M* NaCl, and one of the peripheral barrels, if used for current neutralization, is filled with 3 *M* NaCl. The remaining 3 peripheral barrels are filled with drug solution.

Diffusion of the solutions into the pipette tip is facilitated by centrifugation at 2000 rpm for 30 minutes. For this purpose, we have used a specially designed, plastic device for holding the pipettes firmly in place in standard centrifuge cups, with the tips facing centrifugally. Finally, the electrode tip is manually bumped under microscopic control to a total outer diameter of 3–6 μm. This allows a recording barrel size small enough (2–4 megohms resistance) to isolate action potentials from single neurons while keeping drug barrel resistances low enough for constant current ejection.

The pH of the drug solution is usually adjusted to produce maximal ionization and solubilization of the agent. For example, when cyclic AMP is dissolved as the free acid, it is dissolved in 0.25 *N* sodium hydroxide to yield a 0.5 *M* solution with a pH of 6–8. Several neurohormones may be used as either cations or anions, depending on the pH. In such instances, use as a cation is preferable since interaction with the zeta potential of the glass pipette tip promotes ejection. Microiontophoresis of angiotensin, for example, is effective only if cationic currents are used[5]; no responses are seen if anodal currents are applied to the pipette.

In general, holding currents range from 10 to 40 nA and ejection cur-

[5] R. Nicoll and J. Barker, *Nature* (*London*) **233**, 172 (1972).

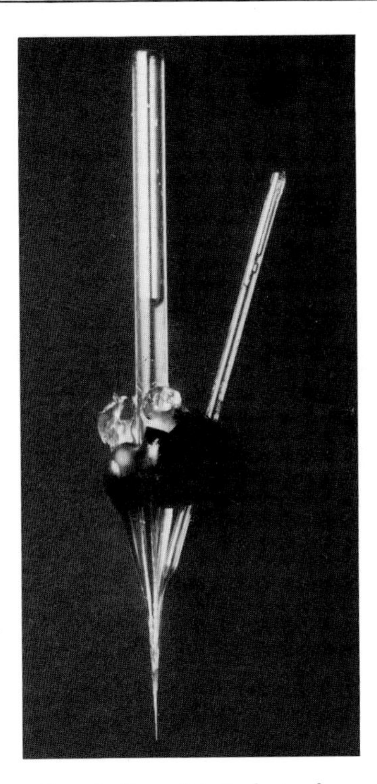

FIG. 4. Completed electrode.

rents may be as high as 200 nA. If cells are particularly sensitive to a particular substance, significant responses may result simply by removing the retaining current and allowing the drug to diffuse from the pipette tip.

Electrodes for recording transmembrane potentials during extracellular drug ejection can be constructed by cementing a fine single-barrel recording pipette (tip diameter $<0.5\ \mu m$) parallel with a standard multibarreled drug micropipette.[6] Best results are usually obtained if the tip of the single-barreled pipette is 20–40 μm in front of the multibarreled pipette (Fig. 4).

It is of utmost importance that the tip of the drug electrode rests firmly against the shaft of the recording electrode. If there is any space between them, they will often separate as they pass through nervous tissue, considerably lessening the chance of obtaining drug responses from the recorded neuron.

[6] A. P. Oliver, *EEG Clin. Neurophysiol.* **31**, 284 (1971).

Recording Techniques

Extracellular Recording

The electrophysiological effects of putative neurohormones on neurons, evaluated by extracellular recording of action potentials, can usually be causally related to changes in spontaneous, or drug or stimulus-induced discharge rate. It must be emphasized, however, that increases or decreases in discharge rate do not, by themselves, unequivocally indicate excitation or inhibition at the membrane level (see below). In order to simplify data collection and analysis of spontaneous activity, the extracellular action potentials under study are usually displayed on an oscilloscope and simultaneously fed into a "window" discriminator to separate them from base-line noise and action potentials of neighboring neurons. The discriminator output is then converted to fixed amplitude pulses, which are subsequently integrated over a preselected interval (usually 1 second). The fixed amplitude pulses may also be displayed on a second channel of the oscilloscope for comparison with the original action potentials. The output of the integration is displayed on one channel of an ink writer, to provide a continuous analog record of the discharge rate. A comparison of this display with the original action potential record is shown in Fig. 5.

In many cases, neurons show very low spontaneous discharge rates. It is often advantageous, in such cases, to induce a steady and regular background discharge to facilitate testing excitatory and inhibitory agents. Applications of small amounts of excitatory amino acids, such as glutamate or DL-homocysteate, from another barrel of the pipette is often used to induce such discharge.[7] Alternatively, a digital computer-based signal averaging paradigm can be used with regularly repeated drug pulses as the "stimulus." This latter technique has the advantage of eliminating the potential artifacts induced by direct interaction of the specific agonist with the excitatory amino acids.

The effects of locally applied hormones on driven as well as spontaneous activity may also be studied. Antidromic spikes and orthodromic excitation and inhibition[8] may all be used. Such studies help to determine whether drugs act on postsynaptic membrane in general, at postsynaptic receptors, or at the presynaptic level. Both pre- and postsynaptic action of many putative transmitters and drugs have been demonstrated[8] in this fashion.

Differentiating pre- and postjunctional action of drugs is also facili-

[7] J. Godfraind, K. Krnjevic, and R. Pumain, *Nature (London)* **228**, 675 (1970).
[8] A. Tebecis and A. DiMara, *Exp. Brain Res.* **14**, 480 (1972).

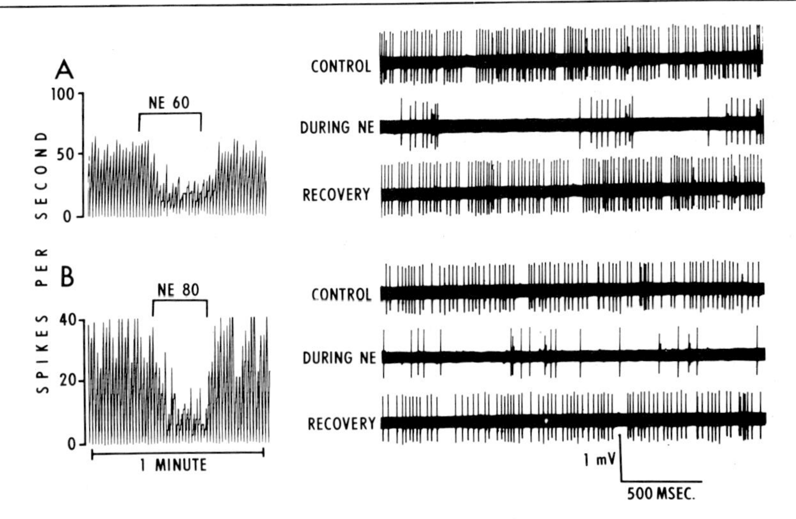

FIG. 5. Effects of microiontophoresis of norepinephrine (NE) on spontaneous Purkinje cell discharge rate. (A) and (B) illustrate two different Purkinje cells. Left half of figure shows integrated polygraph record. Right half of figure shows action potentials themselves. Numbers after drug indicate ejection currents in nanoamperes, and brackets indicate duration of drug ejection.

tated by combining microiontophoresis with either surgical or chemical lesions of afferent pathways. The recent discovery of drugs which destroy specific classes of fibers, such as 6-hydroxydopamine for adrenergic terminals,[9] may permit even more detailed analysis of presynaptic drug actions.

Intracellular Recording

Most intracellular studies of transmitter actions on neurons emphasize effects on membrane potentials, with a depolarizing response indicative of an excitatory action and a hyperpolarization, inhibitory action. However, underlying these potential changes, and perhaps responsible for their generation, may be changes in membrane conductance or permeability to one or more ion species. In this section, techniques for measurement of transmembrane potentials and conductance during microiontophoresis of neurohormones and cyclic nucleotides will be presented.

Since only one pipette can be inserted into most mammalian neurons, the most common procedure for determining membrane ionic conductance utilizes a Wheatstone bridge circuit for passage of current pulses through

[9] F. E. Bloom, B. J. Hoffer, and G. R. Siggins, *Biol. Psychiat.* **4**, 157 (1972).

the same recording pipette used to measure membrane potential. A good example of such a circuit is the one designed by Araki and Otani.[10] The high resistance leg of the bridge (10^8 to 10^9 ohms) is often provided in many commercially available cathode followers.

In brief, the usual procedure for recording membrane potential and determining input resistance (inversely proportional to the sum of all ionic conductances) is as follows: (1) Continually pass constant current pulses of about 20–100 msec duration and 10^{-8} to 10^{-10} amperes intensity (these parameters will depend upon the electrical properties of the specific neuronal membrane and should be empirically determined) through the recording pipettes. (2) With the pipette in the brain but still extracellular, constantly adjust the variable bridge resistor until the bridge is balanced (i.e., the current between the leg with microelectrode and that with the variable resistor is zero). Ideally, the dc potential will then be flat except for the fast "on" and "off" transients of the pulse. (3) With small movements of the micromanipulator, a sudden negative deflection of the dc potential may indicate penetration of a neuron. This can be verified by the size of the potential (—30 to —70 mV), by the presence of large (40–150 mV) positive-going action potentials, and a sudden unbalancing of the bridge. The magnitude of the unbalanced current pulse should now reflect the input resistance of the membrane, provided that the on/off phases of the pulse do not rise so abruptly as to appear discontinuous (a slowly rising or falling potential reflects membrane capacitance). The input resistance is calculated by Ohm's law, dividing the voltage deflection of the pulse by the pulse current. (4) After verifying that the membrane potential is large (—40 to —70 mV) and steady (indicating relatively little injury), determine the effect of drug applications, with a suitable control period after the drug effects have worn off. (5) Note the effect on input resistance of artificial (current-induced) changes in membrane potential (see below). (6) Withdraw the pipette from the cell, note the zero potential, and verify that the Wheatstone bridge is again null-balanced.

It may be helpful to first determine whether it is possible to obtain uninjured intracellular records from a given neuron type with single-barrel pipettes. The attachment of additional barrels to the recording pipette, for microiontophoresis of drugs, often greatly reduces the chances of obtaining good records. In our laboratory, it has been determined that two extracellular barrels are optimal for extracellular application of drugs during intracellular recording of transmembrane potentials in cerebellar Purkinje neurons. The double extracellular barrels are heated together,

[10] T. Araki and T. Otani, *J. Neurophysiol.* **18,** 472 (1955).

twisted once, and then pulled to a fine tip.[6] They are then bumped, glued to the recording pipette, and filled.[6] One extracellular barrel usually contains a drug solution, while the other has 3 *M* NaCl for use as a "balance" barrel to neutralize electrotonic effects. The intracellular recording electrode is usually filled with 3 *M* KCl, K_2SO_4, or potassium citrate. While KCl is most often used, K citrate should be tested if it is suspected that leakage of Cl or SO_4 into the neuron might alter the response under investigation.

Such electrode assemblies make it possible to measure the effects of neurotransmitters and cyclic nucleotides on resting membrane potential, membrane conductance, and postsynaptic potentials (Fig. 6). These analyses are necessary to determine the electrophysiological mechanism of

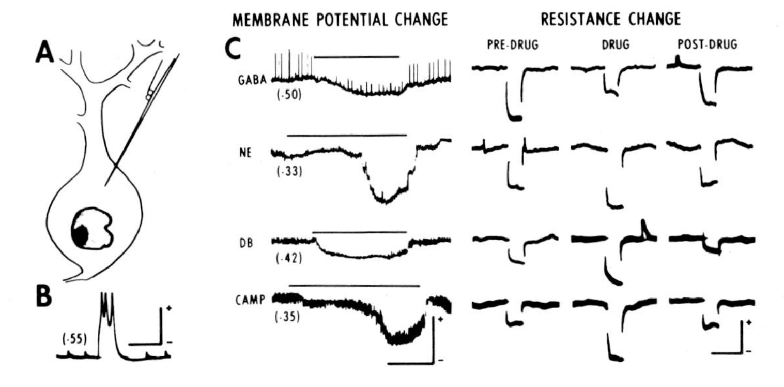

FIG. 6. Intracellular recordings from rat cerebellar Purkinje cells. (A) Schematic representation of a three-barrel micropipette with a Purkinje cell. The intracellular electrode protrudes beyond the orifices of the two extracellular microelectrophoresis barrels. (B) Multispiked spontaneous climbing fiber discharge obtained during intracellular recording from a Purkinje cell. Number in parentheses is resting potential in millivolts (mV); calibration bars are 20 msec and 25 mV. (C) Changes in membrane potential and membrane resistance of four different Purkinje cells in response to γ-aminobutyrate (GABA), norepinephrine (NE), dibutyryl cyclic AMP (DB cyclic AMP), and cyclic AMP. All specimens in each horizontal row of records are from the same cell. Solid bar above each record indicates the extracellular electrophoresis of the indicated drug (100–150 nA). Number in parentheses below each recording is resting potential in millivolts; calibration bar under membrane potential records is 10 seconds and 20 mV for NE, DB cyclic AMP, and cyclic AMP, and is 5 seconds and 10 mV for GABA. The effective input resistance was judged by the size of pulses resulting from the passage across the membrane of a brief constant current (1 nA) pulse before, during, and after electrophoresis of the respective drugs (1 mV = 1 megohm). Discontinuities in the fast transients of the pulses result from the loss of high frequencies (>10 kHz) and from the chopped nature of the frequency-modulated magnetic tape recording used. All "pulse" records were graphically normalized to the same base-line level. Calibration bar on right indicates 80 msec and 15 mV for all pulse records.

drug action and, hopefully, to eventually correlate the electrophysiological and biochemical effects of these substances.

Problems with Use of the Iontophoretic Technique

Quantitating Release

The great variation in transport number, from one pipette to another, has already been discussed. Because of this problem, it is virtually impossible to predict the amount of drug released by a given ejection current. Moreover, the determination of the effective concentration of the drug at the membrane receptors is still further removed, since the exact distance of the electrode from the neuron is unknown, and the assumption of free diffusion through a homogeneous medium is certainly untenable to the central nervous system.

Artifactual Responses

The major artifacts associated with extracellular recording and iontophoretic administration of neurohormones involve pH effects, local anesthetic action, and electrotonic effects of the iontophoretic current. As noted above, the pH of drug solutions is often manipulated to promote ionization and ejection. Although some studies[11,12] have suggested that solutions of biogenic amines with pH of less than 4.0 may elicit artifactual excitation, recent work from several laboratories have provided convincing evidence against this simplistic view. Cortical neurons have been shown to possess identical responses to norepinephrine at pH 3.0 and pH 5.0. As a further control, some drugs may be ejected as cations from low pH solutions and as anions from high pH solutions. An identity of action would argue against pH effects.

Many neuroactive drugs possess powerful local anesthetic side effects and can thereby reduce discharge rates by inactivating the membrane spike generating mechanism rather than by producing physiological hyperpolarization and conductance changes. Several types of controls are useful in distinguishing local anesthetic action from inhibition. First, local anesthetic action usually elicits a marked and progressive diminution of the amplitude of the action potentials prior to their disappearance, whereas inhibition usually does not. In the case of norepinephrine effects of cerebellar Purkinje cells, the inhibition may actually be associated with an increase in action potential size. Second, during a true, non-

[11] R. C. Frederickson, L. M. Jordan, and J. W. Phillis, *Brain Res.* **35**, 556 (1971).
[12] L. M. Jordan, R. C. Frederickson, J. W. Phillis, and N. Lake, *Brain Res.* **40**, 552 (1972).

anesthetic inhibition, neurons can usually be excited by microiontophoretic administration of a depolarizing ionic current or an amino acid, such as glutamate or DL-homocysteate, from another barrel of the same pipette. Cells which have been locally anesthetized do not often respond to such amino acid or cathode applications (Fig. 7).

Electrotonic effects of the iontophoretic currents are another potential source of artifact. These currents can be neutralized by the use of a "balance current," that is, automatic passage, through the sodium chloride barrel, of a current equal in magnitude but opposite in direction to the sum of the currents passing through the other three barrels. When functioning properly, balance currents tend to eliminate any potential difference between the pipette tip and ground. However, if there is still a question of influence of current on responses of a particular neuron, currents equivalent to the magnitude and polarity of the drug current may be passed through the sodium chloride barrel. If the cell then manifests a response to a sodium or chloride ejection similar to that after drug ejection, an electrotonic artifact is a strong possibility.

There are many problems and artifacts associated with single neuron

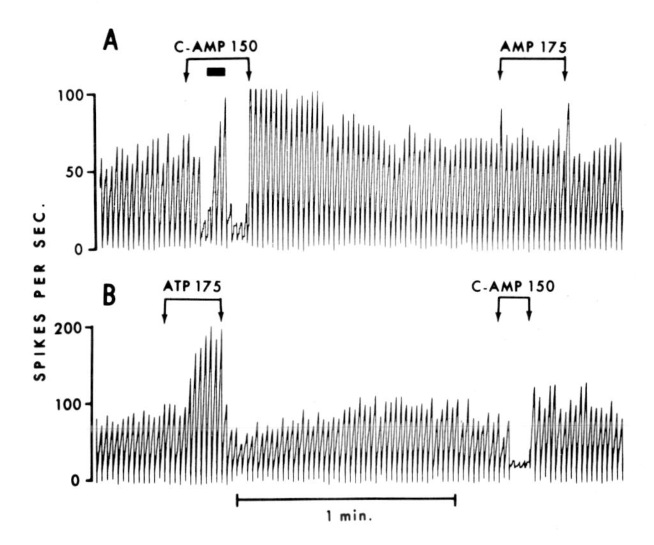

FIG. 7. Effects of microelectrophoretic application of adenine nucleotides on spontaneous Purkinje cell discharge. (A) and (B) illustrate continuous records from the same cell. Duration of drug application is indicated by arrows, and numbers after each drug indicates ejection current in nanoamperes. The black bar in (A) represents application of a cathodal current through the micropipette. The restoration of spontaneous discharge by cathodal current indicates that the slowing seen with cyclic AMP is not secondary to local anesthetic action or hyperdepolarization (cathodal block).

recording from brain and these multiply when microiontophoresis of substances is also employed. One of the most difficult problems is movement, whether of cell or microelectrode. Use of the most stable micromanipulator and stereotaxic apparatus is mandatory. Slight pulsations of nervous tissue, such as arise from respiration and blood pressure, are often troublesome. They may be diminished by pneumothorax and rapid shallow artificial respiration, drainage of the cerebrospinal fluid, and/or application of a "pressor foot" to the exposed brain surface.[13]

When unbalanced iontophoretic current is used to apply drugs extracellularly while recording intracellularly, resistive electrical coupling between recording and drug barrels is a major source of artifact, often producing large changes (5–30 mV) in membrane potential. As a result, proper controls must be performed with the recording electrode outside the cell to determine the voltage-producing effect of iontophoretic currents. These voltage changes must then be algebraicly subtracted from the potentials produced during cell penetration. It is also important to note whether any changes in bridge balance occur with the recording electrode outside the cell during drug ejection. Use of a NaCl-filled extracellular balance barrel usually reduces resistive coupling to less than 1 mV, and thus greatly minimizes this artifactual problem.

Interpretation

One final reservation is in order, relating to the limitations of microiontophoresis. By proper regard for each of the necessary experimental controls, it is possible to get reproducible effects of many neurohormones on the electrophysiological activities of neurons. However, such data alone do not indicate that these responses are a reflection of an underlying hormonal or cyclic nucleotide-chemical input to the cell under study. Critical evidence to corroborate this inference involves selective activation of endogeneous afferent fibers or hormonal inputs to an identifiable cell type[14] and histochemical demonstration of synapses containing the neurohormone under study.[15] With regard to circulating hormones, it should be determined that the endogenous hormone actually reaches the neuron under study. When used in such an interdisciplinary fashion, microiontophoresis may become a powerful analytical tool for evaluating the electrophysiological responses to putative humoral substances in the central nervous system.

[13] B. J. Hoffer, G. R. Siggins, and F. E. Bloom, *Brain Res.* **25,** 523 (1971).

[14] G. R. Siggins, B. J. Hoffer, A. P. Oliver, and F. E. Bloom, *Nature* (*London*) **233,** 481 (1971).

[15] F. E. Bloom, S. Algeri, A. Groppetti, A. Revuelta, and E. Costa, *Science* **166,** 1284 (1969).

Section VII

Exocrine Tissue

[37] Techniques for Studying Development of Normal Mammary Epithelial Cells in Organ Culture

By YALE J. TOPPER, TAKAMI OKA, and BARBARA K. VONDERHAAR

The mammary gland is a useful model system for the study of developmental biology. Normal differentiation of mammary epithelial cells can be simulated *in vitro* using chemically defined media. This facilitates control of the cellular environment and eliminates certain problems of analysis that are inherent in experiments performed *in vivo*. Furthermore, the fully developed cells produce specific milk products which are convenient markers of functional differentiation.

The purpose of this chapter is to describe procedures that have been used for the culture of mammary gland of the mouse and rat and to provide pertinent information on, and references to, analytical techniques used in the assessment of mammary gland development. The methods to be described are intended primarily for the use of investigators who wish to study the murine mammary system, specifically.

Developmental Characteristics in Vivo

As a prelude to discussion of culture procedures, it is useful to outline some of the characteristics of the mouse mammary epithelium at various stages of development *in vivo*. These are summarized in the table.

Mammary Gland Culture

Developmental Stage of Donor Animal

In the organ culture system epithelial cells in mammary explants from nonlactating mice are stimulated to differentiate and produce milk products in the presence of the appropriate hormones. However, the kinetics of response varies depending upon the developmental stage of the donor animal. For example, the cells in the gland of virgin C3H/HeN animals are initially unresponsive to mitogens such as insulin and serum, and acquire responsiveness after about 24 hours in culture.[1,2] This results in delayed cell-proliferation and elaboration of milk products. In general, the duration of the unresponsiveness period is fairly constant for 3- to 5-month-old C3H/HeN mice. However, various environmental factors,

[1] S. Friedberg, T. Oka, and Y. J. Topper, *Proc. Nat. Acad. Sci. U.S.* **67**, 1493 (1970).
[2] T. Oka and Y. J. Topper, *Proc. Nat. Acad. Sci. U.S.* **69**, 1693 (1972).

DEVELOPMENTAL CHARACTERISTICS OF MOUSE MAMMARY GLAND *in Vivo*

Developmental stage	Age	Morphology	Ultrastructure	Cell proliferation	Sensitivity[a] to insulin and serum	Casein synthesis	α-Lactalbumin activity
Embryo	16–21 days of gestation	Primary ducts	Nonsecretory	Active	—	ND[b]	—
Immature virgin	3–10 weeks	Ducts and end buds	Nonsecretory	Active	Not sensitive	ND	ND
Mature virgin	3–5 months	Ducts and alveoli	Nonsecretory	Inactive	Not sensitive	Low	ND
Mid-pregnancy	—	Ducts, alveoli and some lobules	Largely nonsecretory	Active	Sensitive	Higher	+
Mid-lactation	—	Highly lobulo-alveolar	Secretory	Inactive	Sensitive	Highest	++++
Multiparous nonpregnant	—	Ducts and alveoli	Nonsecretory	Inactive	Not sensitive	Low	ND

[a] See section on cell proliferation.
[b] Not detectable.

such as crowding, frequency of handling, bedding, can alter the length of this period, so that occasionally partial responsiveness is observed in freshly isolated tissue, and shortened lag periods occur. Also, the length of the lag period may vary as a function of the age and strain of the animal. We house no more than 5 mice in one cage (10.5 × 6.5 inches) in an animal room which is illuminated from 8:00 AM to 5:00 PM.

Since mouse mammary development is rapid during pregnancy, the study of tissue from pregnant animals requires proper staging. This can be done on the basis of the morphological characteristics of the fetuses,[3] or by counting days after the appearance of the vaginal plug. Generally, use of first-pregnancy animals is recommended.

Culture Media

Chemically defined synthetic media have been used successfully for organ culture of mammary gland. Medium 199, Trowell T8, Weymouth MB752/1, and NCTC 109, supplemented with Hanks' or Earle's salt mixture, are among those used. The composition of these media are given in Biology Data Handbook.[4] Complex supplements, such as blood sera and lactalbumin hydrolyzate, have been employed in particular investigations.[2,5-7] Generally, penicillin G is added to a final concentration of 35 μg/ml. Sterilization of the media is effected by passage through Millipore filters (0.45 μm) contained in Swinnex-13 filter units attached to syringes.

Hormones

Addition of insulin, a glucocorticoid, and either prolactin or placental lactogen to the culture system promotes the conversion of nonsecretory epithelial cells, in explants of mammary tissue derived from animals in various developmental stages, into secretory cells. Partial development can be effected by addition of incomplete complements of these hormones.[8,9] Beef and porcine insulin are about equipotent. The most frequently used glucocorticoid is hydrocortisone, but a number of others

[3] H. Gruneberg, *J. Hered.* **34,** 88 (1943).

[4] Biology Data Handbook (P. L. Altman and D. S. Dittmer, eds.), p. 530. FASEB, Bethesda, Maryland, 1964.

[5] G. C. Majumder and R. W. Turkington, *Endocrinology* **88,** 1506 (1971).

[6] B. I. Balinsky, *Trans. Roy. Soc. Edinburgh* **62,** 1 (1950).

[7] F. J. A. Prop, *Pathol. Biol.* **9,** 640 (1961).

[8] E. S. Mills and Y. J. Topper, *J. Cell Biol.* **44,** 310 (1970).

[9] T. Oka and Y. J. Topper, *J. Biol. Chem.* **246,** 7701 (1971).

have also been shown to have the requisite biological activity.[10] Both ovine and bovine prolactin are effective; human placental lactogen is about 80% as active, on a weight basis, as prolactin in those instances in which it has been tested.[11]

Although these hormones have been shown to produce their characteristic effects when used at levels between 10^{-9} and 10^{-8} M,[1,12] they are usually employed at considerably higher concentrations, i.e., 5 μg/ml each. Appropriate amounts of freshly prepared stock solutions of each hormone are added directly to Medium 199 before sterilization. Insulin stock: 1 mg per 0.5 ml of 5×10^{-3} HCl (2 μg/μl). Prolactin stock: 1 mg per 0.5 ml of 10^{-4} *M* NaOH (2μg/μl). Hydrocortisone stock: 1 mg per 0.2 ml of anhydrous ethanol (5 μg/μl). When systems to which hydrocortisone is added are to be compared to systems without added glucocorticoid, a corresponding amount of ethanol should be added to the latter. The final concentration of alcohol in the media should be no more than 0.5%.

Preparation of Siliconized Lens Paper

In most of the studies on mammary gland explants, the tissue is supported at the surface of the medium. A common support is lens paper, which is siliconized to enable it to float. Prior to siliconization the paper is suspended in ethyl ether for 30 minutes and then the ether is removed by aspiration. This is done 3 times. The same procedure is then performed with 95% ethyl alcohol. Finally, the paper is washed 4 times for 15-minute intervals with quartz-distilled water, and dried at 37°. Siliconization is accomplished by submerging the dried paper in Siliclad (Clay Adams)–quartz-distilled water (1:1000) for 10 minutes at room temperature. After removal of the Siliclad solution by aspiration, excess silicone is eliminated by treatment with water as described above. Again the papers are dried at 37°.

A convenient type of container for culture of the explants is the sterile and disposable Falcon microdiffusion dish (Catalog No. 7004) whose dimensions are 65×15 mm. Accordingly, the dry lens paper is cut into pieces corresponding in size to that of the inner well of these dishes. The pieces are placed in covered, glass petri dishes and heated at 150° for 1 hour in a dry oven. This serves two purposes. The lens paper is sterilized, and the impregnated silicone is modified so that it now permits the paper to float.

[10] R. W. Turkington, W. G. Juergens, and Y. J. Topper, *Endocrinology* **80,** 1139 (1967).

[11] R. W. Turkington and Y. J. Topper, *Endocrinology* **79,** 175 (1966).

[12] F. E. Stockdale, W. G. Juergens, and Y. J. Topper, *Develop. Biol.* **13,** 266 (1966).

Preparation of Tissues

The explantation technique employed is determined by the developmental stage of the donor animal, and by the particular interests of the investigator. Thus, in some instances the whole gland is cultured as a unit, while in others the gland is cut into smaller pieces before explantation.

Embryonic Rudiments. Fetal mouse and rat mammary rudiments are detectable by day 13 of gestation. The effects of androgens on whole rudiments *in vitro* have been studied by Kratochwil.[13] Ceriani[14] has reported that whole fetal rat rudiments can be stimulated to produce casein in the presence of the appropriate hormones.

Whole Glands from Postnatal Mice. For studies on lobuloalveolar development it is advantageous to maintain the epithelial architecture intact. To this end, Prop[7] and Ichinose and Nandi[15] have developed a method for the culture of whole thoracic mammary glands from 3-week-old mice.

Gland Fragments. Most frequently the abdominal glands are used in organ culture studies. Limitations in the rate of diffusion of gases and components from the media necessitate the use of small explants prepared from these glands.[16] Using instruments sterilized in 70% ethanol and rinsed in boiling water, the glands are isolated aseptically after the animals have been killed by cervical dislocation. Isolation is facilitated by placing the animal on a dissection board, ventral surface up, and securing this position with pins through the appendages. After carefully shaving the abdomen and thighs with a sharp razor blade, loose hair is removed by swabbing with a wet cloth. Four incisions, through the skin only, are then made with a scissors. One is in the middle of the abdomen from the genitalia to the diaphragm. Another intersects the first at its upper extremity, completely transversing the ventral surface. Incisions are then made from the lower end of the median cut diagonally across each thigh. Skin flaps can then be peeled from the abdominal wall with forceps. These are pinned to the dissection board, exposing the mammary glands on the inner skin surface. Lymph nodes, one or two in number, are extirpated with forceps from each gland. The glands are then excised with forceps and scissors, removing as much connective tissue as possible, and placed in a sterile microdiffusion dish containing 1–2 ml of sterile medium 199.

It is desirable that the explants to be prepared contain approximately

[13] K. Kratochwil, *J. Embryol. Exp. Morphol.* **25**, 141 (1971).
[14] R. L. Ceriani, *Develop. Biol.* **21**, 530 (1970).
[15] R. R. Ichinose and S. Nandi, *Science* **145**, 496 (1964).
[16] J. J. Elias, *Proc. Soc. Exp. Biol. Med.* **101**, 500 (1959).

equal numbers of epithelial cells. Since the extent to which the ductal–alveolar system extends into the mammary fat pad is a function of both the species and the developmental state of the animal, preparation of equivalent explants from a given animal may require selective utilization of certain portions of the gland. Mature virgin and multiparous nonpregnant mice (approximate weight per gland, 100–150 mg): discard peripheral portion, constituting about 20% of the gland. Pregnant (approximate weight per gland, 80–100 mg) and lactating (approximate weight per gland, 200–300 mg) mouse gland: use entire gland. Mature virgin, multiparous nonpregnant, and 1–15-day pregnant rat gland: use inner core, constituting about 30% (250–350 mg) of the entire gland. Late pregnant and lactating rat gland: use entire gland. These precautions will assure that explants from a given type of animal will contain approximately equal numbers of epithelial cells.

Base-line activities and quantitative responses to hormones may vary somewhat from animal to animal. In experiments in which it is necessary to use pooled tissue from more than one animal it is desirable to employ equal numbers of explants from each animal in every determination.

The situation is different with the abdominal gland of the 3-week-old mouse, however. In this animal the epithelial cells are concentrated in a small area at the level of the nipple. For operational reasons it is desirable to use all the explants derived from one gland for each determination. The contralateral gland may serve as a control.

The explants are commonly prepared directly in the microdiffusion dish in which the glands and medium 199 were placed. The tissue is gently cut into small pieces (0.5–1.0 mg wet weight). Remove as much connective tissue as possible. Transfer of the explants to the culture dishes which have been preloaded with 2 ml of medium and siliconized lens paper is facilitated by first aspirating away most of the medium 199 in which the explants were prepared. Using a fine forceps, the explants may then be transferred to the floating lens paper, taking care not to carry over too much medium with the tissue fragments. The number of explants needed for each experimental point is determined by the parameter being studied. Up to about 25 explants can be accommodated in each dish.

Maintenance of Cultures

Incubation of the explants is carried out in a closed system. The microdiffusion dishes are stacked in heavy-duty polycarbonate boxes containing an open water container for maintenance of high humidity. Each box has, in addition to a hinged top, a closable opening at either end for introduction of 5% CO_2–95% air. The gas mixture is first bubbled through sterile distilled water, and is then permitted to flow into the box for a predetermined time at a predetermined rate such that the pH of the media is brought into range 7.3–7.6. This equilibration may take up to an hour after the gas flow has been stopped and the box has been sealed. The temperature is maintained at 37° in a water-jacketed incubator. Media should be replaced at least every 48 hours. Old media may be removed by aspiration, using sterile Pasteur pipettes, and 2 ml of fresh media introduced with sterile pipettes. The lens papers are easily refloated under these conditions. In some cases it is desirable to transfer the lens paper containing the explants to a fresh dish preloaded with medium. Gassing is then repeated.

To terminate the culture, the lens paper is removed from the dish and gently blotted to remove adhering medium. Explants are then carefully collected with a fine forceps.

Responses of Mammary Epithelial Cells

Morphological Methods

Whole Mounts. Whole mounts prepared from mammary gland according to the method of Lyons[17] provide insight into the architecture of the epithelial component of the whole tissue. Preparations from animals in different developmental states dramatically illustrate the changing ratios of epithelial cells to mammary fat, and the increased branching of the epithelial tree, as maturation proceeds.

Histology. The preparation, fixation, and staining of secretions from cultured explants have been performed by using standard histological techniques.[16] Examination of histological sections, prepared for light microscopy, from explants has provided information not obtainable by other methods. Insulin-induced increase in the number of epithelial cells per average alveolar cross section has been determined.[8] Also, alterations in the spatial arrangement of the alveolar cells and in the size of alveolar lumina consequent to hormone treatment have been reported.[12] Additional observations made on such sections have supplemented determinations made by other methods. Secretory material in the lumina is stainable with eosin. Increased nucleic acid content in the cytoplasm is reflected by enhanced basophilic character.[12]

Electron Microscopy. Visualization of most cellular organelles, how-

[17] W. R. Lyons, R. E. Johnson, R. D. Cole, and C. H. Li, *in* "The Hypophyseal Growth Hormone" (R. W. Smith, O. H. Gaebler, and C. N. H. Long, eds.), p. 461. McGraw-Hill (Blakiston), New York, 1955.

ever, requires electron microscopy. Details of the methodology employed in studies on mammary gland have been presented.[8,18-21] Morphological effects of insulin, hydrocortisone, and prolactin on explants from mid-pregnancy C3H/HeN mice have been studied,[8] using a multiple incubation technique.[22]

Biochemical Methods

The mammary gland contains two major cell-types, epithelial and fat. The proportions of these vary widely as a function of the developmental state of the animal. Changes in the synthesis of certain products, such as casein and α-lactalbumin, obviously reflect response of the epithelial cells. In other instances one cannot directly ascertain whether a response of explants is ascribable to the component epithelial or fat cells. The following techniques have been helpful in making this distinction.

The small clump of epithelial cells concentrated in the nipple area can be removed from the 3-week-old animal by cauterization.[23] As the animal matures this gland remains essentially devoid of epithelial cells, and explants prepared from it can be used to determine fat-cell responses. The unoperated abdominal gland can be used as a contralateral control.

Epithelial cells, free from fat cells but contaminated with a small percentage of connective tissue cells, can be prepared by treating finely minced mammary gland or explants with crude collagenase obtained from Worthington.[1,24] The mince is suspended in a 0.15% solution of collagenase in Medium 199 containing 4% bovine serum albumin, adjusted to pH 7.4. The suspension is shaken at 37° for 40 minutes. At intervals of about 5 minutes the suspension is aspirated into Pasteur pipettes. Use of a series of these with progressively smaller bore-size aids in the dispersion of the epithelial cells and their separation from fat cells. Fibrous tissue which adheres to the tips of the pipettes can be removed. After centrifugation (200 *g*; room temperature for 2 minutes) the epithelial pellet is washed 4 times with Medium 199. Residual fibrous tissue can be removed mechanically during this series of resuspensions and centrifugations. The yield of epithelial cells is about 90%, based on thymidine-^{3}H-labeled DNA.

[18] W. Bargman and A. Knoop, *Z. Zellforsch. Mikrosk. Anat.* **49**, 344 (1959).

[19] K. H. Hollman, *J. Ultrastruct. Res.* **2**, 23 (1959).

[20] K. K. Sekhri, D. R. Pitelka, and K. B. DeOme, *J. Nat. Cancer Inst.* **39**, 459 (1967).

[21] S. R. Wellings, R. A. Copper, and E. M. Rivera, *J. Nat. Cancer Inst.* **36**, 657 (1966).

[22] R. W. Turkington, D. H. Lockwood, and Y. J. Topper, *Biochim. Biophys. Acta* **148**, 475 (1967).

[23] K. B. DeOme, L. J. Faulkin, H. A. Bern, and P. B. Blair, *Cancer Res.* **19**, 515 (1959).

[24] E. Y. Lasfargues, *Anat. Rec.* **127**, 117 (1957).

The collagenase technique has been applied in many studies in which the tissue was first cultured as explants. Determinations on the residual pellet provide direct information on the response of the epithelial cells. Use of this method as a means of preparing epithelial cells which are to be cultured subsequently has been less successful. The isolated cells appear to have limited ability, compared to the cells within explants, to respond to the hormones.

Cell Proliferation. The epithelial cells in mammary explants derived from animals in different physiological states make DNA and divide under the influence of insulin,[25] various blood sera,[5] and epidermal growth factor.[26] The cells from mature C3H/HeN virgin mice and Sprague-Dawley rats are insensitive to insulin and serum factor(s) in terms of DNA synthesis.[1,2] They are also insensitive to insulin in terms of transport of α-aminoisobutyric acid (AIB), and the synthesis of glucose-6-phosphate dehydrogenase and gluconate-6-phosphate dehydrogenase.[1] After about 24 hours of culture the explants acquire these sensitivities. No exogenous hormones are required for the acquisition of these sensitivities. Cells in explants from mid-pregnancy mouse and late-pregnancy rat mammary gland, on the other hand, respond to these agents promptly after isolation.

The proliferative response can be quantitated in a number of ways:

1. The extent of epithelial DNA synthesis can be determined by comparing the epithelial DNA content of freshly isolated explants with that after culture, using the method of Dische[27] as modified by Burton.[28] This determination is performed on the cell-pellet obtained after treating the mouse tissue with collagenase. Minimum tissue requirement: 40–60 mg mid-pregnancy; 60–75 mg mature virgin.

2. DNA synthesis can also be estimated by incorporation of [^{3}H]thymidine into an acid-insoluble fraction. Since fat cells do not make DNA under the conditions used[12] treatment with collagenase is not necessary. Tissue requirement: 4–8 mg.

3. In some instances it may be desirable to know which epithelial cells within explants have synthesized DNA. Such information can be obtained by autoradiography of histological sections prepared from [^{2}H]thymidine-labeled tissue.[12]

4. The number of cells which are dividing at any given time during culture can be estimated by counting mitotic figures in histological sections prepared from explants. The results are usually expressed as mitotic in-

[25] F. E. Stockdale and Y. J. Topper, *Proc. Nat. Acad. Sci. U.S.* **56,** 1283 (1966).
[26] R. W. Turkington, *Exp. Cell Res.* **57,** 79 (1969).
[27] Z. Dische, *Mikrochemie* **8,** 4 (1930).
[28] K. Burton, *Biochem. J.* **62,** 315 (1956).

dices, i.e., the ratio of the number of cells in mitosis (M) to the total number of cells counted. The total number of cells which have entered M during a particular period can be determined by culturing in the presence of a mitotic inhibitor such as colchicine. The ratio of mitotic figures to total cells counted reflects the percentage of cells which have entered M during the culture period in question.

5. The increase in the number of cells can be directly determined, after treatment of the tissue with collagenase, by counting the suspended epithelial cells in a hemacytometer.[29] Counting is facilitated by brief treatment of the cell suspension with 0.5% trypsin solution (GIBCO) to prevent clumping. Minimum tissue requirement: 75 mg, mid-pregnancy mouse; 100 mg, mature virgin.

An important consideration in differentiation is whether or not certain developmental processes are dependent upon cell proliferation. Inhibitors of proliferation are useful tools for attacking this question experimentally. Agents which arrest cells in the mitosis (M) phase of the cell-cycle are usually not satisfactory because cells in M have a reduced general capacity for macromolecular synthesis.[30,31] Agents which prevent cells from entering the DNA-synthesis phase (S), such as 1β-D-arabinosylcytosine and fluorodeoxyuridine, can be fruitfully employed. Under these circumstances one may regard the inhibited cells as normal G_1 cells if proper controls are unaffected. In order to minimize the possibility of toxic side effects the smallest concentration of agent giving maximal inhibition should be used; this may vary as a function of the hormones added, and the developmental state of the donor animal. It is also necessary to establish that the DNA-inhibitor does not affect the developmental process in question when the agent is added to postmitotic cells.[32]

Assay of Products Made Uniquely by Mammary Gland. The formation of the following secretory products is a reliable indication that the mammary epithelial cells have attained a highly differentiated state. Development of these functional capacities may not be synchronous *in vitro*.[29]

1. Casein. Casein is a family of phosphoproteins. The isolation and characterization of the material synthesized by mammary explants is facilitated by incorporation of $^{32}P_i$ or labeled amino acids added to the culture system. After suitable labeling periods the phosphoproteins are precipitated by the addition of Ca^{2+} and rennin to the soluble fraction

[29] B. K. Vonderhaar, I. S. Owens, and Y. J. Topper, *J. Biol. Chem.* **248**, 467 (1973).
[30] H. Fan and S. Penman, *Science* **168**, 135 (1970).
[31] D. Martin, Jr., G. M. Tomkins, and D. Granner, *Proc. Nat. Acad. Sci. U.S.* **62**, 248 (1969).
[32] I. S. Owens, B. K. Vonderhaar, and Y. J. Topper, *J. Biol. Chem.* **248**, 472 (1973).

of the tissue.[33] If no further fractionation is to be performed, carrier bovine casein is added prior to homogenization. If the precipitate is to be resolved into separate casein polypeptides, carrier mouse casein, rather than bovine casein, is added prior to homogenization.[34,35]

The precipitate contains casein and other phosphoproteins. Although not all this material is authentic casein,[22] measurement of its isotope content represents a fairly reliable and rapid method for determining the relative rate and extent of casein synthesis by explants.

More precise quantitation of casein synthesis is accomplished by electrophoresis of the labeled phosphoproteins on starch gels or polyacrylamide gels.[34,36] Individual casein polypeptides can be visualized by staining or autoradiography. The individual casein bands can then be cut out and counted. Minimum tissue requirement: 4–8 mg per determination. The synthesis of casein by different systems within a given experiment may often be compared by expressing results as count per minute per unit wet weight tissue. However, in some instances it is necessary to express results as counts per minute per unit of epithelial DNA.[32] Similar considerations apply to the study of other functional parameters.

2. α-Lactalbumin. α-Lactalbumin is the B-protein component of the lactose synthetase system.[37] The ability of α-lactalbumin to enable the A-protein of lactose synthetase to utilize glucose as an acceptor of the galactose moiety of UDP-Gal, with consequent formation of lactose, is the basis for its assay. A modification[29] of the method of Brew[38] has been used for determination of α-lactalbumin activity in homogenates from cultured explants. Minimum tissue requirements: 50–80 mg of mature virgin or 30–40 mg mid-pregnancy tissue homogenized in 300 μl of buffer; 20–30 mg of lactation tissue.

3. Lactose. The incorporation of radioactive galactose into lactose during culture[39] has also been used to study the control of lactose synthetase activity.[40]

4. Fatty acids. Up to 70% of the fatty acids in rabbit milk trigly-

[33] W. G. Juergens, F. E. Stockdale, Y. J. Topper, and J. J. Elias, *Proc. Nat. Acad. Sci. U.S.* **54,** 629 (1965).

[34] R. W. Turkington, W. G. Juergens, and Y. J. Topper, *Biochim. Biophys. Acta* **111,** 573 (1965).

[35] A. E. Voytovich and Y. J. Topper, *Science* **158,** 1326 (1967).

[36] R. W. Turkington and M. Riddle, *Endocrinology* **84,** 1213 (1969).

[37] U. Brodbeck, W. L. Denton, N. Tanahashi, and K. E. Ebner, *J. Biol. Chem.* **242,** 1391 (1967).

[38] K. Brew, T. C. Vanaman, and R. L. Hill, *Proc. Nat. Acad. Sci. U.S.* **59,** 491 (1968).

[39] J. C. Bartley, S. Abraham, and I. L. Chaikoff, *J. Biol. Chem.* **241,** 1132 (1966).

[40] R. D. Palmiter, *Biochem. J.* **113,** 409 (1969).

cerides is comprised of octanoic and decanoic acids. The hormone dependence of the synthesis of these characteristic products by explants from rabbit mammary gland has been studied.[41]

Other Metabolic Activities of Mammary Gland in Vitro. Other functional parameters of mouse mammary gland which have been studied by the organ culture method, in attempts to understand more fully the development of this tissue, are listed:

1. Accumulation of α-aminoisobutyric acid[1,42]
2. Glucose uptake and total fatty acid synthesis[43,44]
3. Synthesis of glucose-6-phosphate dehydrogenase and gluconate–6-phosphate dehydrogenase[1,2,45]
4. RNA synthesis[12,46–48]
5. Formation of rough endoplasmic reticulum.[8,9]
6. DNA polymerase activity[49]
7. Synthesis and acetylation of histones[50–52]

[41] I. A. Forsyth, C. R. Strong, and R. Dils, *Biochem. J.* **129**, 929 (1972).
[42] T. Oka and Y. J. Topper, *Proc. Nat. Acad. Sci. U.S.* **68**, 2066 (1971).
[43] R. L. Moretti and S. Abraham, *Biochim. Biophys. Acta* **124**, 280 (1966).
[44] R. Mayne and J. M. Barry, *J. Endocrinol.* **46**, 61 (1970).
[45] D. P. Leader and J. M. Barry, *Biochem. J.* **113**, 175 (1969).
[46] R. W. Turkington and O. T. Ward, *Biochim. Biophys. Acta* **174**, 291 (1969).
[47] M. R. Green and Y. J. Topper, *Biochim. Biophys. Acta* **204**, 441 (1970).
[48] M. R. Green, S. L. Bunting, and A. C. Peacock, *Biochemistry* **10**, 2366 (1971).
[49] D. H. Lockwood, A. E. Voytovich, F. E. Stockdale, and Y. J. Topper, *Proc. Nat. Acad. Sci. U.S.* **58**, 658 (1967).
[50] W. F. Marzluff, Jr., K. S. McCarty, and R. W. Turkington, *Biochim. Biophys. Acta* **190**, 517 (1969).
[51] W. F. Marzluff, Jr. and K. S. McCarty, *J. Biol. Chem.* **245**, 5635 (1970).
[52] P. Hohmann and R. D. Cole, *J. Mol. Biol.* **58**, 533 (1971).

[38] Androgen Metabolism and Receptor Activity in Preputial Glands and Kidneys of Normal and Androgen-Insensitive tfm Rats and tfm/y Mice[1]

By C. Wayne Bardin, Leslie P. Bullock, and Irene Mowszowicz[2]

In recent years, many laboratories have examined the pretranscriptional events involved in testosterone activation of target cells, and the

[1] Supported in part by PHS Grant No. HD05276.

[2] On leave of absence from Faculte de Médecine, Pitié-Salpetrière, Service de Biochimie, Paris.

following steps are believed to be important: reductive metabolism of testosterone; binding of testosterone or one of its metabolites to cytoplasmic proteins or receptors; and transfer of this steroid–protein complex to the nucleus of the cell where DNA and RNA synthesis are initiated. It was reasoned that studies of genetic mutants with abnormal responses to androgenic steroids would be useful for the further elucidation of hormone action. Animals with inherited end organ insensitivity to androgens (testicular feminization) seemed ideal for this type of investigation.

Testicular feminization has been described in the rat (tfm)[3,4] and mouse (tfm/y).[5,6] In these species, the disorder is transmitted by the female to one-half of her male offspring in a pattern consistent with either a sex-limited autosomal dominant or an X-linked recessive gene mutation. The affected males have a female phenotype with an XY karyotype and a chromatin negative nuclear sex. These animals have no gross evidence of a reproductive tract other than bilateral inguinal testes. The absence of wolffian development and of masculinization of the external genitalia can be attributed to the lack of the differentiating influence of androgens, whereas the absence of mullerian derivatives is in accord with the postulated inhibitory influence of the testis on the development of the female gonaducts. A variety of studies have demonstrated that all organs in these rodents are insensitive to physiological doses of testosterone, dihydrotestosterone and other androgens.[7] Since these animals lack prostate and seminal vesicles, it was necessary to examine the early events of androgen action in organs other than those of the reproductive tract. The preputial glands in the rat and kidneys in the mouse were selected for intensive study. The procedures developed to investigate androgen metabolism and receptor binding activity in these organs are detailed below.

Procedure for Estimating *in Vitro* 5α-Reduction of Testosterone in Rat Preputial Glands

In vivo studies have demonstrated that preputial glands of normal rats convert testosterone to dihydrotestosterone and concentrate the latter seroid in nuclei.[8] To study the *in vitro* rate of testosterone conversion

[3] A. J. Stanley and L. G. Gumbreck, Program of the Endocrine Society, p. 40 (1964).

[4] C. W. Bardin, L. P. Bullock, G. Schneider, J. E. Allison, and A. J. Stanley, *Science* **167**, 1136 (1970).

[5] M. F. Lyon and S. G. Hawkes, *Nature (London)* **227**, 1217 (1970).

[6] S. Ohno and M. F. Lyon, *Clin. Genet.* **1**, 121 (1970).

[7] C. W. Bardin, L. P. Bullock, R. J. Sherins, I. Mowszowicz, and W. R. Blackburn, *Recent Progr. Horm. Res.* **29**, 65 (1973).

[8] L. P. Bullock and C. W. Bardin, *J. Steroid Biochem.* **4**, 139 (1973).

to dihydrotestosterone and other 5α-reduced metabolites, tissue minces and cell fractions of preputial glands are incubated at 37° in a Dubnoff incubator under 95% O_2:5% CO_2. Samples are preincubated for 5 minutes prior to the addition of substrate. Reactions are stopped by freezing in dry ice; acetone and samples are held at —16° until further analysis is performed.

Studies in Preputial Gland Minces. Preputial glands from 3–5 animals are pooled and aliquoted (80–100 mg) into 2 ml of buffer (0.14 *M* NaCl, 8.8 m*M* Na_2HPO_4, 2.7 m*M* KCl, 1.5 m*M* KH_2PO_4, 0.7 m*M* $CaCl_2$, 0.5 m*M* $MgCl_2$, 6.7 m*M* glucose, pH 7.6). [^{3}H]Testosterone (140,000 cpm) is added in 10 μl of ethanol and the tissue is incubated for 15–240 minutes. Testosterone reduction to dihydrotestosterone increases linearly with substrate concentrations between 50 n*M* and 5 μM. A substrate concentration of 0.5 μM is, therefore, used in most incubations. Dihydrotestosterone formation from testosterone is linear during 1 hour of incubation. The 5α-reductase activity of preputial gland minces is expressed as picomoles of dihydrotestosterone or the sum of dihydrotestosterone plus 5α-androstanediol formed per unit weight per unit time.[8]

Studies with Preputial Gland Cytoplasm and Nuclei. From 0.8 to 1.0 g of preputial glands are finely minced with iris scissors and homogenized in 4 ml of buffer containing 0.32 *M* sucrose, 30 m*M* Tris, 3 m*M* $MgCl_2$, pH 7.6. Homogenates are filtered through flannel and centrifuged at 800 *g* for 10 minutes. Supernatant is decanted and designated cytoplasm. The crude nuclear pellet is suspended in 1 ml of homogenization buffer, dispersed with a Dounce homogenizer, and sedimented (50,000 *g*, 60 minutes) through 2.2 *M* sucrose, 30 m*M* Tris, 1 m*M* $MgCl_2$, pH 7.6. Purified nuclei ($1.1–1.6 \times 10^7$) are added to 2.0 ml of Tris buffer containing an NADPH generating system and 1.5 ml of cytoplasm (8–10 mg of protein)are mixed with 0.5 ml of a similar incubation mixture concentrated 4-fold. [^{3}H]Testosterone (5 μM) is added to 10 μl of ethanol and incubated for 7.5 minutes to 2 hours.

Dihydrotestosterone and 5α-androstanediol accumulation in cytoplasm is a function of testosterone concentration up to 5 μM. At higher substrate concentrations, dihydrotestosterone and 5α-androstanediol formation plateaus. When the 5α-androstanediol isomers are resolved by paper chromatography, 85–90% is as the 3β,17β-isomer.[8] When testosterone is incubated with isolated preputial gland nuclei, dihydrotestosterone formation increases linearly with number of nuclei (1×10^6 to 1×10^7), time (15–120 minutes), and substrate concentration (10 n*M* to 5 μM testosterone).[8]

Isolation of Radioactive Steroid Metabolites. Testosterone, dihydrotestosterone, and 5α-androstanediol (200–900 cpm ^{14}C) are added to each sample to correct for procedural losses. Preputial gland incubates are then extracted with 6–8 volumes of methylene chloride, and extracts were washed successively with 0.1 *N* NaOH, 6% acetic acid, and water. Extracts are dried and spotted on silica gel thin-layer plates and developed in benzene to remove nonpolar lipids. Samples are then compressed to a fine line at the origin with acetone and developed in chloroform:methanol, 49:1. This latter chromatographic system resolves each sample into 3 major fractions (testosterone, 5α-androstanediol, and dihydrotestosterone plus androsterone) which are acetylated. Testosterone acetate and androstanediacetate are rechromatographed on thin-layer plates in benzene: ethyl acetate, 4:1 and benzene:ethyl acetate, 9:1, respectively. This latter system is also used to resolve the androsterone and dihydrotestosterone acetates. The overall recovery of [^{14}C]steroid markers is 71, 73 and 76% for testosterone, dihydrotestosterone, and 5α-androstanediol, respectively.

Testosterone metabolism by preputial gland minces as well as by cytoplasmic and nuclear fractions from male, female, and tfm animals has been compared. These experiments indicate that there is no significant difference in androgen metabolism in the preputial glands from these three groups. These studies, as well as those on other tissues, indicate that decreased 5α-reductase activity cannot explain the decreased dihydrotestosterone retention in preputial gland nuclei of tfm rats after testosterone administration.

Procedure for Estimating Androgen Receptors in Rat Preputial Gland

Preputial glands from 4–6 rats are removed in the cold room and dissected free of connective tissue. The glands are then cut into a fine paste with a Sorvall TC-2 tissue sectioner. It is essential that all further steps are performed at 1°. The tissue is homogenized in 2 volumes of buffer TEDG (0.05 *M* Tris, 0.1 m*M* EDTA, 10% glycerol containing 0.25 m*M* dithiothreitol added fresh daily) for 5 seconds with a Polytron ST-10 homogenizer at a rheostat setting of 5. The homogenate is centrifuged for 1 hour at 120,000 *g*. The resulting supernatant (cytosol) has a protein concentration of 8–14 mg/ml.

Portions (0.5 ml) of this preparation are incubated with [^{3}H]dihydrotestosterone (40,000 cpm, 2 n*M*) for 1–2 hours and mixed with protein markers (sheep[^{131}I]γ-globulin and rat [^{14}C]progesterone-cortisol binding globulin). A fraction of each sample (0.3 ml) is layered on 5 to 20% sucrose gradients made in buffer TEDG and sedimented in an SW 56 rotor at 300,000 *g* for 15 hours. Gradient tubes are emptied from the bottom and aliquoted into counting vials.

With this technique, an 8 S androgen receptor is demonstrated in preputial gland cytosol which is similar to that of prostate. This binding protein has a high affinity for dihydrotestosterone rather than testosterone and can be demonstrated in preputial glands from normal male and females but not from tfm rats.[9]

Procedure for Mouse Kidney 3-Ketoreductase Activity

In vivo studies demonstrated that kidneys of normal mice concentrated testosterone rather than dihydrotestosterone in nuclei.[10] The fact that 5α-reductase activity in mouse kidney is very low suggests that very little testosterone can be converted to dihydrotestosterone. In addition, the accumulation of this latter steroid may be precluded by its rapid metabolism to 5α-androstanediols by 3-ketoreductase (EC 1.1.1.50). The activity of 3-ketoreductase is measured as follows[11]:

[^{3}H]Dihydrotestosterone (2×10^5 dpm, 0.7 nmole) is dissolved in 0.1 ml of buffer (0.32 *M* sucrose, 30 m*M* Tris, 3 m*M* $MgCl_2$, pH 7.4) containing 10 μg of NADPH per milliliter. At time 0, 0.1 ml of the solution to be assayed is added, and the sample is incubated for 30 minutes at 22°. The final incubation volume is 0.2 ml with a substrate concentration of 3.5 m*M* and an NADPH concentration 4 μ*M*. Incubations are stopped by addition of methanol, and the samples are extracted from 20% methanol with methylene chloride after addition of ^{14}C-labeled and cold steroid carriers. Untransformed substrate (dihydrotestosterone) and products (5α-androstanediol) of the reaction are separated by thin-layer chromatography in chloroform:methanol, 97.5:2.5. The purity of isolated fractions was further checked by acetylation (acetic anhydride:pyridine, 1:1) and chromatography in benzene:ethyl acetate, 90:10. Since no further purification is achieved by acetylation, the second chromatographic step is usually omitted in the routine assay. Results are expressed as nanomoles of androstanediol formed per milligram of protein in 30 minutes.

Salient features of this enzyme are that: 95% is located in cytosol; it requires NADPH as cofactor; it has a high K_m; it prefers 5α- to 5β-substrates. The enzyme may exist in two molecular forms which can be separated by polyacrylamide gel electrophoresis. Forms A and B have mean molecular radii which correspond to molecular weights of 38,700 and 28,700, respectively. Sufficient 3-ketoreductase activity is present in

[9] L. P. Bullock and C. W. Bardin, *J. Clin. Endocrinol. Metab.* **35**, 935 (1972).
[10] L. P. Bullock, C. W. Bardin, and S. Ohno, *Biochem. Biophys. Res. Commun.* **44**, 1537 (1971).
[11] I. Mowszowicz and C. W. Bardin, *Steroids* **23**, 793 (1974).

cytosol at 0° to reduce physiological concentrations (2 nM) of dihydrotestosterone without addition of cofactor. 3-Ketoreductase activity is higher in male than in female mice and decreases after orchiectomy. Androgen treatment of castrate male and female mice increases enzyme activity to that seen in intact males. 3-Ketoreductase activity in tfm/y mice is similar to that of females but does not respond to testosterone administration.

Procedures for Estimating Androgen Receptor Activity in Mouse Kidney

Tissue Preparations. Kidneys are removed in the cold room, and all further processing is done at 1°. The tissue is weighed, finely minced and homogenized in 2 volumes of buffer TEDG (see Preputial Gland Procedure) for 5 seconds with a Polytron homogenizer at a rheostat setting of 5. If only cytosol is needed, the tissue homogenates are spun at 220,000 g for 1 hour at 1°. Resulting supernatant (cytosol) has a protein concentration of 14–17 mg/ml. When nuclear studies are to be done, the tissue homogenate is first spun at 800 g for 10 minutes. The supernatant is then centrifuged to obtain cytosol, and the nuclear pellet is washed successively in buffer STM (0.32 M sucrose, 30 mM Tris·HCl, 3 mM $MgCl_2$, pH 7.4), buffer STM containing 0.5% Triton-X and twice more in buffer STM. This results in a nuclear preparation with only a small amount of cellular debris and nuclei devoid of cytoplasmic tags by light microscopy.

Sucrose Gradients. Cytosol and 0.5 M KCl nuclear extracts prepared from mouse kidneys removed from animals 0.5 hour after an intravenous injection of [^{3}H]testosterone (50 μCi) contain androgen binders that sediment as 7.9 S and 3.6 S macromolecules, respectively.[12] Both these binders are specific for testosterone. The cytoplasmic binder can also be studied *in vitro* by incubating 0.5 ml of cytosol with [^{3}H]steroid (40,000 cpm, 2 nM) for 2 hours. The labeled cytosol is mixed with marker proteins and sedimented on linear sucrose gradients as described above for the preputial gland.

Dextran-Coated Charcoal Assay. High-affinity androgen binding activity in mouse kidney cytosol is measured with a dextran-coated charcoal assay similar to that of Korenman.[13] For estimating the dissociation constant (K_d) and the concentration of binding sites (n), cytosol is diluted with buffer TEDG and duplicate 0.2-ml samples (8–10 mg of protein per milliliter) are incubated with increasing amounts of [^{3}H]tes-

[12] L. P. Bullock and C. W. Bardin, *Endocrinology* **94**, 746 (1974).

[13] S. G. Korenman, *J. Clin. Endocrinol. Metab.* **28**, 127 (1968).

tosterone (0.4–7 nM) for 2 hours. Blanks are prepared in a similar fashion but contain an additional 500-fold excess of unlabeled testosterone. To absorb free steroids, 0.5 ml of dextran-coated charcoal [0.5 ml of 50 mM Tris, 0.1 mM EDTA, 10% glycerol containing 0.5% charcoal (Norit-A) and 0.05% dextran] are added to each sample. The tubes are vortexed for 5 seconds and centrifuged at 1000 g for 2.5 minutes. Bound steroid is calculated from the radioactivity in the supernatant minus the blank. Results are analyzed by the method of Scatchard.[14]

Sephadex G-200 Column Chromatography. The androgen receptor in mouse kidney cytosol can also be studied using gel exclusion chromatography. One milliliter of [^{3}H]testosterone-labeled cytosol is applied to a G-200 column (3 × 33 cm) equilibrated in buffer TEDG along with marker proteins of known Stokes radii. The elution volumes of the known proteins and the androgen receptor are determined. It is then possible to calculate the Stokes radius and axial ratio (f/f_0) of the cytosol receptor.[15]

The physical properties of the androgen receptor in the mouse kidney cytosol are shown in the table. It is significant that this binder has a

PHYSICAL PROPERTIES OF THE MOUSE KIDNEY CYTOSOL ANDROGEN RECEPTOR

Molecular weight	270,000
Sedimentation coefficient	7.9×10^{-13} sec
Dissociation constant (K_d)	1.7×10^{-9} M
Sites (per mg cytosol protein)	5.6×10^{-14} M
Stokes radius	82 Å
Axial ratio (f/f_0)	1.88

high affinity for testosterone both *in vivo* and *in vitro*. The binding affinity of this protein for dihydrotestosterone has not been determined since the 3-ketoreductase in mouse kidney cytosol rapidly reduces dihydrotestosterone even at 0°. Androgen binding activity is not demonstrable in kidney cytosol from androgen-insensitive tfm/y mice with either sucrose gradient or dextran-coated charcoal assays.[12]

[14] G. Scatchard, *Ann. N.Y. Acad. Sci.* **51,** 660 (1949).

[15] W. I. P. Mainwaring, *J. Endocrinol.* **45,** 531 (1969).

[39] Enzyme Secretion and K^+ Release Mediated by Beta and Alpha Adrenergic Receptors in Rat Parotid Slices

By MICHAEL SCHRAMM and ZVI SELINGER

Slice System for Measuring Enzyme Secretion Induced by Activation of the Adrenergic β-Receptor and by Butyryl Cyclic AMP

A single parotid gland from a 200-g rat fasted overnight weighs about 150 mg and contains roughly 20 mg of protein of which half is exportable protein. Amylase represents a considerable fraction of the exportable protein (30%, 5000 u per gland) and conveniently serves to monitor secretion. DNase which is also present in large amounts can serve equally well as a measure of secretion. Gland slices secrete 80–90% of the exportable protein in 90 minutes. Such a high yield of secretion is only obtained when the β-adrenergic receptor is activated while K^+ release, which is induced through the α-adrenergic receptor, does not occur. For this purpose the following alternatives are available: isoproterenol in Krebs-Ringer bicarbonate medium which contains 6 m*M* K^+; epinephrine and norepinephrine in the same medium but without Ca^{2+} or with Ca^{2+} but in the presence of 50 m*M* K^+. In the latter system the high K^+ replaces an equivalant amount of Na^+. Efficient secretion is also obtained in the regular Krebs-Ringer bicarbonate medium if the adrenergic receptors are bypassed using butyryl derivatives of cAMP as inducers.[1–3]

Equipment

- Rotary shaker bath
- Polytron homogenizer (Kinematica)
- Gas cylinder containing 95% O_2/5% CO_2
- Stainless steel single edge razor blades
- One beaker 250 ml, 4 beakers 100 ml
- Cellulose nitrate tubes 25 × 75 mm, each perforated 10 mm from the top to introduce a gassing tube. The tare weight of each tube is recorded
- Polyethylene vials 25 × 50 mm of the type used for scintillation counting

[1] S. Batzri, and Z. Selinger, *J. Biol. Chem.,* **248,** 356 (1973).

[2] S. Batzri, Z. Selinger, M. Schramm, and M. R. Robinovitch, *J. Biol. Chem.,* **248,** 261 (1973).

[3] Z. Selinger, S. Batzri, S. Eimerl, and M. Schramm, *J. Biol. Chem.,* **248,** 369 (1973).

Micropipettes, 50 μl
Small test tubes containing 0.45 ml of amylase dilution buffer
Cotton
Ether
Wooden board, tacks, and a dissection set
Gas distribution system: The gas is passed first through a washing bottle containing water at 37° and then through an empty suction flask to trap excess water. The flask is connected to a manifold from which polyethylene tubing (1.5 mm diameter) leads the gas into four individual reaction mixtures.

Reagents

Fortified Krebs-Ringer bicarbonate medium (m*M*): NaCl, 118; KCl, 5; $MgSO_4$, 1; $CaCl_2$, 2.5; KH_2PO_4, 1; $NaHCO_3$, 25; β-hydroxybutyrate, 5. Nicotinamide, inosine, and adenine[1] are not routinely added since they have been found to increase secretion in the absence of added hormone. The salt mixture without bicarbonate, calcium, and the organic components can be kept in the refrigerator.

Reagents for amylase determination: phosphate buffer pH 6.9, 20 m*M* containing NaCl 7 m*M* and bovine serum albumin fraction V, 1 mg/ml. This buffer serves for dilution and assay of amylase according to Bernfeld.[4]

Preparations

Rat Parotid Slices. Male albino rats, 200 g, kept at 26°, are transferred to a cage without food but with ample supply of water 18–20 hours before the experiment. The fasting time should coincide with the night (dark) period during which the rats usually eat. To minimize differences between individual animals, at least 8 glands from 4 animals are pooled in each experiment. The animal is placed in a large glass jar which is saturated with ether. After 3 minutes of anesthesia, the animal is bled by piercing the chest and cutting through the heart with scissors. The skin of the neck area is lifted and carefully cut at the median line. The glands are exposed by further lateral cuts of the skin which is stretched by tacks to each side of the neck. Connective tissue and fat which lies over the glands is carefully removed with the aid of fine forceps and scissors. The parotid gland is recognized by its almost white color. In contrast to the other salivary glands and lymph nodes which are quite compact, the parotid lobules extend from the ear cartilage to the large capsule of the

[4] P. Bernfeld, this series, Vol. 1, p. 149.

submaxillary gland and also to the skin of the neck, which has been pulled aside. The glands are dissected free of fat, connective tissue, and buff-colored lymph nodes and are immediately placed in the Krebs-Ringer bicarbonate medium at 37°, through which a 95% O_2–5% CO_2 mixture is continuously bubbled. When eight glands have been collected, they are placed in a single pile on 4 layers of Parafilm and are cut into slices of 1–2 mm^3 with the aid of two razor blades which are moved in opposite directions. The slices are washed for 5 minutes in a 100-ml beaker containing 40 ml of medium through which the O_2–CO_2 mixture is continually bubbled. The medium is removed by decantation, the slices are poured into the back of a polythene petri dish, drained by tilting and divided onto four equal portions. Slices are further processed immediately.

Determination of Initial Rates of Secretion. Each portion of slices equivalent to 2 glands is placed in a cellulose nitrate tube containing 4 ml of medium at 37°. The tubes are placed in groups of 2 in 100-ml beakers which are fixed in the rotary shaker bath. Gassing tubes introduced through the hole in the cellulose nitrate tubes are quickly adjusted so that the tip is just under the surface of the medium and gassing is continued. Manipulation of the gassing tubes is necessary to avoid excessive foaming due to protein secretion. Shaking is set to a rate that causes a definite movement of the slices and preincubation is continued for at least 15 minutes to boost the intracellular ATP level.[1] At any chosen time point the test tube is removed from the shaking beaker, is tilted and a sample of medium of 50 μl is taken with a micropipette for amylase assay. The sample is delivered into the small tubes containing 0.45 ml of the phosphate-albumin buffer kept at 0°. The samples can be frozen maintaining undiminished amylase activity. To measure initial rates of secretion two samples are withdrawn at 3-minute intervals, an inducer of secretion is then added and samples are again taken at 3-minute intervals. If only a single system is tested, it is possible to obtain even more detailed kinetics by withdrawing samples every minute; 10–15 minutes after onset of secretion the tube is removed from the bath and weighed. The tissue together with the medium is cooled in ice and homogenized twice, each time for 15 seconds with a Polytron homogenizer set at 5 (half of maximum). The homogenate is quantitatively transferred to a storage tube by washing the original tube and homogenizer with the phosphate buffer to bring the final volume of the homogenate to 15 ml. If amylase is not determined on the day of the experiment, samples should be stored at −20°.

Amylase is assayed in the homogenate and in the samples removed from the medium during the experiment.[4] The amount of enzyme in the homogenate plus the amount removed from the medium during the experiment represent the total amount initially present in the slices. In order to calculate the amount of enzyme in the medium, the volume of medium must be known. It is obtained from the total weight of the tubes minus their tare weight. The specific gravity of the medium is assumed to be 1.0 and the weight of the glands is not subtracted since it is small and most of it is due to extracellular fluid. The amount of amylase present at any time in the medium is expressed as percent of total.

The maximal initial secretion rate of amylase is produced by 10 μM catecholamine and is 1.5–3% per minute. The same rate is obtained with 1 mM mono- or dibutyryl cyclic AMP after a lag period of about 10 minutes. The rate of amylase release without added inducer of secretion is about 0.05% per minute. Amylase release in the controls without hormone is reduced to 0.02% per minute if a nutrient-rich medium F12[5] is used instead of Krebs-Ringer bicarbonate at all stages of the experiment. Enzyme secretion induced by the hormone is as sufficient as in Krebs-Ringer bicarbonate.

Long-Term Incubation. Because some evaporation and foaming occur with the above procedure, a somewhat different system from that described above is used when longer incubation periods are desired. Slices equivalent to one gland are placed in a polyethylene vial containing 2 ml of medium at 37°C. An inducer of secretion and other substances are added, the vial is gassed for 30 seconds with the O_2–CO_2 mixture and is rapidly and tightly closed with a rubber stopper. The vials are incubated in the shaker bath as above. Every time a sample of medium is removed for amylase assay, but at least once every 30 minutes, the vial must be flushed with the O_2–CO_2 gas mixture as above. If the system contains catecholamine, a fresh amount, equivalent to that initially present is also added every 30 minutes. Incubation can be carried out for 2 hours. At the end of the experiment a 50-μl sample of medium is withdrawn for amylase assay and the rest of the medium with the slices is homogenized and processed as described above under determination of initial rates of secretion.

Slice System for Measuring K^+ Release Induced by Activation of the Adrenergic α-Receptor

The α-adrenergic response is measured as the extent of K^+ released from the slices. The final extent represents a steady state between the efflux of K^+ due to the action of the α-adrenergic receptor and the influx K^+ due to the action of the Na^+, K^+-activated ATPase. Epinephrine, 0.1 mM, causes release of 20–30% of total cellular K^+. Initial rates of K^+

[5] R. G. Ham, *Proc. Nat. Acad. Sci. U.S.* **53**, 288 (1965).

release are more variable than the final extent, probably owing to limitations in the rate of diffusion from the slices.

Equipment. The apparatus for slice preparations, gassing, and shaking is described above for the system for measuring secretion. In addition the following are needed:

- Atomic absorption spectrophotometer.
- Micropipettes, 25 μl.
- Cellulose nitrate tubes of known weight, 15 × 75 mm, each perforated 10 mm from the top to introduce a gassing tube. The weight of the tube is recorded before the experiment.

Reagent

Fortified Krebs-Ringer bicarbonate medium as described above, but with the KCl component reduced from 5 to 3 mM (total K^+ = 4 m*M*)

Operation. Slices are prepared as described above for amylase secretion. Immediately after cutting the tissue, portions equivalent to 1.5 glands each are placed in 3 ml of medium in the tubes which stand in a rack in a bath at 37°. Tubes are gassed and shaken as already described. After 10 minutes of preincubation, the medium from each tube is removed by a Pasteur pipette and is replaced by 1 ml of fresh medium and the experiment is started.

Aliquots of 25 μl are taken at suitable times and added to 5 ml of distilled water to give a final dilution suitable for measurement of K^+ by atomic absorption. Five minutes after onset of incubation a sample is removed for assay (zero time), epinephrine, 0.1 m*M*, is immediately added, and samples are again taken every minute. The release reaction reaches steady state in less than 6 minutes. When the experiment is terminated, the tubes are weighed, 2 ml of medium are added, and the mixture is homogenized by the Polytron instrument for 30 seconds on a setting of 5 at room temperature. Crude particles are removed by centrifugation for 10 minutes at 3000 *g*. Duplicate aliquots of the homogenate diluted as above serve for determination of total K^+ introduced by the slices into the system at zero time. K^+ is also determined in the reagent medium without addition of slices.

Calculations. The concentration of K^+ determined on the homogenate minus the concentration in the reagent medium without slices, multiplied by the total volume gives the total K^+ introduced by the slices at zero time. The volume of medium is calculated as already described above. The amount of K^+ released from the slices by hormone action is derived as follows. The K^+ concentration determined on aliquots of the medium during the experiment multiplied by the volume of the reaction medium gives the total K^+ in the medium. Subtraction of the K^+ which was in the medium just prior to hormone addition yields the amount of K^+ released into the medium through hormone action. Under the conditions described there is only negligible K^+ release in the absence of epinephrine during the period of the experiment. K^+ release is expressed as percentage of the total introduced by the slices at zero time. The maximal extent of K^+ release is obtained with 0.1 m*M* epinephrine or norepinephrine. Isoproterenol which activates only β-receptors does not cause K^+ release.

[40] An Insect Salivary Gland: A Model for Evaluating Hormone Action

By WILLIAM T. PRINCE, MICHAEL J. BERRIDGE, and HOWARD RASMUSSEN

The complex anatomical arrangement of many tissues can severely alter the interpretation of experiments investigating how hormones and neurotransmitters act at the cellular and subcellular levels. For example, histochemical studies on the liver have suggested that epinephrine acts on parenchymal cells whereas glucagon acts on the reticuloendothelial cells which comprise approximately 30% of all liver cells.[1] The responsiveness of blood vessels to various agents has complicated studies on the control of lipogenesis by adipose tissue. Many transport systems such as the frog skin, toad bladder, gastric mucosa, mammalian salivary glands, and the pancreas are notorious for their structural complexity and varying cell types.[2] In an attempt to study precisely how a hormone acts on a single cell type, some ingenious methods have been devised to try and separate out homogeneous populations of cells. Frog skins and toad bladder have been split, pure suspensions of adipose cells have been isolated, and special neoplasmic lines have been introduced to magnify the numbers of a given cell type. The latter technique has been particularly useful in studying insulin secretion by β-cells.[3] However, an inherent danger in all these cases is that the artificial "purification" procedure severely alters basic aspects of the system by disrupting the normal cellular relationships. Nature with her large reservoir of diverse organisms

[1] L. Reik, G. L. Petzold, J. A. Higgins, P. Greengard, and R. J. Barrnett, *Science* **168,** 382 (1970).

[2] M. J. Berridge and J. L. Oschman, "Transporting Epithelia." Academic Press, New York, 1972.

[3] I. D. Goldfine, J. Roth, and L. Birnbaumer, *J. Biol. Chem.* **247,** 1211 (1972).

often provides material uniquely suited to study any given problem. The importance of the giant axon of the squid in our understanding of the action potential readily comes to mind, exemplifying that invertebrates are not to be ignored as potential sources of experimental systems for solving fundamental problems. The ideal system for the study of hormone action would consist of a homogeneous population of cells that maintain their normal cell-to-cell relationships when studied *in vitro*. The salivary glands of the adult blowfly, *Calliphora erythrocephala,* come close to this ideal, and they offer many advantages, especially for studying the way in which ion and water transport is regulated.

These insect salivary glands are paired tubular structures that run the length of the fly from the abdomen, through the thorax, to connect to a common duct which opens into the mouth (Fig. 1A).[4] Each gland consists of two discrete regions: a secretory region that secretes isotonic potassium chloride together with an, as yet, unidentified protein (possibly amylase); and a reabsorptive region which removes potassium chloride so that the final saliva secreted by the fly is hypotonic. The abdominal region of the gland is used in the experiments described in this article. This region is entirely secretory and consists of a tubular epithelium (about 1 cm long and 100 μm in diameter) one cell layer thick and is comprised of only one cell type (Fig. 1B). It lies free of musculature, vasculature, nerves and a connective tissue. These features facilitate the isolation of the gland, the performance of various experimental procedures and the interpretation of the results. It is true, however, that the tubular structure of the gland complicates the interpretation of some of the electrophysiological data and, also, that the gland is only suited to those biochemical techniques that can be adapted to suit small amounts of tissue as one gland weighs approximately 0.1 mg (wet weight). Nevertheless, the opportunity of studying a simple epithelium composed of a single cell type has many advantages when attempting to analyse the intracellular events associated with the action of a hormone.

The salivary glands of *Calliphora* are stimulated by 5-hydroxytryptamine (5-HT).[5] A detailed study of the structure–activity relationships of the 5-HT receptors in this tissue have shown that they are similar to the receptors of other 5-HT-stimulated systems.[6] This fact together with the sensitivity of the glands to 5-HT and the identification of 5-HT as the active factor in the brain of the fly suggest that 5-HT is the natural secretogogue and that 5-HT is not just mimicking the effect of a larger

[4] J. L. Oschman and M. J. Berridge, *Tissue Cell* **2,** 281 (1970).

[5] M. J. Berridge and N. G. Patel, *Science* **162,** 462 (1968).

[6] M. J. Berridge, *J. Exp. Biol.* **53,** 171 (1970).

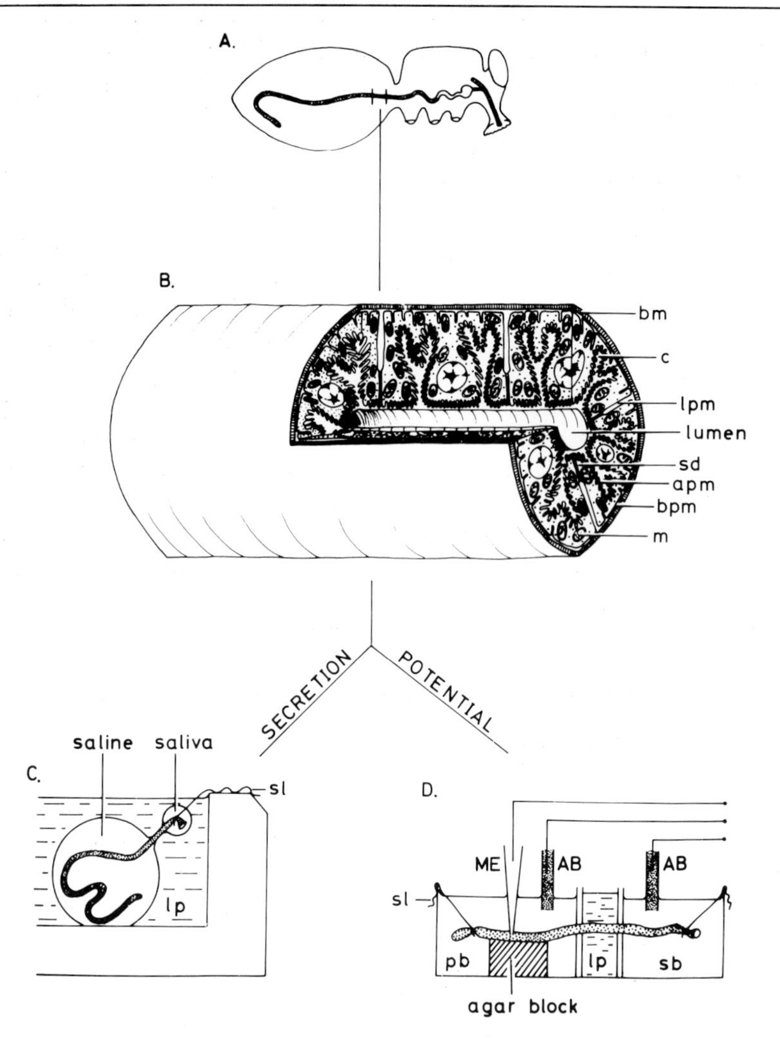

FIG. 1. The salivary gland of the adult fly. (A) The arrangement of the gland in the whole fly. Only one of the paired glands is shown. (B) A short segment from the salivary gland lying in the abdominal region. Large canaliculi (c), which open into the lumen, are formed by infoldings of the apical plasma membrane (apm). Both the canaliculi and free apical surfaces are lined with sheetlike microvilli. The basal plasma membrane (bpm) has a few short infoldings, whereas the lateral plasma membrane (lpm) is straight, and adjacent cells are linked together by septate desmosomes (sd) in the apical region. Large mitochondria (m) are scattered throughout the cytoplasm. (C) The technique used for measuring the rate of fluid secretion. A silk ligature (sl) pulls the cut end of the gland out into the liquid paraffin (lp). Secretion forms at the cut end of the gland and is collected for measurement. (D) The arrangement of isolated salivary glands in a Perspex

polypeptide hormone as it does in *Rhodnius* Malpighian tubules.[5,7] There is some evidence that suggests that 5-HT may be the active agent in other insects such as the cockroach but here again the evidence is not conclusive.[8]

In common with other homeostatic hormones the action of 5-HT is mediated by both calcium and adenosine 3′,5′-monophosphate (cyclic AMP).[6,9-11] The structural simplicity and uniformity of these insect salivary glands present unique opportunities not only for studying the primary interaction of a hormone with its receptor, but also for analyzing the intracellular sequence of events linking receptor activation with a change in cellular activity. Since the main function of salivary glands is to secrete fluid these glands are also a convenient model system for studying how hormones influence ion movement and, in particular, a potassium transporting system. The following techniques have been employed in the study of how 5-HT stimulates fluid secretion: (1) the collection of saliva and measurement of secretory rate; (2) the measurement of transepithelial and intracellular potential; (3) the measurement of calcium fluxes and cyclic AMP concentration.

The Collection of Saliva and Measurement of Secretory Rate

The secretory rates of isolated salivary glands can be made by dissecting the abdominal portion of the gland into 20 μl of *Calliphora* saline under liquid paraffin in a watch glass (Fig. 1C). The open end of each gland is ligatured with a fine silk thread and pulled out of the saline into the liquid paraffin. The gland is then nicked near the ligature so that the fluid secreted can escape and collect at the end of the

[7] S. H. P. Maddrell, *J. Exp. Biol.* **51**, 71 (1969).
[8] A. T. Whitehead, *J. Morphol.* **135**, 483 (1971).
[9] M. J. Berridge and W. T. Prince, *Advan. Cyclic Nucleotide Res.* **1**, 137 (1972).
[10] M. J. Berridge and W. T. Prince. *Advan. Insect Physiol.* **9**, 1 (1972).
[11] W. T. Prince, M. J. Berridge, and H. Rasmussen, *Proc. Nat. Acad. Sci. U.S.* **69**, (1972).

perfusion chamber during measurement of transepithelial and intracellular potentials. The gland lies in a groove connecting the perfusion bath (pb) and saliva bath (sb), which are separated by a middle bath containing liquid paraffin (lp). Agar bridges (AB) inserted into the two outer baths are connected to calomel electrodes to record transepithelial potentials. A microelectrode (ME) inserted into the gland was used to record the potential either across the basal or the apical surface of the gland. Agar blocks support the gland during penetration with the microelectrode. Taken from M. J. Berridge and W. T. Prince, *Advan. Cyclic Nucleotide Res.* **1**, 137 (1972).

gland. Drops of secretion are periodically taken off, fine glass capillary pipettes being used. They are then expelled from the pipettes under liquid paraffin and their diameter is measured. Conveniently, the secretory rates of 12 glands can be measured at the same time. A maximum of 3 glands is placed in a drop of saline. If there are more, the glands soon become oxygen deficient and the secretory rate slows. The medium bathing the glands can be changed by simply removing the drop bathing the glands and replacing it with another. When the composition of the bathing medium is changed the glands are washed at least three times in the new solution.

As experiments can be performed with a number of glands at the same time, the effect of a large range of molecules or different solutions can be screened rapidly and effectively, and statistically significant results obtained. The onset and recovery of secretion in response to a molecule such as 5-HT is very rapid and repeatable.[6] Also, the viability of glands *in vitro* is extremely good (they will survive for hours at room temperature and may be left overnight in the refrigerator) so that a number of experiments may be performed with the same group of glands. There is also an advantage in that the gland is responsive to exogenous cyclic AMP so that if a molecule or treatment is apparently inhibiting the response to 5-HT the receptor can be bypassed with cyclic AMP and the viability of the gland still tested, thus examining the possibility of an effect at a site other than the 5-HT receptor.

The fluid which is collected during measurements on the rate of secretion can be analyzed to determine its composition. In 15 minutes a single gland secretes sufficient fluid for analysis by flame photometry and conventional colorimetric techniques. The major ions secreted by the gland are potassium and chloride.[4,12] Some understanding of the mechanism of secretion can be obtained by studying how the composition of the saliva varies with changes in the ionic composition of the bathing medium. Radioactive tracers can also be employed to study the net flux of various substances across the salivary gland epithelium. However, in all these experiments only the net flux can be determined and this provides little information on the relative rates of movement across the two surfaces of the cell. In common with all transporting epithelia, therefore, the problem of understanding hormone action is to determine the relative effects at both cell surfaces. This objective was achieved by studying both intracellular and transepithelial potentials.[9,10,12-14]

[12] W. T. Prince and M. J. Berridge, *J. Exp. Biol.* **58**, 367 (1973).
[13] M. J. Berridge and W. T. Prince, *J. Exp. Biol.* **56**, 139 (1972).
[14] W. T. Prince and M. J. Berridge, *J. Exp. Biol.* **56**, 323 (1972).

The Measurement of Potential and Resistance

The tissue bath and perfusion system were designed so as to achieve rapid changes of perfusion medium with minimal fluid perturbations in the vicinity of the gland so that intracellular microelectrodes are not dislodged. The ability to record intracellular potentials for more than 1 hour, while frequently altering the bathing medium, attests to the success of the design. The gland is set up in a Perspex perfusion chamber divided into three separate sections[9,10,13] (Fig. 1D). The two outer baths contain saline, and the middle bath is filled with liquid paraffin. The salivary gland lies across the three baths in a groove in the partitions of the chamber. It is held by silk ligatures, tied at both ends, inserted into petroleum jelly seals at the ends of the two outer baths. The lumen of the gland is open to the saliva bath through a nick made immediately behind the silk ligature. Agar blocks are placed between the grooves in the tissue chamber to support the gland during penetration of the gland with microelectrodes and to improve the circulation of saline within the perfusion bath.

The bath containing the closed end of the gland is perfused continuously with saline during experiments. A six-way ball and socket tap[15] is placed near the bath so that the solutions bathing the gland may be changed rapidly. Solutions perfuse at 5 ml per minute, and the dead space between the tap and bath is <0.3 ml. Using the intracellular resting potential of the salivary gland cells and salines of different potassium concentrations, it has been possible to establish that new solutions enter the bath within 1 second and a new equilibrium is established in the bath in 5–10 seconds. Since the cost of perfusing high concentrations of certain substances, such as cyclic AMP, is prohibitive, an alternative technique was used to apply such agents. Perfusion is stopped and the compound is injected directly into the bath while draining the control saline to maintain a constant level. Removal of such compounds is achieved simply by restarting perfusion of control saline.

Transepithelial potential recordings are made by recording the potential difference between the two calomel cells connected to both outer baths by agar bridges. Potential changes are monitored on an oscilloscope and recorded on a pen recorder. Intracellular potentials are also measured using microelectrodes so that the potential changes recorded between the outer baths of the tissue chamber can be interpreted more lucidly. The glass microelectrodes (filled with 3 *M* KCl) used to monitor intracellular potentials always have a resistance of 15–25 MΩ (tip potential <5 mV),

[15] D. B. Sattelle, The ionic basis function in fresh-water gastropods. Ph.D. thesis, Univ. of Cambridge, 1971.

as these are found to allow easy penetration and maintainance of intracellular potentials. The criteria used for a successful penetration are that the potential should appear suddenly, should be maintained, and should fall to zero when the electrode is withdrawn.[14] The microelectrode is connected, through a calomel cell, to an impedance converter, which in turn is connected to the oscilloscope and pen recorder. When the microelectrode is in an intracellular position, recordings can be made across either the basal membrane using the perfusion bath as a reference or the apical membrane by reference to the saliva bath. Potentials recorded by the microelectrode and by the calomel cells connected to the two outer baths can be monitored continuously so that simultaneous recordings can be made across the whole epithelium and across either the apical or basal membrane. Also, by inserting a microelectrode in the lumen the validity of the method for measuring transepithelial potential can be tested.

Recently, techniques have been developed (in collaboration with Dr. B. Lindley) to measure the resistance and short-circuit current of this epithelial tissue. To measure the resistance of the epithelium, an electrode is placed in the lumen of the gland near to the meniscus between the saline in the perfusion bath and the liquid paraffin. Pulses of constant current are passed between the two outer baths of the tissue chamber, and the microelectrode then measures the voltage deflection produced by this pulse across the whole epithelium with reference to the perfusion bath.

The interpretation of the measurements are complicated by the fact that the space constant of this tissue is short. That is, the pulse recorded by the microelectrode decays away quickly as the electrode is moved away from the meniscus. However, the space constant can be accurately measured, and as long as the meniscus does not move during experiments reliance can be placed on the results obtained. Confirmation of the measurements of the space constant have been made by passing current through one microelectrode in the lumen away from the meniscus and examining the decay of the current pulses by using a second microelectrode inserted into the lumen at different distances from the current passing electrode. Results using this technique agree with those obtained using just one microelectrode and passing the current across the whole epithelium. Also resistance measurements can be made across the basal surface by placing the microelectrode in an intracellular position and measuring the voltage change that a current pulse produces across the basal membrane. However, here again the space constant of the basal surface must be determined. It must not be forgotten that changes in resistance, such as are caused, for example, by 5-HT will also change the space constants recorded across the basal surface and the whole epithelium and that a change in resistance of the apical surface may apparently change the resistance of the basal surface.

The potential recorded by a microelectrode in the lumen can be short-circuited (i.e., voltage clamped at 0) bypassing a current between the two outer baths of the chamber such that the potential recorded by the microelectrodes (one in the perfusion bath, the other in the lumen of the gland) are compared by differential feedback amplifier, which then passes current into the saliva bath so that any voltage difference between the two microelectrodes is nulled. The magnitude of this current is measured by the output of a second feedback amplifier connected to the reference electrode in the perfusion bath which holds the saliva bath at virtual earth. By introducing pulses of constant voltage into the input of the differential amplifier the conductance of the tissue can be measured by the change in clamping current that such pulses produce. The conductance measurements obtained using this technique are in good agreement with the resistance measurements obtained using the technique described previously. It should be mentioned here that we have found that unlike some other tissues such as frog skin or toad bladder the short-circuit current of this tissue is not always equivalent to the net ion transport across this epithelium.

The Measurement of Calcium Fluxes and Cyclic AMP Concentration

Calcium Uptake

Calcium uptake has been measured in salivary glands as a function of time.[11] It has been shown that uptake is increased by 5-HT but not by cyclic AMP. Calcium uptake is measured using experimental samples containing 20 salivary glands per tube in 500 μl of control saline (pH 7.4, 0.5 mM calcium) at 27°C. To each tube 10 μCi of ^{45}Ca is added 5 minutes after the addition of the compound, whose effect on calcium uptake is to be tested. The time allowed between addition of the compound to be tested and ^{45}Ca is chosen by the time taken for the secretory rate of glands tested by the method described in the section on collection of saliva to reach a maximum level. At appropriate times after the addition of ^{45}Ca, 5 ml of cold saline ($<4°$), without ^{45}Ca, are added, and then the glands in each tube are washed four times. After the final wash the volume is reduced to 100 μl, and 10 ml of scintillation fluid (1000 ml of toluene, 500 ml of Triton X-100, and 6 g of butyl PBD) are added directly to each tube. The contents are then transferred to scintillation vials and counted in a Packard TriCarb scintillation counter.

Calcium Efflux

Calcium efflux can be measured from glands that have been prevously labeled in ^{45}Ca. This is done by incubating overnight tubes containing 20 salivary glands per tube in 0.5 ml of saline containing 50 μCi calcium and 0.5 m*M* calcium at 4°. The tubes are then warmed to 27° for a further 30 minutes, after which time they are once more cooled to 4°. Each tube is then washed four times with ^{45}Ca-free saline. Finally, the glands are suspended in 0.8 ml of the appropriate experimental medium and transferred to a shaking water bath at 27°. Beginning 15 minutes later, 50-μl samples of the supernatant are removed from each tube at 5-minute intervals and transferred to 10 ml of toluene scintillation fluid. After six such samples have been taken, 0.5 ml of experimental solution (e.g., control, 5-HT, or cyclic AMP) is added. Then 100-μl samples are taken at 5-minute intervals for counting.

Cyclic AMP

5-HT-induced increases in cyclic AMP have been demonstrated in salivary glands using three techniques.[11] The most successful of these techniques has been that originally developed by Gilman.[16] This technique uses a cyclic AMP binding protein and measures absolute concentrations. Only the preparation of tissue samples will be described here since the assay technique is well documented.[16a]

Samples of glands (30/tube) are washed twice in 10 ml of normal *Calliphora* saline and the volume reduced to 100 μl. Then each sample is washed twice with 1 ml of experimental saline (e.g., control, 5-HT- or LSD-containing saline). Finally, the volume is reduced to 100 μl and the tubes are incubated for the required time. After incubation, 5 μl of 1 *N* HCl are added to each tube, and the tube is boiled for 5 minutes when 5 μl of 1 *N* NaOH are added and the tube is allowed to cool. The tubes are centrifuged, and 10-μl samples are taken for assay. With this technique, 3–4-fold increases in cyclic AMP can be seen, e.g., with 10 m*M* 5-HT or LSD.[11,17]

Some insect tissues are known to have guanosine 3′,5′-monophosphate (cyclic GMP) in concentrations 2–3 times higher than cyclic-AMP.[18] It seems unlikely that cyclic GMP is the active intracellular intermediary in this tissue, as Gilman has shown that his assay has 1000 times the

[16] A. G. Gilman, *Proc. Nat. Acad. Sci. U.S.* **67**, 305 (1970).

[16a] See volume 38, this series [7].

[17] M. J. Berridge and W. T. Prince, *Brit. J. Pharm.* in press.

[18] E. Ishikawa, S. Ishikawa, J. W. Davis, and E. W. Sutherland, *J. Biol. Chem.* **244,** 6371 (1969).

sensitivity for cyclic-AMP than cyclic-GMP. Secretion studies have shown that the cyclic-AMP receptor can accommodate certain changes in the base region; indeed cyclic tubercidin 3′,5′-monophosphate is more active than cyclic AMP, but cyclic GMP is totally inactive.[9] Nevertheless to be quite certain that it is cyclic AMP that is the active nucleotide, a particulate hormone-sensitive adenyl cyclase has been looked for, and changes in cyclic AMP have been measured by means of another technique.

The methods of preparation of a particulate adenyl cyclase preparation are now standard and will not be described here. Suffice it to say that a NaF-sensitive adenyl cyclase has been found that is stimulated 3–4-fold by 5-HT.

Cyclic AMP has also been measured by the technique of Krishna *et al.*[19] The adenine pool of the cell is labeled with [^{3}H]adenine and conversion into ^{3}H-labeled cyclic AMP measured by separation of labeled cyclic AMP on a Dowex column. Samples containing 50 glands per tube were incubated in 0.5 ml saline with 100 μCi of ^{3}H-labeled adenine for 30 minutes at 27°. After this time the glands are briefly washed in the presence or in the absence of a stimulant (10 m*M* 5-HT) in 0.5 ml saline containing 10 m*M* theophylline (to inhibit phosphodiesterase). After 10 minutes, the glands are boiled for 5 minutes to extract the cyclic AMP. On cooling, the tubes are centrifuged and 50 μl of the supernatant are taken for assay. The labeled compound formed under these conditions has been subjected to chromatography with two different solvents known to separate cyclic AMP from cyclic GMP. Over 90% of the radioactivity recorded behaved like cyclic AMP.

The use of several different techniques described so far has elucidated our knowledge of the physiology of this insect salivary gland and provide a useful model of hormone action. Secretion studies have identified specific 5-HT and cyclic AMP receptors associated with this tissue.[6,10] The convenience of this technique for studying a large range of different compounds has facilitated these structure-activity studies thus allowing a model of these two receptors to be formulated. Also such studies have shed new light on the mode of action of various hallucinogenic molecules such as LSD, mescaline, and various amphetamines.[20]

While the secretory studies have proved extremely useful for studying the nature of the receptors on this tissue, the electrical and biochemical measurements have enabled a more detailed study of the mechanism by which a 5-HT-receptor interaction at the cell surface is translated into

[19] G. Krishna, B. Weiss, and B. B. Brodie, *J. Pharmacol. Exp. Ther.* **163**, 379 (1968).
[20] M. J. Berridge and W. T. Prince, *Nature (London)*. **243**, 283 (1973).

increases in the intracellular concentrations of calcium and cyclic AMP within the cell and how these intracellular messengers induce secretion. For example the potential measurements showed that 5-HT and cyclic AMP produce similar effects on the potential only when calcium or chloride is omitted from medium bathing the gland.[10,12,13] A theory was proposed in which activation of the receptor by 5-HT not only activated adenyl cyclase but also stimulated calcium entry into the cell independently of the rise in cyclic AMP concentration. The potential measurements indicated that the calcium influx increased anion movements while cyclic AMP stimulated a cation pump. More recently the effects of 5-HT on cyclic AMP and calcium influxes have been measured directly.[11]

The ability to distinguish between the action of each second messenger on the basis of their effects on cation and anion transport has revealed that calcium and cyclic AMP act within the cell with markedly different time courses.[10] Also, it has been shown that there is a complex relationship between cyclic AMP and calcium in the cell in addition to their more obvious actions on ion transport.[11,12] For example, cyclic AMP appears to increase the release of calcium from the mitochondria and calcium can inhibit adenyl cyclase. The precise nature of the actions of these two intermediaries is at present under investigation. As yet, a cyclic AMP-dependent protein kinase has been isolated and certain possible natural substrates have been examined.[21]

Despite the small size, the salivary glands of *Calliphora* are thus amenable to a wide range of experimental techniques. The information gathered so far can be integrated into a simple model to explain how an external signal (5-HT) is detected by the cells and translated into a dramatic alteration of both membrane resistance and rate of ion translocation. The involvement of calcium and cyclic AMP in the transduction of information from receptor to effector is an almost universal feature of the action of many homeostatic hormones. The ease with which the effect of calcium and cyclic AMP can be monitored within the salivary gland cell provides a unique opportunity for probing the complex feedback relationships which modulate the activity of these two second messengers.

[21] H. Rasmussen, W. T. Prince, and M. J. Berridge, in preparation.

Section VIII

Vascular Tissue

[41] Preparation of Isolated Epididymal Capillary Endothelial Cells and Their Use in the Study of Cyclic AMP Metabolism

By J. J. KEIRNS, R. C. WAGNER, and M. W. BITENSKY

Isolation of Capillaries

Principle. Several investigators[1] have prepared pure fractions of fat cells by digesting epididymal fat with collagenase. Upon centrifugation of the dissociated adipose tissue, adipocytes collect at the top and the vascular and stromal tissue is pelleted. The isolation procedure[2] described below involves further purification of this pellet to obtain a relatively homogeneous preparation of intact capillaries.

Reagents

Tyrode's balanced salt solution. One liter contains 8.0 g of NaCl, 0.2 g of KCl, 0.05 g of $NaH_2PO_4 \cdot H_2O$, 0.1 g of $MgCl_2 \cdot 6H_2O$, 1.0 g of $NaHCO_3$, 1.0 g of glucose, and 0.2 g of $CaCl_2$, pH 7.2.

Collagenase, crude (Worthington Biochemical Corp.)

Bovine serum albumin

During isolation only plasticware or glassware siliconized with Siliclad (Clay Adams) was used.

Instructions. The distal portion of epididymal fat pads from Sprague-Dawley rats (250–300 g) are removed, placed in ice cold Tyrode's balanced salt solution, and thoroughly minced. Three fat pads are placed in a siliconized vial with 6 ml of Tyrode's containing 40 mg of collagenase and 6 mg of bovine serum albumin. The mixture is incubated at 37° for 45 minutes on a wrist-action shaker. It is then centrifuged at 30 *g* for 2 minutes. The pellet, which contains larger blood vessels, stromal cells, and attached fat cells, is discarded. The supernatant solution containing the fat cake is decanted and centrifuged at 800 *g* for 2 minutes. The vascular pellets from the second centrifugation are washed twice in Tyrode's and collected by centrifugation at 800 *g* for 2 minutes. The final pellet is pink. It contains fragments of intact capillary networks with minimal contamination by stromal elements and with no fat cells as judged by phase contrast and electron microscopy.

[1] M. Rodbell, *J. Biol. Chem.* **239**, 375 (1964).

[2] R. C. Wagner, P. Kreiner, R. J. Barrnett, and M. W. Bitensky, *Proc. Nat. Acad. Sci. U.S.* **69**, 3175 (1972).

Measurements of Adenyl Cyclase and Cyclic AMP Phosphodiesterase. These enzymes were assayed using radioactively labeled substrate isolating labeled product by thin-layer chromatography.[3] Homogenized (Potter-Elvehjem) or sonicated preparations of the capillaries show modest adenyl cyclase activity (about 30% the specific activity of fully stimulated liver), but this activity is stimulated at most only 15% by hormones or prostaglandins. The cyclase is inhibited at least 95% by 5 m*M* alloxan. The homogenates contain an active cyclic AMP phosphodiesterase, which exhibits the usual inhibition by aminophylline or papaverine.

Effect of Hormones or Prostaglandins on the Cyclic AMP Levels of Intact Endothelial Cells. Cyclic AMP content was measured by a modification[2] of the radioimmunoassay method of Steiner.[4] The cyclic AMP content in the absence of added stimulator was 1.2 pmoles of cAMP per milligram of protein. The levels observed in the presence of stimulators are indicated in terms of percentage of basal activity (100% is no stimulation). The most potent stimulators are catecholamines: 20 μM norepinephrine (333%), 50 μM epinephrine (383%), 20 μM isoproterenol (460%), and 20 μM phenylephrine (200%). Phentolamine and propranolol (at 20 μM), which are without effect by themselves, reduce the stimulation by norepinephrine to 225% and 200%, respectively. Among other hormones tested, only 20 μM glucagon (200%) and 50 μM vasopressin (250%) have any significant effect. Hormones without effect include MSH, ACTH, insulin, parathormone, bradykinin, and histamine. Prostaglandins (at 30 μM) significantly stimulate, although the efficacy varies through the prostaglandin series: E_1 (292%), E_2 (284%), F_1 (175%), F_2 (183%), A (150%), and B (158%).

Cytochemical Localization of Adenyl Cyclase

Principle. The method of Reik *et al.*[5] is based on the appearance of lead precipitates of phosphate or pyrophosphate which are produced by hydrolysis of ATP. The present method[2] contains two modifications to increase the specificity. First, instead of ATP, 5′-adenylylimidodiphosphate, which is a substrate for cyclase but not for any other known membrane ATPases, is used. Second, the specific cyclase inhibitor, alloxan,[6]

[3] J. J. Keirns, M. A. Wheeler, and M. W. Bitensky, *Anal. Biochem.*, (1974) in press.

[4] A. L. Steiner, D. M. Kipnis, R. Utiger, and C. Parker, *Proc. Nat. Acad. Sci. U.S.* **64,** 367 (1969). See also this series, Vol. 38, [13].

[5] L. Reik, G. L. Petzold, J. A. Higgins, P. Greengard, and R. J. Barnett, *Science* **168,** 382 (1970).

[6] K. L. Cohen and M. W. Bitensky, *J. Pharmacol. Exp. Ther.* **169,** 80 (1969).

is used to prevent appearance of precipitate assigned to the action of cyclase.

Reagents

Tris-maleate buffer (0.05 *M* each of Tris and maleic acid), pH 7.4, 18% (w/v) dextran (molecular weight 250,000, Pharmacia)
$Pb(NO_3)_2$, 40 m*M*
$MgSO_4$, 40 m*M*
Aminophylline
5′-Adenylylimidodiphosphate (AMP-PNP) (ICN Corp., Irvine, California)
ATP
Alloxan, 0.1 *M*
Tyrode's cacodyate buffer, pH 7.4. One liter contains: 6 g of NaCl, 0.2 g of KCl, 10.7 g of sodium cacodylate, 0.1 g of $MgCl_2 \cdot 6H_2O$, 1.0 g of $NaHCO_3$ 1.0 g of glucose, 0.2 g of $CaCl_2$. Adjust pH with HNO_3
25% glutaraldehyde in water
OsO_4, 4%

Instructions. Fresh unfixed capillaries are suspended in incubation medium made up as follows: 4 ml of Tris-malate buffer; 4 ml of dextran; 0.5 ml of $Pb(NO_3)_2$, 1.0 ml of $MgSO_4$, and either 0.5 ml of alloxan or 0.5 ml of water; 21 mg of aminophylline is added, and 2 mg ATP, 2 mg AMP-PNP, or no substrate. Incubations (in 0.1 ml total volume) are carried out at 25° for 5 minutes (with ATP) or for 25 minutes (with AMP-PNP). Incubation is stopped by dilution with 10 ml of Tris-malate buffer and centrifugation at 5000 *g* for 10 minutes. The capillaries are washed with Tris-maleate buffer and then fixed for 30 minutes at 25° in Tyrode's cacodyate buffer containing 2% glutaraldehyde. The capillaries are sedimented and washed twice with Tyrode's cacodylate buffer. After the final wash the tissue is spun for 5 minutes in a Beckman 152 microfuge to form a tightly packed pellet. The pellets are fixed again in a 1:1 mixture of Tyrode's cacodylate buffer and 4% OsO_4, dehydrated in ethanol, embedded in Epon, and sectioned with an LKB 1 ultramicrotome. Figure 1 shows that reaction product formed in the presence of AMP-PNP is localized on the lumenal membrane, in intercellular junctions, and in attached micropinocytic vesicles and vesicles within the cytoplasm. The reaction product is totally absent in the presence of alloxan or in the absence of substrate. Reaction product is not formed spontaneously in the absence of tissue. If ATP is used as substrate, the distribution of reaction product is much more generalized, and its appearance is not prevented by alloxan.

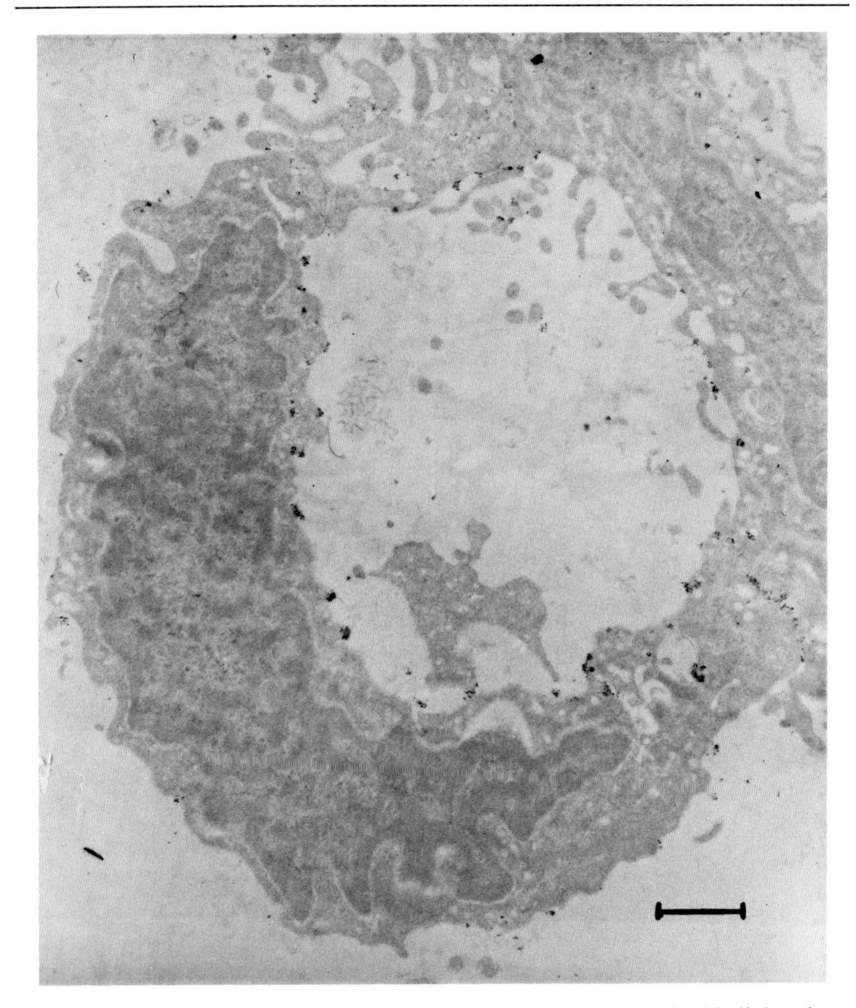

FIG. 1. Cross section of a capillary treated with 0.5 mM adenylylimidodiphosphate for 25 minutes at room temperature (not counterstained). $\times$ 24,000; scale marker = 0.5 μm.

Section IX

Unicellular Organisms

[42] The Growth and Utilization of *Dictyostelium discoideum* as a Model System for Hormone or Protohormone Action

By MICAH I. KRICHEVSKY and LESLIE L. LOVE

Maintenance of Stock Cultures

Stock cultures of *Dictyostelium discoideum* are grown on an agar surface with a bacterial associate as part of the food supply unless axenic cultures are desired (see below). The species of bacteria chosen are usually *Escherichia coli* or *Aerobacter aerogenes*. Other species may be used if convenient.

To minimize bacterial contamination, a highly penicillin-resistant strain of *A. aerogenes* may be used as the bacterial associate. In this case, penicillin is incorporated in the medium.

The following medium[1] is used to maintain stock cultures as well as grow large amounts of amoebae on solid medium (g/liter): yeast extract, 1.0; Bacto peptone, 10.0; glucose, 10.0; $MgSO_4$, 1.0; K_2HPO_4, 0.096; KH_2PO_4, 0.146; agar (as needed) 20.0. When penicillin-resistant *A. aerogenes* is the bacterial associate add 1×10^6 units/liter of penicillin G, i.e., about 1.0 g of 1625 units/mg. Dispense 600 ml in each 1 liter Erlenmeyer flask. Use a cotton stopper. Autoclave 15 minutes.

Procedure. The following procedure is used with either *E. coli* or *A. aerogenes*. A 25-ml Erlenmeyer flask containing 10 ml of the above medium (without agar) is inoculated with an "inoculating needle full" of bacteria from a stock slant of bacterial stock culture. The flask is incubated on a reciprocating shaker at 37° for 6–8 hours. Pour plates of sterile medium, 15 ml in 100×15 mm type petri dishes (plastic dishes of 90 mm actual diameter serve very well). With a sterile pipette, transfer 0.4 ml of liquid culture to the surface of each agar plate.

Touch a flamed, hot inoculating needle to the surface of the agar plate to both cool and wet it. Touch the wet needle to 4–6 spore heads (chosen as rising well into the air to minimize chance surface contamination). The spores should stick to the needle through surface tension. The needle is touched to the drop of liquid culture on the agar surface. A sterile glass L-shaped spreader (made from 5 mm diameter glass rod 20 cm long; the bend approximately 6 cm from one end) is used to distribute the mixed bacteria and amoeba spore suspension over the surface of the plate.

[1] J. T. Bonner, *J. Exp. Zool.* **106**, 1 (1947).

The plates are incubated at 22° for 2–3 days. After sorocarp (or fruit) formation is complete, the plates are stored at 2–5°.

Repeat weekly.

Growth and Harvesting of Large Amounts of Amoebae (Solid Surfaces)

Aluminum baking sheets with raised edges (about $15 \times 11 \times 1$ inches) are fitted with aluminum tops with 0.25-inch bent sides. Each baking sheet has approximately the same surface area as 20 petri dishes.

A 250-ml flask, containing 125 ml of the above medium (liquid), is inoculated with *E. coli* and incubated at 37° for 7 hours.

Autoclaved baking sheets are filled with 600 ml of the above solid medium. For 20 baking sheets, prepare the following inoculum. Select 6 fresh (less than 1 week storage) stock plates. Wash the spores from the surface of each plate by flooding the surface of the plate with 15 ml of the *E. coli*-containing medium. The medium is expelled forcibly onto the agar surface from a 10-ml pipette to dislodge the spores. The resulting suspension is returned to the original 250-ml flask.

Each baking sheet is inoculated with 5.0 ml of the combined *E. coli–D. discoideum* suspension. A glass spreader is used to distribute the inoculum by first spreading in the long direction, then in the short direction. The baking sheets are incubated at 22° until sporulation is complete, if this is the desired result. Such preparations are used for isolating steroid or polysaccharides, etc.

When the desired result is a suspension of cells, predominantly at any given stage of differentiation, the following method is used. Three batches (5 in each) of baking sheets are moved from 22° to 18° at 18, 21, and 24 hours after inoculation. At 42 hours the cells are harvested if predominantly vegetative amoebae are desired. (Longer incubation times are used to produce cells at later stages. The harvesting procedures are identical, irrespective of the stage of development.)

Five baking sheets are selected which show a predominance of the desired stage of morphogenesis. The amoebae are floated from the agar surface with a total of 30 ml of ice cold distilled water per sheet. The cells are washed by centrifugation in 50-ml round-bottom tubes using a Sorvall RC-2B centrifuge (Ivan Sorvall, Inc., Newton, Connecticut) with a SS-34 angle head (i.e., 4.25 inch diameter rotor to outer edge of tube). The centrifuge temperature is 5°. The speed is brought to 4500 rpm and allowed to coast to a stop. (Centrifugal forces greater than approximately 800 g disrupt the cells.) The supernatant suspension is carefully decanted, and the cells are resuspended in 30 ml of cold distilled water for each tube. This procedure is repeated as many as 8 times or until microscopic inspection shows essentially no residual bacteria (complete removal is not a practical possibility by this method). For most experiments,[2-4] the cells are finally suspended in 5 ml of cold distilled water per original baking sheet.

Growth and Harvesting Small Amounts of Amoebae (Solid Surfaces)

Vegetative amoebae for morphogenetic experiments[5] are obtained from an agar culture prepared by the same method described for stock cultures. As many as 20 petri dish cultures are incubated at 22°. At 18, 21, and 24 hours, groups of plates are moved to 18°. After 42 hours, select a culture which shows essentially all the bacteria gone but no evidence of aggregation patterns. The amoebae are floated off the agar surface with 10 ml of ice cold distilled water. The cell suspension is poured into a 50-ml round-bottom centrifuge tube after making the total volume up to 40 ml with more water. The cell suspension is centrifuged as described above. The cells are washed 3 times in the same way and inspected microscopically for essential absence of residual bacterial cells. The final volume is 10 ml.

Growth in Liquid Culture with Bacteria

Depending on need, one of the following two methods for growing amoebae with *E. coli* as the primary food supply have been found useful. If radioactive medium components are used, the first method will yield amoebae which are uniformly labeled with the chosen isotope.

The medium contains (g/liter): K_2HPO_4, 7; KH_2PO_4, 3; $(NH_4)_2SO_4$, 1; $MgSO_4 \cdot 7H_2O$, 0.1; glucose, 1–2. The medium is either filter-sterilized or autoclaved (the glucose is autoclaved separately). One-liter flasks, containing 250 ml of medium, are inoculated with *E. coli*. The flasks are incubated on a reciprocating shaker (112 stroker per minute, 3.5 inches stroke length) at 37° until growth stops. The flasks are cooled to room temperature.

Each flask is inoculated with amoebae from one petri dish (obtained by washing with sterile cold distilled water as described above). The flasks are incubated at room temperature (22–25°) on a reciprocating shaker as above. Growth is followed either microscopically or by an auto-

[2] M. I. Krichevsky and L. L. Love, *J. Gen. Microbiol.* **41,** 367 (1965).

[3] M. I. Krichevsky and L. L. Love, *J. Gen. Microbiol.* **50,** 15 (1968).

[4] M. I. Krichevsky, L. L. Love, and B. M. Chassy, *J. Gen. Microbiol.* **57,** 383 (1969).

[5] M. I. Krichevsky and B. E. Wright, *J. Gen. Microbiol.* **32,** 195 (1963).

matic cell counter. For example, the cells may be counted with a model "B" Coulter Counter (Coulter Electronics, Hialeah, Florida) using a 100 μm orifice. The lower threshold is set to minimize the error due to coincident intake of the autoclaved *E. coli* particles with the amoeba. Control counts have shown that the error is about 2000 counts/ml, which is about 50% of the total counts obtained at the time of inoculation. No upper threshold is used. All raw counts should be corrected for coincidence loss, using the data provided by the instrument manufacturer. The cultures are counted by taking 0.1 ml of the culture, which is diluted to 50 ml with 0.085% NaCl. To avoid lysis, growth should not be allowed to proceed to a stationary phase.

The second method is based on that of Gerisch.[6] The medium consists of *E. coli* cells and sodium phosphate buffer (20 m*M* phosphate, pH 6.2). Frozen, packed *E. coli* cells are obtained commercially (Grain Processing, Muscatine, Iowa). For each liter of medium desired, 20 g of cells are placed in buffer at room temperature. The mixture is stirred for 15 minutes. The suspension is dispensed in 1-liter flasks and autoclaved for 15 minutes. After cooling, the flasks are inoculated and incubated as described in the first method.

Axenic (Bacteria-Free) Growth

The germination medium contains (g/liter): (A) Proteose peptone, 14; yeast extract, 7; (B) glucose, 1.2; penicillin G (potassium), 0.5; streptomycin sulfate, 0.5. Solution A is autoclaved. Solution B is filter-sterilized. To a 25-ml, sterile Erlenmeyer flask, add 10 ml of solution A and 0.3 ml of solution B.

From a fresh sporulated stock plate (i.e., less than a week old), add 3 or 4 spore masses to each flask (see Maintenance of Stock Cultures, above; cool and wet inoculating wire in liquid of flask). Allow the flasks to stand at 22° for 3–5 days. During this time, check the spore suspension microscopically to see whether the spores have germinated. After germination is well under way, add the contents of each flask to a 250-ml Erlenmeyer flask containing 100 ml of the medium described below.

Various strains (AX-1 but not NC-4) of *D. discoideum* can be grown axenically as follows.

Medium

Solution A: Proteose peptone, 20 g/liter; glucose, 2 g/liter; Na_2HPO_4, 14.2 mg/liter; K_2HPO_4, 13.6 mg/liter

[6] G. Gerisch, *Naturwissenschaften* **46**, 654 (1959).

Solution B: $Fe(NH_4)_2(SO_4) \cdot 6H_2O$, 77 mg/liter; $CaCl_2$, 30 mg/liter; $MgCl_2 \cdot 6H_2O$, 30 mg/liter; NaCl, 1.2 g/liter

Solution C (mg/0.1 liter): biotin, 0.1; calcium pantothenate, 30; folic acid, 30; riboflavin, 30; pyridoxine, 330; nicotinic acid, 330; inositol, 330; thiamine, 330; *p*-aminobenzoic acid, 330. The vitamin mixture is solubilized by addition of small amounts of solid $NaHCO_3$.

To each liter of solution A is added 1 ml of solution B and 5 ml of solution C. The complete growth medium is filter sterilized and dispensed (100 ml/250-ml flask). After inoculation with the germinated cell suspension, the cells are incubated by the same procedure as that outlined for "Growth in Liquid Culture with Bacteria" (second method).

Assay for Effects on the Course of Differentiation

Stimulation.[4,5] In testing the effect of a substance or substances on the developmental process, the material is incorporated into 10 ml of 2.5% Difco Noble agar. On the day preceding the actual experiment, the agar is poured into a 9 cm in diameter petri dish. The vegetative amoebae obtained from one petri dish culture, prepared as outlined above, are suspended in a final volume of 10 ml of cold distilled water. Six droplets of 0.01 ml each are placed on the agar surface of each test plate without being spread. Thus, the cell population and density is maintained constant. The pH is initially 6.2 except where pH is a variable being tested.

The test plates are incubated at 22° for a time period (usually 20–24 hours) such that only a small fraction of the ultimately possible mature sorocarps form on the surface of the control plates (i.e., those with nothing added to the 2.5% agar). After incubation, the number of sorocarps as well as the total number of bodies at all stages of development (from definite aggregation through mature sorocarp formation) may be counted in each spot and totaled for each plate. However, in most cases studied, the stimulatory effect of a given material was qualitatively the same regardless of whether the total number of mature sorocarps, the total number of developing centers, or the ratio of sorocarps to centers was considered. When this constancy of ratio does not exist (for example, when high levels of glucose are used[5]) changing efficiency of aggregation (i.e., aggregation territory size) or pseudoplasmodium breakup is indicated.

Inhibition.[7] The usual method is to incubate all test plates and a control plate containing plain 2.5% agar until the control plate is about to finish forming its first complete sorocarp. If one waits long enough, all

[7] M. I. Krichevsky and L. L. Love, *J. Gen. Microbiol.* **37**, 293 (1964).

plates again would contain approximately equal numbers of mature sorocarps. Inhibition is measured as a decrease in the relative number of mature sorocarps at a given time when compared to control plates.

Assay of Biochemical Changes in Cell Suspensions

For each incubation mixture desired,[3] the amoebae from one baking sheet of solid growth medium are harvested just before complete ingestion of bacterial growth, washed free from residual *E. coli* and finally suspended in ice cold distilled water to a final volume of 5 ml. The cell concentration is adjusted to an extinction so that a $\frac{1}{15}$ dilution = 0.55 at 660 nm; 1 cm light path. Each incubation mixture has 4.8 ml of the suspended amoebae in 48 ml final volume. The incubations are done in standard 1-liter beakers on a six-place magnetic stirrer (Associated Biomedic Systems, Inc., Buffalo, New York) so that all suspensions are stirred at identical rates at ambient temperature (about 22°). By the use of such incubation suspensions the effect of various materials on the cells may be tested.

Cell-free filtrates of the incubation mixtures are obtained by vacuum filtration of 2-ml samples using Millipore filters of pore size 0.45 μm and 47 mm diameter (Millipore Filter Corporation, Bedford, Massachusetts; catalog No. HAWPO 4700). Such filtrates may be assayed for released enzymes, other macromolecules, or small molecules. The cells retained on the filters may be assayed for pool contents, etc. The incubation mixtures are usually sampled at 0, 1, 2, 3, and 4 hours. Four-hour incubations of this sort do not affect the potential of the cells to differentiate.

Assay for Chemotaxis

The method used to test the effect of materials on the rate of morphogenesis will also detect any chemotactic potential of the materials. So far 3′,5′-cyclic nucleotides[8,9] and folic acid[10] (and some analogs) have been found to give chemotactic activity. This method yields approximately the same results as the "Cellophane" square test.[11]

For example, on agar with 1 m*M* cyclic AMP, a peripheral ring is

[8] T. M. Konijn, J. G. C. van de Meene, J. T. Bonner, and D. S. Barkley, *Proc. Nat. Acad. Sci. U.S.* **59**, 1152 (1967).

[9] B. M. Chassy, L. L. Love, and M. I. Krichevsky, *Proc. Nat. Acad. Sci. U.S.* **64**, 296 (1969).

[10] P. Pan, E. M. Hall, and J. T. Bonner, *Nature (London) New Biol.* **237**, 181 (1972).

[11] J. T. Bonner, E. M. Hall, W. Sachsenmaier, and B. K. Walker, *J. Bacteriol.* **102**, (1970).

clearly visible after incubation overnight at room temperature.[4,9] The diameter of the original droplet containing the amoebae is approximately half the diameter of the peripheral ring; that is, the amoebae migrate outward. This ring of outward migration is a good direct test of chemotactic activity of cyclic AMP. If both cyclic AMP and folic acid are incorporated in the agar, two concentric rings are formed. Unless a chemotactic substance is present, no ring of outward migration is visible.

Another direct assay of chemotactic activity can be performed by placing streaks of amoebae opposite streaks of test compound.[11] After 18 hours, a clearly defined line of single amoebae and groups migrates toward the streak of cyclic AMP. In a control experiment with 5-AMP, substituted for cyclic AMP, there is no such attraction. Instead, these cells begin to aggregate within the original boundaries of the streak of amoebae.

Method for Changing Conditions during Morphogenesis

Since the normal development of *D. discoideum* requires geometric integrity among the cells themselves as well as between the cells and the substratum, it is not possible to change the environment of the cells by simple addition or dilution of liquids. In order to pulse-label cell components, assess efflux of enzymes or cell constituents, determine reversibility of inhibition of development, etc., a filter membrane can be interposed between the cells and the substratum. Thus, the cells on the membrane can be moved from condition to condition without perturbing the geometry. This point is illustrated by the following method for determining the reversibility of inhibition of morphogenesis.[12]

The procedures described above for ascertaining the effects of materials (stimulation or inhibition) are followed with the exceptions that follow. Instead of petri dishes containing agar, 60 $\times$ 15 mm style tissue culture dishes are used as containers for: first, a prefilter pad such as a Millipore White Filter Pad AP1004700, 47 mm diameter soaked with 2.5 ml of the material to be tested and, second, the wet pad is overlaid with a filter membrane such as a Millipore White HAWP 04700, 0.45 μm pore size, 47 mm diameter. Samples of cell suspension are placed on the filter membranes as before and incubated for 24–48 hours until inhibition of morphogenesis is evident when compared to samples placed on filter membranes lacking the test material in the soaking solution. After the initial incubation period, some of the filter membranes supporting

[12] E. A. Goidl, B. M. Chassy, L. L. Love, and M. I. Krichevsky, *Proc. Nat. Acad. Sci. U.S.* **69**, 1128 (1972)

inhibited amoebae are moved onto support prefilter pads soaked in more control solution from which the test material was omitted. Incubation is continued for at least 5 days. If the inhibition is reversible, this fact is normally observable within that time span.

Acknowledgment

The method for axenic growth was generously contributed by Dr. Arthur Washington.

Section X

Substrate and Ion Fluxes

[43] Modulations of Substrate Entry into Cells

By DAVID M. REGEN

Every pathway of cell metabolism begins with substrate entry. This step seems to be of regulatory significance in fatty acid synthesis and triglyceride synthesis in adipocytes, glycogen storage and glycolytic ATP synthesis in muscle, amino acid catabolism in liver, protein balance in liver and muscle, incorporation of precursors into nucleic acids, and incorporation of labeled phosphate into a variety of cell components. This article is concerned with the problems of recognizing and characterizing a transport modulation and with assessing its magnitude and significance to the regulation of metabolism. It deals with the principles of experimental design and interpretation rather than analytical details.

General Considerations

Crossover Analysis. For the present purposes, a transport modulation is a change in substrate entry rate due to some factor other than the concentrations of substrate outside and inside the cell, e.g., due to a hormone, disease, drug, sex, age, pH, or cellular energy sufficiency. The crossover-analysis[1] concept is applicable here. Thus, if a treatment increases substrate uptake, U_g, decreases external substrate concentration, $[G_e]$, and increases intracellular level, G_i, then entry has been enhanced and this may be partly responsible for the faster uptake. Figure 1 explains these symbols. If one makes such observations under steady-state conditions [absolute value $d(G_o + G_i)/dt \ll$ absolute value dG_e/dt], then one can judge the significance of the transport enhancement by measuring substrate uptake and cellular substrate levels in the treated cells in the presence of lower external substrate concentrations (see Fig. 2). By interpolation it can be determined what the uptake rate would be in the treated cells when intracellular substrate is that of the untreated cells with the higher external substrate concentration. The difference, between this rate and that of the treated cells with the higher external concentration, is the effect of the transport modulation on the steady-state substrate uptake. If this interpolated rate differs from that of the untreated cells with the higher external concentration then a modulation of at least one subsequent reaction is demonstrated.

Steady-state intracellular substrate levels, G_i, are calculated by an equation which purports to subtract interstitial substrate, G_o, from total

[1] J. H. Exton, this series, vol. 37, [23].

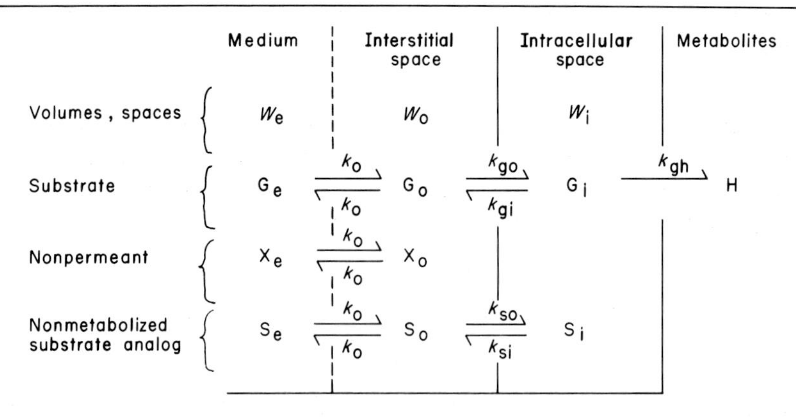

FIG. 1. Diagram of an ideal tissue exposed to a metabolized permeant (G), a nonpermeant (X) of similar size, and a nonmetabolized permeant (S) which enters the cell by the same mechanism as does G. W_e is the volume of the medium, W_o is the interstitial space, and W_i is the intracellular space accessible to S and G. The symbol H represents products of G metabolism. The k symbols are coefficients which, multiplied by appropriate concentrations give the flux rates for the designated events. For example, $d(S_o + S_i)/dt = -dS_e/dt = ([S_e] - S_o/W_o)k_o$, and $dS_i/dt = -d(S_e + S_o)/dt = S_o k_{so}/W_o - S_i k_{si}/W_i$, and $dG/dt = -G_i k_{gh}/W_i$.

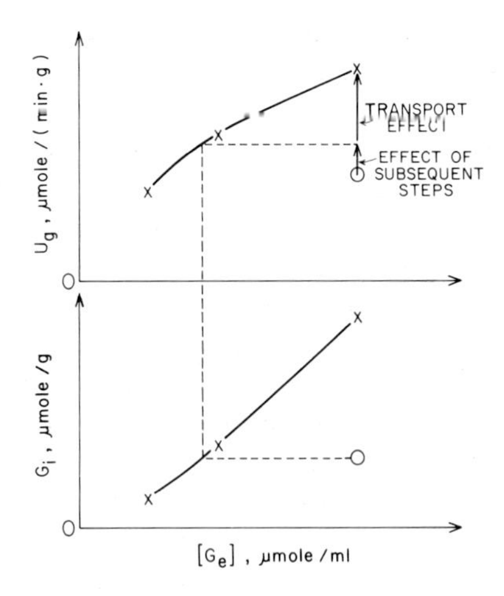

FIG. 2. Diagram illustrating the significance of transport enhancement on steady-state nutrient uptake. The open circle in each panel represents the control data at "normal" external concentration. At this concentration, the modulation (X data) resulted in higher G_i and U_g. As judged from the dashed-line interpolation of the X data at lower $[G_e]$, the modulation would have increased U_g even if G_i had not increased (effect of subsequent steps, i.e., G_i disposal). However, the major increase in U_g was due to the increase in G_i (transport effect).

tissue substrate, G_b:

$$G_i = G_b - G_o \tag{1}$$

These are expressed as amount per tissue unit (e.g., per gram wet or dry tissue or per milliliter of cell water). Interstitial substrate is calculated from the average interstitial concentration, $[G_o]$, and the interstitial space, W_o, expressed as milliliters per tissue unit.

$$G_o = [G_o]W_o \tag{2}$$

Interstitial space is the equilibrium nonpermeant space:

$$W_o = \text{equilibrium } X_b/[X_e] \tag{3}$$

Average interstitial concentration is difficult to estimate. If one can assume that interstitial gradients are small, one may use the approximation

$$[G_o] \leq [G_e] \tag{4}$$

All the errors are on the side of underestimating G_i, the more so when uptake is faster. Any delay in stopping (by freezing, heating, poisoning) enzyme activity after separation of tissue from medium will result in a fall in G_b. The negative term will be too large by the factor $[G_e]/[G_o]$, average $[G_o]$ being less than $[G_e]$ owing to the gradients between medium and cells. This error is not likely to be serious in perfused organs or tissues *in vivo* if A-V difference is a small fraction of serum concentration. Raising the serum substrate level improves this relationship. The error is reduced also if the A-V average is taken as $[G_e]$. In incubated solid tissues, the error is more likely to be serious; but higher medium concentrations will reduce the error, as will use of a thinner preparation and lower temperature. A means of taking interstitial gradients into account will be described later. Any entry of the nonpermeant into cells will make the negative term too large. It should be noted that a free-cell suspension has a pseudointerstitial space, that portion of the medium which cannot be removed from the cells prior to analysis. Equation (1) is, therefore, needed for these systems; and here the main source of error would come from delay in stopping enzyme activity—the pseudointerstitial space having substrate at the same concentration as does the medium.

Since the underestimation of G_i is worse with faster uptake, crossover analysis will never mistakenly identify transport as a site of pathway regulation, but it may fail to disclose a transport modulation. If a nonmetabolized substrate analog, S, is available which enters by the same mechanism as does the substrate, then studies of S transport (described later) may reveal a suspected modulation not seen by the above procedure. This is especially successful if S is transported more slowly than G.

Flux Coefficients. The above measurements do not quantify the transport modulation. This requires some characterization of the transport process in the control and modulated states. The characterization consists of estimates of the influx coefficient, k_{go}, and the efflux coefficient, k_{gi}, in the presence of various substrate concentrations outside, $[G_o]$, and inside, $[G_i]$. The influx coefficient is the extracellular volume containing the amount fluxing into the cells of a tissue unit each minute. Thus, $U_{g+} = [G_o]k_{go}$, and $k_{go} = U_{g+}/[G_o]$. Likewise for efflux, $U_{g-} = [G_i]k_{gi}$, and $k_{gi} = U_{g-}/[G_i]$. Net uptake is the difference between influx and efflux, $U_g = U_{g+} - U_{g-}$. If the substrate enters by simple diffusion (by dissolving in the lipid phase of the membrane or by passing certain kinds of pores), then the flux coefficients may be independent of substrate concentration, and may, therefore, be flux constants. Likewise, if transport is mediated, but saturable only with concentrations much higher than met *in vivo* and *in vitro*, then the same applies. In either case, the changes in k_{go} and/or k_{gi} suffice to quantify the modulation. Any modulation that would increase the entry rate in the presence of an unchanged or lower external concentration and an unchanged or higher internal concentration would constitute an enhancement of transport. A rise in k_{go} with no more than a proportional rise in k_{gi} would meet this criterion, as would a fall in k_{gi} with no fall in k_{go}. There are gray areas outside these limits which may satisfy the criterion if other conditions are specified. For example, if steady-state $[G_i]/[G_o]$ is sufficiently less than k_{go}/k_{gi}, then U_g and G_i would be increased by a rise in k_{go} despite a more than proportional rise in k_{gi}; for, under this circumstance, efflux is of little consequence. Also, if steady-state $[G_i]/[G_o]$ is almost as great as k_{go}/k_{gi}, then a fall in k_{gi} could increase U_g and G_i if an accompanying fall in k_{go} is less than proportional; for, under this circumstance, the ratio k_{go}/k_{gi} is more important than the value of either flux constant.

Passive Carrier Transport. Substrate entry by passive, carrier-mediated transport (facilitated diffusion) presents another element of complexity; for the flux coefficients will not be constants. "Passive" in this case means that the equilibrium ratio $[G_i]/[G_o] = 1$. Substrate entry will obey the law:

$$U_g = \frac{[G_o]F_g(1 + [G_i]/R_g) - [G_i]F_g(1 + [G_o]/R_g)}{1 + [G_o]/K_{go} + [G_i]/K_{gi} + [G_o][G_i]/R_gB_g} \tag{5}$$

The theoretical basis for this equation is presented elsewhere.[2] It applies to a noncooperative carrier, i.e., one obeying classical Michaelis kinetics. The equation does not use V_{max} because this is not a constant. The activity constant, F_g, is the V_{max}/K_m ratio, which is a constant. It is the value of k_{go} and k_{gi} observed at very low substrate level. The positive

[2] D. M. Regen and H. E. Morgan, *Biochim. Biophys. Acta* **79**, 151 (1964).

and negative numerator terms, respectively, divided by the denominator are influx and efflux rates; hence $k_{go} = F_g(1 + [G_i]/R_g)$/denominator, and $k_{gi} = F_g(1 + [G_o]/R_g)$/denominator.

Equation (5) contains only three independent constants besides F_g; for, B_g is a derivative of the other three:

$$1/B_g = 1/K_{go} + 1/K_{gi} - 1/R_g \quad (6)$$

The carrier hypothesis predicts Eq. (6), and any one of its constants may be calculated from experimental evaluation of the other three. If one evaluates all four experimentally and is convinced that Eq. (6) cannot be satisfied, then the carrier hypothesis is insufficient and another model must be sought. There are four conceptually simple experiments to evaluate separately the four constants of Eq. (6) and F_g. These will be referred to as the F_g-K_{go} procedure, F_g-K_{gi} procedure, F_g-B_g procedure, and R_g procedure.

F_g-K_{go} procedure: Measure the initial substrate entry, U_g, immediately upon exposure of the cell to external substrate (labeled or unlabeled). This initial U_g is estimated at several widely differing concentrations of extracellular substrate, $[G_o]$. Under these conditions, the G_i terms of Eq. (2) drop out and the equation reduces to $U_g = [G_o]F_g/(1 + [G_o]/K_{go})$. A plot of $[G_o]/U_g$ (i.e., $1/k_{go}$) vs $[G_o]$ will have a y intercept of $1/F_g$ and an x intercept of $-K_{go}$, the Michaelis constant for substrate entry. If the plot is horizontal, entry is unsaturable and k_{go} is an influx constant.

F_g-K_{gi} procedure: Cells are initially loaded with labeled or unlabeled substrate. The medium is changed to one containing no substrate. One then estimates exit, $-U_g$, under conditions where $[G_o]$ is very low. The rate law is then $-U_g = [G_i]F_g/(1 + [G_i]/K_{gi})$. A plot of $-[G_i]/U_g$ (i.e., $1/k_{gi}$) vs $[G_i]$ will have intercepts at $1/F_g$ and $-K_{gi}$, the latter being the Michaelis constant for exit. A horizontal plot means that k_{gi} is an efflux constant.

F_g-B_g procedure: This is the most precise evaluation of the four but is possible only if substrate metabolism is sufficiently slow that the steady-state substrate ratio $[G_i]/[G_o]$ is near the equilibrium ratio. This will obtain if $k_{gh} \ll k_{gi}$; and, under this circumstance, steady-state influx, U_{g+}, will be only slightly greater than efflux, U_{g-}. One may estimate this exchange flux by addition of tracer to the medium of cells already in approximate equilibrium with unlabeled substrate and measurement of U_{g+}, the early rise in cell radioactive substrate. Alternatively, one may equilibrate with labeled substrate, then change the medium to one containing only unlabeled substrate at the original concentration and estimate U_{g-}, the efflux of labeled substrate from the cells. One does this experiment with several widely differing concentrations of unlabeled substrate. The rate law is $U_{g-} \leq U_{g+} = [G_o]F_g/(1 + [G_o]/B_g)$. A plot of $[G_o]/U_{g+}$ (i.e., $1/k_{go}$) or $[G_o]/U_{g-}$ vs $[G_o]$ will have intercepts of $1/F_g$ and $-B_g$, the latter being the apparent dissociation constant of the substrate-carrier complex. A horizontal plot means that k_{go} and k_{gi} are constants.

Flux-ratio–R_g procedure: The flux-ratio constant, R_g, is evaluated by an estimate of influx and efflux under conditions where $[G_i]/[G_o]$ is not near the equilibrium ratio. If, for example, k_{gh} is not small relative to k_{gi}, then the experiment would be to measure steady-state uptake, U_g, of unlabeled substrate and, during this steady-state, to add tracer substrate to the medium for a measurement of influx. Efflux, U_{g-}, is calculated by the relation $U_g = U_{g+} - U_{g-}$. From the numerator of Eq. (5), it can be shown that the ratio of influx to efflux coefficients is given by the following:

$$\frac{U_{g+}/[G_o]}{U_{g-}/[G_i]} = \frac{k_{go}}{k_{gi}} = \frac{1 + [G_i]/R_g}{1 + [G_o]/R_g} \tag{7}$$

which can be solved for R_g. If k_{gh} is very small, one must estimate the uptake, U_g, by the rise in G_i after exposure to unlabeled substrate (or substrate labeled with isotope a). When the accumulation is about ⅓ of the steady-state, one must estimate influx, U_{g+}, by addition of substrate tracer (or substrate labeled with isotope b).

The equations for the above four procedures are simplifications of Eq. (5). There are numerous derivations from Eq. (5) which are not simple and allow interpretation of fluxes and rates when G_o and G_i are both present at a variety of concentrations. In each case, if one has two constants evaluated, two more may be determined by a slope intercept plot, after which Eq. (6) may be used to complete the evaluations. The following three procedures are especially useful.

F_g-$R_g \rightarrow K_{gi}$-B_g net procedure:

$$\frac{1 + [G_o]/R_g}{[G_o] - [G_i]} - \frac{F_g}{U_g} = \frac{1}{K_{gi}} - \frac{[G_o](1 + [G_i]/R_g)}{[G_o] - [G_i]} \frac{1}{B_g} \tag{8}$$

Plot the left-hand complex against the complex factor on $1/B_g$. The y intercept will be $1/K_{gi}$ and the slope will be $1/B_g$.

F_g-$B_g \rightarrow K_{go}$-R_g net procedure:

$$\frac{F_g}{U_g} - \frac{1 + [G_i]/B_g}{[G_o] - [G_i]} = \frac{1}{K_{go}} + \frac{[G_i](1 + [G_o]/B_g)}{[G_o] - [G_i]} \frac{1}{R_g} \tag{9}$$

Treat as with Eq. (8).

F_g-$R_g \rightarrow K_{go}$-B_g flux procedure:

$$\left(\frac{F_g}{k_{go}} - 1\right)\frac{1 + [G_i]/R_g}{[G_o] - [G_i]} = \frac{1}{K_{go}} + \frac{[G_i](1 + [G_o]/R_g)}{[G_o] - [G_i]}\frac{1}{B_g} \tag{10}$$

The procedures described above may be extremely difficult if transport so limits uptake that only barely detectable amounts of intracellular substrate are present in the steady state. In this case, one may be able to carry out the F_g-K_{go} procedure estimating U_g as the steady-state uptake at various $[G_o]$, or estimating the initial rise in cell labeled substrate, $G_i{}^*$, plus all labeled products, H^*, from external labeled substrate, $G_o{}^*$. The other procedures would be difficult and less relevant, since they are concerned with the presumed physiologically insignificant effects of G_i on net transport. However, if one has a nonmetabolized substrate analog, S, which shares the substrate's entry mechanism, then all the equations and procedures described for G transport apply also to S transport, and the transport process can be characterized with the analog.[2] (One should substitute the letter S for G and s for g in the equations to make clear that the constants and coefficients are analogs of the G transport constants and coefficients.) The absence of S metabolism makes characterization of S transport simpler and more definitive. It is worthwhile to consider the effects of substantial G levels on the movements of S presented at tracer levels. The interaction is described by the equation[3]

$$U_s = \frac{[S_o]F_s(1 + [G_i]/R_g) - [S_i]F_s(1 + [G_o]/R_g)}{1 + [G_o]/K_{go} + [G_i]/K_{gi} + [G_o]\,[G_i]/R_gB_g} \tag{5a}$$

One can identify four procedures from this equation that may help in evaluation of G constants.

F_s-K_{go} procedure: If transport so limits G uptake that steady-state G_i is negligible, then S influx in the presence of various $[G_o]$ will obey the law, $U_{s+} = [S_o]F_s/(1 + [G_o]/K_{go})$. A plot of $[S_o]/U_{s+}$ (i.e., $1/k_{so}$) vs $[G_o]$ will have the intercepts $1/F_s$ and $-K_{go}$. This inhibitory effect is the usual basis for saying that S uses the saturable G transport mechanism. This value for K_{go} will be more accurate than one based on steady-state U_g, since even low levels of G_i may affect net G transport significantly but are not likely to affect k_{so}.

Counterflow-R_g procedure: Simply allow the above experiment to continue until $dS_i/dt = 0$. From the numerator of Eq. (5a), it is seen

[3] P. M. Buschiazzo, E. B. Terrell, and D. M. Regen, *Amer. J. Physiol.* **219,** 1505 (1970).

that this "equilibrium" is influenced by G as follows:

$$\text{Equilib. } \frac{[S_i]}{[S_o]} = \frac{k_{so}}{k_{si}} = \text{steady-state } \frac{1 + [G_i]/R_g}{1 + [G_o]/R_g} \tag{7a}$$

The carrier hypothesis predicts the relation $F_sR_s = F_gR_g$. If one is convinced that this relation cannot be satisfied, then the carrier hypothesis is not sufficient and another model must be sought.

F_s-B_g procedure: If G metabolism is so slow that steady-state $[G_i]/[G_o]$ is near equilibrium, then the counterflow-R_g procedure would fail. The F_s-K_{go} procedure would involve simultaneous presentation of S and G and measurements of U_{s+} only prior to significant G_i accumulation. If one presents S after $[G_i]/[G_o]$ is near equilibrium then the rate law is $U_{s+} = [S_o]F_s/(1 + [G_o]/B_g)$. If one preloads with S in the presence of G, then changes to a medium with the same level of G (but no S) and studies S efflux, the rate law is $U_{s-} = [S_i]F_s/(1 + [G_o]/B_g)$. A plot of $[S_o]/U_{s+}$ (i.e., $1/k_{so}$) or $[S_i]/U_{s-}$ (i.e., $1/k_{si}$) vs $[G_o]$ will have intercepts of $1/F_s$ and $-B_g$.

F_s-$R_g \rightarrow K_{go}$-B_g flux procedure: If, for some substantial substrate concentrations, steady-state $[G_i]/[G_o]$ is distinctly less than the equilibrium ratio but more than negligible, then one can evaluate all the constants of Eqs. (5a) and (6) from measurements of k_{so} (i.e., $U_{s+}/[S_o]$) and k_{so}/k_{si} (i.e., equilibrium $[S_i]/[S_o]$) at a variety of substrate levels (in the steady state). R_g will be evaluated several times by Eq. (7a). F_s is k_{so} in the absence of G. K_{go} and B_g are evaluated by Eq. (10) with F_s/k_{so} in place of F_g/k_{go}. K_{gi} is then evaluated by Eq. (6).

Active Transport. The equilibrium ratio $[G_i]/[G_o]$ will be unity unless some form of energy other than the substrate gradient is dissipated by substrate movement in one direction. A gradient in a substance (e.g., electrical charge, H^+, Na^+, K^+) which moves with or against the substrate in a coupled manner or a nonequilibrium chemical reaction which is coupled to substrate movement will cause the equilibrium substrate ratio $[G_i]/[G_o]$ to assume some value, E_g, other than unity.[3] If entry is nonsaturable, k_{go}/k_{gi} is always equal to E_g. With mediated transport, $k_{go}/k_{gi} = E_g$ only if equilibrium is attained or if very low substrate concentrations are used relative to R_g. One can convert Eqs. (5), (7), (8), (9), (10), (5a), and (7a) to the corresponding ones for "active" carrier-mediated transport simply by dividing $[G_i]$ by E_g and $[S_i]$ by E_s wherever these inside concentrations appear. It should be pointed out that E_g and E_s will be constants only if the energy source is not taxed by the transport.

The intracellular concentrations $[G_i]$ and $[S_i]$ in the above equations are in fact calculated as G_i/W_{gi} and S_i/W_{si}, respectively, where G_i and S_i are intracellular amounts per unit of tissue, W_{gi} and W_{si} being the substrate-accessible and analog-accessible intracellular spaces per unit of tissue. These spaces may be difficult to determine, if the cells contain several compartments or if circumstances cannot be found where $[G_i] = [G_o] = [G_e]$ or $[S_i] = [S_o] = [S_e]$; for, it can only be calculated on the assumption of even distribution by an equation such as: $W_{gi} = G_i/[G_e] = G_b/[G_e] - X_b/[X_e]$. If W_{gi} or W_{si} is not known, one must substitute G_i/W_{gi} or S_i/W_{si} for the intracellular concentrations in all equations. The intracellular amounts, G_i and S_i, then become the variables and W_{gi} or W_{si} becomes a constant in each intracellular term. It will be constant only if osmolality of medium and cells are balanced. If W_{gi} and W_{si} are not known, then E_g and E_s cannot be proved to be unity, and vice versa. The following equations[3] illustrate how these uncertainties affect some of the equations already given. Others will be given later. Equation (5a) would read

$$U_s = \frac{[S_o]F_s(1 + G_i/W_{gi}E_gR_g) - S_iF_s(1 + [G_o]/R_g)/W_{si}E_s}{1 + [G_o]/K_{go} + G_i/W_{gi}E_gK_{gi} + [G_o][G_i]/W_{gi}E_gR_gB_g} \qquad (5a')$$

The $F_s - K_{go}$ and $F_s - B_g$ (by analogy the $F_g - K_{go}$ and $F_g - B_g$) rate laws are unaffected. The $F_s - K_{gi}$ rate law would read

$$U_{s-} = S_iF_s/[W_{si}E_s(1 + G_i/W_{gi}E_gK_{gi})]$$

A plot of S_i/U_{s-} (i.e., W_{si}/k_{si}) vs G_i will have intercepts of $W_{si}E_s/F_s$ and $-W_{gi}E_gK_{gi}$. Equation (7a) would read

$$\text{Equilib. } \frac{S_i}{[S_o]} = \frac{k_{so}}{k_{si}/W_{si}} = \text{steady-state } W_{si}E_s \frac{1 + G_i/W_{gi}E_gR_g}{1 + [G_o]/R_g} \qquad (7a')$$

Equilibrium $S_i/[S_o]$ in the absence of G is $W_{si}E_s$. If $S_i/[S_o]$ is measured in the presence of several steady-state levels of $[G_o]$ and G_i, then a plot of $W_{si}E_s[S_o]/[G_o]S_i - 1/[G_o]$ vs $G_i[S_o]/S_i[G_o]$ will have an intercept of $1/R_g$ and a slope of $-W_{si}E_s/W_{gi}E_gR_g$.

If, in addition, k_{so} is evaluated in these experiments, then one or all of the remaining constants can be calculated as described in the $F_s - R_g \rightarrow K_{go} - B_g$ flux procedure. The equation for this is as follows:

$$\left(\frac{F_s}{k_{so}} - 1\right)\frac{1 + G_i/W_{gi}E_gR_g}{[G_o] - G_i/W_{gi}E_g} = \frac{1}{K_{go}} + \frac{G_i(1 + [G_o]/R_g)/W_{gi}E_g}{[G_o] - G_i/W_{gi}E_g}\frac{1}{B_g} \qquad (10a)$$

The intracellular space may be divided among two or more compartments. Sometimes these compartments can be manifest in a compartmental analysis.[4] If such compartments actively concentrate or extrude substrate, this has the effect of making W_{gi} greater or smaller than the physical space accessible to substrate. This is of no consequence to the equations described so far, since intracellular space is in any case not a real volume but an empirical, imaginary volume (apparent mediumlike solvent volume in the cells). It is not evaluated as such, but in combination with other constants. For example, k_{si}/W_{si} is the fraction of S_i fluxing out per minute, $W_{gi}E_g$ is the equilibrium value of $G_i/[G_o]$. The equations and the use of W_{gi} or W_{si} as constants in them are valid for initial fluxes and for steady-state phenomena, if it can be assumed that the fraction of intracellular substrate in each compartment in the steady state is independent of substrate concentration. One can test this assumption by compartmental analysis of labeled substrate exit from cells in the steady state with respect to unlabeled substrate at several concentrations. The concepts of compartmentation and graphical analysis have been dealt with elsewhere.[4,5] This author prefers to use the analog computer for compartmental analysis.

Experimental Design

Up to this point we have considered the kinds of data that can characterize a substrate entry process. In many tissue–substrate systems the characterization will be only partial, and that partial characterization will be compromised by uncertainties about interstitial concentrations and their changes with time. Nevertheless, the ideal procedures suggested above provide a standard which the experimenter should approach as best he can with his particular system. A matter too complex and varied to discuss here is the use of integrated transport equations to deal with transport data when differential rates are difficult to measure. In some cases integrated equations enable one to get more information about the transport kinetics than is contained in the initial rate.

It can be seen that labeled substrate plays a major role in these procedures. The choice of label and the separations needed prior to counting depend on the kind of data sought. One should, if possible, choose a label that is converted to an easily separable product at or adjacent to the first irreversible metabolic step. If the substrate changes from nonvolatile to volatile (^3H_2O or $^{14}CO_2$) or vice versa, from nonionic to ionic or vice versa, from water-soluble to fat-soluble or vice versa, from acid-soluble to acid-precipitable or vice versa, then the separation will be simple and the appearance of such products can assist in the assessment of U_g. Inter-

[4] S. M. Skinner, R. E. Clark, N. Baker, and R. A. Shipley, *Amer. J. Physiol.* **196,** 238 (1959).

[5] A. B. Borle, this volume [44].

mediates prior to this step must also be separated, shown to have negligible radioactivity, or calculated out if they are a constant fraction of total cell radioactivity. Labeled intermediates prior to the first irreversible step can be considered compartments of G_i, for these intermediates behave dynamically like intracellular compartments. If this is done, then W_{gi} is artificially enlarged and one must expect $W_{gi} > W_{si}$.

The following discussion of tissue preparations will provide some suggestions as to how one may make the measurements of rates and/or coefficients called for in the previous section.

Cell Suspensions. The most satisfactory transport studies are carried out on erythrocytes well stirred in an isotonic buffer. In such a system, concentrations of substances at the outer cell surface are nearly the same as in the medium, there being no interstitial space between medium and cells. The cell volume is largely water, and, therefore, can contain large amounts of permeant. Cells can be serially sampled and readily separated from the medium for analysis. There are no organelles, hence the cell contents can be considered essentially evenly distributed in the cell water. To measure influx, one mixes the suspension of cells (of known extracellular and intracellular volume) with an isotonic solution containing the permeant, G or S (and a nonpermeant, X). If nothing is known of the time course, then samples should be taken at increasing intervals such as 0.1, 0.3, 1, 3, 10, etc., minutes. If a rough idea of the time course is established, then it is best to obtain a series of samples when the smallest measurable fraction of the steady-state accumulation has occurred and one or two samples after the steady-state is reached. The early rise in intracellular concentration is the entry rate, which reveals the initial influx coefficient (k_{go} or k_{so}). The steady-state intracellular to extracellular concentration ratio will be the final ratio, $k_{go}/(k_{gi} + k_{gh})$ or k_{so}/k_{si}. One can calculate $k_{gi} + k_{gh}$ or k_{si} from these data only under those conditions where k_{go} or k_{so} is constant in time.

A variety of methods are available to separate the cells from the medium for analysis of cell contents. Rapid filtration followed immediately by acid extraction should be effective but may not be the best procedure, since substantial amounts of medium will remain with the filter. The amount of nonpermeant in the extract provides the information needed to correct for the permeant in the extract derived from the contaminating medium [pseudointerstitial space of the filtered cells, see Eq. (1)]. If transport and metabolism can be sufficiently slowed by cooling (and addition of a transport inhibitor) then the sample may be mixed with ice cold saline (with the inhibitor) above the filter.[5,6] This

[6] D. M. Miller, *Biophys. J.* 8, 1329 (1968).

will greatly reduce the correction for contaminating medium. If the transport rate and metabolism are very slow in this solution, then separation can be effected by centrifugation through dense serum albumin (or its equivalent) layered under the mixture (the albumin being dissolved in the same cold saline with its additives), followed quickly by removal of supernatant and acid extraction.[7]

Cells are prepared for an efflux experiment by preincubation with the permeant (preferably radioactive) for a period sufficient to attain the desired internal concentration (preferably steady state). The cells are then centrifuged, medium removed, and cells resuspended in medium containing a nonpermeant but lacking the permeant (or containing the permeant but not radioactive permeant).[2,6,7] If it was necessary to cool the cells for centrifugation (to slow utilization of the substrate), then some means of attaining the desired temperature immediately upon resuspension must be used. For example, the vessel containing the cells could be incubated briefly just before resuspension, or a small volume of cold cells could be transferred to a vessel at the desired temperature together with a medium warmed above the incubation temperature, such that, upon mixing, the desired temperature is attained. Efflux should be followed by analyses of both cells and medium.

Other cell types in suspension are handled similarly, but data analysis may be more difficult and less definitive. Intracellular compartments create uncertainty as to the concentration of permeant in the cytosol. This is dealt with by use of the expression $G_i/W_{gi}E_g$ or $S_i/W_{si}E_s$ as described earlier. Adipocytes present a special technical problem in that the cytoplasm constitutes only a small fraction of the cell volume.[8] Hence initial influx rates will involve measurements of very small amounts of permeant in the cell and may be obscured by modest amounts of contaminating medium in the cell sample. Adipocytes, preloaded and centrifugally floated in the loading medium for an efflux experiment will be accompanied by large amounts of medium relative to cell water. This may interfere with estimates of k_{gi} from appearance of G_e, but it should not interfere greatly with estimates of k_{si} based on the loss of S_i. If transport and metabolism are slowed sufficiently by cooling, the cells can be floated from dense albumin through a layer of less dense albumin.

Perfused Organs. Next to suspended cells, perfused organs are the best objects of transport studies. The perfusion device should be so designed that the permeant and nonpermeant can be presented to the tissue in a sudden step wave and shut off in a step wave. This means having a

[7] R. T. Hoos, H. L. Tarpley, and D. M. Regen, *Biochim. Biophys. Acta* **266**, 174 (1972).

[8] J. R. Carter and O. B. Crofford, this series, Vol. 37, [22].

stopcock to switch from one medium to another just above the tissue. Or it means having an infusion line (for sudden introduction of permeant and nonpermeant) entering the inflowing medium just before it reaches the tissue. In the latter case, a square-edged constriction between infusion point and tissue will aid mixing. It should be possible to collect the effluent medium immediately as it leaves the tissue rather than after it has traversed a long tube. If it must traverse a tube, then the effluent should be pulled at 2 or 3 times the perfusion flow through the tube to the collecting vessel by a pump, a Y-tube being in the line near the tissue to let air in (to segment the effluent in the tube). The perfusion flow must be fast enough to provide O_2 to the cells but may need to be faster as described below. Several features of experimental design must be based on a pilot study carried out as follows. Perfuse for a few minutes with permeant and nonpermeant at low chemical concentrations but sufficient radioactivities that they count at 1000 or more times background. Switch to the nonradioactive influent and collect effluent over very short intervals at first, increasing the intervals as the experiment proceeds. For these experiments, a few more definitions are needed: $[G^*_{load}]$ = influent labeled substrate concentration during loading; $[G^*_{inf}]$ = influent labeled substrate concentration at a given time; $[G^*_{eff}]$ = effluent labeled substrate concentration at a given time; H^* = all labeled products of G^* metabolism. Plot $[G^*_{eff}]/[G^*_{load}]$ and $[X_{eff}]/[X_{load}]$ on semilog paper against time, as illustrated in Fig. 3. The nonpermeant plot will begin at 1.0 and show a delayed and then hopefully straight decline for about two cycles. The washout time, at which effluent nonpermeant falls to 0.01 or ceases its rapid fall, is the "nonpermeant equilibration or washout time." If perfusion flow is sufficient, the substrate plot will begin just slightly below 1.0 and follow the nonpermeant closely during its early straight-line course. If the substrate plot falls initially more gently than the nonpermeant plot, then future perfusion flow must be faster. At some point while the nonpermeant plot is straight, the substrate plot will cross the nonpermeant plot and become distinctly gentler. This is the "beginning" of cell substrate washout. The total washout time necessary for the effluent substrate to fall below 1% of its value at the beginning of cell washout is the minimum time needed in further experiments to achieve a steady state. Later studies with higher substrate concentrations may show longer steady-state times and the loading periods should be appropriately adjusted.

Solid-tissue experiments are much more laborious and much less precise than cell-suspension experiments, and one should confine effort to the most rewarding procedures. If the tissue is reasonably homogeneous (e.g., liver), steady-state washout experiments will be the best to begin

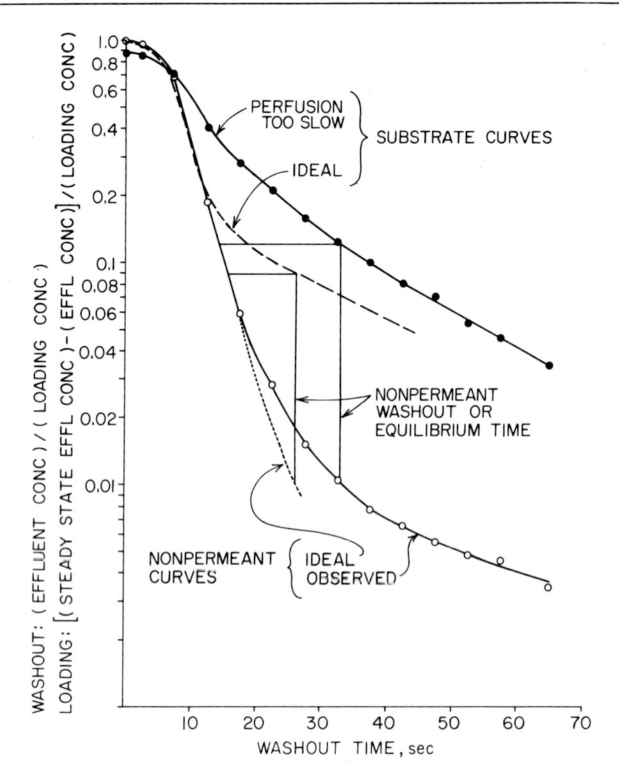

FIG. 3. Steady-state washout of substrate (30 mM [^{3}H]D lactate, ●) and nonpermeant ([^{14}C]-sucrose, ○) from perfused liver. The same curve would be obtained if effluent concentrations of a steady-state loading experiment were plotted as suggested on the ordinate. The effluent data of both kinds of experiment contain the same information, the loading data being more noisy toward the end, owing to the small differences between effluent concentration and steady-state effluent concentration. The observed data were far from ideal, despite a flow rate of 2 ml/min g. A flow rate of 10 ml/min g and much shorter sampling intervals would probably be needed to obtain a near ideal substrate curve. The later, gentler slope of the nonpermeant curve shows that some extracellular space of the system washed out slowly. When the nonpermeant curve falls to 0.01 (nonpermeant washout time), one can assume that the ratio G_i^*/G_o^* is within 1% of its steady value.

with. For these, load the tissue to the steady state by perfusion with labeled substrate and nonpermeant, then switch to unlabeled substrate at the same concentration. This is repeated at a wide variety of substrate concentrations. Follow $[G^*_{eff}]$, $[H^*_{eff}]$, and $[X_{eff}]$ and obtain some steady-state values of $[G_{eff}]$. Terminate some perfusions at the nonpermeant washout time and others shortly thereafter, for analysis of tissue weight, tissue water, G_b^*, H_b^*, X_b, G_b, and $[G_{inf}]$. Some experiments will have to terminate prior to washout, for analysis of $[G^*_{load}]$, $[X_{load}]$, equilibrium X_b, and steady-state G_b^*.

Estimates of $[G_o^*]$, G_i^* and U_g^* are neither simple nor theoretically perfect. Net labeled substrate entry (negative entry in this case) may be calculated by either or both of the arguments of Eq. (11):

$$\begin{aligned} U_g^* &\simeq ([G^*_{inf}] - [G^*_{eff}]) \times \text{flow}/(\text{tissue weight}) - \Delta(G_o^* + G_p^*)/\Delta t \\ &\simeq \Delta(G_b^* + H_b^*)/\Delta t - \Delta G_o^*/\Delta t + [H^*_{eff}] \times \text{flow}/(\text{tissue weight}) \end{aligned} \tag{11}$$

where G_p^* is the substrate in the effluent tubing or $[G^*_{eff}] \times$ (tube volume)/ (tissue weight).

$$G_o^* = [G_o^*]W_o \tag{2a}$$

$$W_o = \text{equilib. } X_b/[X_{inf}] \tag{3a}$$

$$G_i^* = G_b^* - G_o^* \tag{1a}$$

Interstitial labeled substrate can be calculated by Eq. (4a) if interstitial gradients are small, or by Eq. (12) if interstitial gradients are larger:

$$[G_o^*] \simeq ([G^*_{inf}] + [G^*_{eff}])/2 \tag{4a}$$

$$\frac{[G^*_{inf}] - [G^*_{eff}]}{[G^*_{inf}]W_o - G_o^*} \simeq \text{washout } \frac{[X_{eff}]}{X_b} \tag{12}$$

Steady-state $[G_o]$, G_i, and U_g are calculated by the same principles as are $[G_o^*]$, G_i^*, and U_g^*. However, U_g can be calculated from the flux coefficients once they are determined:

$$U_g = [G_o]k_{go} - G_i k_{gi}/W_{gi} = G_i k_{gh}/W_{gi} \tag{13}$$

These flux coefficients can be determined by simultaneous solution of the following three equations:

$$U_g^* = [G_o^*]k_{go} - G_i^* k_{gi}/W_{gi} \tag{14}$$

$$\begin{aligned} \text{Steady-state } [G_o]/G_i &= \text{steady-state } [G_o^*]/G_i^* \\ &= k_{gi}/W_{gi}k_{go} + k_{gh}/W_{gi}k_{go} \end{aligned} \tag{15}$$

$$\frac{\Delta H^*/\Delta t}{G_i^*} = k_{gh}/W_{gi} = \text{steady-state } U_g/G_i \tag{16}$$

If the tissue is not so homogeneous (e.g., heart containing chambers, or kidney containing different layers), or if steady-state G_i/G_o is very low, then a steady-state loading experiment may be preferred to the washout experiment. To do this, we simply reverse the order of the two media used. In principle, both steady-state experiments are the same, the calculations and information being identical. One must decide whether the loading or washout data are more precise.

Evaluation of transport constants from steady-state experiments is as follows: A plot of $1/k_{go}$ vs $[G_o]$ will intercept the y axis at $1/F_g$. If the plot is horizontal, then k_{go} is a constant and entry is unsaturable. A plot of $k_{gi}/W_{gi}k_{go}$ vs $[G_o]$ will intercept the y axis at $1/W_{gi}E_g$. These plots are not necessarily straight, and a smooth curve should be drawn through them for the above extrapolations. If $G_i/[G_o]$ was near $W_{gi}E_g$ for any of the perfusions, draw a straight line through those $1/k_{go}$ points and the $1/F_g$ intercept. It will have an x intercept near $-B_g$. If $G_i/[G_o]$ was negligible relative to $W_{gi}E_g$ for any of the experiments (but $[G_o]$ was high enough that k_{go} was significantly less than F_g), then a straight line drawn from the corresponding $1/k_{go}$ points through the $1/F_g$ intercept will intercept the x axis near $-K_{go}$. For those experiments where $G_i/[G_o]$ was high enough for accurate estimate of k_{gi}/W_{gi} but significantly less than $W_{gi}E_g$, calculate R_g by solution of the following equation:

$$\frac{k_{go}}{k_{gi}/W_{gi}} = W_{gi}E_g \frac{1 + G_i/W_{gi}E_gR_g}{1 + [G_o]/R_g} \tag{7b}$$

If the R_g calculation was possible, then it is also possible to determine K_{gi} and B_g as described in the F_g-$R_g \rightarrow K_{gi}$-B_g net procedure or the F_g-$R_g \rightarrow K_{go}$-B_g flux procedure. The equation for the F_g-$R_g \rightarrow K_{gi}$-B_g net procedure will be

$$\frac{1 + [G_o]/R_g}{[G_o] - G_i/W_{gi}E_g} - \frac{F_g}{U_g} = \frac{1}{K_{gi}} - \frac{[G_o](1 + G_i/W_{gi}E_gR_g)}{[G_o] - G_i/W_{gi}E_g}\frac{1}{B_g} \tag{8a}$$

If the steady-state experiments provide data for these procedures then no other experiments are necessary. If $G_i/[G_o]$ was negligible at all levels of $[G_o]$, then the steady-state experiments constitute an F_g-K_{go} procedure and nothing can be learned from further experiments (with only G). If, however, steady-state $G_i/[G_o] \simeq W_{gi}E_g$ for all levels of $[G_o]$, then the steady-state experiments constitute an F_g-B_g procedure, and one may be able to determine the remaining constants by performing a pseudo F_g-K_{go} and/or a pseudo F_g-K_{gi} procedure. For these non-steady-state loading or washout experiments, one uses empty medium (lacking unlabeled substrate) in place of unlabeled substrate during the preperfusion or washout period. Calculations of U_g^*, $[G_o^*]$, G_i^*, $[G_o]$, and G_i are as described above (U_g being calculated from U_g^* and specific activity), but one cannot estimate flux coefficients. Since F_g and B_g are known before one undertakes these experiments, one can estimate the remaining unknown constants by the F_g-$B_g \rightarrow K_{go}$-R_g net procedure using the following form of Eq. (9):

$$\frac{F_g}{U_g} - \frac{1 + G_i/W_{gi}E_gB_g}{[G_o] - G_i/W_{gi}E_g} = \frac{1}{K_{go}} + \frac{G_i(1 + [G_o]/B_g)/W_{gi}E_g}{[G_o] - G_i/W_{gi}E_g}\frac{1}{R_g} \tag{9a}$$

Incubated Tissues. In order for an incubated tissue to be comparable to a perfused organ as an object for transport investigation, it must contain mainly intact cells and be thin enough so that oxygen requirements are met at its center. It must also be sufficiently thin[9] that the half-time for washout of the interstitial space (nonpermeant) is very short compared to the half-time for intracellular permeant washout as indicated in pilot experiments. In a sense, a thinner tissue is equivalent to faster perfusion flow. If entry at the cell membrane is mediated, then a lower temperature can slow it, thereby improving the half-time relationship. An improvement can also be made by use of a substrate analog with inherently slow transport.

If the half-time relations are right, one can get about the same information from incubated tissues as from perfused organs. The concepts of experimental design and data interpretation are the same. However, the experimental maneuvers and data are somewhat different. A loading time course must be followed by measurement of G, G*, X, and H* in tissue and medium until X_b is steady for a time. Calculations of $[G_o^*]$, G_i^*, and U_g^* are similar to those of the perfusion experiment except that no effluent data are available.

$$U_g^* = -\Delta G_e^*/\Delta t - \Delta G_o^*/\Delta t$$
$$= \Delta(G_b^* + H_b^*)/\Delta t - \Delta G_o^*/\Delta t + \Delta H_e^*/\Delta t \quad (11a)$$

$$\frac{\Delta(G_b^* + H_b)/\Delta t + \Delta H_e^*/\Delta t}{[G_e^*]W_o - G_o^*} \simeq \frac{\Delta X_b/\Delta t}{[X_e]W_o - X_b} \simeq \frac{-\Delta G_e^*/\Delta t}{[G_e^*]W_o - G_o^*} \quad (12a)$$

For Eq. (12a) it is best to choose nonpermeant data at a time when $\Delta(X_b/[X_{load}])/\Delta t$ has a value near the final value of $\Delta(G_b^*/[G_{load}^*] + H_b^*/[G_{load}^*])/\Delta t + \Delta(H_e^*/[G_{load}^*])/\Delta t$ or $\Delta(G_e^*/[G_{load}^*])/\Delta t$. Note that G_e^* and H_e^* should be expressed as amount per tissue unit just as G_o^*, G_b^*, H_b^*, etc.

Labeled substrate exit is studied by washout from tissues preloaded to a steady state with labeled permeant and nonpermeant. The tissue is transferred serially through a series of vessels containing empty medium or unlabeled permeant (non-steady-state or steady-state procedures, respectively). Duration in each vessel will be at first very short and then gradually longer. The total labeled permeant found in each washout vessel divided by time in the vessel and by tissue weight is egress rate in that vessel. These are plotted on semilog paper as stairsteps of width equal to time in vessel. A smooth line drawn through these steps should at each interval have the area of the step. If the substrate egress data were divided by $[G_{load}^*]$ and the nonpermeant egress data by $[X_{load}]$, the

[9] L. S. Weis and H. T. Narahara, *J. Biol. Chem.* **244**, 3084 (1969).

data would resemble those of Fig. 3, except that both plots would begin at the same value and fall without lag. Once the nonpermeant washout time is known, experiments should terminate at this time for tissue analysis. The equations described for calculation of loading data are equally valid for washout data.

The equations [(11), (4a), (12), and (12a)] suggested for evaluating $[G_o]$, G_i, and U_g in solid tissues are approximations, which, at least, illustrate some of the problems one faces in working with solid tissues. There are methods for analyzing washout transients, which are theoretically capable of evaluating transport coefficients (k_{go}, k_{gi}/W_{gi}, etc.) more precisely, if these are constant in time (as in steady-state experiments). Use of these methods requires participation of a trained biomathematician. Moment analysis[10] may be useful if only one cellular compartment is involved. The modeling approach used by Goresky *et al.*[11] is much more powerful and can be adapted to systems of more than one cellular compartment.

In Vivo Studies. Transport studies *in vivo* cannot be as thorough as those *in vitro*. Nevertheless, some information as to the mechanism and rates of permeation can be obtained. One may be confined to studies of substrate (or analog) entry and steady-state distribution. If it is not possible to present tissues with a step wave of radioactive substrate, the variation with time of serum concentration must be taken into account mathematically or with an analog computer model. We have studied sugar transport into muscle[12] and brain.[9,13,14] The brain is particularly suited to *in vivo* transport measurements because *in vitro* brain preparations deteriorate rapidly and because the important transport membrane is often the blood-brain barrier, which separates about 97% of the water (W_i or the parenchymal water) from 3% of the water (W_o or the vascular water). In such cases, the "interstitial space" is very small and turns over rapidly and does not interfere with estimates of parenchymal content.

If the tissue cells (or parenchyma in the case of brain) can be assumed to behave as a single compartment and the interstitial space assumed to equilibrate instantly with the central circulation, then the accumulation of labeled substrate in the tissue (dG_b^*/dt) can be described by the equation

$$dG_b^*/dt = dG_i^*/dt = [G_e^*]k_{go} - G_i^*(k_{gi} + k_{gh})/W_{gi} \qquad (17)$$

[10] L. B. Wingard, Jr., T. Chorbajian, and S. J. Galla, *J. Appl. Physiol.* **33**, 264 (1972).

[11] C. A. Goresky, G. G. Bach, and B. E. Nadeau, *J. Clin. Invest.* **52**, 991 (1973).

[12] H. E. Morgan, D. M. Regen, and C. R. Park, *J. Biol. Chem.* **239**, 3084 (1964).

[13] W. A. Growdon, T. S. Bratton, M. C. Houston, H. L. Tarpley, and D. M. Regen, *Amer. J. Physiol.* **221**, 1738 (1971).

[14] T. J. Moore, A. P. Lione, D. M. Regen, H. L. Tarpley, and P. L. Raines, *Amer. J. Physiol.* **221**, 1746 (1971).

The concentration in the serum can be fit to a series of exponential terms:

$$[G_e^*] = L_1 e^{-t\lambda_1} + L_2 e^{-t\lambda_2} + \cdots \quad (18)$$

The coefficients (L) and exponent constants (λ) are the intercepts and fractional slopes of a semilog analysis[15] of the $[G_e^*]$ time course. The combined and integrated form of these equations is

$$G_b^* = [G_e^*]W_o + k_{go}\left(\frac{L_1[\operatorname{aln}(-t(k_{gi}+k_{gh})/W_{gi}) - \operatorname{aln}(-t\lambda_1)]}{\lambda_1 - (k_{gi}+k_{gh})/W_{gi}} + \cdots\right) \quad (19)$$

where the triple periods indicate more terms differing only in the subscript on L and λ, as many terms as needed in Eq. (18) to mimic the $[G_e^*]$ time course. One determines k_{go} and $(k_{gi} + k_{gh})/W_{gi}$ by a trial and error fitting of Eq. (19) to the G_b^* time course.[11,12] Equation (19) will be strictly applicable only if entry is nonsaturable or if unlabeled substrate is approximately in a steady state.

Interpretation of Carrier Transport Constants

The mechanistic interpretations of transport constants and their modulations are theoretical problems too involved to be discussed here. The approach which we find most revealing has been published recently[16] and is compatible with others presented earlier.[2,17] This field is in its infancy in the sense that no transport modulation has been investigated in sufficient depth to allow a mechanistic interpretation. The procedures described here are intended to provide the necessary information.

[15] E. Gurpide, J. Mann, and E. Sandberg, *Biochemistry* **3**, 1250 (1964).
[16] D. M. Regen and H. L. Tarpley, *Biochim. Biophys. Acta* **339**, 218 (1974).
[17] P. Geck, *Biochim. Biophys. Acta* **241**, 462 (1971).

[44] Methods for Assessing Hormone Effects on Calcium Fluxes *in Vitro*[1]

By André B. Borle

The measurement of calcium fluxes in tissues or in isolated cell systems is immensely complicated by several facts: (1) calcium is distributed among several external and internal pools, the glycocalyx, the

[1] This work is supported by a grant from the National Institutes of Health, U.S. Public Health Service No. AM07867.

plasma membrane, the cytoplasm, the mitochondria, the endoplasmic or sarcoplasmic reticulum; (2) calcium exists in several exchangeable or unexchangeable forms, in free or ionized form, in diffusible but complexed form, in bound form and precipitated in mineral form which may or may not be exchangeable. The total cell calcium, which is perhaps the only reliable measurement that can be made, often bears no relationship with the physiologically or biochemically important fractions of cellular calcium. For instance, the total calcium content of striated muscle is around 2×10^{-3} mole/kg of tissue,[1a] whereas intracellular ionized calcium is in the range of 10^{-8} M.[2] It is obvious that a significant change in intracellular free calcium concentration will hardly affect the total cell calcium. But the converse may not be necessarily true; a significant change in total cell calcium is likely to be reflected at the subcellular level. Standard cell fractionation by ultracentrifugation is not very useful to study cellular calcium distribution because an important redistribution of calcium occurs between the various subcellular fractions during the fractionation process. There is no available method to measure the free calcium in the various cellular compartments. Some methods can detect relative changes in calcium activity of the cytoplasm in specialized cell systems. One of them utilizes the properties of the photoprotein aequorin, which is extracted from the photogenic organs of the hydromedusa *Aequorea forskalea*. Ashley and Ridgway[3] applied the properties of aequorin, which emits light only in presence of calcium ions, to study the relationship between the rise in calcium activity and tension development in barnacle muscle. Enzyme activity profiles have been used by Rasmussen and his collaborators to relate the changes in gluconeogenesis of renal cells and the presumed shifts in cytoplasmic calcium activity.[4-6] However neither method gives any information on the absolute level of calcium activity or about calcium fluxes. Kinetic analyses of calcium-45 distribution in isolated cells or in tissues may bring some indication of the size of exchangeable calcium pools and about their rate of exchange. But compartmental analyses of calcium-45 movements also have serious limitations. The size of any exchangeable calcium pool brings no direct information on its physiologically active ionized calcium fraction. The identification of a particular pool is also difficult; the accumulation of indirect evidence

[1a] D. Gilbert and W. O. Fenn, *J. Gen. Physiol.* **40**, 393 (1957).
[2] S. Winegrad, *in* "Mineral Metabolism" (C. L. Comar and F. Bronner, eds.), Vol. 3, p. 191. Academic Press, New York, 1969.
[3] C. C. Ashley and E. B. Ridgway, *J. Physiol. (London)* **209**, 105, (1970).
[4] N. Nagata and H. Rasmussen, *Biochemistry* **7**, 3728 (1968).
[5] N. Nagata and H. Rasmussen, *Proc. Nat. Acad. Sci. U.S.* **65**, 368 (1970).
[6] K. Kurokawa, T. Ohno, and H. Rasmussen, *Biochim. Biophys. Acta* **313**, 32 (1973).

may only increase the confidence of the investigator in assuming that his identification is correct, but it cannot prove it. Nevertheless this type of study can detect which compartment is affected by a particular hormone whether a particular flux is increased or decreased and the possible influence of a subcellular compartment on another.

Measurement of Total Cell Calcium

The chemical measurement of total calcium in isolated cell systems should be a relatively simple procedure, compared to the more cumbersome method required for tissues or tissue slices. Indeed the correction for interstitial calcium which requires the use of extracellular markers is not necessary. However a major fraction of the total cell calcium resides in the glycocalyx, the cell coat which contains proteins and mucopolysaccharides.

Enzymatic digestion of this extracellular coat by collagenase, trypsin, hyaluronidase, or complexing of calcium by EDTA are the standard procedures used to isolate cells from living tissues. This will obviously remove significant amounts of calcium from the glycocalyx and lower the total cell calcium. Using cell cultures of established cell lines does not eliminate the problem. The standard procedures of suspending cells from monolayer cultures also call for trypsin, for EDTA, or for the mixture of both.[7] Figure 1 shows the influence of the trypsinization time on the total calcium of HeLa cells grown as monolayers.[8] The only cell systems that may be immune from this problem are cells cultured as suspensions or cells that are already in suspension when collected: blood cells, thymocytes, or Ehrlich ascites tumor cells. Cells from monolayer cultures should be harvested by scraping them off the culture vessel with a policeman, if one is to avoid any loss in total cell calcium. Total calcium loss is not the only problem of cell isolation: EDTA also increases calcium efflux from the cells and may deplete some intracellular calcium pools. Figure 2 shows the effect of 0.1 m*M* EDTA on the efflux rate coefficient of calcium from isolated kidney cells (LLC MK_2). Even in the absence of EDTA, a calcium-free medium will double the efflux rate of calcium out of isolated kidney cells (Fig. 3).

The total calcium concentration of a cellular system depends not only on the method of preparation, but also on the ionic composition of the isolating or incubating medium.[9] The total cellular calcium of kidney

[7] D. J. Merchant, R. H. Kahn, and W. H. Murphy, "Handbook of Cell and Organ Culture." Burgess, Minneapolis, 1964.

[8] A. B. Borle, *J. Cell Biol.* **36**, 567 (1968).

[9] A. B. Borle *in* "Phosphate et métabolisme phosphocalcique" (D. J. Hioco, ed.), p. 29. Expansion Scientifique Française, Paris, 1971.

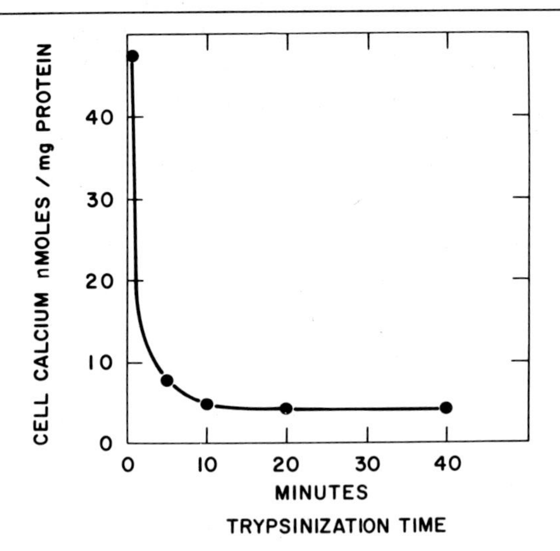

FIG. 1. Effect of trypsin and EDTA as a function of time on the total cell calcium in HeLa cells. The concentrations of trypsin and EDTA were, respectively, 0.125% and 0.01%. From A. B. Borle, *J. Cell Biol.* **36**, 567 (1968). Reproduced with the permission of the *J. Cell Biol.*

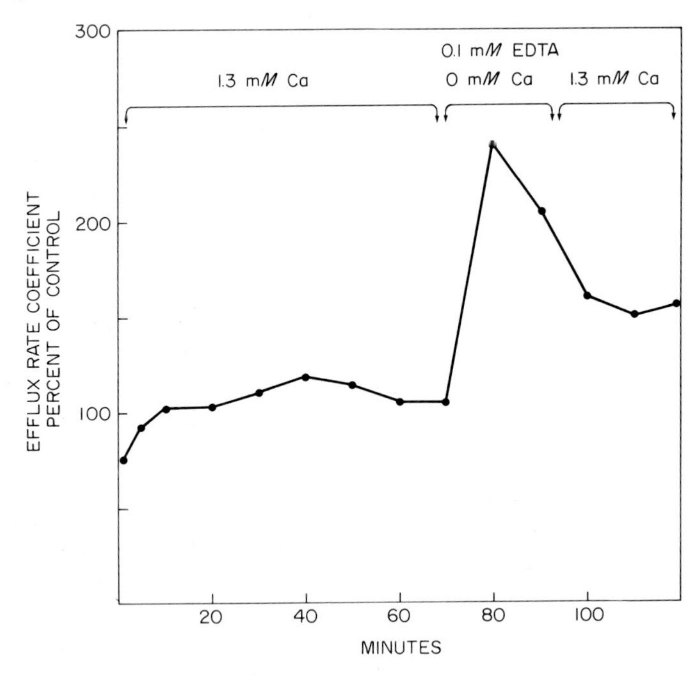

FIG. 2. Effect of 0.1 m*M* EDTA on the efflux rate coefficient of calcium from isolated kidney cells.

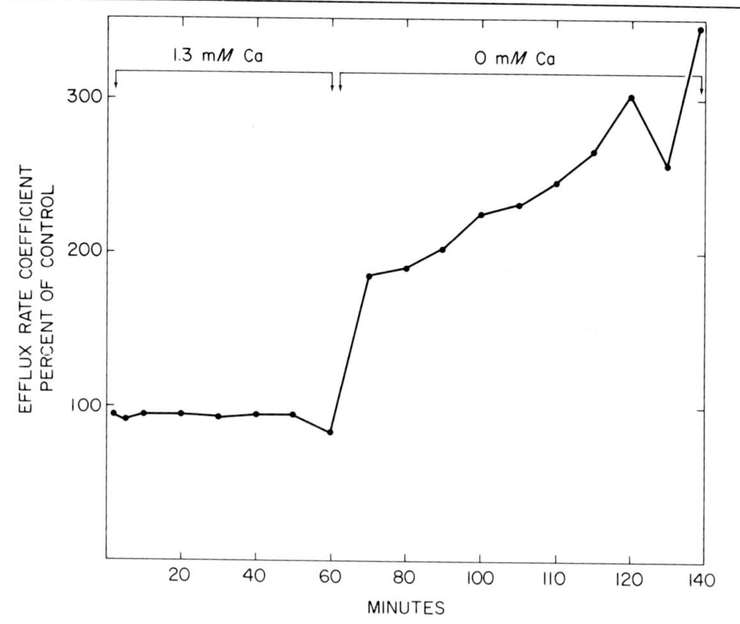

FIG. 3. Effect of a calcium-free medium on the efflux rate coefficient of calcium from isolated kidney cells.

cells, for instance, increases when the phosphate concentration of the medium is raised from 0 to 20 mmoles/liter (Fig. 4). It is also dependent on the medium calcium concentration (Fig. 5). Between 0 and 2 m*M* calcium, the total cellular calcium seems to be directly proportional to the concentration of calcium in the medium, but the slope of the proportionality curve depends on the medium phosphate: as phosphate increases, the cell is more sensitive to the changes in extracellular calcium concentration (Fig. 6). Clearly, the composition of the incubation medium and the buffer chosen, whether bicarbonate or phosphate, will have great influence on the absolute levels of the total cellular calcium concentration. Consequently, the value obtained for total cell calcium depends to a large extent on the method of isolation and on the composition of the suspending buffer.

Most methods available for the determination of small amounts of calcium in biological material suffer from interference by many substances present in the sample. Sodium, potassium, and phosphate interfere with flame photometry[10,11]; sodium, potassium, phosphate, and proteins with

[10] P. S. Chen, Jr. and T. Y. Toribara, *Anal. Chem.* **25**, 1642 (1953).
[11] R. P. Geyer and E. J. Bowie, *Anal. Biochem.* **2**, 360 (1961).

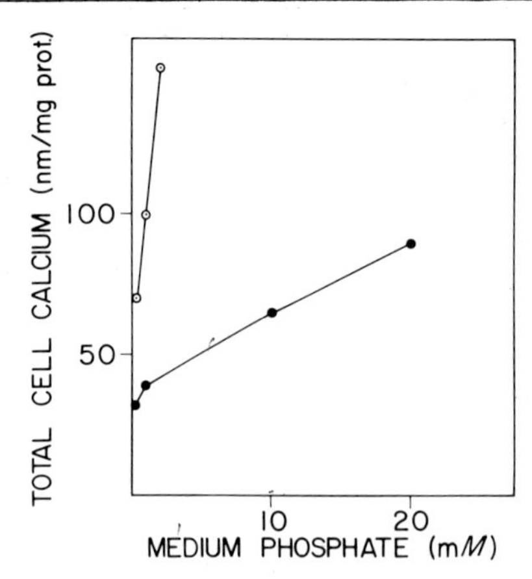

FIG. 4. Effect of the medium phosphate concentration on the total calcium of isolated kidney cells. The medium calcium concentration was either 0.4 mM (filled circles) or 1.3 mM (open circles). From A. B. Borle *in* "Phosphate et Métabolisme Phosphocalcique" (D. J. Hioco, ed.), p. 29. Expansion Scientifique Française, Paris, 1971. Reproduced with the permission of l'Expansion Scientifique Française.

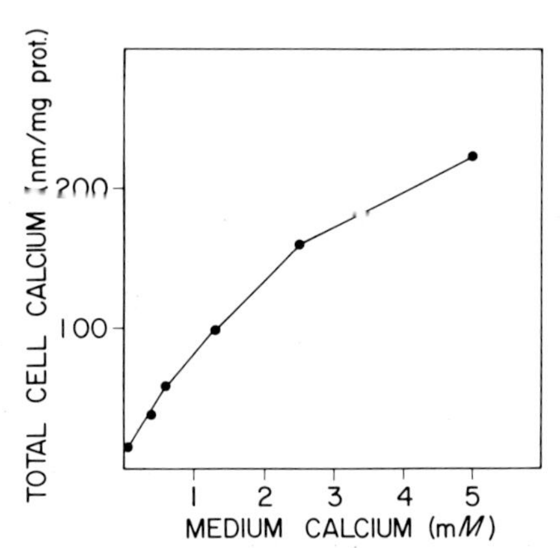

FIG. 5. Effect of the medium calcium concentration on the total calcium of isolated kidney cells. The medium phosphate concentration was 1 mM. From A. B. Borle *in* "Phosphate et Métabolisme Phosphocalcique" (D. J. Hioco, ed.), p. 29. Expansion Scientifique Française, Paris, 1971. Reproduced with the permission of l'Expansion Scientifique Française.

atomic absorption[12]; and magnesium, ATP, and nonspecific fluoresent substances with direct fluorometry.[13,14] Complexometric titration of a

[12] J. B. Willis, *Spectrochim. Acta* **16,** 259 (1960).

[13] D. F. H. Wallach and T. L. Steck, *Anal. Chem.* **35,** 1035 (1963).

[14] H. B. Collier and G. Dudson, *Anal. Biochem.* **15,** 367 (1966).

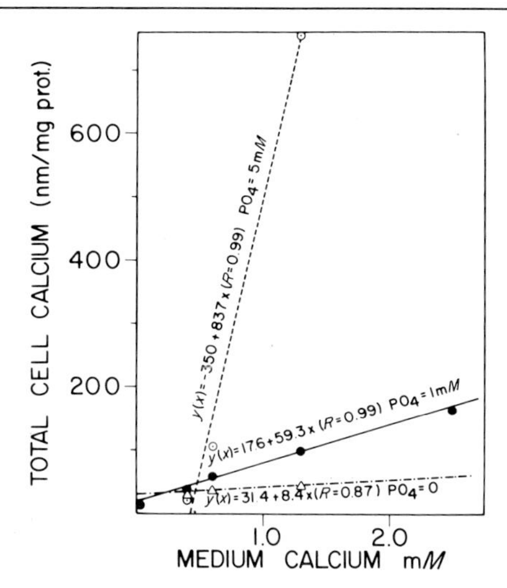

FIG. 6. Relative influence of the medium concentration of phosphate and of calcium on the total calcium of isolated kidney cells. The medium phosphate concentration was respectively 0 m*M* (open triangles), 1 m*M* (filled circles), and 5 m*M* (open circles). The equation of each regression line is given and the regression coefficient *R* is included in parentheses. From A. B. Borle *in* "Phosphate et Métabolisme Phosphocalcique" (D. J. Hioco, ed.), p. 29. Expansion Scientifique Française, Paris, 1971). Reproduced with the permission of l'Expansion Scientifique Française.

fluorescent calcium–calcein complex seems to be one of the best methods for the microdetermination of calcium in biological material.[15,16] The method developed by Borle and Briggs is extremely sensitive and can detect 1 nmole (0.04 μg) of calcium with a coefficient of variance of less than 2%.[16] It is highly specific and free of interference, since magnesium, phosphate, and ATP do not interfere up to molar ratios of 200, 20,000, and 40,000, respectively. Proteins such as albumin, troponin, tropomyosin, actin, and cell sonicates do not interfere with the titration up to weight ratio of 36,000. In terms of sensitivity and lack of interference fluorometric titration has an advantage over atomic absorption. Other methods may require the separation of calcium by oxalate precipitation,[17,18] a procedure which is not immune from significant losses of calcium; other techniques may require the elimination of organic material from the sample by ashing procedures. Dry ashing requires (1) the drying of the sample in a drying oven at 100°; (2) ashing for 3–12 hours, depending on

[15] P. Cartier and J. Clement-Metral, *Clin. Chim. Acta* **4**, 357 (1959).
[16] A. B. Borle and F. N. Briggs, *Anal. Chem.* **40**, 339 (1968).
[17] T. Y. Toribara and L. Koval, *J. Lab. Clin. Med.* **57**, 630 (1961).
[18] T. Y. Toribara and L. Koval, *Talanta* **7**, 248 (1961).

the amount of material, in a muffle furnace at 500°, and (3) the dissolution of the ash with 0.1 *N* HCl. Regardless of the method, the determination of total calcium in tissues of cell systems is not as simple a procedure as it first appears. Finally, exceptionally high values of total cell calcium should be considered suspect. In our hand, one of the first signs of cellular "distress" is an immediate increase in the permeability of the plasma membrane for calcium, quickly followed by a massive accumulation of calcium by the cell. This accumulation is probably due to the uptake of calcium by the cell mitochondria.[19,20] In many respects, a tissue or cell preparation which has lost its membrane integrity resembles a mitochondrial preparation.

Measurement of Single Time Point Cellular Calcium Uptake

The determination of calcium uptake by tissues or cell preparations by radioisotope techniques is a simple experimental procedure. The information obtained, however, can be easily misinterpreted. Technically the method is not different from the determination of other labeled ions or molecules taken up by tissues or cell suspensions. After a period of preincubation to achieve steady-state conditions, carrier-free calcium-45 is added to the incubation mixture and the cells are incubated for a fixed period of time, predetermined by the investigator. Several methods are available for separating the tissue or the cells from the radioactive medium at the end of the incubation period. The simplest can be used with tissue slices or with cell monolayers which adhere to their culture vessel. Six consecutive rinses with an ice-cold and unlabeled buffer solution are usually used to wash the extracellular radioactivity from the cells.[8] With suspensions of isolated cells, two different techniques are available: (1) separation with a Millipore filter with a pore diameter approximately 3 times smaller than the cell diameter; a pore size of 5 μm can be used for epithelial cells with diameters of 12–20 μm; (2) separation by centrifugation.[21,22] By this method, a 1-ml aliquot of cell suspension is added to 40 ml of ice-cold unlabeled buffer contained in a centrifuge tube, and centrifuged for 45 seconds at 2000 *g*. This immediately slows down exchange processes and dilutes the radioactive medium 40-fold. The buffer is decanted, and the cell pellet is homogenized in distilled water with an ultrasonic probe. Several determinations can be made to estimate the cell mass: a cell count by hemacytometer or by electronic cell counter

[19] R. E. Thiers, E. S. Reynolds, and B. L. Vallee, *J. Biol. Chem.* **235,** 2130 (1960).
[20] E. S. Reynolds, R. E. Thiers, and B. L. Vallee, *J. Biol. Chem.* **237,** 3547 (1962).
[21] A. B. Borle, *J. Gen. Physiol.* **55,** 163 (1970).
[22] C. Levinson and L. E. Blumenson, *J. Cell. Physiol.* **75,** 231 (1970).

in a replicate sample of the suspension; DNA or cell protein determination of the cell sonicate.[7,23] The radioactivity and the total calcium of the incubating medium and of the cell sonicate are measured by standard techniques. From these measurements several parameters can be calculated: (1) the activity of the medium, α_m; (2) the specific activity of the medium, sp. α_m; (3) the total cell calcium, Ca_c, expressed per milligram of protein, per microgram of DNA or per million of cells; (4) the cell activity, $^{45}Ca_c$, expressed on the same basis of reference; and (5) the cell specific activity, sp. α_c. The cellular calcium uptake can be derived as shown in Eq. (1).

$$\text{Calcium uptake} = \frac{^{45}\text{Ca}_c}{\text{sp. } \alpha_m} \tag{1}$$

The results will be expressed, for instance, in nanomoles of calcium per milligram of cell protein.

The cell specific activity relative to the medium specific activity (relative specific activity) will give an indication of the turnover rate or, in other words, of the fraction of the total cellular calcium which has exchanged with the radioactive medium at that particular time point:

$$\text{Relative specific activity} = (\text{sp. } \alpha_c)/(\text{sp. } \alpha_m) \tag{2}$$

This parameter is often expressed as percent exchange:

$$\text{Percent exchange} = \text{relative specific activity} \times 100 \tag{3}$$

The percent exchange never reaches 100% since there is a significant pool of cellular calcium which is totally unexchangeable.[22,24]

Calcium uptake by cells or by tissues is often misinterpreted as an indication of the rate of calcium influx from the medium to the cell interior. The amount of radioactive calcium present in the tissue at any given time between 1 minute and isotopic steady state is determined by several independent variables: the amount of isotope taken up by the tissue, the amount of isotope released by the tissue, and the size of the cellular exchangeable calcium pools. Calcium uptake being by definition a single time point determination cannot discriminate between an increased entry of calcium into the cell, a decreased release, or a larger exchangeable pool. As an example, Table I shows that 0.1 μM cyclic AMP increases the calcium uptake of isolated kidney cells 40%. This simple measurement gives absolutely no indication whether this increase is due to a greater influx, a larger exchangeable pool or a decreased efflux. The kinetic analysis of the calcium uptake curve shown in Fig. 7 and in Table

[23] V. I. Oyama and H. Eagle, *Proc. Soc. Exp. Biol. Med.* **91**, 305 (1956).
[24] A. B. Borle, *J. Gen. Physiol.* **53**, 43 (1969).

TABLE I

EFFECT OF 0.1 μM CYCLIC AMP ON CALCIUM UPTAKE BY ISOLATED KIDNEY CELLS, AFTER 60 MINUTES OF INCUBATION

	Number of experiments	Calcium uptake[a] (nmoles/mg protein)
Control	10	3.83 ± 0.08
cAMP, 0.1 μM	6	5.39 ± 0.44[b]

[a] Values are the mean ±SE.
[b] $p < 0.001$.

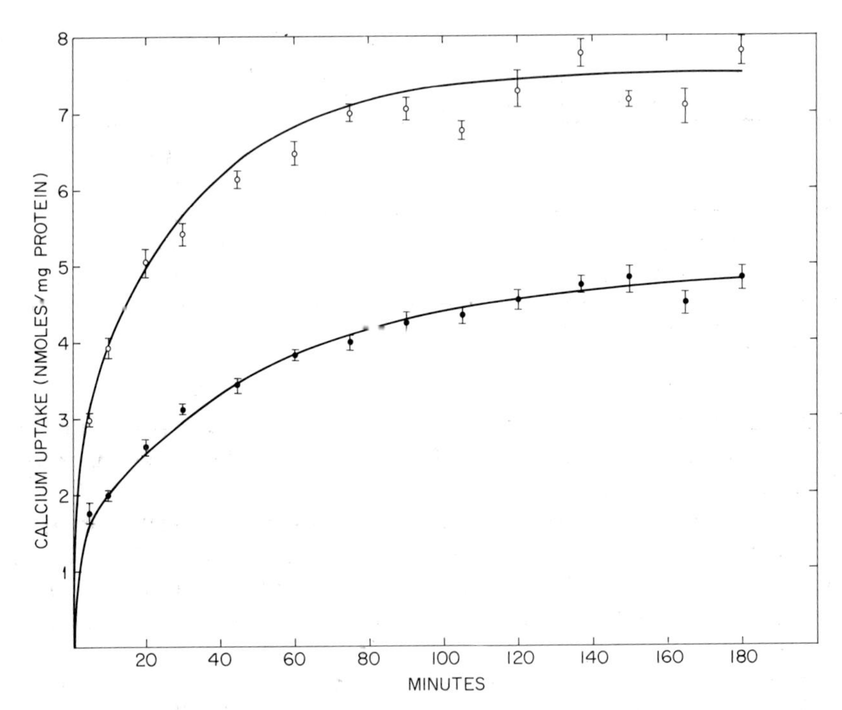

FIG. 7. Effect of 0.1 μM cyclic AMP on the calcium uptake curve of isolated kidney cells. The medium was a Krebs-Henseleit bicarbonate buffer containing 1.3 mM calcium and 1.0 mM phosphate, at pH 7.4. The values are the mean ± the standard error of 6 experiments in each group. The difference between groups at each time point is statistically significant at the 0.01 level. The solid lines represent the theoretical curves reconstructed from the respective differential equations presened in the figure. ●——●, Control, $(dR/dt)/E = 70e^{-0.02t} + 1300e^{-0.65t}$; ○——○, 0.1 μM cAMP, $(dR/dt)/E = 160e^{-0.032t} + 2340e^{-0.64t}$.

TABLE II
EFFECT OF 0.1 μM CYCLIC AMP ON THE SLOW PHASE OF CALCIUM INFLUX IN ISOLATED KIDNEY CELLS

	Control	0.1 μM cAMP
Calcium influx (pmoles mg protein^{-1} min^{-1})	70	160
Calcium pool (nmoles mg protein^{-1})	3.5	5.0
Rate constant (min^{-1})	0.020	0.032
Half-time (minutes)	34.6	21.7

TABLE III
EFFECT OF VITAMIN D DEFICIENCY ON CALCIUM UPTAKE BY ISOLATED CHICK INTESTINAL CELLS, AFTER 60 MINUTES OF INCUBATION

	Number of experiments	Calcium uptake[a] (nmoles/mg protein)
Control	7	7.57 ± 0.54
Vitamin D deficiency	7	5.85 ± 0.43[b]

[a] Values are the mean ±SE.
[b] $p < 0.05$.

II reveals that this increased calcium uptake is due to the doubling of calcium influx into the cell and a 50% increase in the intracellular exchangeable calcium pool.

Conversely a decreased calcium uptake does not necessarily mean a decreased calcium influx. For instance, many investigators have found that calcium uptake by the intestinal mucosa was significantly decreased in vitamin D deficiency. The same observation can be made in isolated intestinal cells, as shown in Table III. It is usually assumed that this decreased uptake reflects a depressed calcium influx. However, kinetic analyses of the calcium uptake curves shown in Fig. 8 indicate that this is entirely due to a decreased intracellular exchangeable calcium pool.[25] Indeed the parameters derived from these curves and presented in Table IV, show that calcium influx per se is not affected in spite of the low calcium uptake.

Another difficulty in assessing the significance of calcium uptake is

[25] A. B. Borle, *J. Membrane Biol.* **16**, 207 (1974).

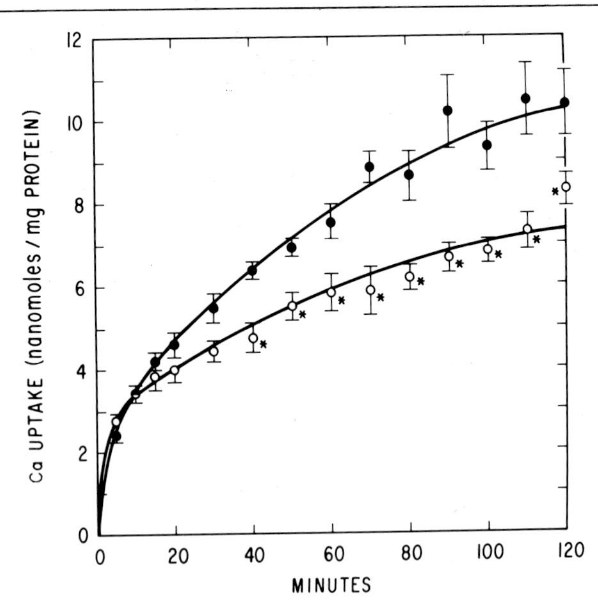

FIG. 8. Difference in the calcium uptake of intestinal cells isolated from control (●) and vitamin D-deficient (○) chicken. Intestinal cells were isolated from the duodenum by hyaluronidase (0.1 mg/ml). The values are the mean ± the standard error of 7 controls and 6 experimentals. The medium composition was identical to that described in Fig. 7. From A. B. Borle, *J. Membrane Biol.* **16,** 207 (1974).

TABLE IV
EFFECT OF VITAMIN D DEFICIENCY ON THE SLOW PHASE OF CALCIUM INFLUX IN ISOLATED INTESTINAL CELLS[a]

	Control[b] (7)	Vitamin D deficiency[b] (6)
Calcium influx (pmoles mg protein^{-1} min^{-1})	144 ± 14	144 ± 26
Calcium pool (nmoles mg protein^{-1})	7.40 ± 0.91	5.57 ± 0.46[c]
Rate constant (min^{-1})	0.0185 ± 0.0026	0.0227 ± 0.0035
Half-time (minutes)	37.4 ± 5.4	30.4 ± 4.7

[a] From A. B. Borle, *J. Membrane Biol.* **16,** 207 (1974).
[b] Values are mean ±SE. Numbers in parentheses indicate number of experiments.
[c] $p < 0.05$.

TABLE V
CALCIUM UPTAKE FROM THEORETICAL CURVE OF FIG. 9

	Calcium uptake (nmoles/mg protein)		
	20 min	30 min	60 min
Control	2.70	3.16	3.96
Hormone X	2.54	3.19	4.24

the presence of an extracellular calcium pool, the glycocalyx. This compartment may be as great or greater than the intracellular calcium pool. Its rate of exchange is about 20–30 times faster than the exchange of calcium between the inner and outer phase of the cell. A small change in this extracellular compartment may completely obscure a significant rise or fall in calcium influx into the intracellular pool. Conversely a calcium uptake significantly different from control may be due exclusively to a modification of the glycocalyx pool.

For instance it has been shown that parathyroid hormone increases calcium uptake in the HeLa cells and in isolated kidney cells.[8,26] This increased uptake turned out to be due to an increased calcium influx and to a larger intracellular exchangeable calcium pool.[21] However depending on some unpredictable variable, for instance on the method of cell preparation, the effect of any hormone X on calcium uptake may be completely obscured, although kinetic analyses can still detect very clearly an increased influx and an increased intracellular pool. Table V shows that at 20 minutes calcium uptake is less than control, at 30 minutes it is identical, and at 60 minutes it is only slightly increased. It would be impossible to draw any conclusions from such data. The uptake curves of Fig. 9 do not appear very different either. However, when these curves are dissected into their two kinetic components it is clear that the hormone X has a significant effect on the slower phase of calcium influx which represents transfer into the cell (Fig. 10 and Table VI). This effect was completely obscured by a decreased calcium exchange in the glycocalyx.

It is equally conceivable that a significant increase in calcium uptake may represent only a larger extracellular exchangeable calcium pool (fast phase) while calcium influx into the cell and the intracellular calcium pool remain unchanged or actually decrease.

Finally isotopic uptake by cells and tissues is sometimes used to study the active or passive nature of the process transporting a substance from the incubating medium to the inner phase of the cells. The validity of

[26] A. B. Borle, *Endocrinology* **83**, 1316, (1968).

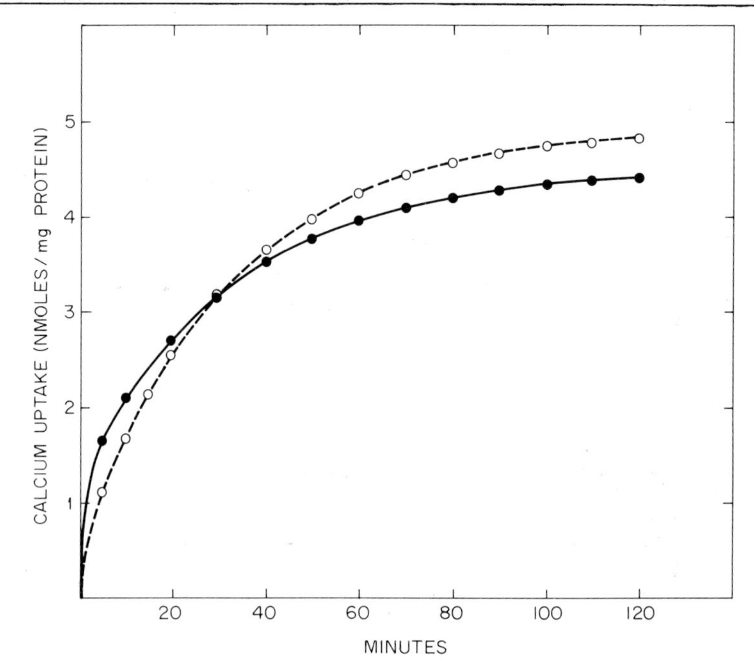

FIG. 9. Theoretical uptake curves showing little difference between control conditions (●——●) and calcium uptake in presence of a hormone X (○---○).

this method applied to calcium transport is extremely doubtful. Recently, it became apparent that calcium-45 uptake represents three different pro cesses: the binding of calcium to an extracellular compartment, the transport of calcium into the cell, and the sequestration of calcium in a subcel-

TABLE VI
PARAMETERS OF CALCIUM INFLUX DERIVED FROM THE THEORETICAL CURVES OF FIG. 9

	Control	Hormone X
Fast phase (glycocalyx)		
Calcium influx (pmoles mg protein^{-1} min^{-1})	1000	500
Calcium pool (nmoles mg protein^{-1})	1.66	0.70
Rate constant (min^{-1})	0.6	0.7
Slow phase (intracellular)		
Calcium influx (pmoles mg protein^{-1} min^{-1})	100	140
Calcium pool (nmoles mg protein^{-1})	3.3	4.5
Rate constant (min^{-1})	0.030	0.031

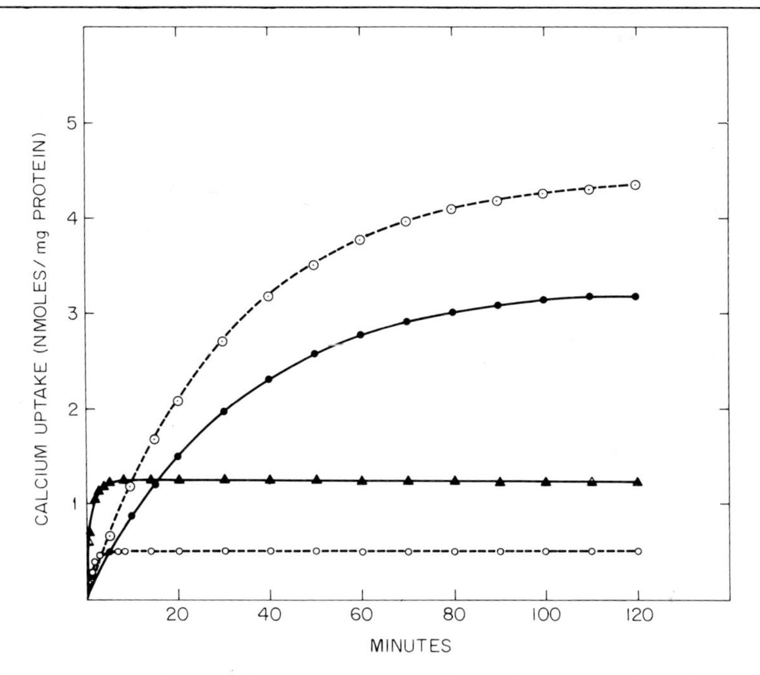

FIG. 10. Fast and slow kinetic components dissected from the theoretical curves of Fig. 9. These curves were reconstructed from the coefficients and exponential constants of the differential equations derived from the curves in Fig. 9 and presented in Table VI. ●——●, Control, slow phase; ▲——▲, control, fast phase; ⊙---⊙, hormone X, slow phase; ○---○, hormone X, fast phase.

lular compartment which has been identified with the mitochondria.[27-29] Single time point uptake experiments are unable to distinguish between these three separate transport processes. Even kinetic analyses of calcium uptake curves cannot demonstrate the existence of two intracellular pools because the rate of calcium uptake by mitochondria is about 1000 times greater than the rate of calcium influx through the plasma membrane. Consequently the two intracellular compartments are seen as one unique intracellular pool. It is obvious that the effects of metabolic inhibitors on calcium uptake will be very difficult to interpret, depending on the contribution of the mitochondria to the overall calcium uptake. A decreased calcium uptake by an inhibitor does not necessarily prove that calcium influx into the cell is a metabolically dependent process. Con-

[27] A. B. Borle, *in* "Cellular Mechanisms for Calcium Transfer and Homeostasis" (G. Nichols, Jr. and R. H. Wasserman, eds.), p. 151. Academic Press, New York, 1971.

[28] A. B. Borle, *Fed. Proc., Fed. Amer. Soc. Exp. Biol.* **32,** 1944 (1973).

[29] A. B. Borle, *J. Membrane Biol.* **10,** 45 (1972).

versely, the lack of demonstrable inhibition even with specific mitochondrial inhibitors does not necesarily exclude the participation of mitochondria to the overall process of uptake. Indeed the energy for calcium accumulation by mitochondria can be derived from 3 sources: substrate oxidation, NADH oxidation, and ATP. Each energy source is inhibited by different metabolic inhibitors or uncouplers.[30] Thus depending on the inhibitor, the exogenous substrate available, the endogenous substrate of the cell and the ATP levels of the mitochondria, calcium uptake may or may not be affected. It would be inappropriate, on the basis of such results alone, to draw definite conclusions concerning the active or passive nature of calcium fluxes into the cell.

Despite all the limitations of the method, the measurement of a single time point cellular uptake of calcium-45 may still be useful. First, it is technically easy and rapid. Second, although negative results cannot exclude the possible effect of a hormone on a particular calcium pool, positive results indicate that one of the three calcium compartments is affected. Third, potentiation or inhibition by other factors of the positive effects can be easily screened. But regardless of the results, the interpretation of the data can only be limited and tentative.

Measurement of Calcium Influx by Kinetic Analysis

Kinetic analysis of radiotracer movements requires different mathematical treatments depending on the experimental conditions (for instance open or closed systems), on the number of compartments assumed to be present, and on their relationship (series or parallel case). Several articles dealing with the theory of compartmental tracer kinetics and with their mathematical treatment are available in the literature and will not be covered here.[31-34]

Methods

A homogeneous population of cells, in suspension or in monolayer cultures, can be used to measure calcium influx by tracer kinetics. Tissues are not as easy to study because several additional compartments increase the complexity of the system: vascular space, interstitial space,

[30] A. B. Borle, *Clin. Orthop.* **52**, 267 (1967).
[31] C. W. Sheppard, "Basic Principles of the Tracer Method." Wiley, New York, 1962.
[32] A. K. Solomon, *in* "Mineral Metabolism (C. L. Comar and F. Bronner, eds.), Vol. 1, Part A, p. 119. Academic Press, New York, 1960.
[33] J. S. Robertson, *Physiol. Rev.* **37**, 133 (1957).
[34] J. S. Robertson, D. C. Tosteson, and J. L. Gamble, *J. Lab. Clin. Med.* **49**, 497 (1957).

the presence of different cell species with different kinetic behavior will introduce many more kinetic components, which must be taken into account. Since a homogeneous cell population already has three recognized calcium compartments, any additional pool will increase the difficulty of resolving the calcium uptake curve by compartmental analysis. It is also easier to study the kinetic behavior of radiocalcium in steady-state conditions, although the study of non-steady-state compartmental systems is possible.[31]

Technically, the study of calcium influx in cellular systems is essentially a multiple determination of single time points calcium-45 uptake. Calcium-45 uptake is determined from time zero until the system reaches or approaches isotopic steady state. A period of preincubation without isotope should precede the experimental period to allow the system to reach steady state. The length of the preincubation period depends on the composition of the incubating medium, especially if it is very different from the composition of the growth medium in which the cells were cultured or collected. In this respect, the ionic concentration of calcium, phosphate, sodium, and magnesium, the pH, the buffer chosen, the kind of substrate and its concentration are all important variables to be taken into consideration. More importantly, if a hormone is expected to alter cellular calcium metabolism, a period of preincubation should allow the cell to reach the new steady state before the isotope is introduced into the system. Unfortunately there is no good criterion for deciding whether a cell suspension has reached steady state. A stable total cell calcium is a prerequisite, but it is not a guarantee that steady state has been achieved since it has little relation to the much smaller intracellular exchangeable calcium pools. A stable relative specific activity is a much better criterion, but it requires several pilot experiments.

The method for the measurement of calcium influx in monolayer cultures is relatively simple.[24] Cells are grown on a flat culture vessel (Bellco T-60 flask, Falcon plastic flask 75 cm^2, or milk dilution bottles) until they reach confluence and their stationary phase of growth. The growth medium is decanted, and the cells are preincubated in the experimental buffer containing a metabolic substrate. At time zero, 20–100 μCi of carrier-free calcium-45 is added to each flask and the cells are incubated at 37° on a rocking platform (Fig. 11). The cells are harvested at various time intervals from 1 minute or less to 3 or 4 hours or more to approach or reach isotopic steady state. Harvest involves decanting the radioactive medium, washing the monolayer 6 times with ice cold isotopic buffer, adding a few milliliters of distilled or deionized water, and scraping the cells off the vessel surface with a rubber or plastic policeman. The cells are then homogenized with a high intensity ultrasonic probe, and aliquots

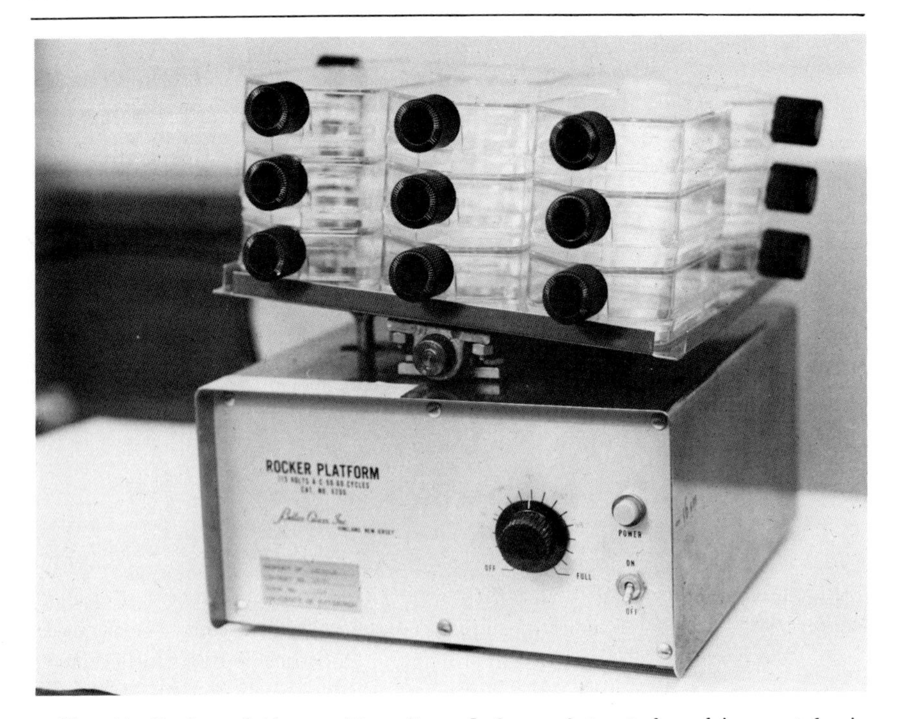

FIG. 11. Rocker platform with culture flasks used to study calcium uptake in monolayer cultures. Rocking rates should be adapted to the experimental conditions. The experimental set-up should be placed in a 37° incubator.

of the homogenate are analyzed for total calcium, calcium-45 activity, and protein (or DNA). The radioactive incubating medium is also analyzed for total calcium and calcium-45 radioactivity.

The method of choice, however, utilizes cells in suspension.[21] The experimental set-up is simple, as shown in Fig. 12. Cells are suspended in 60 ml of buffer in a cylindrical glass, plastic, or Lucite vessel of 100-ml capacity. The amount of cells in 60 ml should be equivalent to 40 or 60 mg of cell protein or to 2 ml of cells packed at 100 *g*. The cells are kept in suspension by a Teflon-coated magnetic bar driven by an immersible magnetic stirrer. After the period of incubation and the addition of 50–100 μCi of calcium-45, 1-ml aliquots of the cell suspension are taken with an Eppendorf pipette, injected into 40 ml of ice-cold, unlabeled buffer in a 50-ml centrifuge tube and centrifuged for 45 seconds at 2000 *g*. The cell pellet is immediately separated from the buffer by decanting and the centrifuge tube kept inverted to drain. The internal wall of the centrifuge tube is wiped dry with a cellulose tissue, 4 ml of distilled water are added to the cell pellet, and the cells are homogenized with an ultra-

FIG. 12. Experimental set-up to study calcium uptake by isolated cells in suspension. The water bath is kept at 37°. The suspension vessel is exposed to 5% CO_2 in air from a gas cylinder. The gas is warmed, washed, and humidified by passing through a washing bottle. The cells are kept in suspension by a Teflon-coated magnetic bar driven by an immersible magnetic stirrer. To the right of the water tank are the Eppendorf pipettes used to sample the cell suspension and the centrifuged tubes containing 40 ml of ice cold unlabeled medium (usually on ice).

sonic probe. Total calcium, calcium-45 activity, and cell proteins are measured in the homogenate and on replicate samples of the radioactive medium separated from the cells by centrifugation. A cell suspension of 60 ml allows for the determination of 40 different time points of calcium uptake (or 20 duplicates) and 6 determinations of the medium specific activity. A 15- to 20-ml volume of cell suspension should be left in the incubation vessel at the end of the experiment to cover the magnetic bar. Our experience shows that if the Teflon bar is exposed to the air, the stirring becomes turbulent and has a deleterious effect on the cells in suspension. This manifests itself by a rise in the tail end of the uptake curve.

Separation of the cells from the radioactive medium can also be achieved by filtration through Millipore filters as outlined in the previous section. Finally, Langer and collaborators have developed a very elegant method for monitoring calcium exchange in myocardial cell monolayers by direct, continuous counting of cellular radioactivity.[35] The cells are grown as monolayer on a glass scintillator slide. The $50 \times 20 \times 1.5$ mm rectangular slide is made of NE 901 glass (Nuclear Enterprises Ltd., San

[35] G. A. Langer, E. Sato, and M. Seraydarian, *Circ. Res.* **24,** 589 (1969).

Carlos, California). Both sides of the slide are covered by the monolayer by inoculating the culture on two successive days on each side of the slide. This places the monolayer directly in contact with the radioisotopic detector. The glass scintillator slide is then placed in a specially designed flow cell which fits the well of a scintillation spectrometer (Fig. 13). Calcium-45 uptake or washout can both be continuously monitored. The flow cell chamber is made of Lucite plastic with polished inner and outer surfaces. The fluid capacity of the chamber with slide in place is 5.4 ml.

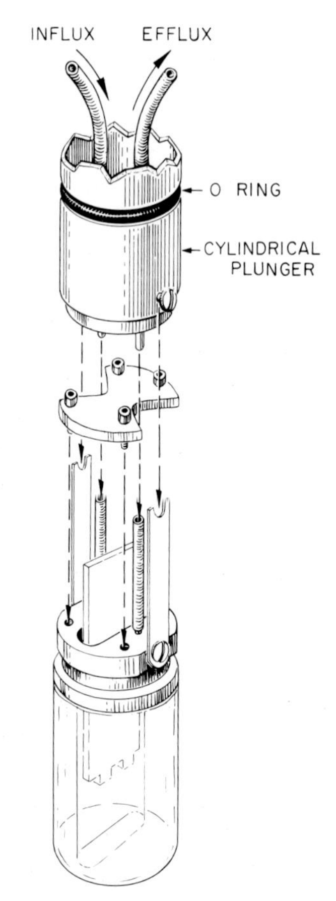

FIG. 13. Flow cell and scintillator slide. The slide is shown during insertion into the flow cell. After insertion, the cap (with four screws) is tightened in place, the tubular fittings on the plunger are inserted in the flow cell tubes, and the plunger is attached to metal strap-fittings on each site of the cell. The flow cell is then ready for insertion into the spectrometer well. From G. A. Langer, E. Sato, and M. Seraydarian, *Circ. Res.* **24,** 589 (1969). Reproduced with the permission of The American Heart Association, Inc.

Fluid entering the chamber is directed against the slide, passes under the bottom of the slide which divides the chamber longitudinally and then flows upward toward the chamber outlet. Monitoring of the radioactivity may be done either by alternating periods of counting and printout (respectively, 6 and 7.2 seconds in a Beckman model LS 200 B spectrometer) or by continuous recording of the count rate if a rate meter can be adapted to the spectrometer. According to Langer, the counting efficiency of glass scintillator NE 901 for calcium-45 placed on the surface of the glass varies between 22% and 28% when the scintillator is placed in the chamber filled with fluid. The minimum flow rate of fluid through the chamber should be determined not to be a limiting factor for the washout of radioactivity from the slide. According to Langer, flow rates exceeding 24 ml per minute do not increase further the calcium-45 washout rate and can be used in his system. In these studies the characteristics of calcium-45 uptake or washout to and from the slide without attached cells (blank) should always be determined to estimate the participation of the glass to the whole system. The cellular uptake, for instance, is obtained by subtracting the blank curve (slide alone) from the curve obtained with the whole system (cell plus slide). This is shown in Fig. 14. There is one possible pitfall in this method. Not all cell monolayers adhere to their glass substrate as tightly as myocardial cells. Cell detachment produced by a stream of fluid flowing on a monolayer is not unusual. If the cells were detaching at a more or less fixed rate, it could produce spurious results and create a totally artificial kinetic compartment. Consequently before using such a method for cell strains other than myocardial cells, it will be essential to ascertain that no cell detachment occurs.

Calculations

The specific activity curve of calcium uptake by isolated cells can be adequately described by a double exponential equation.[21,22] We can therefore assume that the simplest compartmental model for the system includes two cellular compartments plus the radioactive buffer phase. The exchange rates, and the compartment size of a three-compartment steady-state closed system can be readily calculated from the solution to the general equations published by Robertson *et al.*[34] The system can be represented schematically as follows:

$$S_1 \underset{k_{21}}{\overset{k_{12}}{\rightleftarrows}} S_2 \underset{k_{32}}{\overset{k_{23}}{\rightleftarrows}} S_3$$

The symbols used in this analysis are defined in Appendix I. According to Robertson *et al.*,[34] the differential equation for the behavior of the

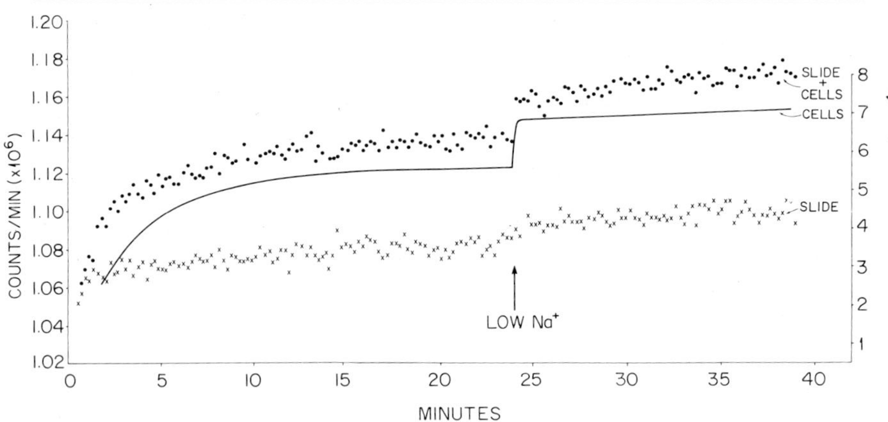

FIG. 14. Calcium-45 uptake by scintillator slide with and without cells. Calcium-45 activity (left ordinate) is plotted against time for the slide with cells (●) and for the slide without cells (×). The cellular component (solid line) was obtained by subtraction (right ordinate). Low sodium solution was substituted at the twenty-fourth minute in both labeling procedures. From G. A Langer, E. Sato, and M. Seraydarian, *Circ. Res.* **24,** 589 (1969). Reproduced by permission of The American Heart Association, Inc.

tracer in compartment 1 in steady state situation is

$$dR_1/dt = -J_{12}X_1 + J_{12}X_2 \tag{4}$$

If one assumes that the two cellular compartments are in parallel, S_2 represents the medium and $S_1 + S_3$ the two cellular pools. If one assumes that the two cellular pools are in series, S_1 becomes the medium and S_2 and S_3 the cellular compartments. By measuring the radioactivity of the cells, the calcium-45 of the two cellular compartments $R_1 + R_3$ is determined experimentally in the parallel case, or $R_2 + R_3$ in the series case. If one assumes a parallel case, the increase in tracer activity of the cells which comprises compartment 1 and 3 is equal to its disappearance from the medium compartment 2, after the tracer is added to the medium.

Therefore

$$-dR_2/dt = dR_1/dt + dR_3/dt = dR_{1+3}/dt \tag{5}$$

Division of the terms of Eq. (4) by S_1 yields Eq. (6), expressed in terms of specific activities. Analogous procedures yield the equations for the second and third compartment:

$$dX_1/dt = -k_{12}X_1 + k_{12}X_2 \tag{6}$$

$$dX_2/dt = k_{21}X_1 - (k_{21} + k_{23})X_2 + k_{23}X_3 \tag{7}$$

$$dX_3/dt = k_{32}X_2 - k_{32}X_3 \tag{8}$$

The general solutions of the system of Eqs. (6), (7), and (8) are also given by Robertson *et al.*[34] as:

$$X_1 = E + C_{11}e^{-\lambda_1 t} + C_{12}e^{-{}_2\lambda t} \tag{9}$$

$$X_2 = E + C_{21}e^{-\lambda_1 t} + C_{22}e^{-\lambda_2 t} \tag{10}$$

$$X_3 = E + C_{31}e^{-\lambda_1 t} + C_{32}e^{-\lambda_2 t} \tag{11}$$

The three parameters, calcium flux J_{ij}, rate constant k_{ij}, and compartment size S_i, can be calculated for the series or parallel case. In the series case, Eq. (9) (with numerical values for $X_1 0$, E, C_{11}, C_{12}, λ_1, and λ_2) is obtained by experiment. The rate constants, compartment sizes, and fluxes are obtained by using Eqs. (12) to (20).

$$k_{12} = \frac{C_{11}\lambda_1 + C_{12}\lambda_2}{X_1 0} \tag{12}$$

$$k_{21} = \frac{k_{12}(\lambda_1 + \lambda_2 - k_{12}) - \lambda_1\lambda_2[1 - (E/X_1 0)]}{k_{12}} \tag{13}$$

$$k_{32} = (E\lambda_1\lambda_2)/(X_1 0 k_{21}) \tag{14}$$

$$k_{23} = \frac{\lambda_1\lambda_2[1 - (E/X_1 0)]}{k_{12}} - k_{32} \tag{15}$$

$$S_1 = R_1 0/X_1 0 \tag{16}$$

$$\rho_{12} = S_1 k_{12} \tag{17}$$

$$S_2 = \rho_{12}/k_{21} \tag{18}$$

$$\rho_{23} = S_2 k_{23} \tag{19}$$

$$S_3 = \rho_{23}/k_{32} \tag{20}$$

In the parallel case, access to compartment 2 only is assumed, so Eq. (10) is obtained by experiment. Rate constants, compartment size, and fluxes are calculated from Eqs. (21) to (29):

$$k_{12} = (-[C_{21}\lambda_1 + C_{22}\lambda_2 - X_2 0(\lambda_1 + \lambda_2)] - \{[C_{21}\lambda_1 + C_{22}\lambda_2 - X_2 0(\lambda_1 + \lambda_2)]^2 - 4X_2 0 E\lambda_1\lambda_2\}^{1/2})/2X_2 0 \tag{21}$$

$$k_{32} = (E\lambda_1\lambda_2)/(X_2 0 k_{12}) \tag{22}$$

$$k_{21} = \frac{k_{12}(\lambda_1 + \lambda_2 - k_{12} - k_{32}) - \lambda_1\lambda_2 + k_{12}k_{32}}{k_{12} - k_{32}} \tag{23}$$

$$k_{23} = (C_{21}\lambda_1 + C_{22}\lambda_2)/X_2 0 - k_{21} \tag{24}$$

$$S_2 = R_2 0/X_2 0 \tag{25}$$

$$\rho_{12} = S_2 k_{21} \tag{26}$$

$$\rho_{23} = S_2 k_{23} \tag{27}$$

$$S_1 = \rho_{12}/k_{12} \tag{28}$$

$$S_3 = \rho_{23}/k_{32} \tag{29}$$

A computer program using PIL/X (Pitt interpretive language) for a DEC system-10 computer is given in Appendix II and can be used to calculate these parameters. If the investigator does not know whether the system under consideration belongs to the parallel or to the series case, he can attempt to make the distinction by assuming first the one case and then the other. The results of compartmental analysis of calcium in isolated cells appears to fit either assumption.

Programs for the numerical solution by digital computer of differential-difference equations for a three compartment system have been published.[31] However, it is possible to calculate by graphical analysis of hand-drawn curves the best equation fitting the experimental data. In this case, one should always reconstruct a theoretical curve derived from the equation to make sure that it actually fits the experimental data. However, even a very good fit between the theoretical curve and the experimental data does not prove the validity of the model or of its underlying assumptions; it is only consistent with them.

A typical experiment can be analyzed as follows: Figure 15 shows the specific activity curve of calcium-45 uptake by isolated kidney cells in a low calcium medium. From the experimental points a first approximation of the specific activity curve is drawn by hand. The specific activ-

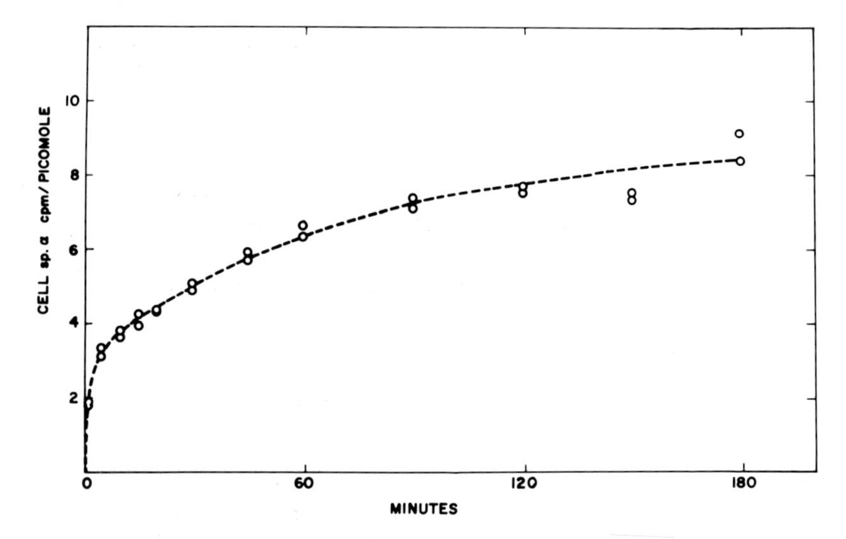

FIG. 15. Kidney cells specific activity plotted against time, during a calcium-45 uptake experiment. The exponential equation was obtained by graphical analysis as shown in Table VII and Fig. 16. The open circles are the experimental points. The dashed line is the curve reconstructed by computer from the coefficients and exponential constants of the differential equation $dSA/dt = 3.10e^{-0.718t} + 0.085e^{-0.0144t}$. The medium calcium concentration was 13 μM.

TABLE VII
GRAPHICAL ANALYSIS OF THE SPECIFIC ACTIVITY CURVE OF FIG. 15

1	2[a]	3	4	5[b]	6	7[c]	8
Time (min)	X_{1+3} (cpm/pmole)	ΔX_{1+3}	$\frac{\Delta X_{1+3}}{\text{min}}$	$\lambda_2 t$	$e^{-\lambda_2 t}$	$C_2 e^{-\lambda_2 t}$	$C_1 e^{-\lambda_1 t}$
1	1.60	1.60	1.60	0.0144	0.985	0.0837	1.52
2	2.40	0.80	0.80	0.0288	1.029	0.0874	0.71
3	2.85	0.45	0.45	0.0432	1.044	0.0887	0.36
4	3.10	0.25	0.25	0.0576	1.059	0.0900	0.16
5	3.25	0.15	0.15	0.0720	1.074	0.0913	0.056
10	3.75	0.50	0.10	0.1440	1.154	0.0981	0.002
15	4.10	0.35	0.07				
20	4.40	0.30	0.06				
30	5.00	0.60	0.06				
40	5.50	0.50	0.05				
50	5.95	0.45	0.045				
60	6.35	0.40	0.040				
70	6.65	0.30	0.030				
80	6.95	0.30	0.030				
90	7.20	0.25	0.025				
100	7.40	0.20	0.020				
110	7.60	0.20	0.020				
120	7.75	0.15	0.015				

[a] Cell specific activity.
[b] $\lambda_2 = 0.0144$.
[c] $C_2 = 0.085$.

ity of the cells at fixed time intervals is recorded from the curve as shown in the first two columns of Table VII. The difference between each value is calculated and then expressed per unit time, usually per minute (columns 3 and 4 of Table VII). The values of column 4 are then plotted on a multicycle semilogarithmic plot as shown in Fig. 16. The curve follows the general equation:

$$X_{1+3} = C_1 e^{-\lambda_1 t} + C_2 e^{-\lambda_2 t} \tag{30}$$

The asymptote of the slowest component of the curve with the lesser slope is drawn and extended to the origin. This can be done by finding the best fit visually or by the method of the least squares. However, since the earlier points should be given a much greater weight than the later points, the least-square method may be erroneously biased by the tail end of the asymptote.[36] The value of the asymptote at zero time is 0.085 which represents C_2. Its slope is obtained by dividing 0.693 by the half-

[36] W. E. Roseveare, *J. Amer. Chem. Soc.* **53,** 1651 (1931).

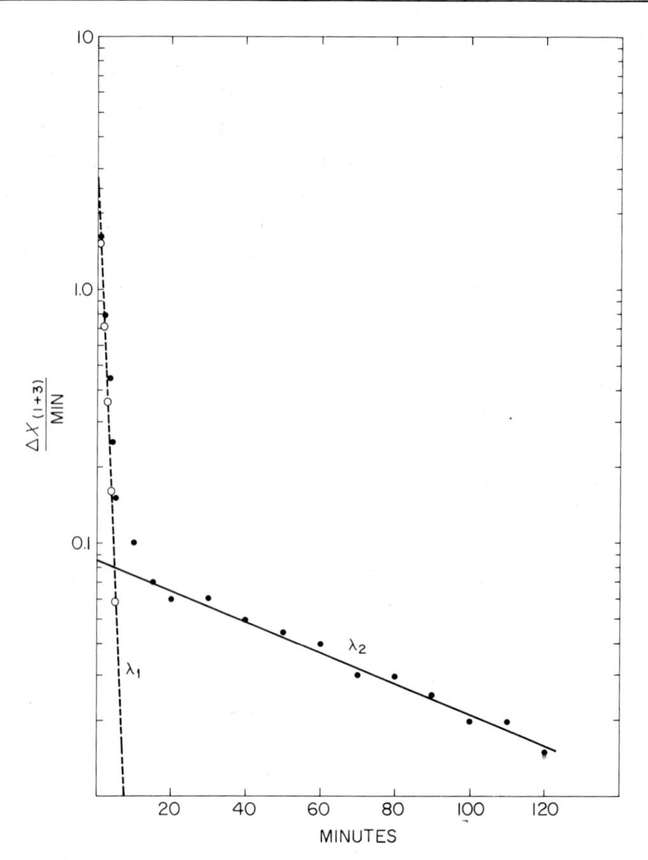

FIG. 16. Graphical analysis of the specific activity curve shown in Fig. 15. The points are taken from Table VII (column 4). The slopes λ_1 and λ_2 and the intercepts C_1 and C_2 are used to reconstruct the theoretical curve shown as a dotted line in Fig. 15. ○---○, fast phase, $C_1 = 3.10$, $\lambda_1 = 0.718$; ●——●, slow phase, $C_2 = 0.085$; $\lambda_2 = 0.0144$.

time of the decay line: the asymptote falls to a value of 0.085/2, i.e., 0.0425 at 48.1 minutes. The slope λ_2 is thus $0.693/48.1 = 0.0144$. Knowing C_2 and λ_2, the value $C_2e^{-\lambda_2 t}$ can be calculated for various time points t. The various values of $e^{-\lambda_2 t}$ are obtained from natural exponential function tables or by computer. This is shown in columns 5–7 of Table VII. Finally $C_2e^{-\lambda_2 t}$ is subtracted from the first fast component of Fig. 13 (column 4 minus column 7 of Table VII) and the values plotted on the same semilogarithmic plot (open circles, Fig. 16). The asymptote of the fast component is drawn and extended to zero time. In this particular case, the intercept, C_1, is found to be 3.10 and the slope, λ_1, 0.718.

Thus the curve can be represented by Eq. (31).

$$dX_{(1+3)}/dt = 3.10e^{-0.718t} + 0.085e^{-0.0144t} \quad (31)$$

To be sure that the derived curve actually fits the experimental data, it should be reconstructed by calculation or by computer and superimposed on the experimental data points. This is shown in Fig. 15, where the dashed line represents the theoretical curve derived above. If the curve does not fit the data, a new derivation or some correction of the coefficients and the exponential constants must be made until a good fit is obtained.

Three other parameters are needed—X_20, R_20, and E—to calculate the fluxes, the rate constants, and the compartment sizes according to Eqs. (12) to (29). The pertinent data concerning the experiment shown in Fig. 15 and to solve the equations are presented in Tables VIII and IX. The results for the series or parallel cases can be obtained by the digital computer program given in Appendix II or by simple stepwise solution of Eqs. (12) to (29). The results are presented in Table X. It is obvious that there is no significant difference between the compartment sizes and the fluxes between the parallel and the series case. Indeed, because the fast phase exchanges so rapidly that it becomes almost immediately part of the medium compartment, the parameters of the slow component can be calculated as in a two-compartment system or as in the parallel case of a three-compartment system. Huxley, in an appendix to Solomon's article,[32] points out that when one is faced with two exponen-

TABLE VIII
BASIC DATA FROM THE EXPERIMENT SHOWN IN FIG. 15[a]

Medium	
Calcium concentration (mM)	0.013
Medium calcium (pmoles/ml)	13,000
^{45}Ca radioactivity (cpm/ml)(R_20)	1,980,000
Specific activity at 0 time (cpm/pmole)(X_20)	152.3
Specific activity at ∞ time (cpm/pmole)(E)	145
Cells	
Cell calcium (pmoles/ml)	10,800
Cell protein (mg protein/ml)	1
Cell ^{45}Ca activity at ∞ time (cpm/mg protein)	88,500
Cell specific activity at ∞ time (cpm/pmole)	8.2
Relative specific activity (cell sp α/medium sp α) at ∞ time	0.0566
Exchangeable cell calcium (pmoles/mg protein)	610

[a] In a medium of physiological concentration of calcium, 1 mM, the difference between X_20 and E is negligible and can hardly be detected. E can be calculated by subtracting the cell activity at ∞ time from the medium activity at zero time R_20.

TABLE IX

DATA FOR THE DETERMINATION OF THE PARAMETERS OF A THREE-COMPARTMENT CLOSED SYSTEM ACCORDING TO EQS. (12) TO (20) (SERIES CASE) AND (21) TO (29) (PARALLEL CASE) FROM THE SPECIFIC ACTIVITY CURVE SHOWN IN FIG. 15 AND DESCRIBED IN EQ. (31)[a]

C_{11} (cpm/pmole)	3.10
C_{12} (cpm/pmole)	0.085
λ_1 (min^{-1})	0.718
λ_2 (min^{-1})	0.0144
X_20 (cpm/pmole)	152.3
E (cpm/pmole)	145.0
R_20 (cpm/ml)	1,980,000

[a] C_{11}, C_{12} are the coefficients of the fast and slow exponential terms and λ_1, λ_2 the exponential constants derived from Eq. (31) describing the cell specific activity curve. X_20 is the medium specific activity at 0 time, E the specific activity at ∞ time, and R_20, the amount of tracer in the medium at 0 time.

tial terms describing the movement of an isotope into two compartments such as:

$$R_1 + R_3 = Ae^{-\lambda_1 t} + Be^{-\lambda_2 t} \tag{32}$$

it is incorrect to assume that

$$R_1 = Ae^{-\lambda_1 t} \tag{33}$$

and

$$R_3 = Be^{-\lambda_2 t} \tag{34}$$

TABLE X

PARAMETERS OF THE THREE-COMPARTMENT CLOSED SYSTEM OBTAINED FROM THE DATA OF TABLE IX ACCORDING TO EQS. (12) TO (29)

Rate constant (min^{-1})	Compartment size (pmoles/mg protein[a])	Flux pmoles/(mg protein minute)[a]
Parallel case		
k_{21} 0.0142	S_1 262	J_{21} 184.61
k_{12} 0.7037	S_2 13,000	J_{23} 5.48
k_{23} 0.0004	S_3 392	
k_{32} 0.0139		
Series case		
k_{12} 0.0146	S_1 13,000	J_{12} 190.10
k_{21} 0.6838	S_2 277	J_{23} 5.41
k_{23} 0.0194	S_3 376	
k_{32} 0.0143		

[a] The units in the solution of the equations are actually picomoles per milliliter of cell suspension, but since there is 1 mg of cell protein per milliliter of cell suspension, the results can also be expressed on a cell protein basis.

since both compartments contain terms with each of the rate constants:

$$R_1 = a_1 e^{-\lambda_1 t} + b_1 e^{-\lambda_2 t} \tag{35}$$

and

$$R_3 = a_3 e^{-\lambda_1 t} + b_3 e^{-\lambda_2 t} \tag{36}$$

This assumption often leads to a gross overestimate of the slow compartment. However, this error is often negligible when the time constants of the two phases are widely different as they are in the above experiment. The error can be estimated by the equation given by Huxley

$$R_3 0 = \frac{AB(\lambda_1 - \lambda_2)^2}{A\lambda_1^2 + B\lambda_2^2} \tag{37}$$

where $R_3 0$ is the radioactivity of the slow compartment at time 0 and A, B, λ_1, and λ_2, the coefficients and exponential constants of the differential equation. According to Table IX

$$A = 36.47B \tag{38}$$

and

$$\lambda_1 = 49.86\lambda_2 \tag{39}$$

where $A = C_{11}$ and $B = C_{12}$. Substituting the values of Eqs. (38) and (39) in Eq. (37) shows that

$$R_3 0 = 0.975B \tag{40}$$

Therefore, accepting Eqs. (33) and (34) as an approximation of the system would overestimate the slow compartment by only 2.5%, which is an acceptable error for studies of this type. These considerations are not purely academic. The possibility to analyze the data as a parallel system offers great advantages and will allow a much simpler analysis of the data. There are indeed several difficulties in the type of kinetic analysis described above. For instance, the very small difference between the medium specific activity at zero time ($X_2 0$) and the specific activity at infinite time (E) has to be known with great accuracy, and the smallest imprecision will result in very great errors. In addition, errors of a few percent in the determination of the cell specific activity at infinite time may produce differences of up to 25% in the estimate of the fast component flux and compartment size, with little effect, however, on the parameters of the slow phase. Such errors are most likely to occur in experiments performed at low medium calcium concentrations because longer time of incubation may be required to reach steady-state values for the specific activity of the cells.

However Eq. (40) shows that if the exponential constants λ_1 and λ_2 differ by more than one order of magnitude, the error introduced in the analysis by assuming a parallel system will be very small. In such a parallel three-compartment closed system, several additional assumptions can be made:

1. Since the compartment size of the medium S_2 is at least 20 times larger than the two cellular compartments (in a buffer containing 1 mM calcium S_2 is more than 1000 times larger than $S_1 + S_3$) and since the fluxes in and out of the medium are so small as to be negligible, S_2 and X_2 (respectively, the pool size and the specific activity of the medium) can be considered constants. This assumption is supported by actual measurements of the medium specific activity.

2. From assumption 1, it follows that

$$X_2 0 = X_2 = E \tag{41}$$

3. The system is in a steady state, therefore

$$J_{12} = -J_{21} \tag{42}$$

and

$$J_{23} = -J_{32} \tag{43}$$

and all fluxes are constants.

4. The change in specific activity in each compartment can be described by the following exponential function:

$$1 - (X_i/E) = e^{-\lambda_i t} \tag{44}$$

5. After the tracer is added in compartment S_2, the increase in tracer activity of the cells which comprise compartment S_1 and S_3 is equal to its disappearance from compartment S_2. Therefore:

$$-dR_2/dt = dR_1/dt + dR_2/dt \tag{45}$$

or

$$-dR_2/dt = dR_{1+3}/dt \tag{46}$$

We can state

$$dR_{1+3}/dt = (J_{21} + J_{23})X_2 - J_{12}X_1 - J_{32}X_3 \tag{47}$$

Since, from Eq. (44)

$$X_i = E(1 - e^{-\lambda_i t}) \tag{48}$$

we can write

$$dR_{1+3}/dt = (J_{21} + J_{23})X_2 - J_{12}[E(1 - e^{-\lambda_1 t})] - J_{32}[E(1 - e^{-\lambda_3 t})] \tag{49}$$

Since $X_2 = E$ = constant and since all fluxes J's are constants

$$J_{12}E + J_{32}E - (J_{21} + J_{23})X_2 = \text{constant} \tag{50}$$

or

$$E(J_{12} + J_{32} - J_{21} - J_{23}) = C \tag{51}$$

At steady state $C = 0$, therefore

$$d(R_{1+3}/E)/dt = -J_{12}e^{-\lambda_1 t} - J_{32}e^{-\lambda_3 t} \tag{52}$$

A curve relating the cell calcium-45 activity/medium specific activity (R_{1+3}/E) with time can be drawn. Figure 17 presents the same experiment as that shown in Fig. 15 but plotted as R_{1+3}/E. From the hand-drawn curve, $d(R_{1+3}/E)/dt$ is determined, and the coefficients and exponential functions are calculated as shown in Table XI and Fig. 18. The equation obtained is

$$d(R_{1+3}/E)/dt = 190e^{-0.683t} + 5.41e^{-0.0144t} \tag{53}$$

Since λ_1 is approximately 50 times greater than λ_2, we can solve the equation as though the system was in parallel and we can accept the approximation shown in Eqs. (33) and (34) without introducing an error greater than a few percents.

The calculation of the compartment size is simplified by three previous assumptions: (1) the three compartments can be analyzed as though they were in parallel, (2) the medium compartment is very large compared to the exchangeable cellular pools, especially at a normal calcium concentration of 1 mM, and (3) the rate of calcium exchange between

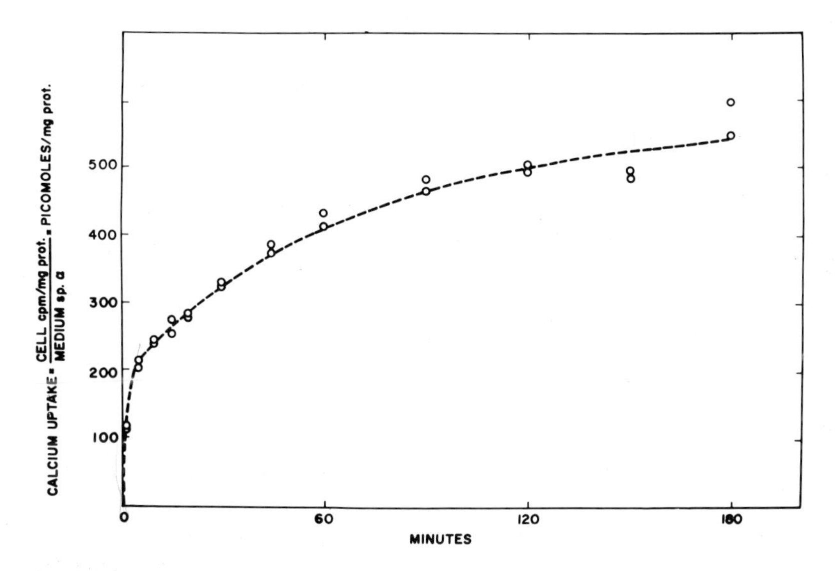

FIG. 17. Calcium-45 uptake in isolated kidney cells. The exponential equation was obtained by graphical analysis as shown in Table XI and Fig. 17. The open circles are the experimental points. The dotted line is the curve reconstructed from the coefficients and exponential constants of the differential equation $(dR/E)/dt = 190e^{-0.683t} + 5.41e^{-0.0144t}$. The medium calcium concentration was 0.013 mM.

TABLE XI
GRAPHICAL ANALYSIS OF THE CALCIUM UPTAKE CURVE OF FIG. 17

1	2	3	4	5[a]	6	7[b]	8
Time (min)	R_{1+3} (pmoles/mg prot)	ΔR_{1+3}	$\frac{\Delta R_{1+3}}{\text{min}}$	$\lambda_2 t$	$e^{-\lambda_2 t}$	$C_2 e^{-\lambda_2 t}$	$C_i e^{-\lambda_i t}$
1	101	101	101	0.0144	0.985	5.33	95.7
2	154	53	53	0.0288	1.029	5.57	47.4
3	184	30	30	0.0432	1.044	5.65	24.4
4	202	18	18	0.0576	1.059	5.73	12.3
5	213	11	11	0.0720	1.074	5.81	5.19
10	243	30	6	0.1440	1.154	6.24	0
15	266	23	4.6				
20	287	21	4.2				
30	324	37	3.7				
40	357	33	3.3				
50	385	28	2.8				
60	409	24	2.4				
70	430	21	2.1				
80	448	18	1.8				
90	464	16	1.6				
100	478	14	1.4				
110	490	12	1.2				
120	500	10	1.0				

[a] $\lambda_2 = 0.0144$.
[b] $C_2 = 5.41$.

the medium and the cellular compartments is very small when compared to the medium compartment. In these conditions, the exponential constant λ of the equation approaches the value of the rate constant k of the transport process as in a single compartment. The justification for this assumption is presented in Appendix III.

Consequently, from an equation such as

$$d(R_{1+3}/E)/dt = -Ae^{-\lambda_\alpha t} - Be^{-\lambda_\beta t} \tag{54}$$

we can then assume that $A = J_{12}$, $B = J_{32}$, $\lambda_\alpha = k_{12}$, and $\lambda_\beta = k_{32}$ and solve for the fluxes, rate constant, and compartment size of the two phases, knowing that $J_{ij} = k_{ij}S_i$. The results are presented in Table XII. The results are practically identical to those obtained from the specific activity curve and calculated according to Robertson *et al.*[34] (Table XIII)

The main advantage of this method is that it requires only the determination of the cell calcium-45 radioactivity and of the cell protein. The change in the medium specific activity is so small that it can be neglected,

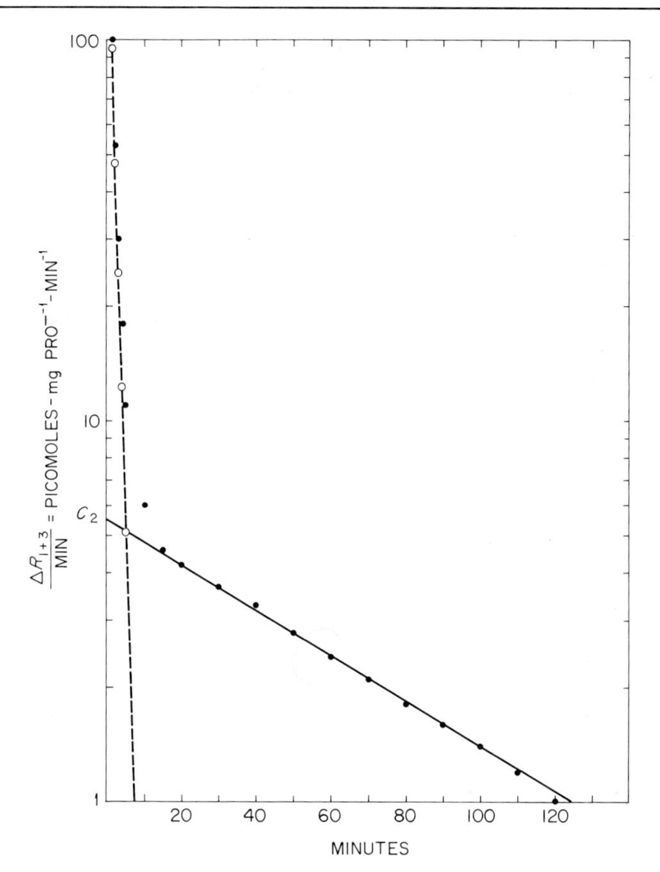

FIG. 18. Graphical analysis of the calcium-45 uptake curve shown in Fig. 17. The points are taken from Table XI (column 4). The slopes λ_1 and λ_2 and the intercepts C_1 and C_2 are used to reconstruct the theoretical curve shown as a dotted line in Fig. 17. ○---○, fast phase, $C_1 = 190$, $\lambda_1 = 0.683$; ●——●, slow phase, $C_2 = 5.41$, $\lambda_2 = 0.0144$.

and this important parameter can be considered a constant. However, this method of calculation has its limitations, which should be clearly underlined. The system can be treated as a parallel one, only because the fast phase exchanges so rapidly that it becomes almost immediately part of the medium compartment. It should be emphasized that this third method will not satisfactorily describe a series model in general, but only the special case in which the rate constant of the fast component is at least one order of magnitude larger than the rate constant of the slow component. All three methods may produce errors in the estimate of the fast phase. However, the accuracy of the slow phase values is very good

TABLE XII
PARAMETERS OF THE THREE COMPARTMENT CLOSED SYSTEM CALCULATED FROM EQS. (41) TO (53)

Rate constant (min^{-1})	Compartment size (pmoles/mg protein[a])	Fluxes pmoles/(mg protein min)
k_{21} 0.0146	S_1 279	J_{21} 190
k_{12} 0.6830	S_2 13,000	J_{23} 5.41
k_{23} 0.0004	S_3 377	
k_{32} 0.0144		

[a] Picomoles per milliliter; see Table X, footnote *a*.

and independent of the inherent errors in the fast component. If one is mostly interested in the slow compartment and if the conditions outlined above are met, the third method provides a simple means to measure calcium fluxes in isolated cells.

Levinson and Blumenson have used different mathematical models to evaluate statistically the model's performance when fitted to the experimental data.[22] They also concluded that calcium uptake in Ehrlich mouse ascites tumor cells can be fitted best to a model predicting two cellular exchangeable compartments.

TABLE XIII
RESULTS OBTAINED BY THE THREE DIFFERENT METHODS OF CALCULATION: (1) SERIES CASE (2) PARALLEL CASE OF SPECIFIC ACTIVITY CURVE (FIG. 15 AND TABLE X), AND (3) PARALLEL CASE OF CALCIUM UPTAKE CURVE (FIG. 17 AND TABLE XII)[a]

	Specific activity curve		Calcium uptake curve
	Series case	Parallel case	parallel case
Compartment size (pmoles/ml)			
Medium	13,000	13,000	13,000
Fast compartment	277	262	279
Slow compartment	376	392	377
Fluxes [pmoles/(mg protein minute)]			
Fast component	190	184	190
Slow component	5.41	5.48	5.41

[a] Values can be expressed either as picomoles per milliliter of medium or picomoles per milligram of cell protein since there is 1 mg of cell protein per milliliter of medium.

Measurement of Calcium Efflux by Kinetic Analyses

Another method for study of the distribution and the fluxes of calcium in tissues or cellular systems uses isotopic desaturation of calcium-45 from prelabeled cells. The complexity of the method depends on the number of exchangeable calcium pools in the system under study. With whole tissues, calcium desaturation curves include vascular and interstitial spaces which add several kinetic components to those of the cells themselves. In addition, the inhomogeneity of the cell population will increase the number of compartments if different cells have different calcium transport rates and different compartment sizes. Although it is technically easier to use whole tissues, the data obtained are more difficult to interpret. On the other hand, cell suspensions give results that are easier to interpret because the system is homogeneous and it contains fewer kinetic compartments. Technically, however, cell suspensions are more troublesome.

Methods

The determination of calcium efflux by calcium-45 desaturation from prelabeled tissues has been used extensively, and the methods described are essentially a variation of the same basic technique. It has been used in crayfish nerve, squid giant axon, and vagus nerve,[37-41] in myocardial, skeletal, and smooth muscle,[42-50] as well as in liver slices.[51] The method consists of prelabeling the tissue with calcium-45 for a fixed period of time. The tissue is then rinsed rapidly in an unlabeled buffer to clear the extracellular space and superficial binding sites of the isotope. Desaturation of the tracer from the tissue can be performed in several ways: (1) by successive soaking of the labeled tissue in a series of test

[37] S. Soloway, J. H. Welsch, and A. K. Solomon, *J. Cell Comp. Physiol.* **42,** 471 (1953).
[38] A. L. Hodgkin and R. D. Keynes, *J. Physiol.* (*London*) **138,** 253 (1957).
[39] M. Luxoro and S. Rissetti, *Biochim. Biophys. Acat* **135,** 368 (1967).
[40] E. Rojas and C. Hidalgo, *Biochim. Biophys. Acta* **163,** 550 (1968).
[41] P. Kalix, *Pfluegers Arch. Gesamte Physiol. Menschen Tiere* **326,** 1 (1971).
[42] S. Winegrad and A. M. Shanes, *J. Gen. Physiol.* **45,** 371 (1962).
[43] G. A. Langer and A. J. Brady, *J. Gen. Physiol.* **46,** 703 (1963).
[44] G. A. Langer, *Circ. Res.* **15,** 393 (1964).
[45] H. Reuter and N. Seitz, *J. Physiol.* (*London*) **195,** 451 (1968).
[46] C. P. Bianchi, *J. Gen. Physiol.* **44,** 845 (1961).
[47] A. Isaacson and A. Sandow, *J. Gen. Physiol.* **50,** 2109 (1967).
[48] A. Isaacson and A. Sandow, *J. Pharmacol. Exp. Ther.* **155,** 376 (1967).
[49] C. Van Breemen and D. Van Breeman, *Biochim. Biophys. Acta* **163,** 114 (1968).
[50] J. Nagasawa and T. Suzuki Tohoku, *J. Exp. Med.* **102,** 1 (1970).
[51] S. Wallach, D. L. Reizenstein, and J. V. Bellavia, *J. Gen. Physiol.* **49,** 743 (1966).

tubes containing nonisotopic buffer solutions, for periods of several hours. The successive washout periods may vary from a few minutes to half an hour; (2) by leaving the tissue attached in the same vessel and replacing the buffer at fixed intervals, or (3) by perfusing the labeled tissue attached in a perfusion chamber and continuously monitoring the radioactivity of the outflowing perfusate. The perfusion method has also been used by Langer to monitor the radioactivity of the tissue itself by placing a Geiger tube in its close proximity.[43] However the measurement of tissue calcium-45 activity by this method may not be reliable. Cosmos and Harris[52], pointed out that calcium-45 emits weak beta particles so that the emission reaching the counter tube derives only from the surface of the tissue. Besides making the readings a measure of surface calcium rather than of total calcium, the values obtained are subject to errors depending on the variable thickness of the tissue and of the water film existing between the counter and the tissue. The tracer activity of the effluent collected during the washout is plotted semilogarithmically against time, and the curves are analyzed graphically according to the method outlined above.

A more elegant method which combines the advantages of a tissue attached to a substrate and that of a homogeneous population of cells without interstitial and vascular spaces has been published by Langer *et al.*[35] and has been described in the preceding section. It consists of culturing cells on a glass scintillator slide which can be introduced in a specially designed flow cell fitting the well of a scintillation spectrometer (Fig. 13). The cells are labeled and desaturated in the flow cell and the cells' radioactivity is continuously monitored. This technique requires the determination of the kinetic characteristics of the slide itself without the cells attached, and the cellular component is obtained by subtracting the slide contribution to the total system (Fig. 19). Whether the blank is an accurate reflection of the slide contribution when most of its surface area is covered by the cell monolayer is unclear and should be carefully evaluated.

Finally cell suspensions have been successfully used for calcium-45 desaturation experiments.[25,29] A suspension consisting of 100–200 mg wet weight of cells (equivalent to 10–20 mg of cell protein or to 0.3–0.6 ml of packed cells for mammalian kidney cells) in 6 ml of buffer are incubated at 37° in 75 $\times$ 120 mm polycarbonate or polyethylene tubes. The cell suspension is kept dispersed with a minute Teflon-coated magnet bar 10 mm long and 3 mm in diameter (No. 8546, Cole Parmer, Chicago, Illinois) and an immersible magnetic stirrer (MS-7 Tri-R Instruments,

[52] E. Cosmos and E. J. Harris, *J. Gen. Physiol.* **44,** 1121 (1961).

Jamaica, New York 11435) (Fig. 20). After a period of preincubation to allow the cells to achieve steady state, 50–100 μCi of calcium-45 are added to the suspension, and the cells are labeled for a predetermined period of time (10 minutes, 1, 2, or 3 hours). At the end of the labeling period, usually 1 hour, the cells are centrifuged at 2000 g for 45 seconds, and the radioactive medium is decanted. The cells are resuspended and washed in unlabeled medium to clear the trapped extracellular fluid and superficial binding sites from the isotope and centrifuged once again. The cells are then dispersed in 6 ml of unlabeled buffer, divided in two separate fractions of 3 ml each into clean polycarbonate tubes. Both fractions are desaturated simultaneously. At 1, 5, 10, and 20 minutes and every 10 minutes thereafter, the medium is separated from the cells by centrifugation at 2000 g. The Teflon-coated stirrer is removed from the suspension so as not to disturb the cell pellet during the centrifugation and the decanting of the medium. The 3 ml of medium are decanted into

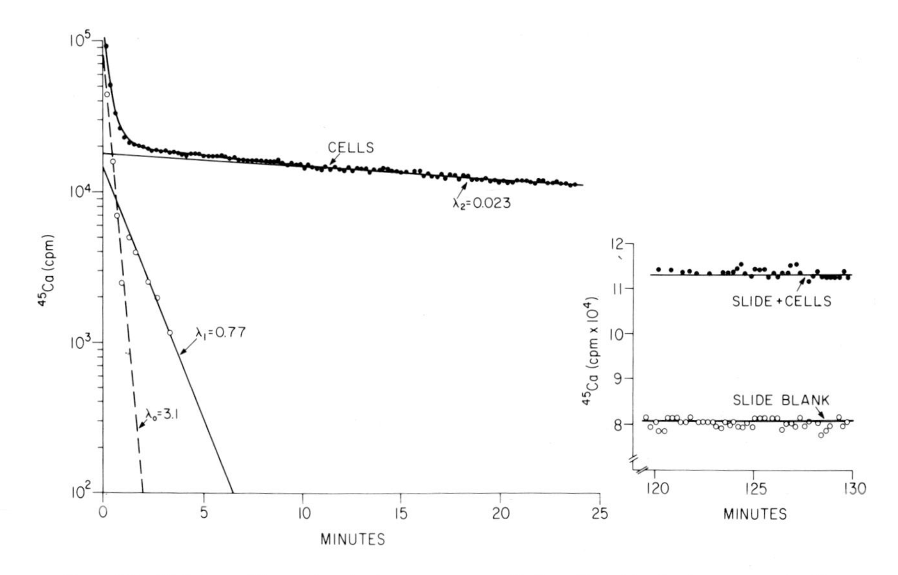

FIG. 19. Calcium-45 washout from myocardial cells attached to a scintillator slide. The last 10 minutes of a 130-minute labeling period is indicated in the inset at right. Subtraction of the activity of the blank from the slide indicates the counts per minute attributable to the cell layer. The contribution of the blank has been subtracted from the washout curve (on left) so that the curve described by the solid circles represents the cells. It is graphically resolved into two exponential components (solid lines) with the slopes λ_1 and λ_2. The broken line with a slope λ_0 represents the washout of the calcium-45 perfusate from the flow cell chamber. From G. A. Langer and J. S. Frank, *J. Cell Biol.* **59**, 441, 1972. Reproduced with the permission of *J. Cell Biol.*

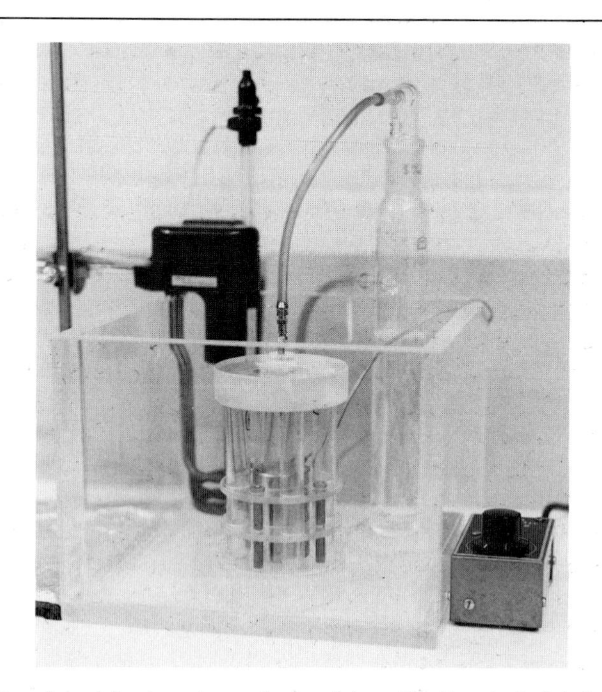

FIG. 20. Experimental set-up to perform calcium-45 efflux in isolated cell suspensions. The Lucite tube holder holds 4 tubes, 2 containing the cell suspensions and 2 containing the subsequent washout medium, which is thus preequilibrated with regard to the temperature bath and the CO_2 concentration. The cells are dispersed by minute Teflon-coated magnets driven by a submersible magnetic stirrer. A humidified and warmed gas phase is perfused through the Lucite top resting on the test tubes.

scintillation vials containing 10 ml of scintillation solution.[53] Three milliliters of unlabeled fresh medium are added to the cell pellet, and the cells are replaced in the incubating bath and dispersed with the stirrer. The centrifugation and the change of medium should not last more than 1 minute. At the end of the washout, the cell pellet is homogenized in dis-

[53] Several scintillation solution formulas will accept 20–30% aqueous solutions. "Cocktail D" recommended by Beckman is a cocktail for ambient temperature liquid scintillation spectrometry. It will accept up to 25% of aqueous solutions. Its composition is as follows: PPO 5 g/liter, naphthalene 100 g/liter, dioxane to 1 liter. Aquasol (New England Nuclear, Boston, Massachusetts) is a xylene-based scintillation solution with exceptional miscibility with water. A 10-ml amount of Aquasol with 3 ml of pure water will form a stiff cloudy gel; with 3 ml of physiological buffer ($\mu = 0.16$) it forms a cloudy solution. A stiff gel can be obtained by acidifying the buffer solution as follows: 0.2 ml of 6 N HCl, 3 ml of physiological buffer and 10 ml of Aquasol.

tilled water with an ultrasonic probe, and aliquots are used to determine the cell protein, the total calcium, and the calcium-45 left in the cells.

Calculations

The sum of all the radioactivity in each wash plus the radioactivity left in the cells at the end of the experiment is taken as the total radioactivity present in the cells at the beginning of the desaturation period. The radioactivity remaining in the cells at each time point is obtained by sequentially subtracting the radioactivity of each wash. The results can be expressed as counts per minute per total cell mass or can be converted to percent of initial radioactivity. Table XIV shows the results of a typical desaturation experiment. A computer program for these calculations is given in Appendix IV. The calcium-45 in the cells, expressed as percent of the initial radioactivity is drawn as a semilogarithmic plot as shown in Fig. 21. Three different phases can be obtained by graphical analysis and the slope and intercept of each component can be readily calculated, as in previous section. The plot of the slowest phase should be consistently linear for several hours to provide clear evidence that the last component is a single exponential.

Several assumptions can be made in the analysis of the data. First,

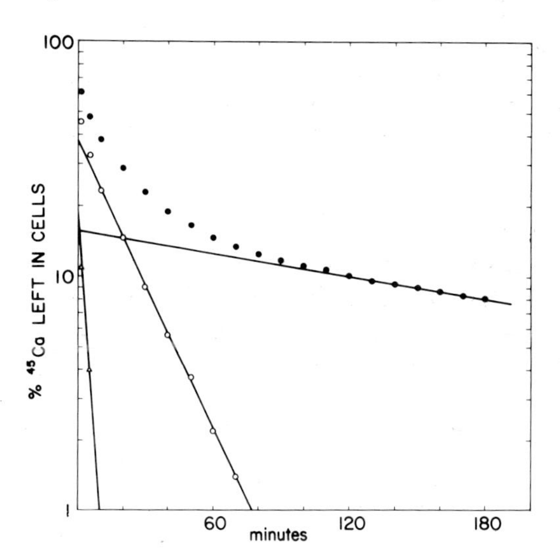

FIG. 21. Calcium-45 efflux from isolated kidney cells. The data are plotted as the percent of the initial (time 0) radioactivity left in the cells, calculated according to Table XIV (column 5). The various kinetic phases (solid lines) are obtained by graphical analysis. From A. B. Borle, *J. Membrane Biol.* **10**, 45 (1972). Reproduced with the permission of *J. Membrane Biol.*

TABLE XIV

CALCULATION OF A ^{45}Ca DESATURATION EXPERIMENT IN KIDNEY CELLS[a]

3 ml wash sample No.	Time (min.)	cpm	^{45}Ca left in cells	Percent ^{45}Ca left in cells
1	1.0	174,417	220,969	55.89
2	5.0	53,393	167,576	42.38
3	10.0	39,477	128,099	32.40
4	20.0	34,955	93,144	23.56
5	30.0	17,915	75,229	19.03
6	40.0	9725	65,504	16.57
7	50.0	5803	59,701	15.10
8	60.0	3860	55,841	14.12
9	70.0	2905	52,936	13.39
10	80.0	2193	50,743	12.83
11	90.0	1548	49,195	12.44
12	100.0	1353	47,842	12.10
13	110.0	1074	46,768	11.83
14	120.0	795	45,973	11.63
15	130.0	758	45,215	11.44
16	140.0	626	44,589	11.28
17	150.0	582	44,007	11.13
18	160.0	545	43,462	10.99
19	170.0	488	42,974	10.87
20	180.0	453	42,521	10.75
21	100.0	395	42,126	10.65
22	200.0	405	41,721	10.55
23	210.0	304	41,417	10.48
24	220.0	329	41,088	10.39
Cells at end experiment	—	41,088		
Σ = total cpm at time 0		395,386		

[a] The radioactivity of each of the 24 washout samples is given in the third column. The total radioactivity remaining in the cell of the end of the experiment is added to this column to provide the total radioactivity (Σ) present in the cells at the beginning of the desaturation (zero time). The radioactivity left in the cells at the end of each washout period and shown in the fourth column is obtained by subtracting the counts of the first sample from the total radioactivity; the ^{45}Ca left in the cells at 5 minutes is equal to the ^{45}Ca in the cells at 1 minute minus the radioattivity of the second washout sample, and so on. The same results can be expressed in percent of the initial activity by setting $\Sigma = 100\%$ as shown in the fifth column: 395,386 = 100%, 220,969 = 55.89%, 167,576 = 42.38%.

in all cell types studied until now, the time constants of the three phases differ by at least one order of magnitude. Consequently the system can be analyzed as though it consisted of three parallel compartments without introducing an error greater than a few percent[29] (Appendix V). However, since the technique of isotope desaturation described above may not be fast enough to measure with an acceptable precision the first or fastest component of calcium efflux with a time constant of a few minutes, it is advisable to leave this fast component aside and to consider only the second and the third phase of desaturation. Second, if the experiment is carefully designed, the system can be considered to be in steady state. Consequently the respective fluxes, compartment sizes, and rate constants of transfer are constant and equal during the uptake of the isotope and during the desaturation period. During the labeling period, the change in specific activity X_i in each compartment i can be described by the following exponential function:

$$X_i/E = 1 - e^{-\lambda_i t} \tag{55}$$

Where E is the specific activity of compartment i at infinite time or equilibrium specific activity. Since the fluxes in and out of the medium are so small as to be negligible (see Appendix III), the specific activity of the medium is constant during the labeling phase and can be accepted to be equal to E. By definition:

$$X_i = R_i/S_i \tag{56}$$

where R_i is the radioactivity and S_i the amount of calcium in compartment i. Therefore, from Eqs. (55) and (56)

$$R_i/S_i = E(1 - e^{-\lambda_i t}) \tag{57}$$

or

$$S_i = R_i/[E(1 - e^{-\lambda_i t})] \tag{58}$$

At the end of 1 hour of calcium-45 uptake, $t = 60$ minutes. $R_{i(60\,\mathrm{min})}$ and λ_i are obtained from the desaturation curve: $R_{i(60\,\mathrm{min})}$ is the radioactivity of compartment i at the beginning of the washout period calculated from the intercept of each phase with time 0 of the desaturation. λ_i is obtained from the slope of the component i of the desaturation curve. Since during the desaturation period the system is an open one, the slope of each kinetic phase obviously represents the rate constant of efflux k_{ij} from compartment i to the medium j. But in addition, since the medium compartment size is about 3 orders of magnitude larger than the intracellular pools, the slope of each desaturation phase is also identical to λ_i in Eq. (58), which is the exponential constant of calcium influx into each compartment during the cell labeling in a closed system. The justification for this simplification is presented in Appendix V. Consequently the compartment size S_i of each kinetic component can be calculated from Eq. (58) knowing E, the medium specific activity during labeling, $R_{i(60\,\mathrm{min})}$ and λ_i.

Since λ_i is also the rate constant of calcium efflux k_{ij}, the flux in and out of this compartment can be calculated, knowing that

$$J_{ij} = k_{ij}S_i \tag{59}$$

Since most of these calculations depend on the validity of the assumption that the system is in a steady state, it is essential to ascertain that the cells are indeed in steady state before the introduction of the isotope. When a hormone is to be used, it will be necessary to preincubate the cells for several hours before the beginning of an experiment. Any hormone or inhibitor can affect one compartment without affecting the other. Figure 22, for instance, shows the effect of 1 m*M* cyclic AMP on the efflux of calcium-45 from isolated kidney cells. The slow compartment 3 is markedly affected whereas the faster compartment 2 is practically identical to the control component. Experimental conditions can markedly modify the effects of hormones on both compartments, particu-

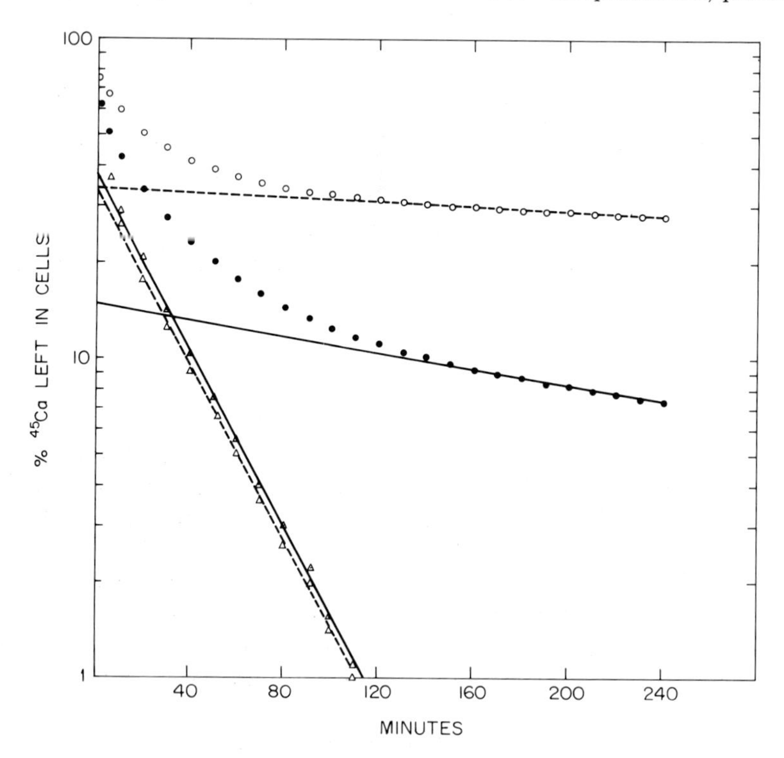

FIG. 22. Effect of 1 m*M* cyclic AMP on the efflux of calcium-45 from prelabeled isolated kidney cells. The slowest kinetic component is markedly affected by cyclic AMP (○---○) whereas the faster second phase is unaltered (△---△) when compared with control cells (●——●, ▲——▲).

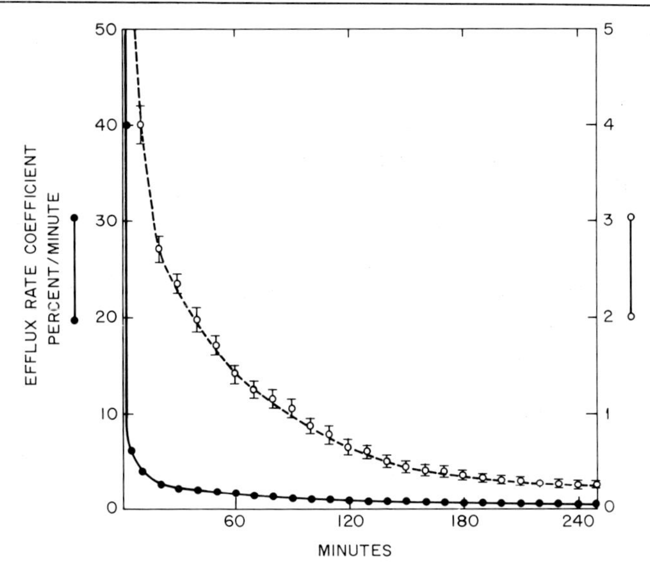

FIG. 23. Efflux rate coefficient (ERC) of calcium-45 from isolated kidney cells as a function of time. The results are the means ± standard deviations of 19 experiments. The same results are graphed on two different scales: the left ordinate relates to the filled circles, the right ordinate to the open circles.

larly the phosphate concentration of the buffer, its ionic composition, and the nature of the buffer system employed (bicarbonate, phosphate, or Tris).[54]

Perturbations of Steady State Fluxes

All the methods described in the previous sections deal with steady-state conditions. Since the intracellular compartments and their rates of exchange are all interrelated, it is often difficult to differentiate the primary effect of hormones from the secondary consequences of their effects. Moreover, these techniques do not provide any information about the rapidly of onset of hormonal action. However, it is possible to obtain this kind of information by perturbing the steady-state fluxes of unstimulated cells. In this type of experiment, it is not possible to learn the magnitude of the fluxes and of the various calcium compartments or the magnitude of the hormonal effect. On the other hand, perturbation of

[54] A. B. Borle, *in* "Calcium, Parathyroid Hormone and the Calcitonins" (R. V. Talmage and P. L. Munson, eds.), p. 484. Excerpta Med. Found., Amsterdam, 1972.

the steady state will show how fast hormones are acting and whether they stimulate or inhibit calcium fluxes.

Methods

Perturbation of steady-state calcium desaturation can be performed in whole tissue or in suspended cells. The experimental procedures are identical to those described in the previous section: the cells are labeled

TABLE XV
COMPUTER OUTPUT GIVING THE RAW DATA OF THE STEADY STATE PERTURBATION EXPERIMENT SHOWN IN FIG. 24[a]

k = "experiment code number"
sigma = 705023.02
Control

Sample	Time	cpm	Ca left in cells	Percent left	Rate coefficient
1	1.0	203,485	501,538	71.14	33.73
2	5.0	76,615	424,923	60.27	4.13
3	10.0	57,830	367,093	52.07	2.92
4	20.0	66,788	300,305	42.60	2.00
5	30.0	45,343	254,962	36.16	1.63
6	40.0	30,323	224,639	31.86	1.26
7	50.0	23,981	200,658	28.46	1.13
8	60.0	19,219	181,439	25.74	1.01
9	70.0	13,844	167,595	23.77	0.79
10	80.0	11,222	156,373	22.18	0.69
11	90.0	9011	147,362	20.90	0.59
12	100.0	7127	140,235	19.89	0.50
13	110.0	5910	134,325	19.05	0.43
14	120.0	5055	129,270	18.34	0.38
15	130.0	4482	124,788	17.70	0.35
16	140.0	3685	121,103	17.18	0.30
17	150.0	3227	117,876	16.72	0.27
18	160.0	2621	115,255	16.35	0.22
19	170.0	2381	112,874	16.01	0.21
20	180.0	2169	110,705	15.70	0.19
21	190.0	1855	108,850	15.44	0.17
22	200.0	1762	107,088	15.19	0.16
23	210.0	1567	105,521	14.97	0.15
24	220.0	1382	104,139	14.77	0.13
25	230.0	1339	102,800	14.58	0.13
26	240.0	1168	101,632	14.42	0.11

TABLE XV (*Continued*)

Experimental
sigmae = 719552.04

Sample	Time	cpm	Ca left in cells	Percent left	Rate coefficient	Percent of control
1	1.00	208,810	510,742	70.98	33.94	100.64
2	5.00	78,649	432,093	60.05	4.17	100.87
3	10.00	58,439	373,654	51.93	2.90	99.33
4	20.00	67,365	306,289	42.57	1.98	99.00
5	30.00	43,634	262,655	36.50	1.53	93.92
6	40.00	32,590	230,065	31.97	1.32	104.61
7	50.00	22,839	207,226	28.80	1.04	92.63
8	60.00	27,318	179,908	25.00	1.41	140.29
9	70.00	25,668	154,240	21.44	1.54	193.67
10	80.00	20,561	133,679	18.58	1.43	206.16
11	90.00	16,925	116,754	16.23	1.35	227.80
12	100.00	13,753	103,001	14.31	1.25	252.54
13	110.00	10,709	92,292	12.83	1.10	254.75
14	120.00	6723	85,569	11.89	0.76	197.11
15	130.00	4192	81,377	11.31	0.50	142.33
16	140.00	3183	78,194	10.87	0.40	133.10
17	150.00	2636	75,558	10.50	0.34	126.97
18	160.00	2083	73,475	10.21	0.28	124.32
19	170.00	1878	71,597	9.95	0.26	124.03
20	180.00	1694	69,903	9.71	0.24	123.40
21	190.00	1481	68,422	9.51	0.21	126.72
22	200.00	1218	67,204	9.34	0.18	110.06
23	210.00	1111	66,093	9.19	0.17	113.09
24	220.00	1034	65,059	9.04	0.16	119.61
25	230.00	934	64,215	8.91	0.14	111.74
26	240.00	845	63,280	8.79	0.13	116.09

[a] The computer program used to calculate the different parameters is given in Appendix IV.

with calcium-45 for a fixed period of time and then desaturated by continuous or sequential washout. At a predetermined time, the composition of the washout medium is altered by changing its ionic composition, or by adding an inhibitor, a hormone, or both. If calcium desaturation is performed simultaneously in two separate cell suspensions with cells uniformly labeled, the perturbation can be done in one group only and the results can be compared to the unstimulated cells. This can easily be done by labeling the cells together in one single tube and then dividing the suspension into two groups to be desaturated simultaneously.

Calculation of the Data

A semilogarithmic plot of the calcium-45 remaining in the tissue or in the cells is often inadequate to demonstrate a hormonal effect. One simple method to evaluate such an effect is to calculate the *efflux rate coefficient* or ERC:

$$\text{ERC} = \text{cpm}_w/[(\text{cpm}_c)_{\text{mean}}\Delta t] \times 100\% \tag{60}$$

Where cpm_w is the radioactivity lost in the washout from the cells during the time interval Δt and $(\text{cpm}_c)_{\text{mean}}$ is the mean radioactivity left in the cells between time t and $t + \Delta t$:

$$(\text{cpm}_c)_{\text{mean}} = (\text{cpm}_{c(t)} + \text{cpm}_{c(t+\Delta t)})/2 \tag{61}$$

The computer program shown in Appendix IV will calculate the ERC of both control and experimental cells and the percent stimulation or inhibition of the experimental cells compared with their controls.

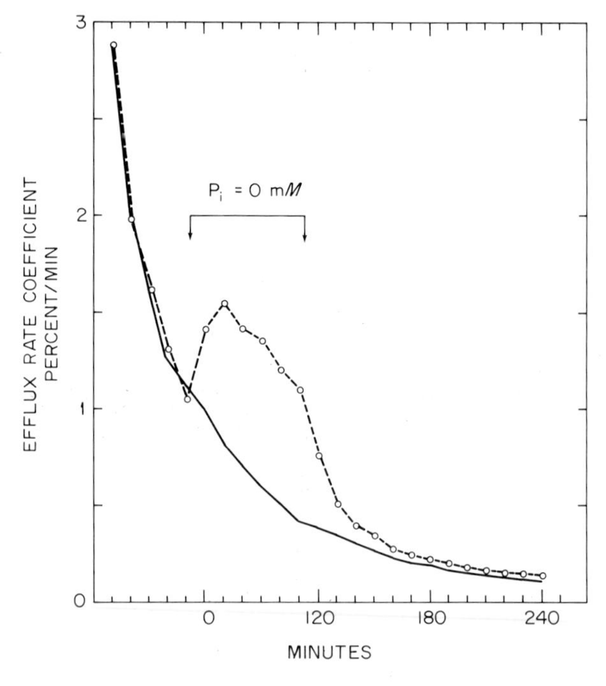

FIG. 24. Effect of a phosphate-free medium on the efflux rate coefficient of calcium-45 from isolated kidney cells. The control medium contains 1 m*M* phosphate. Both control and experimental media contain 1.3 m*M* calcium. ——, control ERC; ○---○, ERC of the experimental group. The results are plotted from Table XV (sixth column).

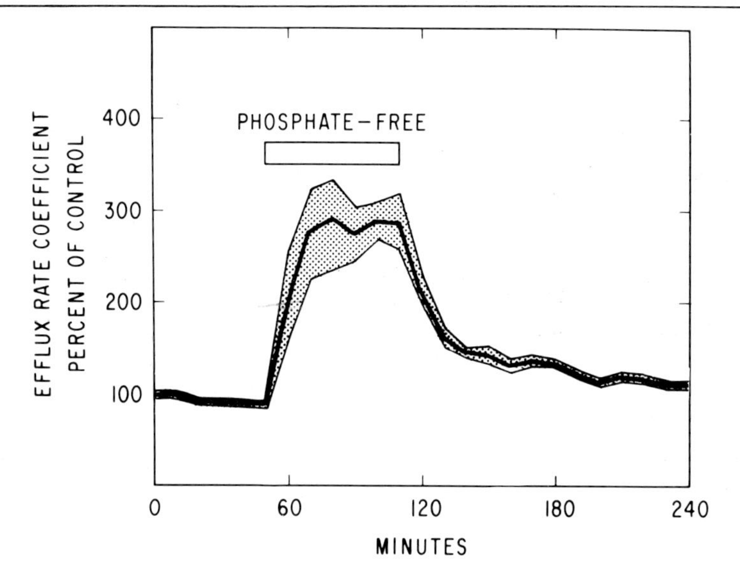

FIG. 25. Effect of a phosphate-free medium on the efflux rate coefficient, expressed as percent of control (column 7 of Table XV). Control: phosphate = 1 m*M*. The results are the mean ± standard error of 5 experiments. From A. B. Borle, *J. Membrane Biol.* **10**, 45 (1972). Reproduced with the pemission of *J. Membrane Biol.*

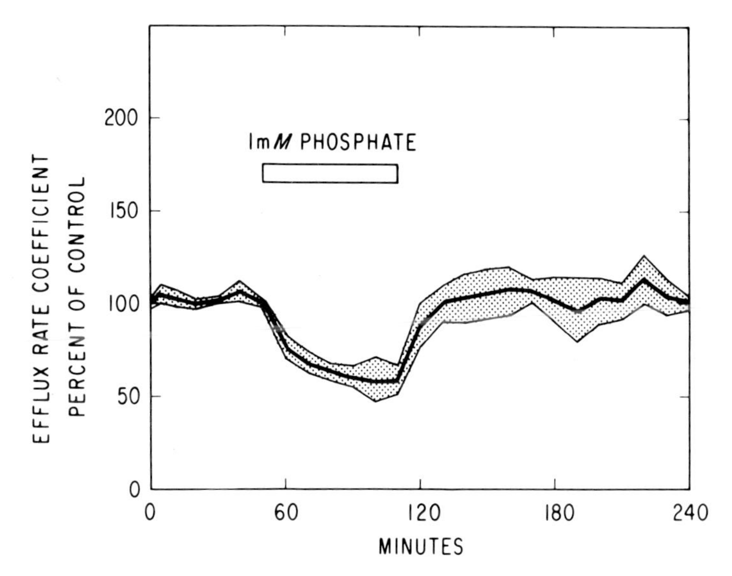

FIG. 26. Effect of a medium containing 1 m*M* phosphate on the efflux rate coefficient of calcium-45 from kidney cells (expressed as percent of control). The control medium contained 1.3 m*M* calcium and no phosphate. The results are the mean ± standard error of 4 experiments. From A. B. Borle, *J. Membrano Biol.* **10**, 45 (1972). Reproduced with the permission of *J. Membrane Biol.*

The efflux rate coefficient is a function of the rate constants of calcium efflux of all the kinetic components of the desaturation curve. Since the half-time of each phase differs by at least one order of magnitude their respective contribution to the ERC will progressively change with time. During the early time points of desaturation, the rate constant of the fastest phase will predominantly determine the ERC, whereas after several hours, the faster phase's contribution will approach zero and ERC will reflect mostly the rate constant of the slowest phase. Consequently, the efflux rate coefficient will constantly decrease with time (Fig. 23). The effects of a change in the composition of the washout medium can be readily demonstrated by plotting the change in efflux rate coefficient. The absolute values of the ERC of the control of experimental groups can be plotted on the same graph or the results can be expressed as the percent change in experimental ERC compared to control. As an example, the effect of the removal of phosphate in the washout medium during a 60-minute period is shown in the computer output presented in Table XV. Column 6 of Table XV, the ERC of control and experimental group, can be plotted against time as shown in Fig. 24. Column 7 of Experimental part Table XV gives the percent change of the ERC during the

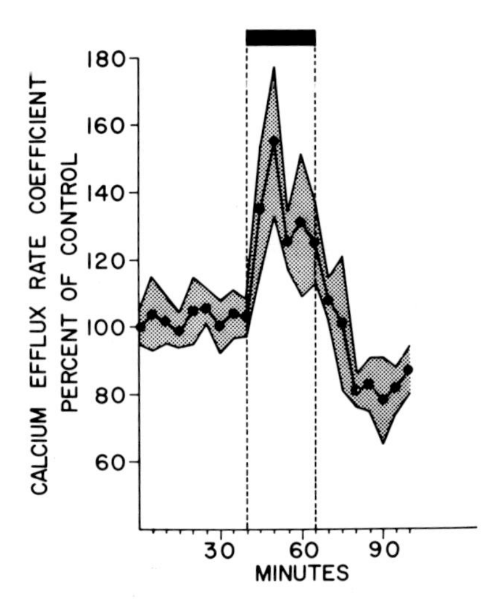

FIG. 27. Effect of 1 m*M* cyclic AMP on the efflux rate coefficient of calcium-45 from isolated kidney cells. The medium contained 1.3 m*M* calcium and 1.0 m*M* phosphate. The results are the mean ± standard error of 3 experiments. From A. B. Borle, *in* "Calcium, Parathyroid Hormone and the Calcitonins" (R. V. Talmage and P. L. Munson, eds.), p. 484. Excerpta Med. Found., Amsterdam, 1972. Reproduced with the permission of Excerpta Medica Foundation.

control period (0 to 50 minutes), during the experimental period (50 to 110 minutes), and during recovery (110 to 240 minutes). These values can be plotted as in Fig. 25, which shows the means and standard deviations of 5 similar experiments. Not only stimulatory, but also inhibitory effects, can be easily demonstrated. For instance, the effect of the addition of 1 m*M* phosphate during a calcium-45 washout performed in a phosphate-free medium can be clearly seen in Fig. 26. The interpretation of data of this type is not easy. Depending upon the experimental conditions, an increase in the rate coefficient of calcium-45 efflux prelabeled cells may reflect one or several of the following processes: (1) an increased calcium influx into the cell, (2) an increased transport rate through the plasma membrane out of the cell, (3) an increased calcium activity in the cell cytoplasm, or (4) an increased release of calcium-45 from subcellular compartments. By carefully selecting different experimental conditions, or by using inhibitors specific for certain transport

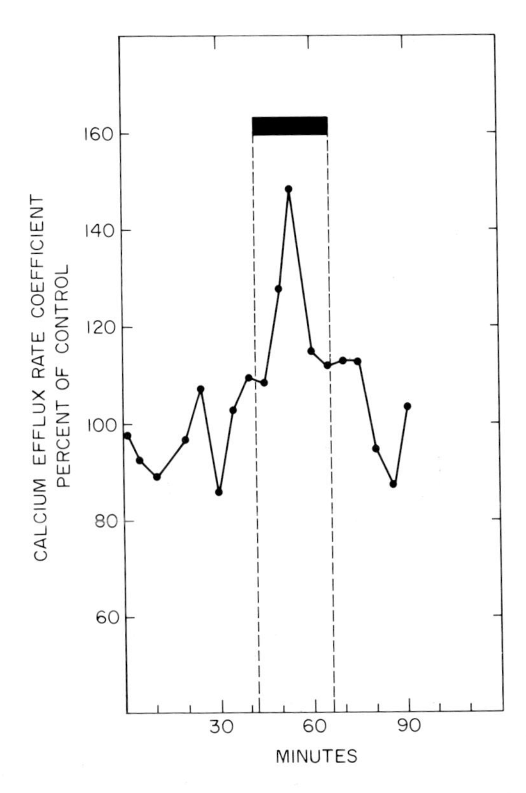

FIG. 28. Effect of 1 m*M* cyclic AMP on the efflux rate coefficient of calcium-45 from isolated kidney cells. The medium contained 1.0 m*M* phosphate and no calcium.

processes, it may be possible to narrow the possibilities by elimination. For instance, cyclic AMP stimulates calcium efflux from prelabeled kidney cells as shown in Fig. 27.[54] This could be due to an increased influx of unlabeled calcium into the cell, displacing the intracellular tracer. However, in some circumstances, the same effect is observed in the absence of extracellular calcium as shown in Fig. 28, which is a strong argument against that possibility. Since intracellular calcium homeostasis is still not clearly understood, it is very difficult to interpret these kinds of data.

Finally, perturbation of steady-state calcium fluxes can also be performed during an uptake experiment. In this case, to see a stimulation of uptake, the perturbation should occur when the cells are near the isotopic steady state (Fig. 29). Conversely, to demonstrate an inhibition

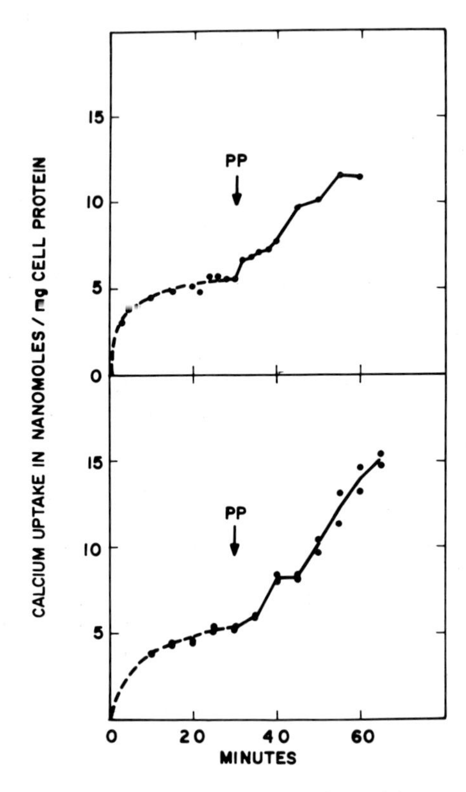

FIG. 29. Effect of 0.3 mM pyrophosphate on the calcium-45 uptake of isolated kidney cells. Pyrophosphate (PP) was added at the time shown by the arrow. From A. B. Borle, *in* "Phosphate et métabolisme phosphocalcique (D. J. Hioco, ed.), p. 29. Expansion Scientifique Française, Paris, 1971. Reproduced with the permission of l'Expansion Scientifique Française.

of calcium uptake, the perturbation should be produced during the early phase of calcium-45 uptake when a break in the uptake curve is more likely to be discernible.

Appendix I

The symbols used in this chapter are defined as follows:

Symbols	Definitions
S_i	Amount of exchangeable calcium in compartment i
R_i	Amount of calcium-45 in compartment i
X_i	Specific activity of calcium-45 in compartment i; $X_i = R_i/S_i$
X_i0	Specific activity of calcium-45 in compartment i at time zero
E	Equilibrium (infinite time) specific activity
J_{ij}	Rate of calcium transport from compartment i to compartment j
k_{ij}	Rate constant of calcium transport from compartment i to compartment j; fraction of S_i transferred to compartment j in unit time, $k_{ij} = J_{ij}/S_i$
A	Coefficient of the first exponential term in the equation describing the behavior of the tracer in a cellular compartment when the tracer is initially in the medium compartment
B	Coefficient of the second exponential term in the equation describing the behavior of the tracer in a cellular compartment when the tracer is initially in the medium compartment
λ_1	Exponential constant of the first exponential term
λ_2	Exponential constant of the second exponential term
ρ	Rate of calcium exchange between compartments i and j in steady-state conditions; $= J_{ij} = -J_{ji}$
ERC	Efflux rate coefficient
cpm_w	Amount of calcium-45 appearing in the medium during one washout period of a desaturation experiment
cpm_c	Amount of calcium-45 remaining in the cells during a desaturation experiment
$(cpm_c)_{mean}$	Mean radioactivity of calcium-45 remaining in the cells during one washout period of a desaturation experiment
Δt	Length of time in minutes of a washout period of a desaturation experiment

Appendix II

Computer program using PIL/X, Pitt interpretative language for a $DEC_{system\text{-}10}$ computer (Computer Center, University of Pittsburgh, Pittsburgh, Pennsylvania, 15213), to calculate the exchange rates, compartment sizes, and rate constants of exchange in a three-compartment, steady-state closed system for the series and parallel cases, according to Robertson *et al.*[34]

Computer symbols	Definition according to Appendix I
MK	Code name or code number of experiment
c21	A
c22	B
l1	λ_1
l2	λ_2
x20	X_20
E	E
R20	R_20
k12	k_{12}
S2	S_2
M12	J_{12}

COMPUTER PROGRAM FOR THE CALCULATION OF THE KINETIC PARAMETERS OF A THREE-COMPARTMENT SYSTEM ACCORDING TO ROBERTSON *et al.*[34]

```
1.1   demand MK,c21,c22,l1,l2,x20,E,R20.
1.2   do part 2
1.25  line.
1.3   do part 3.
1.4   do part 1.
1.5   do part 5.

2.0   set z=(c21*l1)+(c22*l2)-(x20*(l1+l2)).
2.01  set k12=-(z-sqrt of ((((c21*l1)+(c22*l2)-(x20*(l1+l2)))**2)-
        (4*x20*E*l1*l2)))/(2*x20).
2.1   set k32=E*l1*l2/(x20*k12).
2.2   set k21=((k12*(l1+l2-k12-k32))-(l1*l2)+(k12*k32))/(k12-k32).
2.3   set k23=(((c21*l1)+(c22*l2))/x20)-k21.
2.4   set S2=R20/x20.
2.5   set M12=S2*k21.
2.6   set M23=S2*k23.
2.7   set S1=M12/k12.
2.8   set S3=M23/k32.

3.0   type "parameters of three compartment system, parallel case".
3.01  line.
3.02  type MK.
3.1   type in form 1,"k(21)",k21,"S(1)",S1,"M(12)",M12,"k(12)",k12,"S(2)",
        S2,"M(23)",M23.
3.2   type in form 1,"k(23)",k23,"S(3)",S3.
3.3   type in form 1,"k(32)",k32.
3.4   line.
3.5   line.

4.0   set c11=c21.
```

```
4.1   set c12=c22.
4.15  set x10=x20.
4.16  set R10=R20.
4.2   set k12=((c11*l1)+(c12*l2))/x10.
4.21  set k21=((k12*(l1+l2-k12))-((l1*l2)*(1-(E/x10))))/k12.
4.22  set k32=(E*l1*l2)/(x10*k21).
4.23  set k23=(((l1*l2)*(1-(E/x10)))/k12)-k32.
4.24  set S1=R10/x10.
4.25  set M12=S1*k12.
4.26  set S2=M12/k21.
4.27  set M23=S2*k23.
4.28  set S3=M23/k32.

5.0   type "parameters of three compartment system, series case".
5.01  line.
5.02  type MK.
5.1   type in form 1,"k(12)",k12,"S(1)",S1,"M(12)",M12,"k(21)",k21,"S(2)",
        S2,"M(23)",M23.
5.2   type in form 1,"k(23)",k23,"S(3)",S3.
5.3   type in form 1",k(32)",k32.

Form 1.
      #####  _._____!!!!  ####  _____._____!!!!  #####  _____._____!!!!
```

Appendix III

The assumption that the three-compartment system can be analyzed as if it were in parallel, is supported by the fact that the differences between the series case and the parallel case are negligible (Table X). The calculation of the parameters of each compartment can then be derived from the kinetic equations for a two-compartment system, S_1 being the medium and S_2 the unknown compartment.

$$R_1 = X_1 S_1 \tag{62}$$

$$R_2 = X_2 S_2 \tag{63}$$

Since S_1 and S_2 are constant

$$dR_1/dt = (S_1 dX_1)/dt \tag{64}$$

and

$$dR_2/dt = (S_2 dX_2)/dt \tag{65}$$

We can also write

$$dR_1/dt = J_{21}X_2 - J_{12}X_1 \tag{66}$$

and

$$dR_2/dt = J_{12}X_1 - J_{21}X_2 \tag{67}$$

Since the system is in steady state

$$J_{12} = -J_{21} = \rho \tag{68}$$

Therefore

$$dR_1/dt = (S_1 dX_1)/dt = \rho(X_2 - X_1) \tag{69}$$

and

$$dR_2/dt = (S_2 dX_2)/dt = \rho(X_1 - X_2) \tag{70}$$

Combining Eqs. (69) and (70)

$$dX_1/dt - dX_2/dt = -\rho(1/S_1 + 1/S_2)(X_1 - X_2) \tag{71}$$

The rate of change of $X_1 - X_2$ being proportional to $X_1 - X_2$, Eq. (71) yields an exponential relation. Since X_2 ultimately approaches X_1, the terminal value of $X_1 - X_2$ approaches zero. Moreover since all the radioactivity is initially in compartment 1 with a specific activity $X_1 0$, the initial $(X_1 - X_2)0 = X_1 0$. The solution is then

$$X_1 - X_2 = X_1 0 e^{-\rho(S_1+S_2)/S_1 S_2 t} \tag{72}$$

From Eq. (70)

$$X_1 - X_2 = (dR_2/dt)/\rho \tag{73}$$

and combining Eqs. (72) and (73) yields

$$(dR_2/dt)/\rho = X_1 0 e^{-\rho(S_1+S_2)/S_1 S_2 t} \tag{74}$$

Since $X_1 0 = E$

$$(dR_2/dt)/E = \rho e^{-\rho(S_1+S_2)/S_1 S_2 t} \tag{75}$$

Writing Eq. (75) in logarithmic forms gives

$$\log\left(\frac{dR_2/dt}{E}\right) = \log \rho - \frac{\rho(S_1 + S_2)}{S_1 S_2} t \tag{76}$$

A semilogarithmic plot of $(dR_2/dt)/E$ versus time t will yield a straight line with a slope λ where

$$\lambda = [\rho(S_1 + S_2)]/S_1 S_2 \tag{77}$$

The intercept of the regression line with zero time will yield ρ. From Eq. (70)

$$dR_2/dt = -\rho(X_2 - X_1) \tag{78}$$

since at time zero X_1 equals X_10 (or E), and $X_2 = 0$, one can write

$$[(dR_2/dt)/E]_0 = \rho \tag{79}$$

The slope λ can be obtained by dividing 0.693 by the half-time $t/2$ (the abscissal value of time t for which $X_1 - X_2$ is half of X_20). Thus

$$\lambda = 0.693/(t/2) = \rho(S_1 + S_2)/S_1S_2 \tag{80}$$

Knowing λ, ρ, and the compartment size of the medium S_1, the unknown compartment S_2 can be calculated as follows:

$$\lambda = \rho(S_1 + S_2)/S_1S_2 \tag{81}$$

or

$$\rho/\lambda = S_1S_2/(S_1 + S_2) \tag{82}$$
$$S_1S_2 = (\rho/\lambda)S_1 + (\rho/\lambda)S_2 \tag{83}$$
$$S_1S_2 - (\rho/\lambda)S_2 = (\rho/\lambda)S_1 \tag{84}$$
$$S_2(S_1 - \rho/\lambda) = (\rho/\lambda)S_1 \tag{85}$$
$$S_2 = \rho/\lambda[S_1/(S_1 - \rho/\lambda)] \tag{86}$$

Since in all circumstances studied so far ρ/λ is about three orders of magnitude smaller than S_1, $S_1/(S_1 - \rho/\lambda)$ is nearly equal to 1, and the unknown compartment S_2 can be simply obtained by dividing the intercept of the semilogarithmic plot ρ by its slope λ. Table XVI shows that even when the medium calcium concentration is very low, 0.013 mM, the medium compartment S_1 is still far greater than ρ/λ. It is especially true in a physiological calcium concentration of 1.3 mM. The error introduced is in the order of a fraction of a percent (Table XVI).

In summary, if the exponential constants differ by more than one order of magnitude, the three-compartment closed system can be analyzed as a parallel case. If the medium compartment is much greater than the cellular pools and the exchange rates, it is legitimate to assume that the coefficients and the exponential constants of the double exponential equation can be taken for the fluxes and rate constants of their respective kinetic components. Thus in Eq. (54) one can write: $A = J_{12}$; $B = J_{32}$; $\lambda_\alpha = k_{12}$; $\lambda_\beta = k_{32}$.

Appendix IV

Computer program using PIL/X, Pitt interpretative language for a DEC$_{\text{system-10}}$ computer (Computer Center, University of Pittsburgh, Pittsburgh, Pennsylvania 15213) to calculate from a desaturation experiment, the radioactivity left in the cells, the percent calcium-45 left in the

TABLE XVI

PERCENT ERROR INTRODUCED IN THE CALCULATION OF THE COMPARTMENT SIZE BY ASSUMING THAT $\lambda = k$ IN EQ. (76)

	ρ (pmoles/min)	λ (min^{-1})	ρ/λ (pmoles)	S_1 (pmoles)	$\frac{S_1}{S_1(\rho/\lambda)}$ (pmoles)	S_2 (pmoles)	% Error
Medium Ca = 0.013 mM[a]							
Fast phase	190	0.683	279	13,000	1.021	284	2.1
Slow phase	5.41	0.0144	377	13,000	1.029	387	2.8
Medium Ca = 1.3 mM[b]							
Fast phase	1161	0.611	1900	1,300,000	1.0010	1902	0.10
Slow phase	96	0.0299	3210	1,300,000	1.0024	3218	0.24

[a] From Table XIII.

[b] From A. B. Borle, *J. Gen. Physiol.* **55**, 163 (1970).

cells, the efflux rate coefficient, and the percent increase in ERC in an experimental group compared to its control.

Computer symbols	Definitions
k	Code name or code number of the experiment
n	Number of washout periods
cpmcel	Total radioactivity of the control cells at the end of the desaturation
cpmCEL	Total radioactivity of the cells of the experimental group at the end of the desaturation
$x_{1 \text{ to } n}$	Time point (in minutes) of the successive washout collections; x(1) = 1 minute time point, x(2) = 5-minute time point x(3) to x(*n*) = 10-minute time points
f (i)	Total radioactivity of the successive washout collections of the control group
m (i)	Total radioactivity of the successive washout collection of the experimental group
sigma	Total radioactivity of all washout collections plus the final radioactivity of the cells of the control group
sigmae	Total radioactivity of all washout collections plus final radioactivity of the cells of the experimental group
i	Washout sampling number
y (i)	Radioactivity left in the cells of the control group after *i* washout periods
z (i)	Radioactivity left in the cells of the experimental group after *i* washout periods
percent (i)	Radioactivity left in the cells of the control group after *i* washout periods, expressed as percent of the total initial radioactivity (sigma)
perct (i)	Radioactivity left in the cells of the experimental group after *i* washout periods, expressed as percent of the total initial radioactivity (sigmae)
rate (i)	Efflux rate coefficient (ERC) of the control group during the washout period *i*
ratae (i)	Efflux rate coefficient of the experimental group during the washout period *i*
pc (i)	ERC of the experimental group during the washout period *i* expressed as percent of the control ERC during the same period.

Computer Program for Calcium Efflux Calculations

```
1.1    demand k,n,cpmcel,cpmCEL.
1.11   set x(1)=1.
1.12   set x(2)=5.
1.13   for i=3 to n:set x(i)=(i-2)*10.
1.2    type "type cpm control".
1.21   demand in free form,(for i=1 to n:f(i)).
1.22   type "type cpm experimental".
```

(Continued)

Computer Program for Calcium Efflux Calculators (*continued*)

```
1.23   demand in free form,(for i=1 to n:m(i)).
1.24   stop.
1.3    do part 2.
1.31   do part 3.
1.32   do part 4.
1.4    do part 6.

2.3    set sum=0.
2.4    for i=1 to n:set sum=sum + f(i).
2.45   set sigma=sum+cpmcel.
2.5    set y(0)=sigma.
2.6    for i=1 to n: set y(i)=y(i-1)-f(i).
2.65   for i=1 to n:set percent(i)=y(i)*100/sigma.
2.7    set x(0)=0.
2.71   for i=1 to n:set rate(i)=(f(i)/(x(i)-x(i-1)))*100/((y(i)+y(i-1))/2).

3.3    set sume=0.
3.4    for i=1 to n:set sume=sume+m(i).
3.45   set sigmae=sume+cpmCEL.
3.5    set z(0)=sigmae.
3.6    for i=1 to n:set z(i)=z(i-1)-m(i).
3.65   for i=1 to n:set perct(i)=z(i)*100/sigmae.
3.7    set x(0)=0.
3.71   for i=1 to n:set ratae(i)=(m(i)/(x(i)-x(i-1)))*100/((z(i)+z(i-1))/2).

4.0    for i=1 to n:set pc(i)=ratae(i)*100/rate(i).

6.1    type k,sigma.
6.13   type "control".
6.2    type form 12.
6.3    for i=1 to n:type in form 13,i,x(i),f(i),y(i),percent(i),rate(i).
6.31   line.
6.32   type "experimental".
6.322  type sigmae.
6.34   type form 14.
6.4    for i=1 to n:type in form 15,i,x(i),m(i),z(i),perct(i),ratae(i),pc(i).
Form 12.
sample  time  cpm  Ca left in cells  percent left  rate coefficient
Form 13.
__  __._  ____!!!!  ____!!!!  ___.__!!!!  ___.__!!!!.
Form 14.
sample  time  cpm  Ca left in cells  percent left  rate coefficient  percent of control
Form 15
__  __._  ____!!!!  ____!!!!  ___.__!!!!  ___.__!!!!  ___.__!!!!
```

Appendix V

In the calculation of the desaturation experiments, we have assumed that because the time constants of the three components differ by more than one order of magnitude, the differential equation can be analyzed as a parallel case of a multicompartment system. However Huxley, in an appendix to Solomon's article[32] has pointed out that when one is faced with two exponential terms describing the desaturation curve of an isotope leaving two compartments such as

$$R_1 + R_2 = Ae^{-\lambda_1 t} + Be^{-\lambda_2 t} \tag{87}$$

it is incorrect to assume that

$$R_1 = Ae^{-\lambda_1 t} \tag{88}$$

and

$$R_2 = Be^{-\lambda_2 t} \tag{89}$$

This assumption leads to an overestimate of the slow compartment, and the error can be calculated by the equation given by Huxley:

$$R_2 0 = [AB(\lambda_1 - \lambda_2)^2]/[A\lambda_1^2 + B\lambda_2^2] \tag{90}$$

Where $R_2 0$ is the radioactivity of the slow compartment at time 0 and A, B, λ_1, and λ_2 are the coefficients and exponential constants of the differential equation. Experimental data obtained from calcium-45 desaturation experiments performed with isolated kidney cells[29] are shown in Table XVII. In this table

$$A = 0.72B \tag{91}$$

and

$$\lambda_1 = 10.2\lambda_2 \tag{92}$$

Substituting Eqs. (91) and (92) in Eq. (90) shows that

$$R_2 0 = 0.95B \tag{93}$$

Therefore our assumption that the system can be calculated as a parallel case and that the intercept B can be taken for $R_2 0$ overestimates its value by 5%, which is an acceptable error for these studies.

TABLE XVII
COEFFICIENT AND EXPONENTIAL CONSTANT OF PHASES 2 AND 3 OF NINE CALCIUM-45 DESATURATION CURVES[a,b]

Coefficients and exponential constants	Experimental values
A	1.83 ± 0.22
B	2.54 ± 0.43
λ_1	0.0466 ± 0.0021
λ_2	0.00455 ± 0.00085

[a] The values are the mean ± SE of nine experiments.
[b] From A. B. Borle, *J. Membrane Biol.* **10**, 45 (1972).

Another assumption which requires justification is that the rate constant of efflux of each compartment k_{ij} can be substituted for the exponential constant λ_i in Eq. (58). Since during a desaturation experiment the system is an open one, the slope of each kinetic phase represents the rate constant of efflux k_{ij} from compartment i to the medium j. Indeed we can write

$$dR_i/dt = -J_{ij}X_i \tag{94}$$

Since

$$X_i = R_i/S_i \tag{95}$$

$$dR_i/dt = -J_{ij}(R_i/S_i) \tag{96}$$

or

$$dR_i/dt = (J_{ij}/S_i)R_i \tag{97}$$

Since the rate of change dR_i/dt is proportional to R_i we obtain an exponential relation. At time 0 all the radioactivity is in compartment i and equal to R_i0 and, at infinite time, $R_i = 0$. Thus we can write

$$R_i = R_i0e^{-(J_{ij}/S_i)t} \tag{98}$$

Taking the logarithmic of both sides

$$\log R_i = \log R_i0 - (J_{ij}/S_i)t \tag{99}$$

We obtain an expression which is linear with time. In a semilogarithmic plot, the intercept with time zero is R_i0 and the slope is $-(J_{ij}/S_i)$. Since

$$J_{ij}/S_i = k_{ij} \tag{100}$$

the slope of component i represents the rate constant of efflux from compartment i to the medium.

Equation (58), however, is an expression of the labeling period occurring in a closed system

$$S_i = R_i/[E(1 - e^{-\lambda_i t})] \tag{58}$$

and the exponential constant λ_i is not equivalent to a rate constant. In Appendix III the relevant equation describing calcium-45 uptake in such a closed system was Eq. (75) or

$$(dR_i/dt)/E = \rho e^{-\rho(S_j+S_i)/S_jS_it} \tag{101}$$

where ρ is the rate of exchange in a steady state system:

$$\rho = J_{ij} = -J_{ji} = \text{constant} \tag{102}$$

S_j is the medium and S_i the unknown intracellular compartment. The slope of the semilogarithmic plot of Eq. (101) versus time is equal to λ_i as shown previously in Eq. (77) or

$$\lambda_i = [\rho(S_j + S_i)]/S_jS_i \quad (103)$$

Therefore

$$\lambda_i/\rho = (S_j + S_i)/S_jS_i \quad (104)$$

Since at steady state

$$J_{ij} = k_{ij}S_i = \rho \quad (105)$$

Eqs. (104) and (105) give

$$\lambda_i/(k_{ij}S_i) = (S_j + S_i)/S_jS_i \quad (106)$$

or

$$\lambda_i = k_{ij}S_i[(S_j + S_i)/S_jS_i] \quad (107)$$

or

$$\lambda_i = k_{ij}[(S_j + S_i)/S_j] \quad (108)$$

In the cellular suspension system described, S_j, the medium compartment is about 1000 times larger than S_i the intracellular compartment. Consequently the ratio $S_j + S_i/S_j$ is so close to unity that the difference can be neglected, and *in these conditions* only we can write

$$k_{ij} = \lambda_i \quad (109)$$

Table XVIII provides the experimental evidence that the error introduced by this simplification is not greater than a fraction of 1%.

TABLE XVIII
CALCULATION OF THE ERROR INTRODUCED BY THE SIMPLIFIED Eq. (109)[a]

	Medium S_j (nmoles/ml)	Cell Ca compartment S_i (nmoles/ml)	$\frac{S_j + S_i}{S_j}$	λ_i	k_{ij}	% Difference
Medium calcium = 1.3 mM						
Second phase of desaturation	1300	3.05 ± 0.37	1.0026	0.04671	0.0466	0.23
Third phase of desaturation	1300	4.23 ± 0.72	1.0032	0.004565	0.00455	0.32

[a] Data derived from A. B. Borle, *J. Membrane Biol.* **10,** 45 (1972), Table 2. Values are the mean ± SE of nine experiments in which the cell suspension averaged 5 mg of protein per 3 ml of suspension.

Author Index

Numbers in parentheses are reference numbers and indicate that an author's work is referred to, although his name is not cited in the text.

A

B

D

E

F

I

J

K

L

M

N

O

P

Q

R

S

T

U

V

W

Y

Z

Subject Index

A

B

D

H

I

J

K

L

M

N

O

P

R

S

T

U

V

W

X

Y

Z